Performance Limits in Communication Theory and Practice

NATO ASI Series

Advanced Science Institutes Series

A Series presenting the results of activities sponsored by the NATO Science Committee, which aims at the dissemination of advanced scientific and technological knowledge, with a view to strengthening links between scientific communities.

The Series is published by an international board of publishers in conjunction with the NATO Scientific Affairs Division

A Life Sciences **B Physics**	Plenum Publishing Corporation London and New York
C Mathematical and Physical Sciences **D Behavioural and Social Sciences** **E Applied Sciences**	Kluwer Academic Publishers Dordrecht, Boston and London
F Computer and Systems Sciences **G Ecological Sciences** **H Cell Biology**	Springer-Verlag Berlin, Heidelberg, New York, London, Paris and Tokyo

Series E: Applied Sciences - Vol. 142

Performance Limits in Communication Theory and Practice

edited by

J. K. Skwirzynski

Marconi Research Centre,
Great Baddow, Chelmsford,
Essex, U.K.

Kluwer Academic Publishers

Dordrecht / Boston / London

Published in cooperation with NATO Scientific Affairs Division

Proceedings of the NATO Advanced Study Institute on
Performance Limits in Communication Theory and Practice
Il Ciocco, Castelvecchio Pascoli, Tuscany, Italy
July 7–19, 1986

Library of Congress Cataloging in Publication Data

NATO Advanced Study Institute on "Performance Limits in Communication Theory and Practice" (1986 : Castelvecchio Pascoli, Italy)
Performance limits in communication theory and practice / editor, J.K. Skwirzynski.
p. cm. -- (NATO ASI series. Series E, Applied sciences ; no. 142)
"Published in cooperation with NATO Scientific Affairs Division."
"Proceedings of the NATO Advanced Study Institute on 'Performance Limits in Communication Theory and Practice,' Il Ciocco, Castelvecchio Pascoli, Tuscany, Italy, July 7-19, 1986"--T.p. verso.
Includes index.
ISBN 9024736951
1. Telecommunication--Congresses. 2. Information theory--Congresses. 3. Statistical communication theory--Congresses. I. Skwirzynski, J. K. II. North Atlantic Treaty Organization. Scientific Affairs Division. III. Title. IV. Series.
TK5101.A1N394 1986
621.38--dc19 88-3797
CIP

ISBN 90-247-3695-1

Published by Kluwer Academic Publishers,
P.O. Box 17, 3300 AA Dordrecht, The Netherlands.

Kluwer Academic Publishers incorporates the publishing programmes of
D. Reidel, Martinus Nijhoff, Dr W. Junk, and MTP Press.

Sold and distributed in the U.S.A. and Canada
by Kluwer Academic Publishers,
101 Philip Drive, Norwell, MA 02061, U.S.A.

In all other countries, sold and distributed
by Kluwer Academic Publishers Group,
P.O. Box 322, 3300 AH Dordrecht, The Netherlands.

Printed in The Netherlands

TABLE OF CONTENTS

PREFACE

In this volume we present the full proceedings of a NATO Advanced Study Institute (ASI) on the theme of performance limits in communication, theory and practice. This is the 4th ASI organised by me and by my friends in the field of communications. The others were "New Directions in Signal Processing in Communication and Control", published in 1974, "Communication Systems and Random Process Theory", published in 1977, and "New Concepts in Multi-User Communication" published in 1980.

The first part of the present proceedings concentrates on the ultimate physical limits in electronic communication. Here we have three important papers. Professor David. G. Messerschmitt discusses the problem of breaking the recursive bottleneck. He concentrates on high performance implementations of algorithms which have internal recursion or feedback. Next, Professor Daniel V. McCaughan concentrates on optimum scales and limits of integration. He claims that these ultimate limits are, in a sense, invariant, determined by consideration of quantum mechanical relationships between velocity, time, physical dimensions and the uncertainty principle. Finally, Professor Jerome Rothstein discusses the ultimate thermodynamic limitations in communication and computation. His special themes are the thermodynamic limits on computation, and measurement in communications and physics.

In the second part of these proceedings we consider statistical, information, computation and cryptographic limits. First, Professor I. Bar-David considers the capacity peak power constrained Gaussian channels. His main theme is the performance of communication systems when peak power constraints are imposed on transmitted signals. Then Professor Ian F. Blake (and his colleagues) consider the complexity issues for public key cryptography. He presents a description of the public key systems, the complexity of implementation, the quadratic sieve algorithm for integer factorisation, discrete logarithms in characteristic two fields, and comparison of the public key systems. Next is Dr. P. Bridge, who discusses the collaborative coding for optical fibre multi-user channels; he discusses the properties of optical fibre systems, optical multiple access networks, and multiple access protocols and coding. Following that is the paper by Professor Y.G. Desmedt who discusses what has happened with Knapsack cryptographic schemes. He overviews the whole history of this cryptographic scheme, including details of the weakness and of the cryptoanalytic attacks on trapdoor Knapsack schemes. Next is Dr. Robert W. Keyes, who considers the optical logic for computers; he considers problems in information processing, logic with transistors and bistability in physical systems. Following him is Professor Anthony Ephremides who concentrates on the limitations in queueing models in communication networks; he considers the problem in capacity allocation in these networks, and also the problem of stability in such channels. Next is Dr. Gunter G. Weber who discusses the limits to network reliability; his theme is the discussion of fault tree analysis, and the performability of such channels. Then Professor Kenneth Steiglitz discusses two non-standard paradigms for computation in analogue machines and in cellular automata; he measures the complexity of these problems, and gives several examples of results of such measurements.

Then we follow with the paper of Professor J. Pieter M. Schalkwijk on the capacity region of the binary multiplying channel; he considers also the initial thresholds in these channels and illustrates that with many examples. Next is Dr. Fred Piper with his discussion on recent developments in cryptography, considering also block cipher techniques and cipher feedback techniques. Then Professor Thomas M. Cover considers the role of feedback in communication, discussing the feedback in memoryless channels, Gaussian channels with feedback and multiple access channels. Mr. G.A. Karpel discusses the complexities of information transfer with reference to a genetic code model. He considers that this is the replication of genetic material in living cells, for this material contains coded information sufficient to specify the entire host organism. Following that we have Professor Robert J. McEliece, with his collaborator, Dr. Khaled Abdel-Ghaffar, who consider the ultimate limits of information density; they consider binary symmetric channels with noise scaling, the universality of orthogonal codes, and illustrate these with several examples. Then we have the paper by Professor P.G. Farrell and his three colleagues on limits of radio communication with collaborative transmission over cellular radio channels. They discuss the cellular mobile radios, and collaborative transmission over digital cellular channels. Finally we have two papers by Professor J.E. Midwinter on performance boundaries of optical fibre systems, as well as aspects of digital optics and optical computing.

The third and final part of these proceedings is concerned with the limits of modelling and of characterisation of communication channels. The first contribution is by Professor K.W. Cattermole who discusses the robustness and sensitivity of communication models; he considers the defficiency of models, and the sensitivity of models, giving us several examples of these difficulties. Following him is Professor Jack Keil Wolf who considers the modulation and coding for the magnetic recording channels. He discusses modulation codes, and the capacity of these codes with random errors. Next is the paper by Professor F.L.H.M. Stumpers on modelling of communication limits for non-Gaussian noise; he considers information theory approaches to the capacity of such channels, and proposes that adaptive noise can cancel this problem. Following that we have a paper by Dr. Z.C. Zhang on the capability of 144 bits ISDN digital signals with existing systems; he discusses the digital-to-digital crosstalk, and voice-to-digital crosstalk. Then Professor J.L. Massey discusses the channel models for random-access systems. He gives us the definitions of capacity of such channels, particularly with many stations, and discusses both the binary feedback and the ternary feedback. Following that we have a paper by Professor Edward C. van der Meulen on the capacity limits for multiple-access channels without feedback; some of them have correlated sources in the sense of Slepian and Wolf, while others are memoryless and are with arbitrary correlated sources. Finally we have the paper by Dr. W. Kuo on the limits on system reliability improvement; he proposes that reliability can be improved by using simple design, by using simple components and by adding some redundancy.

It took me two years to prepare this important Institute, and here I want to thank my co-directors who helped me a lot in this. They are Professor P.G. Farrell, Professor E.A. Palo, Professor J.P.M. Schalkwijk and Professor K. Steiglitz.

My assistant at this Institute was Mr. Barry Stuart, one of my bridge partners, who did an excellent job manning our office and settling all accounts.

Finally, I wish to thank Dr. John Williams, the Director of our Laboratory, who has allowed me to organise this venture.

Great Baddow, September 1987. J.K. Skwirzynski

Part 1.
Ultimate Physical Limits in Electronic Communication

Breaking the Recursive Bottleneck

by

David G. Messerschmitt
Department of Electrical Engineering and Computer Sciences
University of California
Berkeley, California 94720

1. Introduction

If we are looking for ways to implement high-performance systems, there are several directions we can head. One direction is to use high-speed technologies, such as bipolar or GaAs, which allow us to gain performance without modification to the methods or algorithms. If, in contrast, we wish to exploit one of the low-cost VLSI technologies, particularly CMOS, we can gain much more impressive advantages in performance by exploiting concurrency in addition to speed. This is because, while the scaling of these technologies does naturally result in higher speed (roughly proportional to the reciprocal of the scaling factor), it has a much more dramatic effect on the available complexity (which increases roughly as the square of the speed)[1]. Two other characteristics which lead to high performance implementations should also be kept in mind. First, it is desirable to use structures with *localized communications,* since communications is expensive in speed, power, and die area. Second, it is desirable to achieve *localized timing,* meaning that whenever signals must propagate a long distance there is available a suitable delay time[2].

Two forms of concurrency are usually available, *parallelism* and *pipelining.* By parallelism we mean processing elements performing complete tasks on independent data, whereas in pipelining processing elements are performing a portion of a larger task on all the data. Pipelining in particular is a manifestation of the desirable property of localized timing.

These considerations favor implementations which feature arrays of identical or easily parameterized processing elements with mostly localized interconnection and which have localized timing in the form of pipelining. This has led to an interest in *systolic array* and *wavefront array* structures[2, 3], which have these properties. In most applications we should also be aware of the high design cost of complex circuits, and these type of structures also have the desirable property that they are ameniable to a procedural definition.

In this paper we concentrate on high performance implementations of algorithms which have internal *recursion* or *feedback,* wherein the past outputs of the algorithm are used internally in the algorithm. Examples of such algorithms include *IIR digital filters* and *adaptive filters[4].* (The reader should beware that the term "recursion" is sometimes used to denote an identical algorithm applied to an infinite stream of data[2], and that is not what we mean here). Algorithms which exhibit recursion or feedback are usually considered undesirable for high performance implementation, since the internal feedback of the algorithm usually results in non-localized communication and non-localized timing. Further, as we will see later, for any given realization of a recursive algorithm there is a fundamental limit on the throughput with which it can realized in a given technology. Fortunately, as we will also show for a certain class of recursive algorithms, this limit can be circumvented by changing the structure of the realization -- hence the title of the chapter.

The last point of changing the structure of an algorithm deserves elaboration. In searching for appropriate implementations for high performance VLSI realization, there are two directions we can head, both of them fruitful. One is to search for *new algorithms* for a given application which lead naturally to a desirable implementation structure. In some cases this simply entails finding the most appropriate algorithm from those already known, and in some cases it will be fruitful to design entirely new algorithms which give functionally similar results but lead to more

J. K. Skwirzynski (ed.), Performance Limits in Communication Theory and Practice, 3–19.

desirable implementations. An example of this would be the recent interest in chaotic relaxation, and similar techniques for solution of circuit equations. Another example would be to replace a recursive IIR digital filter with a non-recursive FIR digital filter.

The second direction we can head is to use existing algorithms, but *implement them in new ways.* To be more precise, we do not change the input-output characteristic of the algorithm, but we do change the internal structure of the algorithm which implements this input-output characteristic, thereby impacting the finite-precision effects but nothing else. We will term this option *recasting the structure* of the algorithm. An example of this would be to choose LU decomposition in preference to Gaussian elimination for solution to a set of linear equations, since the latter is less regular and includes more data dependencies. The solution to the equations is of course independent of the method of obtaining that solution, aside from finite precision effects. Another example would be the choice of one digital filter structure (say cascaded second-order sections) in preference to another (say lattice filter).

In this chapter we show examples of both approaches, and show that the choice of an algorithm or recasting of the structure of an algorithm can have a dramatic effect on the performance of the implementation. When we speak of performance in this context, we are referring specifically to the speed, as reflected in the sampling rate or throughput, rather than the functionality of the algorithm. While this exercise is useful as a demonstration of the many ways in which algorithms can be modified to suit implementation constraints, the practical significance of the results we demonstrate here is mostly to high performance applications which demand sampling rates in excess of 10 MHz or perhaps significantly greater. Specifically we show that these demands can be met by low cost implementation technologies, albeit at the expense of possibly a great deal of hardware.

In this chapter we will discuss these issues in the context of specific signal processing algorithms. In particular, we concentrate on two of the simplest and most widespread algorithms: digital filtering and adaptive filtering. We show that in both cases, the algorithms which are commonly used are inappropriate for high performance realizations, but that by designing algorithms specifically with the characteristics of VLSI in mind considerable improvement can be achieved. Specifically we discuss block state realizations of digital filters and vectorized lattice realizations of adaptive filters, and show that both can achieve very high sampling rates with low cost technologies. We also discuss bit-level pipelining of these algorithms. The challenge in achieving a high sampling rate is mostly in recursive systems, since as we will see the recursion tends to negate the most obvious ways of improving performance. After briefly introducing non-recursive systems in Section 2, our odyssey into recursive systems begins in Section 3 by considering a simple, almost trivial case, the first-order system. This simple example illustrates most of the essential ideas. Subsequent sections extend this example in two directions -- generalization to higher order systems is considered in Section 4, and the important (in an application sense) case of adaptive filters, an example of a time-varying linear system, is considered in Section 5.

2. Non-Recursive Systems

It has been known for a long time that non-recursive algorithms, such as an FIR filter, are very natural for high sampling-rate implementations, since many output samples can be computed in parallel. Furthermore, several architectures have been proposed in the literature for implementing such filters using arrays of locally interconnected identical processing elements[3, 5-8]. The basic technique, as shown in Figure 1, is to convert the single-input single-output (SISO) system into a multiple-input multiple-output (MIMO) system. For the example shown, four output samples are computed in parallel, where $L = 4$ is the block size of the implementation. The MIMO system also accepts a block of input samples in parallel (in this case $L = 4$ samples). A serial-to-parallel converter is required at the input and a parallel-to-serial converter is required at the output, and the MIMO system operates at a rate L times slower than the input sampling rate. Hence, if we can find a way to keep the internal sampling rate of the MIMO system constant as we increase L, then the effective sampling rate increases linearly with L, and in principle we can achieve an arbitrarily high sampling rate (within the practical limitations of how fast we can operate the serial-parallel-serial converters).

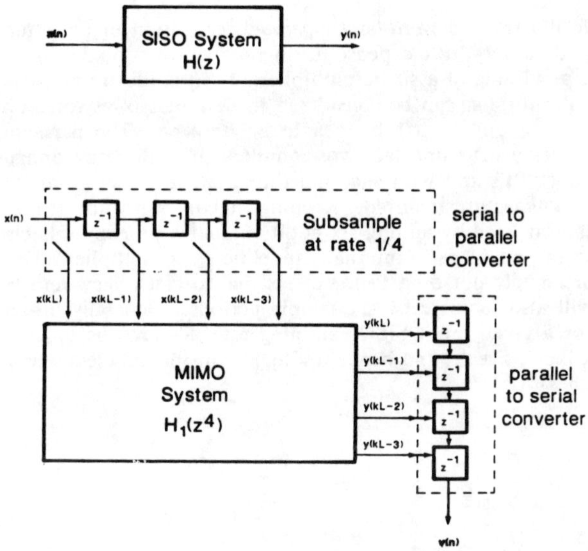

Figure 1. Illustration of single-input single-output (SISO) and multiple-input multiple-output (MIMO) systems for a digital filter.

Figure 2 shows a systolic array implementation of the MIMO realization of an FIR filter for $L = 3$ which realizes the filter

$$y(n) = a_0 u(n) + a_1 u(n-1) + a_2 u(n-2).$$

The technique begins in Figure 2a by laying out the computation (in this case a *matrix-vector multiplication)* on a two-dimensional array of processing elements (PEs). This realizes the overall matrix-vector multiplication in a form where each multiplication and addition is implemented by separate hardware in order to achieve the maximum concurrency, the structure is two-dimensional for mapping onto a silicon die, and the interconnections are local. The next step, shown in Figure 2b, is to "fold" the two-dimensional array to eliminate the trivial processors which multiply by zero while retaining a rectangular array of processors[9]. The two inputs $u(3k-3)$ and $u(3k-4)$ are no longer input, but rather are generated by further delaying the smaller delay inputs. Each delay corresponds to three delays (z^{-3}) at the higher SISO input sampling rate, and a single delay at the lower sampling rate of the MIMO system. The third step in Figure 2c is to add *slice pipelining latches[10],* which are represented by the diagonal lines. Wherever these lines cross a signal path, a delay latch is added to that signal path. Representing these latches as diagonal lines emphasizes that these latches are added to a *feedforward cutset* of the computation graph. Adding one delay to each path at a feedforward cutset implies that the same number of delays is added to each forward path from input to output of the MIMO system, and the operation of the system is not affected except by the addition of a number of delays equal to the number of slice pipelining latches (five for the example). Putting the slice pipelining latches diagonally rather than vertically results in pipelining in both the horizontal and vertical directions. Overall, this pipelining allows the *throughput* of the realization, defined as the rate at which new vectors of samples are accepted by the MIMO system, to equal the rate at which multiplications (with embedded additions) can be performed. The effect of the added pipeline latches is to add *latency* to the algorithm, where latency is defined as the delay between the application of an input vector and the availability of the corresponding output vector. This realization has increased this latency, and has achieved in turn an increased throughput.

Figure 2c illustrates a *word-level pipelined* realization of the filter. The throughput can be further increased, at very little expense in silicon area (due to additional latches), by *bit-level pipelining[9].* The pipelining of a single multiplier[11-16] is illustrated in Figure 3. The technique is similar, since a multiplier can be considered to be a matrix-vector multiplication, where the elements of the vector and matrix happen to be Boolean. The particular realization shown is a three-bit multiplier using truncated twos-complement arithmetic configured as a two-dimensional array of full-adder PEs and with slice pipelining latches. Note that for the wordsize shown, six slice pipelining latches are required. Keeping that in mind, the realization of Figure 2c can be pipelined at the bit level by adding six additional slice pipelining latches for each existing latch, and then using those latches to internally pipeline each multiplier. The throughput will increase, although unfortunately not by a factor of six due to clock skew and latch setup time problems. The latency will also increase by six sample periods. Generally bit-level pipelining is advantageous, since for a given input SISO sampling rate it allows us to use a larger MIMO sampling rate, a smaller block size L, and hence results in considerably less hardware (even considering the additional latches).

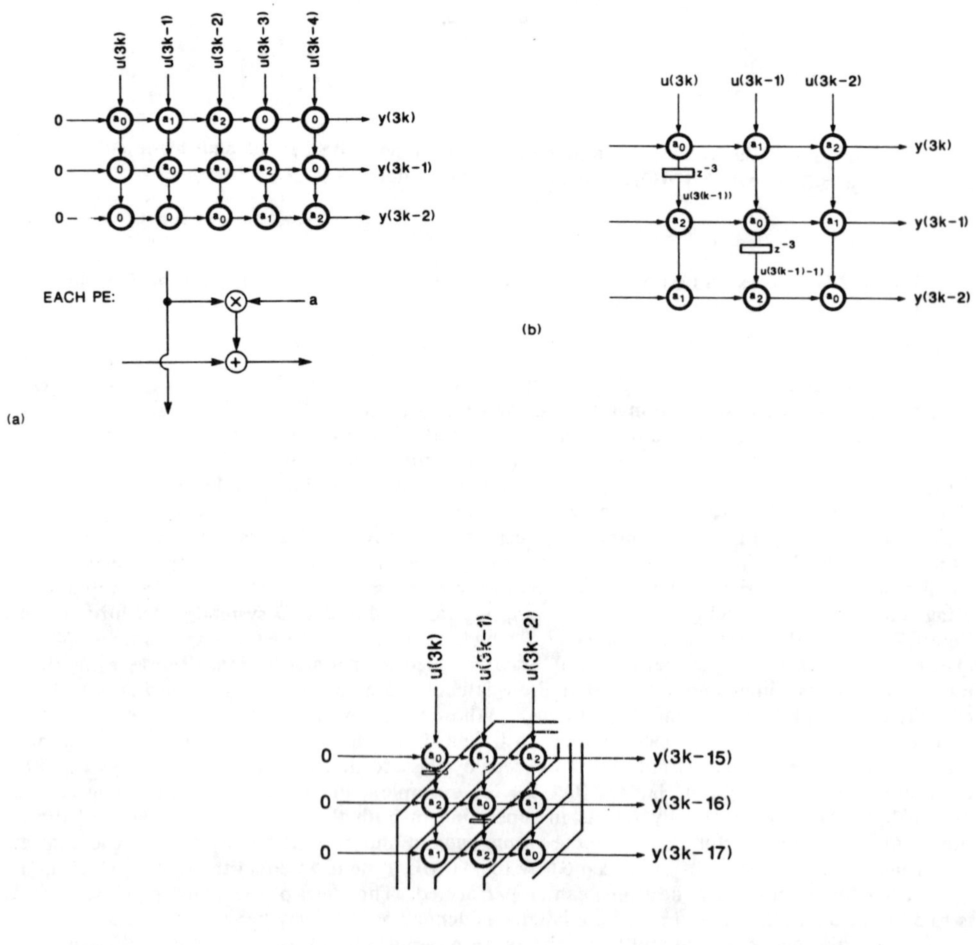

Figure 2. Derivation of a slice-pipelined systolic architecture for an FIR filter.

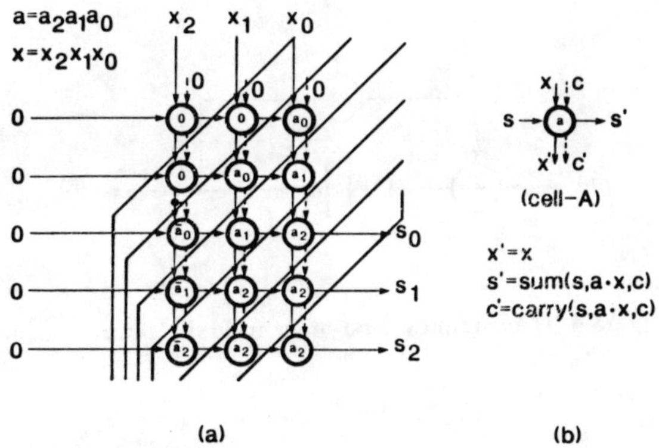

Figure 3. A slice-pipelined constant-coefficient three-bit truncated twos-complement array multiplier.

The approach of Figure 2 can be extended to arbitrarily large L as well as arbitrarily large filter order. The PEs are fully pipelined, implying that aside from potential clock skew problems[1], the realization can be partitioned into multiple chips. Further, a sufficiently small portion of the realization can be mapped onto a single chip so as to accommodate the computation and I/O bandwidth limitations of any given chip technology, implying that a very high sampling rate (or throughput) filter can be realized with a low-speed IC technology as long as the resulting delay (latency) is acceptable.

Unfortunately, many algorithms that we are accustomed to using in signal processing applications are inherently *recursive,* meaning that they have an internal state which depends on the past outputs. Simple but important examples of such systems include IIR digital filters and adaptive filters. In the remainder of this chapter we concentrate on these types of algorithms. Fortunately, these algorithms are also amenable to high speed implementation using locally interconnected processing elements, although this is perhaps not immediately obvious.

3. First-Order Recursive System

Consider the first-order system shown in Figure 4,

$$x(n) = ax(n-1) + u(n). \tag{1}$$

This system holds very little interest except as a simple example to illustrate the essential ideas to follow. Figure 4 is in the form of a *computation graph,* which shows the atomic operations of delay and multiplication and the inherent *feedback* or *recursion* in the algorithm. This example illustrates the inherent (or is it?) limitation on the sampling rate at which a recursive algorithm can be implemented in a given technology[2, 17, 18]. The nature of this limitation is that the multiplication corresponding to time n must be *completed* before the next multiplication at time $n+1$ can be *begun.* The time it takes to perform this multiplication/addition, its *latency,* therefore limits the *throughput* with which the algorithm can be realized.

This example, plus the earlier example of a non-recursive system, illustrates the essential difference between recursive and non-recursive systems: In a non-recursive system, *greater latency* can always be exchanged for *higher throughput.* The example of Figure 2 suggests that this is not the case for recursive systems. If we attempt to increase the throughput of the multiplier by increasing its latency, we actually end up reducing the throughput of the recursive system in

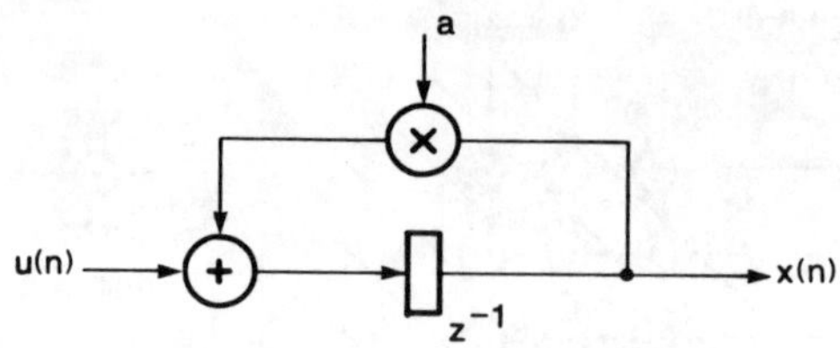

Figure 4. The simplest first-order recursive linear system.

which the multiplier is embedded. Hence, attempting to pipeline the multiplier in the recursive feedback loop is counterproductive -- in fact, since we can only do one multiply at a time, only one stage of the pipelined multiplier will be active at a time, with the remaining stages sitting idle!

Actually looks are deceiving -- we will show that in fact even for this recursive system and a broader class of such systems, greater latency can in fact be exchanged for higher throughput, and further for a given speed of technology an arbitrarily high throughput can be achieved if latency is not a concern.

Before showing this, let us generalize and formalize in the following subsection the relationship between latency for the computation *within* a recursive loop and the throughput with which it can be implemented. In particular, we can obtain a bound on the throughput called the *iteration bound.*

3.1. Iteration Bound

We derive a bound on the throughput for a given computation graph in this subsection[2, 17, 18].

Assume that we have a computation graph which operates at a single sampling rate, like the ones considered in this chapter. As we have seen by example, whenever we have a feedforward cutset for the graph, we can pipeline the left and right subgraphs by introducing pipeline latches on each feedforward path in the cutset. The throughput of the graph is therefore limited by either the left or right subgraph, and it is necessary to consider only subgraphs with no feedforward cutsets from the perspective of limitations on throughput. Every node in such a subgraph is in a directed loop, so we can limit attention to such loops. Consider any particular loop, and let the number of logical delays in this loop be N_{total}, so that the total delay around the loop is $\frac{N_{\text{total}}}{F_s}$. Let T_{total} represent the total latency required for all the computations within the loop.

For every logical delay in the loop that is preceded by a computational element, the actual time delay is less than $\frac{1}{F_s}$; the actual delay is called the *shimming delay.* Let S_{total} denote the total shimming delay in the loop. Then we must have that the total delay equals the computational latency plus the shimming delay,

$$T_{\text{total}} + S_{\text{total}} = \frac{N_{\text{total}}}{F_s}$$

and since the shimming delay must be positive,

$$S_{\text{total}} = \frac{N_{\text{total}}}{F_s} - T_{\text{total}} \geq 0$$

we have the iteration bound on the throughput,

$$F_s \leq \frac{N_{\text{total}}}{T_{\text{total}}}.$$

The available throughput for the entire computational graph is of course bounded by the delay of the directed loop with minimum throughput. The iteration bound suggests that more logical delays in the directed loop are beneficial in increasing throughput (assuming the bound is tight), since they increase the total latency around the loop available for computation.

Applying the iteration bound to the first-order system of Figure 4, we see that as expected since $N_{total} = 1$, $F_s = \frac{1}{T_m}$ where T_m is the latency required for one multiplication/addition. We shouldn't however take this iteration bound too seriously, since it applies only to the realization of Figure 4. In fact, we can recast the algorithm into a new realization for which the bound on throughput is higher -- in fact arbitrarily high! The basic technique for doing this, called *look-ahead computation[15, 16]*, is described in the following subsection.

3.2. Look-Ahead Computation

The technique of look-ahead computation is to iterate an algorithm M times before implementing it. For example, for the first-order system of Figure 1, we can write $x(n)$ in terms of $x(n-j)$ rather than $x(n-1)$, *viz*

$$x(n) = a^j x(n-j) + D(n,j)$$

where it a recursive equation for $D(n,j)$ is easily developed,

$$D(n,j+1) = D(n,j) + a^j u(n-j), \quad D(n,0) = 0.$$

The computation graph for this new algorithm is shown in Figure 5 for $M = 3$. Note that we have not changed the input/output characteristics of the algorithm in any way, but rather we have simply recast its realization. Further note that the values of a^j, $0 \leq j \leq M$ can be *pre-computed* with arbitrary precision, and hence this computation need not burden the implementation of the algorithm. The look-ahead computation of $D(n,M)$ is non-recursive; in fact, it is a vector-vector multiplication which is a degenerate case of the matrix-vector multiplication considered for the FIR filter. Hence, this part of the computation can be pipelined, down to the bit level if desired, at the expense of latency in the manner shown earlier.

The recursive part of the algorithm, represented by

$$x(n+M) = a^M x(n) + D(n,M)$$

is also shown in Figure 5. This recursive part of the algorithm now consists of a directed loop with one multiplication plus embedded addition, and in addition M delays. Hence the iteration bound says that the throughput is bounded by

$$F_s \leq \frac{M}{T_m}$$

which is M times larger than for the original realization! Since we can choose any M we desire, at the expense of a larger look-ahead computation, the iteration bound is arbitrarily large just as in the non-recursive case.

How do we actually achieve the iteration bound? One way would be to replicate the multiplier M times, but a more hardware efficient method is suggested by the derivation of the iteration bound. This more efficient method is called *retiming[2, 19]*, and effectively pipelines the computations within the feedback loop by moving the available delays around. The maximum throughput is achieved when the shimming delay is zero for each logical delay; that is, the computation which precedes the logical delay actually has a latency equal to the full sample period $\frac{1}{F_s}$. We can achieve this ideal if we can divide the total computation, a single multiply/addition, into M parts, each having a latency $\frac{T_m}{M}$, and intersperse them with the delays. In other words, pipeline the multiplier! This is considered in the following subsection.

3.3. Bit-Level Pipelining

Bit-level pipelining of the recursive algorithm is illustrated in Figure 6 for a three-bit multiplier and $M = 7$ [15, 16]. First, in Figure 6a, the computation graph of the recursive part of the

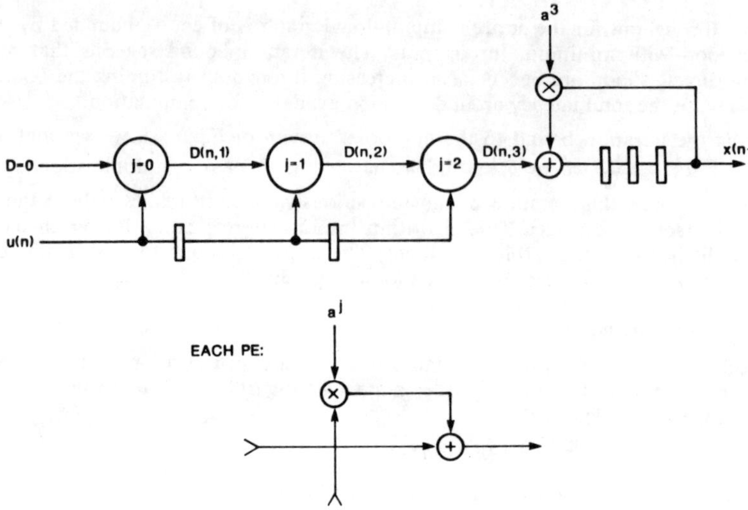

Figure 5. Look-ahead computation version of Figure 4 with $M = 3$.

system is simply replicated, including the details of the multiplier. Note again that the addition comes for free as an embedded part of the multiplier, and also note the seven delay latches which are an inherent part of the algorithm.

We can move the delay latches anywhere we want, effectively making them into slice pipeline latches, as long as we satisfy two constraints. First, each slice pipeline latch must form a cutset of the directed loop, as it is in Figure 6a, so that the recursive computation is unchanged. Second, each slice pipeline latch must form a cutset for the forward path from input to output, as it is in Figure 6a, so that the input computation is not changed. The configuration of Figure 6b satisfies these constraints, and achieves a fully pipelined multiplier. What is the throughput achieved by the configuration of Figure 6b? Unfortunately, it is only five times greater than in Figure 6a, not seven times greater! The reason can be seen upon closer examination -- each directed loop has seven delays but only five full adders, and two of the delays have no associated computational element and hence a shimming delay equal to the reciprocal of the throughput. The configuration of Figure 6b therefore does not achieve the iteration bound.

An alternate configuration which does achieve the iteration bound can be obtained[15, 16] by moving the excess latches outside the feedback loop, thereby retaining the same number of latches in each directed loop and each forward path, but resulting in a *different* number of latches in loops and forward paths. This configuration is shown in Figure 6c. The reader can verify that each forward path has seven latches, and each directed loop has five latches. Every latch in a directed loop has an associated computational element, and hence the iteration bound is achieved for $M = 5$. Hence, Figures 6b and 6c achieve the same throughput, but Figure 6c is more efficient since it has a smaller M and hence a smaller look-ahead computation.

Figure 6c can be obtained by selectively moving delays outside the loops, but it can also be obtained in a more direct fashion, and one which is very useful for application of these techniques. We simply start with the computation graph with no delays, and slice pipeline it as in the non-recursive case ignoring the feedback paths (and insuring that the feedback paths have no latches). We then count the number of latches in one feedback path (which is the same for all paths), and that is the value of M that must be used in the look-ahead computation. The resulting configuration is efficient in the sense that it achieves the iteration bound for that value of M (there are no wasted shimming delays).

Unfortunately the configuration of Figure 6c is not systolic in the sense that potentially long feedback paths exist. It can be made systolic as shown in Figure 6d by including the feedback paths in the slice pipeline latches. This is expensive, however, since the number of latches in a feedback loop is now $M = 16$ and the latency is also higher. Since the feedback paths have no computational elements, they can tolerate more signal path delay, and therefore the large number of latches in the feedback path in Figure 6d is undoubtably not necessary. There are of course intermediate possibilities between the extremes of Figure 6c and 6d which trade fewer latches in the feedback path for reduced look-ahead computation.

The technique we have proposed for increasing the throughput of a recursive first-order system can be described as a combination of *look-ahead computation,* to increase the number of delays in the loop, and *retiming* to reconfigure those delays to effectively pipeline the computation within the loop. The significance of the technique is that a first-order recursive system can be implemented with a throughput equal in theory to the reciprocal of the latency in a full addition (less than that in practice due to clock skew problems). Although we have shown that as M increases, the iteration bound on throughput increases without bound, we have not demonstrated the ability to achieve this bound for an M any larger than that appropriate to fully pipeline a single multiplier. In fact, it is possible to increase the throughput without bound by turning the SISO recursive system into a MIMO system in a manner similar to Figure 1. Rather than illustrate this for a first-order system, we go directly to a higher-order system in the next section.

4. Higher Order Systems

Thus far only the first-order system has been considered, but we have developed all the tools to fully pipeline higher order recursive systems at the word-level[8, 20, 21] or bit-level[15, 16] as well. Consider a general state-space representation of a SISO linear system, where the state update equation is

$$\mathbf{x}(n+1) = \mathbf{A}\mathbf{x}(n) + \mathbf{b}u(n)$$

where $\mathbf{x}(n)$ is a state vector of size N for an N-th order system, and the output equation is

$$y(n) = \mathbf{c}^T\mathbf{x}(n) + du(n).$$

The transfer function of this system is

$$H(z) = \mathbf{c}^T(z\mathbf{I} - \mathbf{A})^{-1}\mathbf{b} + d$$

where the all-important state matrix $\mathbf{A}$ determines the poles of the filter. The state matrix is unique only up to a similarity transformation. As we will see, we can dramatically affect the efficiency with which the system is implemented by the judicious choice of $\mathbf{A}$.

As in the non-recursive case, we can turn this SISO system into a MIMO system by defining a vector of L inputs $\mathbf{u}^{(L)}(kL)$ and L outputs $\mathbf{y}^{(L)}(kL)$, in which case the system becomes

$$\mathbf{x}((k+1)L) = \mathbf{A}^L\mathbf{x}(kL) + \mathbf{B}\mathbf{u}^{(L)}(kL)$$

$$\mathbf{y}^{(L)}(kL) = \mathbf{C}\mathbf{x}(kL) + \mathbf{D}\mathbf{u}^{(L)}(kL)$$

where in addition a look-ahead computation by L samples has been applied to the state equation. This realization is known as a *block state realization[22-29].* Particular expressions, which need not concern us here, can be derived for the matrices **B**,**C**, and **D**. The important point to notice is that the state equation is essentially unaffected except that $\mathbf{A}$ is replaced by $\mathbf{A}^L$, which can again be pre-computed. The structure of the realization is shown in Figure 7, where an *MVM* is a *matrix-vector multiplier* of the type of Figure 2 and the *SUN* is the *state update network* which incorporates the recursive portion of the computation[8]. All the computations are non-recursive, and can be fully pipelined using matrix-vector multipliers, with the exception of the SUN. We will show shortly that this MIMO system can operate at a throughput equal to the reciprocal of the latency of a single full addition, and since the effective throughput of the SISO system is L times as great, the SISO throughput is unbounded as L increases.

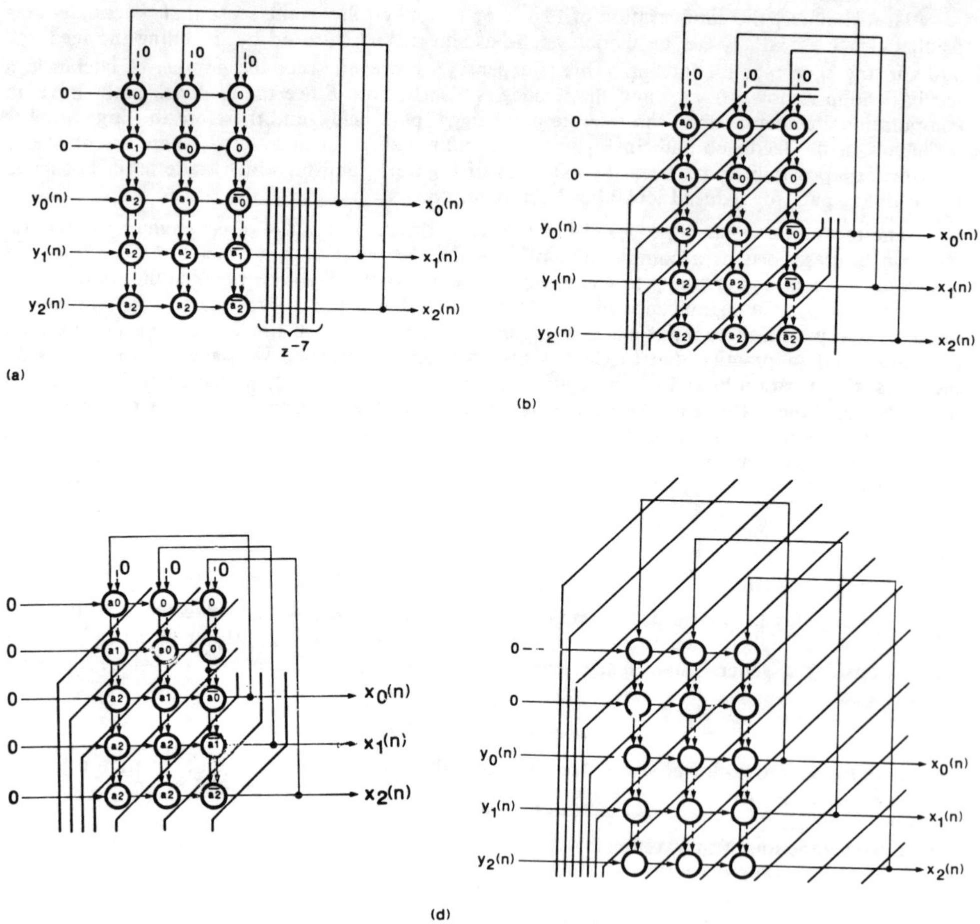

Figure 6. A bit-pipelined version of the recursive portion of a first-order recursive portion with look-ahead computation.

From the perspective of throughput, the only term that concerns us is the recursive system implemented in the SUN, so in the sequel we will focus on only this part. Pipelining of the SUN at the bit level requires, as in the first order case, a look-ahead computation by M samples[15, 16]. Applying this look-ahead to the SUN, it becomes of the form

$$\mathbf{x}((k+M)L) = \mathbf{A}^{ML}\mathbf{x}(kL) + \mathbf{z}(kL).$$

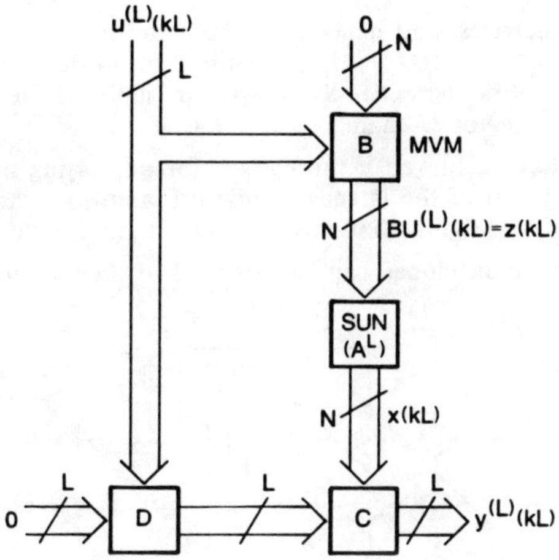

Figure 7. Overall structure of a block-state realization of a recursive digital filter. ***L*** **is the block size and** ***N*** **is the order of the filter. "MVM" refers to a matrix-vector multiplier, and "SUN" refers to a state-update network.**

where the details of the look-ahead computation of $\mathbf{z}(kL)$ need not concern us here, except to note that the complexity of this will increase with M and it is non-recursive and hence can be fully pipelined. Examining the SUN in more detail, the computation is laid out in Figure 8a with word-level pipelining for the special case of $L = 3$ and $M = 5$. The five delays in the loop have been moved so that they constitute slice pipeline latches. In fact, Figure 8a is very similar, at the word pipelined level, to Figure 6b at the bit pipelined level -- topologically the two problems are virtually identical.

Figure 8a allows us to do state updates at a rate equal to the reciprocal of the multiplier latency, and is fully pipelined at the word level. This is at the expense of a look-ahead computation corresponding to $M = 2N - 1$ for an NxN state matrix. Fortunately, if the state matrix is triangular, or close to triangular, the expense in look-ahead computation can be reduced correspondingly. But before looking at this case, first observe that the structure of Figure 8a can easily be pipelined at the bit level. Suppose that bit-level slice pipelining of the multiplier requires W pipeline stages in the sense of Figure 6c, then if M is chosen to be $(2N-1)W$, each of the latches in Figure 8a can be replaced by W latches, which can in turn be used to pipeline the multipliers at the bit level. Further note that a smaller M can be used as in Figure 8b (analogous to Figure 6c) -- this configuration requires $M = N$ for pipelining at the word level or $M = NW$ at the bit level.

The case of a lower triangular state matrix is shown in Figure 8c, where the feedback reduces to feedback around each single multiplier on the diagonal element. Additional slice pipeline latches are added to pipeline the below-diagonal elements at the word- or bit-level. This case requires a much smaller M than the full matrix case, namely $M = W$, and hence a lower look-ahead computation penalty.

Unfortunately, it is not possible to transform any filter transfer function into a similar lower triangular form with real elements. However, it is possible to obtain a *quasi-triangular* state matrix $\mathbf{A}$ by similarity transformation[8]. By quasi-triangular, we mean a matrix with two-by-two sub-matrices along the diagonal and zeros elsewhere above the diagonal (and in fact cascade second-order sections and lattice structures have this form). Further, fortuitously if $\mathbf{A}$ is quasi-triangular, then so is the pre-computed state matrix $\mathbf{A}^{ML}$. The case of quasi-triangular SUN is shown in Figure 8d for $N = 4$. As can readily be seen, the feedback in this case is around the two-by-two diagonal sub-matrices, and hence $M = 2$ is required for word-level pipelining of the recursive portion of the system, and $M = 2W$ is required for bit-level pipelining. Note again that the technique is to slice pipeline the entire SUN structure ignoring the feedback, and then noting after the fact the required value of M (namely 2 or 2W).

To summarize, we have achieved a throughput for processing of blocks of L SISO input samples equal to the reciprocal of the latency of one full addition. As L increases, in principle the effective throughput at the original SISO level grows without bound.

The techniques we have developed can be extended to time-varying systems as described in the following section.

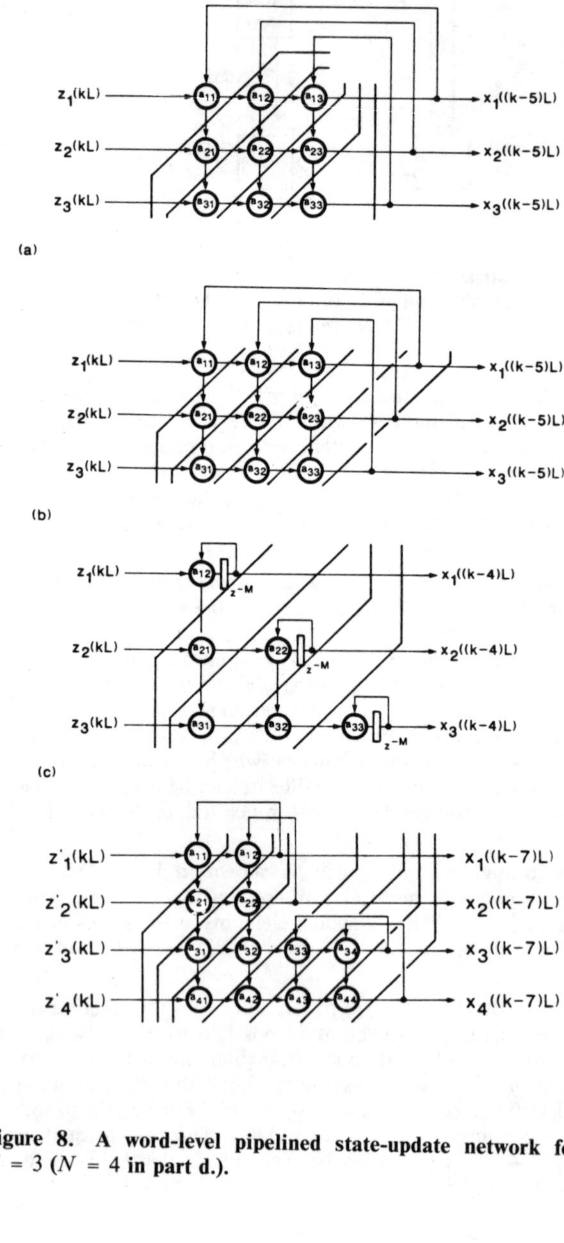

Figure 8. A word-level pipelined state-update network for system of order $N = 3$ ($N = 4$ in part d.).

5. Time-Varying First-Order System

The previous results, relating to general linear time-invariant systems, can be extended to time-varying systems. This opens up a much more interesting class of applications, such as adaptive filters. As in the time-invariant case, most of the insight can be gained by looking at the first-order system case.

5.1. First-Order System

Consider a first order system

$$x(n) = a(n)x(n-1) + u(n) \tag{2}$$

identical to (1) except that the multiplier is now time-varying. This system is subject to look-ahead computation for bit-level pipelining in much the same manner as (1). In fact, (2) reduces to (1) when $a(n)$ is a constant. Writing $x(n)$ as a look-ahead computation,

$$x(n) = C(n,j)x(n-j) + D(n,j)$$

we can easily develop the following recursions for C and D

$$C(n,j+1) = a(n-j)C(n,j), \;\; C(n,0) = 1$$
$$D(n,j) = D(n,j) + u(n-j)C(n,j), \;\; D(n,0) = 0.$$

In particular, we can look-ahead by M samples using

$$x(n) = C(n,M)x(n-M) + D(n,M)$$

as shown in Figure 9 for $M = 3$. The look-ahead computation takes three PEs, each with two multiplies, and each accepting delayed versions of the two input signals $a(n)$ and $u(n)$. This portion of the system can easily be fully pipelined at the bit level. Comparison with the time-invariant case of Figure 5 should be made, where one fewer multiply per PE is required (this is because of the ability to pre-compute $C(n,j)$ for the time-invariant case). The recursive portion of the computation graph has, as in the time-invariant case, one multiply/addition and $M = 3$ delays, and can be fully pipelined at the bit level if M is chosen appropriately[16].

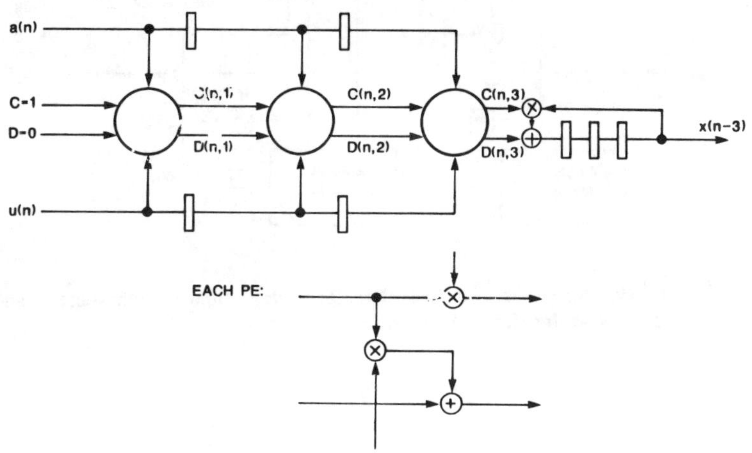

Figure 9. Look-ahead computation version of a first-order time-varying linear system with $M = 3$.

5.2. Example: A Vectorized Lattice Filter

As an example of an application of a time-varying system which can benefit from vectorization and word- or bit-level pipelining, consider an adaptive filter. An adaptive lattice filter[4], as shown in Figure 10, is particularly suitable for this purpose, since the inherent feedback is limited to a single stage of the filter. A joint-process estimator is shown; a predictor is assumed as a special case. A single stage, shown in Figure 10b, accepts as inputs the forward, backward, and joint-process errors of order n and outputs the errors of order n+1. The two coefficients $k_{n+1}(T)$ and $k^c_{n+1}(T)$ characterize the state of this stage at time T. An adaptation algorithm operating on the input signals updates the state, and contains the only recursion in the filter. This adaptation algorithm is of the form of (2),

$$k_{n+1}(T) = a_{n+1}(T)k_{n+1}(T-1) + b_{n+1}(T)$$

with a similar relation for $k^c_{n+1}(T)$, where the two inputs to the state update, $a_{n+1}(T)$ and $b_{n+1}(T)$ are functions of the input signals and depend on their details on the adaptation algorithm used.

The structure of Figure 10 is completely pipelineable using the technique described in the last section[30-32]. At the expense of look-ahead computation and latency, the system can be implemented with a throughput equal in theory to the reciprocal of one full adder time. Furthermore, it can be vectorized, calculating L successive output samples in parallel, to obtain an additional (unbounded) speedup.

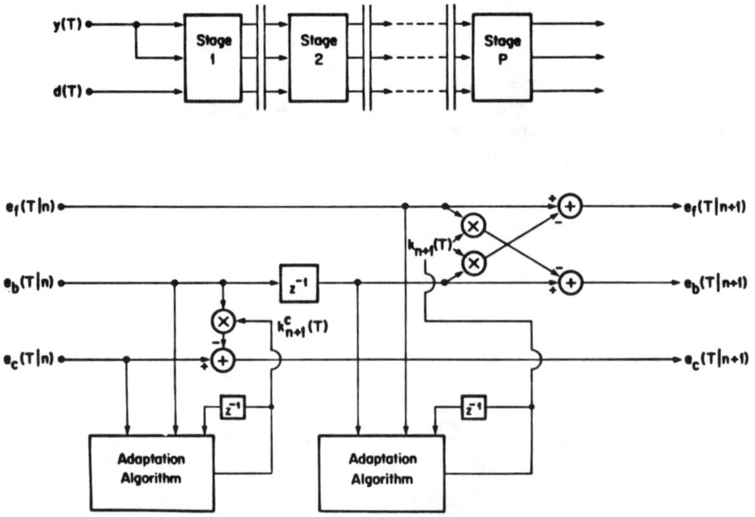

Figure 10. An adaptive lattice filter. The adaptation of each stage is an example of a first-order time-varying linear system.

6. Conclusion

In this chapter we have shown that there are no fundamental limits on the throughput with which a large class of signal processing algorithms can be implemented, assuming that latency is not a consideration. Included are recursive as well as non-recursive algorithms. The techniques which have been used to overcome any potential limits include *pipelining, look-ahead computation,* and *retiming.* Using these techniques, a large class of algorithms can be implemented with the throughput of a full-adder for a vector of L input and output samples. As L increases this throughput increases without bound. The throughput of a pipelined full-adder in currently available CMOS technologies is in the neighborhood of 30 MHz, so that quite high sampling rates can be achieved for modest values of L. This throughput is limited by clock skew problems rather than the speed of the full-adder, so there may be opportunities to increase this throughput further by techniques such as self-timed logic.

In spite of the absence of any fundamental limits on throughput, there are of course practical limitations, a few of which we can mention:

1. In many applications the acceptable latency is limited.
2. Realization of these architectures requires serial-to-parallel conversion of samples starting at the original throughput. This conversion must be performed by a higher-speed technology than that used to perform the signal processing and will present a severe problem for very high sampling rates.
3. Unless major advances are made in the bandwidth on and off a chip, our ability to implement a system of the type described here will soon be limited by I/O bandwidth rather than processing power[1]. Therefore, realizations of many of these systolic systems will necessarily be multi-chip. Fortunately, systolic realizations enable us to divide the system into smaller pieces with reduced I/O requirements, and hence this limitation is actually a further justification of the approaches described here[3].
4. Synchronous interconnection of locally interconnected processing elements leads to problems of clock skew in large chips or especially multi-chip systems[2]. Fortunately there appear to be solutions to this problem short of excessively slowing the clock rate.
5. High speed realizations of signal processing algorithms requires potentially a lot of hardware, which can present a problem in cost-sensitive applications.

7. Acknowledgement

The author is indebted to his colleagues (in alphabetical order) Edward A. Lee, Hui-Hung Lu, Teresa H.Y. Meng, and Keshab Parhi, without whose contributions this chapter would not be possible. This research was supported by the National Science Foundation, Grant ECS-82-11071, the Advanced Projects Research Agency under order number 4031 monitored by the Naval Electronics Systems Command contract number N00039-C-0107, a grant from Shell Development Corporation, and an IBM Fellowship.

---

1. J.W. Goodman, F.J. Leonberger, S-Y Kung, and R.A. Athale, "Optical Interconnections for VLSI Systems," *IEEE Proceedings* **72**(7) p. 850 (July 1984).

2. S. Y. Kung, "On Supercomputing with Systolic/Wavefront Array Processors," *Proceedings of the IEEE* **72**(7)(July 1984).

3. H. T. Kung, "Why Systolic Architectures," *IEEE Computer* **15**(1)(Jan. 1982).

4. M. Honig and D.G. Messerschmitt, *Adaptive Filters: Structures, Algorithms, and Applications,* Kluwer Academic Press, Hingham, Mass. (1985).

5. H. T. Kung and Charles E. Leiserson, "Algorithms for VLSI Processor Arrays," in *Mead and Conway, Introduction to VLSI Systems*, Wesley Publishing Co., Reading, MA (October, 1980).

6. Hassan M. Ahmed, Jean-Marc Delosme, and Martin Morf, "Highly Concurrent Computing Structures for Matrix Arithmetic and Signal Processing," *IEEE Computer* **15**(1)(Jan. 1982).

7. Sailesh K. Rao and Thomas Kailath, *VLSI and The Digital Filtering Problem,* Information Systems Laboratory, Stanford University,, Stanford (1984). Internal Memorandum

8. Hui-Hung Lu, Edward A. Lee, and David G. Messerschmitt, "Fast Recursive Filtering with Multiple Slow Processing Elements," *IEEE Transactions on Circuits and Systems*, (November, 1985).

9. C.W. Wu, P.R. Cappello, and M. Saboff, "An FIR Filter Tissue," *Proc. 19th Asilomar Conf. on Circuits, Systems, and Computers*, (Nov. 1985).

10. J. Robert Jump and Sudhir R. Ahuja, "Effective Pipelining of Digital Systems," *IEEE Trans. on Computers* **C-27**(9)(Sept., 1978).

11. Cappello, Peter R. and Steiglitz, Kenneth, "A VLSI Layout for a Pipelined Data Multiplier," *ACM Transactions on Computer Systems* **1** (2) p. 157 (May 1983).

12. Luk, W. K., "A Regular Layout for Parallel Multiplier of $O(Log sup 2 N)$ Time," pp. 317 in *VLSI Systems and Computations*, , ed. Guy Steele,Computer Science Press, Rockville, Md. (1981).

13. Reusens, Peter, Ku, Walter H., and Mao, Yu-hai, "Fixed Point High Speed Parallel Multipliers in VLSI," pp. 301 in *VLSI Systems and Computations*, , ed. Guy Steele,Computer Science Press, Rockville, Md. (1981).

14. Brent, R.P. and Kung, H.T., "A Regular Layout for Parallel Adders," *Technical Report, Dept of Computer Science, Carnegie-Mellon University* **CMU.CS.79.131** (June 1979).

15. Parhi, Keshab Kumar and Messerschmitt, David G., "A Bit Level Pipelined Systolic Recursive Filter Architecture," *Proceedings of the International Conference on Computer Design*,, (1986).

16. K. Parhi and D.G. Messerschmitt, "Efficient Implementation of Recursive Filters Pipelined at Bit Level," *IEEE Trans. on Acoustics, Speech, and Signal Processing*, ((submitted)).

17. Markku Renfors and Yrjo Neuvo, "The Maximum Sampling Rate of Digital Filters Under Hardware Speed Constraints," *IEEE Trans. on Circuits and Systems* **CAS-28**(3)(March 1981).

18. A. Fettweis, "Realizability of Digital Filter Networks," *Arch. Elek. Ubertrangung* **30** pp. 90-96 (Feb. 1976).

19. Charles E. Leiserson and Flavio M. Rose, "Optimizing Synchronous Circuitry by Retiming," *Third Caltech Conference on VLSI*, (March, 1983).

20. Hui-Hung Lu, *High Speed Recursive Filtering*, University of California, Berkeley (1983). PhD Thesis

21. David G. Messerschmitt, *VLSI Implemented Signal Processing Algorithms*, NATO Advanced Study Institute, Bonas, France (July 1983). Conference

22. B. Gold and K. L. Jordan, "A Note on Digital Filter Synthesis," *Proceedings of the IEEE* **65** pp. 1717-1718 (Oct., 1968).

23. H. B. Voelcker and E. E. Hartquist, "Digital Filtering Via Block Recursion," *IEEE Trans. Audio and Electroacoustics* **AU-18** pp. 169-176 (June 1970).

24. Charles S. Burrus, "Block Implementation of Digital Filters," *IEEE Trans. on Circuit Theory* **CT-18** pp. 697-701 (Nov. 1971).

25. Casper W. Barnes and S. Shinnaka, "Block Shift Invariance and Block Implementation of Discrete-Time Filters," *IEEE Transactions on Circuits and Systems* **CAS-27** pp. 667-672 (Aug., 1980).

26. Sanjit K. Mitra and R. Gnanasekaran, "Block Implementations of Recursive Digital Filters - New Structures and Properties," *IEEE Trans. on Circuits and Systems* **CAS-25** pp. 200-207 (April, 1978).

27. Jan Zeman and Allen G. Lindgren, "Fast Digital Filters with Low Round-off Noise," *IEEE Trans. on Circuits and Systems* **CAS-28** pp. 716-723 (July 1981).

28. D. A. Schwartz and T. P. Barnwell, III, "Increasing the Parallelism of Filters Through Transformation to Block State Variable Form," *Proceedings of ICASSP '84, San Diego*, (1984).

29. Chrysostomos L. Nikias, "Fast Block Data Processing via a New IIR Digital Filter Structure," *IEEE Transactions on ASSP* **32**(4)(August, 1984).

30. Teresa H.-Y. Meng and D.G. Messerschmitt, "Implementations of Arbitrarily Fast Adaptive Lattice Filters With Multiple Slow Processing Elements," *Proc. IEEE ICASSP*, (April 1986).

31. Teresa H.-Y. Meng and D.G. Messerschmitt, "Arbitrarily High Sampling Rate Adaptive Filters," *IEEE Trans. on Acoustics, Speech, and Signal Processing*, ((to appear)).

32. K. Parhi and D.G. Messerschmitt, "Bit Level Pipelined Adaptive Filters," *IEEE Trans. on Acoustics, Speech, and Signal Processing*, ((submitted)).

OPTIMUM SCALES AND LIMITS OF INTEGRATION

Professor Daniel V McCaughan

Marconi Electronic Devices Limited

Introduction

The usefulness and viability of silicon technology is viewed with totally different eyes by the final system user, who has to supply (usually in severe time constraints), systems generated to perform to what are often severe environmental and speed/power limitations, by the inventor or developer of the technology who usually sees its virtues through "rose-tinted" spectacles, and ignores its problems, and by the realist in the middle who is constrained by the conflicting necessities of procurement for forward use, guaranteed sources of supply, and sufficient innovation to ensure technical performance and price competitiveness.

The subject of this paper, in this context, is silicon technology, with particular emphasis on the various MOS technologies, especially CMOS and the lessons we should learn from experience in the use of devices in large scale systems.

Integration and Constraints

To discuss optimum limits of integration we should be clear what we mean by integration; historically the integration of silicon devices on single chips has progressed at a rate unprecedented in any other industry: using single transistors of 1960 vintage for example, one bit of static memory would occupy about $1cm^2$ of printed circuit board i.e. 1Kbit of static RAM would occupy about 1sq ft of board space - making a rudimentary desk-top computer fill about 10000 PCBs of this size. The enormous reduction in power and size brought about by the integration of up to 10^6 components on a single chip by 1985, has made possible great increases in the computational and data storage power of even simple commercial systems, and all the cost reductions associated with this reduction in size. Even such a relatively simple chip as a 4K static Random access memory (SRAM) has perhaps 50,000 connections on it, and is fabricated for less than $1.

Some practical limits have emerged over the last number of years, associated with factors outside the chip itself. Firstly, there are almost no silicon chips made with physical lateral dimensions greater than about 1cm square, while most are smaller. This is related to the physical realities of making the packages as much as any other factor. Secondly the number of connections (package pins) from the outside world seldom exceeds 225 and indeed is much more likely to be between 16 and 84. Thirdly, voltage standards have come down over the years from 12V in the 1960s to 5V now; in practical terms the 5V standard is likely to remain the most common, with implications for physical limitations of electrical performance.

J. K. Skwirzynski (ed.), Performance Limits in Communication Theory and Practice, 21–42.

A factor not generally recognized is the innate conservatism of the silicon integrated circuit industry in general. Aluminium and its simple alloys (with Silicon and Copper) is virtually the only metal used for interconnect wiring on-chip; silicon dioxide and silicon nitride are almost the only insulators used; gate material is primarily polysilicon, with a few variants such as policide (metal-silicon alloys) . A few processes use organic dielectrics for inter-metal insulation in multilevel metal systems, and for top-coat anti-alpha-particle protection for random access memories. Of late, a few circuits using other (refractory) metals have been coming into use; occasionally alternate dielectrics (e.g. Alumina, pioneered by Bell Labs in the 1960s) are now incorporated, particularly in EEPROMS, and silicides of a number of metals have been used as a reduced resistivity interconnect layer. Finally conservatism extends to dopants:- boron, phosphorus, arsenic and antimony dominate; other (possible multi-diffusion) systems, such as Gallium/Germanium, used in power devices, are rare.

This conservatism has its roots in practical realism. Manufacturers have a reluctance to deviate from the well trodden paths which are known to lead to reliable and reproducible circuits. This same reluctance is seen in processing methods : dry plasma etching techniques were common in some major laboratories in the 1960's with "backsputtering" being used as a contact-cleaning technique pre-metallization in (typically) a noble gas environment. The realization in the early 1970's that flourine-and chlorine-containing gases could be used to selectively etch layers of metal, dielectric or silicon was only put to practical production use to any significant extent by the end of the decade and into the 1980's. Similarly, predictions in the 1960's and early 1970's that electron-beam lithography would rapidly replace optical lithography have been unfounded; other laboratory methods such as X-ray lithography have still to make a major impact on device fabrication on a large scale, though both Electron-beam and X-ray methods are now being considered seriously for very fast turnaround prototyping manufacturing systems; these ideas have been in small scale use in major facilities for some time.

Thus, overall, the integrated circuit manufacturing industry is a complex hybrid of very advanced thinking at the academic level, with conservative views in manufacturing and manufacturability.

Limitations from a System viewpoint

Before considering the on-chip practical limits of integration it is worth while considering the large system requirements for processing capability. Large systems are assembled from many chips, usually of many different types, and require a large amount of chip to chip interconnection. Constraints of timing, synchronization, chip to chip driving capability, off-chip signal bandwidths and overall architecture will usually dominate. One can distinguish several generic types of system in this consideration. One is inherently relatively limited in its total requirement; an example of this could be the environmental control of a domestic dwelling or the functional control of a motor vehicle. Here the total number of gates necessary for the total functionality can be computed; at some level of integration all these functions will be contained on one chip and considerations of cost rather than orders of magnitude increases in complexity will dominate. Other systems are not inherently self limiting in this way : the amount of data computation, reception, and transmission required by an office complex could in one sense be unlimited : high bandwidth computing, high bandwidth fibre optic link communication, for both data and voice/vision etc together with all the computational complexity necessary for financial, logistical, and other information systems will require large multichip, multisensor systems for the forseeable future. Similar would be large avionics or flight systems, large radar installations or complex telephone/communications systems. In all such complex systems are one can make some approximations. One is known as "Rent's rule" (1) which relates the number of logic gates in a system to the number of connections that need to be made to it. Approximately:

$C = 2N^{2/3}$--------(1)

where C is number of connections and N the number of circuits per chip. Practically however (2) there is not much data to extrapolate this beyond 1000 connections where there appears to be a 'saturation' at $>10^4$ circuits/chip. If one extrapolates to large biological systems however, the function appears to carry on to $>10^8$ connections in $>10^{12}$ gates equivalent (3).

Similar computations may be done on average length of line on a chip and the number of connections of a given length on a chip. It has been suggested (1) that the average line length increases as the (1/3) power of the number of circuits per chip. As to the distribution of length of wiring on chip, on a typical LSI the function of appearance frequency follows a damped oscillatory pattern out to remarkably long lengths in purely random logic circuits - e.g. up to 14mm on a typical current LSI chip of 10000 transistors (4).

The limiting factors in large systems are thus related to the increasing complexity of chips constituting the system, where gradually the off-chip complexities are being replaced by on-chip complexities and off-chip simplicity as the number of chips/system is decreased. As an example the MEDL family of chips for the 1553B military bus system began as a 5-chip family, with a complex mother board interconnecting the chips made in 5µ technology. This system has recently been supplemented by a two-chip solution in 3 micron technology where one chip performs the data transmission and another the reception; finally a recently announced new version on 3/2.5µm technology puts all the functions of transmission and reception on one chip (5).

A problem, which this paper will not discuss at length, is that of testing. One advantage of a multichip system is that each element may be tested prior to fabrication into the final system. As chip complexity increases and access becomes more difficult the importance of built-in self test (BIST) increases and synchronous systems with scan-path testing methodologies gain in importance.

The (V)LSI advantage in large systems is well illustrated by the discussion above. Clearly off-chip connections between chips represent considerable capacitative loads - a fundamental limitation to the speed of any system is the capacitances that need to be driven within it. The nodal capacitances between chips are very much higher than those internal to a chip and therefore dominate both the off-chip drive capability requirements of any given chip, and thus the overall system speed; indeed within any individual chip the speed is dominated by the necessity to drive local nodal capacitances and in the circuits of practical experience, the whole system speed can be determined by the single slowest node on one single chip in a critical date path. These considerations have also a maximal impact on overall system architectures and from these, the chip architectures themselves lead to a central theme of this paper :

that is, that systolic, pipelined architectures which minimize both internal data paths and the necessity to drive long interconnections, and external data connections in the same way are often the optional choice for computational intensive or signal processing orientated practical systems.

Device Considerations

Before consideration of fundamental technical questions of limitations on device performance we should bear in mind the rate of increase of components (or bits) per chip since the LSI/VLSI era began in the late 1960's. We may tabulate as an example the practical availability of static and random access memories taking prototype sampling as the date of availability :

MOS LSI/VLSI Memory

Year	Memory	
approx 1971	1Kbit DRAM	PMOS
approx 1973/4	4Kbit DRAM	
approx 1976	16Kbit DRAM	NMOS
approx 1979	64Kbit DRAM	
approx 1983	256Kbit DRAM	CMOS
approx 1986	1Mbit DRAM	

Note that (i) there is a log/linear relationship between memory size and year of introduction, still continuing at roughly the same pace - though with indications that the trend is flattening out; and that during the period the technology has moved from PMOS LSI, through NMOS,to current generation CMOS. During the same period of time, line dimensions have followed a similar log-linear trend, with "Rules" of 8μm for the early 1K-bit parts being followed by 5/4μm rules for the 4K-bit parts, 3μm for the 16 - and 64 - Kbit parts and down to 1μm rules for the 1M-bit components now (1986) at early prototype sampling in-house in major companies, (though not yet available generally).

A number of advances have enabled these trends to take place (16); the early 1K-bit memory devices used a three-transistor cell, storing charge in the gate plus wiring capacitance of a MOSFET; later memories (the 4K) used a simple single transistor all with a capacitor to store the data, switched by an actual MOSFET with real sources and drains; later 4K devices used a two level polysilicon process instead of one level, and compressed the cell to be a "virtual" transistor with a capacitor connected to a charge-coupled-device-like drain/source - much more compact and more process tolerant. The later compression of this design by the adoption of common elements between cells and utilization of more complex capacitance - enhancement techniques led to the 256K bit device, and the latest offerings at the 1M-bit level use even more complex isolation methods to compress their cells still further, plus minimizing metal interconnect pitch dimensions.

This illustrates how in practical terms the evolution of IC technology takes place : simple scaling of existing structures is simply not enough to realize the required increases in packing density needed to keep the die size in the chips to managable proportions. It also illustrates a principle : simple consideration of "shrinkage" and "scaling" is not enough - it must be combined with design cleverness if real advances are to be made. In the example of memory devices the evolution from three to one transistor cells was major breakthrough; the use of double-level polysilicon was a second, and enhancement of cell capacitance was a third. These were combined with lithographic and other technological improvements in line width reduction; furthermore the transition to the one transistor cell was accompanied by a reduction in voltage requirement from 12V to 5V, which in turn allowed reduction of diffusion and oxide isolation dimensions, allowing further shrinkage of overall chip size. Finally increased sophistication of sense-amplified design has reduced the detectable charge-packet per cell by orders of magnitude, down to the order of 10^4 - 10^5 electrons per cell in a current 64 - or 256K-bit DRAM, giving a sensing voltage signal swing of approx 200mV in about 100nSec at 75°C. That this is far away from fundamental limits may be seen from the signal packet size of another device using the same basic technology - the Charge-coupled device imager - where, with moderate cooling to reduce background spurious currents, picture element charge packets of 1000 electrons or less are routinely detected.

The number of components (transistors, capacitors etc) per chip has proved to be a good measure over the years of practical progress. Moore (6) proposed that the number of components per chip doubled every year. This was virtually true from 1959 to 1974 but recent years have seen a dramatic slowing of this rate of progress, perhaps to a doubling every 3 to 5 years, with a continuing trend of slowing down (7). Even with this reduction in rate of progress the implication is one of over 4 million components per chip in production in 1991.

Scaling of Devices

The foregoing discussion leads naturally to questions fundamental to an understanding of practical limitations in device technology - i.e. what are the factors that govern technological limits in devices and technology and what methods should be followed in scaling the size of devices, as the n- and p-channel device of Fig 1., which is a cross section of a CMOS device structure fabricated in a LOCOS - type process. In this type of process one starts with a bare silicon wafer, and, on top of a thin SiO_2 layer, deposits a sacrificial layer of silicon nitride. This nitride layer is then patterned so as to leave areas of nitride only over areas of silicon which subsequently form active transistors. Subsequent processing grows thick thermal SiO_2 in every other area of the wafer. As will be seen from the schematic diagram, this results in a SiO_2 layer which tapers towards the active areas. This "birds-beak" is of lesser significance in structures with gate dimensions of 5 or 6 micrometers but as the gate dimensions approach the thickness of the SiO_2 layer itself ($\frac{1}{2}$-1μm) the character of the structure changes so that effects related to these edges can be a major factor in MOS device performance.

Furthermore, bipolar devices laterally isolated by layers of SiO_2 (as are most high-speed devices) will also be fabricated by comparable methods leading to complex structures with performance related closely to edge behaviour. The structure of the oxide will also have significant effects on the capacitances associated with the wiring and interconnection of the devices for both bipolar and MOS.

Fig 1

A typical p-well CMOS structure fabricated in a LOCOS type process.

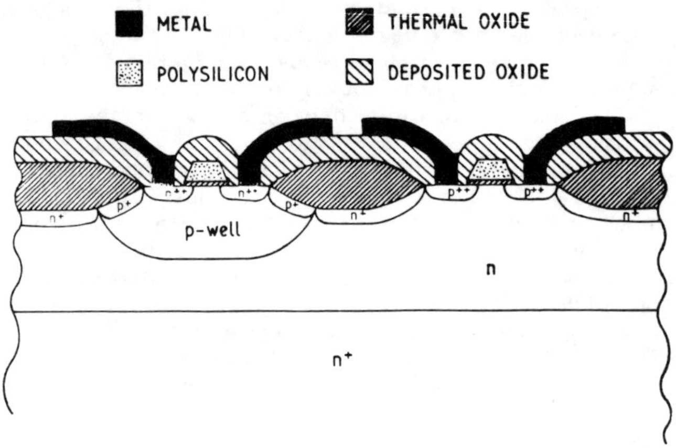

One very practical matter should be kept in mind when reading the vast amount of literature on this subject, and when discussing MOS technologies in generalized terms such as "5μm Technology", "2μm Technology" etc - scaling laws whether for bipolar or MOS technologies usually give consideration to individual device behaviour; practical implementation of these scaling factors in real processes is complicated by considerations of metal pitch, diffusion spacings, contact sizes etc, and also by (especially as device dimensions shrink) factors related to tolerances arising from lithography and etching (spacing round contacts, metal overlaps, nested vs non-nested vias etc). In practical terms therefore one "2μm"technology can easily have a factor of two or three advantage in speed and packing density over another process of supposedly equivalent dimension.

The initial concept of MOS scaling [8] was that co-ordination of changes in dimensions, voltage and doping levels, could produce devices with constant electric field distribution and power density. The following table gives the scaling factor (either assumed initially or derived) for each of a series of parameters, to first order.

PARAMETER	FACTOR
Device Dimension L, W, t_{ox}	1/K
Device Area	$1/K^2$
Junction Depth	1/K
Voltage V	1/K
Current I	1/K
Capacitance EA/1	1/K
Doping N_A/N_D	K
Delay Time CV/I (T_D)	1/K
Power Dissipation IV	$1/K^2$
Power Density IV/A	1
Power - Delay Product	$1/K^3$
Frequency Dependent Power Dissipation (CV^2/T_D)	$1/K^2$

It should be noted however that a number of factors do not scale in a favourable direction, associated particularly with line resistances and contacts (8).

PARAMETER	FACTOR
Line Resistance (pL/Wt)	K
Normalized Voltage Drop (IR/V)	K
Line Response Time ($R_L C$)	1
Line Current Density (I/A)	K
Contact Resistance (Rc)	K^2
Contact Voltage Drop (Vc)	K
Normalized Contact Voltage Drop (Vc/V)	K^2
Normalized Line Response Time	K

The results of these tabulated considerations lead to certain practical conclusions. Firstly there are real advantages in terms of power-delay and density - delay products, which scale as K^3. The power density scaling factor is inverse to the device density factor so the net result is unchanged. The voltage scaling is trickier, however. The 5V standard is so well established in practical terms that the reality of use in systems is that any manufacturer who keeps to the 5V capability has an advantage - hence scaling with constant voltage is also used in practice. This leads to considerable problems with enhanced fields, enhanced subthreshold currents etc. In practical terms some considerable advantages have been gained by scaling the device oxide thickness alone, or with scaling of the transistor lengths, especially when design considerations limit the drive requirements from an individual device; clearly reduction in transistor dimensions reduces its capability to drive large capacitances; if the nodal capacitances are dominated by unscaled terms and interconnect wiring, the gains to be made from transistor shrinkage are very limited. A further factor is that n-channel device gains peak at about 1μm because of considerations of electron velocity saturation in the device channel, and the finite thickness of inversion layers (9).

Practical considerations may be tabulated as follows (10) :

	Parameter	LSI Limits Factor (K)	VLSI Effects reducing Factor K
1.	Feature Size	K	Lithography, Etching limits
2.	Density	K^2	Design, alignment, wiring limits
3.	Channel Length	K	Short Channel limits, sub-threshold currents, reliability
4.	Gate Delay	K	Velocity Saturation, Oxide limits, parasitic resistances, interconnect resistance, capacitance, current drive
5.	Supply Volts	K	Hot Electron effects, Dielectric Breakdown
6.	Power Dissipation	K^2	Subthreshold leakage, drain - barrier lowering
7.	Power - Delay Product	K^3	Parasitic, interconnect limits
8.	Gate Dielectric	K	Need very thin gate oxides with severe limits on reliability
9.	Parasitic factors	-	Contact Resistances do not scale

Taking a Static Ram (by comparison with the early table on dynamic RAM, a static RAM of the same area as a given dynamic RAM will have $\frac{1}{4}$ the bit count) :

APPROX FEATURE SIZE FOR A CMOS PROCESS	METAL PITCH	CELL AREA (6 TRANSISTOR) μm^2
3μm	8μm	1600
2μm	8μm	1300
2μm	6μm	750
1.5μm	6μm	750
1.25μm	4.5μm	420
1μm	3μm	300
1μm	2.5μm	150

Note that advances in device isolation give the packing density increases towards the lower half of the table.

Some of the problem factors associated with decreasing the device dimensions are shown in Fig 2, illustrating the hot carrier effects which are seen as the effective internal fields increase.

Fig 2

Hot Carrier Effects

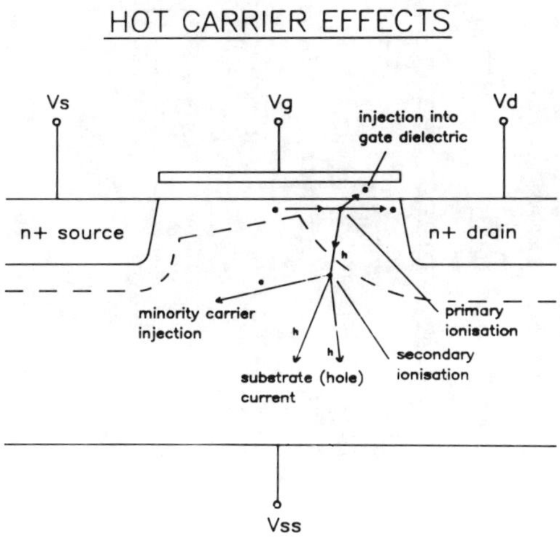

The tables above stress that, in practical terms, scaling of individual devices produces excellent results on unloaded ring oscillators which are not governed by real circuit constraints, but that line and contact resistances, and considerations of electromigration resistance, can dominate real circuit performances. Additional factors relate to sensitivity to radiation, temperature etc, also must be considered.

Non-constant field scaling (11) considerations seem to indicate that as devices drop below 1μm dimensions channel resistance effects and velocity saturation means that p-channel and n-channel device performance converges. Below 0.5μm contact resistance effects almost eliminate any speed improvements whatsoever in conventional circuits. Further details of scaling experiments can be found in ref 7.

A practical example of the size advantage scaling a circuit from one 2.5μm technology to a comparable scale 1.5μm technology is given in Fig 3.

Bipolar Scaling

We consider briefly the effects of scaling on Bipolar devices, in particular I^2L. Scaling of applied voltages in Bipolar technology is usually not applicable, since in practical terms most I^2L technologies operate close to V_{BE} (0.7v). Thus constant voltage scaling is used. (Note that if MOS voltages are scaled to 1v common MOS-Bipolar technologies become easier to implement). Constant voltage bipolar scaling might be accomplished as in the following table:-

Fig 3

A typical logic cell designed in (left) 2.5μm CMOS technology and (right) 1.5μm CMOS technology.

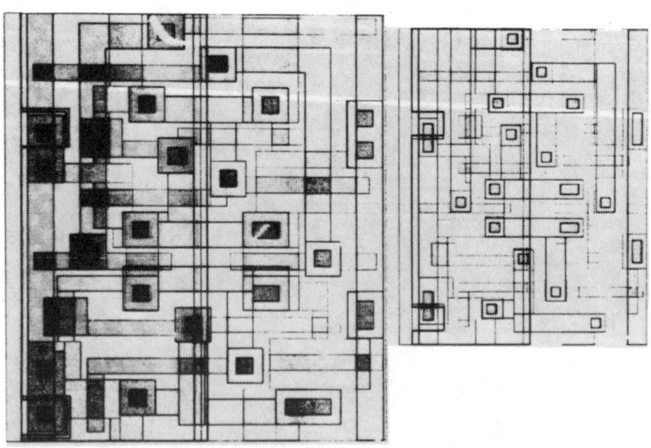

The summary of practical capability in technology for ASICs is shown in the table below. Currently available from vendors is technology between 3 and 1.5μm., in CMOS.

TECHNOLOGY PERFORMANCE

Technology	3μm	2μm	1.5μm	1.0μm	Units
Metal Pitch	8	6.5	5	4.2	μm
Supply Voltage	5	5	5	3.3-5	V
Typical Gate Delay	2	1.2	0.8	0.4	ns
Max. chip speed	50	80	125	250	MHz
Power/gate at max. gate speed	1	0.8	0.8	1.2	mW
Packing density					
Semicustom	160	220	290	350	gates mm^{-2}
Custom (SLM)	320	480	800	1150	gates mm^{-2}
Custom (DLM)			950	1400	gates mm^{-2}

PARAMETER	SCALING FACTOR
Device Dimensions	1/K
Voltage, Current	Not Scaled
Area of Logic Gate	$1/K^2$
Logic Gate Power	Not Scaled
Chip Power Density (IV/A)	K^2
Stored Charge	$1/K^2$
Power - Delay Product	$1/K^2$
Logic Gate Delay	$1/K^2$

Bipolar scaling

Note that interconnections and contact resistances do not scale, just as in the case of MOS devices. Increased currents (which tend to be higher than MOS equivalents in any case) can give rise to increased electromigration problems

Materials Considerations

There are materials considerations at both wafer level and device level which have major impacts on device and circuit performance and on potential for scaling towards practical limits.

Dielectrics

MOS Devices of the 1970's generations had gate oxides of the order of 1000Å thick or greater. The intrinsic dielectric breakdown strength of SiO_2 is of the order of 1.07 x 10^6V/cm so that breakdown voltages, (even allowing for local asperities) were >50V. As oxide dimensions are reduced several practical considerations emerge : control of oxide growth including thickness control and control of interface states and interface change becomes increasingly difficult. Temperature ranges up to 1100°C were used in PMOS processing; recent processes wherein small diffusion (junction depths are paramount, necessitate growth at much lower temperatures (750°C - 900°C) with consequent difficulties in control of the interface parameters and incorporation of chlorine species in the oxide to control mobile charge. A recent development in this context is a two step oxidation/anneal cycle at this different temperatures.

Scaling factors in the tables above give the magnitude of the problem : at constant field a 5μm device will use 1000Å oxide vs 200Å for a 1μm device. One advantage of these much thinner oxides is that devices so fabricated are much less sensitive to threshold shifts due to irradiation (e.g. in space), as shown in Fig 4.

Fig 4

Threshold voltage shift vs thickness of MOS oxide for capacitors and transistors on Silicon on Sapphire.

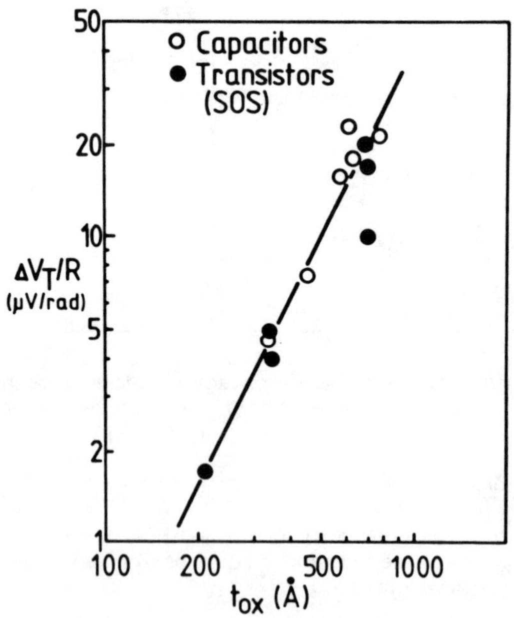

Furthermore, reduction of oxide thickness reduces the effect on V_T (device threshold) of interface charge, and reduces the drain proximity effects. Provided the thin oxide runs over regions of depleted silicon the voltage applied to the structure may be dropped mainly over the silicon depletion region since

$$\frac{V_{OX}}{V_{TOT}} = \frac{3\,T_{OX}}{(3\,T_{OX} + T_{si})}$$ where T_{si} is the depletion width.

In the operation regime of a device, however, where gate dielectric and inversion regions are comparable, the drain current and transconductance can be substantially reduced (by up to 67%) (12).

Another primary effect which must be considered is that of defect density. The fundamental breakdown voltage of a 100Å film of SiO_2 is of the order of 10V-12V. If the device area has a defect of any description this may well reduce considerably, as will also be the effect of polysilicon asperities which give enhanced local fields. The defect densities will be controlled as much by the inherent pre-oxidation process cleaning during manufacture, and by the environment in which the device is kept between oxidation and polysilicon gate deposition. These considerations are hardly fundamental to the oxide properties but are dominant in practice.

Silicon Doping. A paradox (13) is that as device dimensions decrease, each chip, or even each device, will include a larger percentage of the total parametric distribution. Obviously if N_D is scaled by the same scaling factors as the dimensions this factor is reduced - but each device is likely to have, in addition, threshold and punch through control implants at the level of 10^{11} and 10^{12} ions/cm². Indeed at the 1μm level with $10^{11}/cm^2$ implant only 1000 dopant atoms would be incorporated in a device. The device threshold will also be varied by fluctuation in basic wafer doping and by the variations in interface state and fixed interface change density.

These have been tabulated (7) :

PARAMETER	5μ Device	0.5μ Device
tox	1000Å	100Å
Xj	2μm	0.2μm
N_A	2.5×10^{14} cm^{-3}	2.5×10^{15} cm^{-3}
N_{I^2}	2×10^{11} cm^{-2}	1×10^{12} cm^{-2}
V_T	1.25	0.35 - 0.5v
V_{DD}	12v	1.5v - 3v
V_{SUB}	-5v	-1v
ΔV_T	27mv	57mv
$\frac{\Delta I_{DSAT}}{I_{DSAT}}$	1.3%	29%

Thus the effect of parameter fluctuations is very high in the case of the 0.5μm device.

Devices close to one another on the same wafer are likely to have matched threshold voltages but if, as in the case of analogue circuits of very high performance, close matching of parameters across a very large chip is vital, there will be considerably greater problems at the 0.5μm level than at the 5μm level. Local defect densities in the material are, in fact, much reduced by gettering processes during fabrication and by tighter (more expensive !) material specifications during manufacture.

A further consideration is that, for a number of practical reasons, devices of very small dimensions are likely to be made on epitaxial silicon material. The phenomenon of 'latch-up' gets considerably worse as device dimensions are reduced. Considering the device structure of Fig 1 latch up occurs when (i) the gains of the parasitic n+pn transistors inherent in a bulk CMOS structure are such that their product is >1, and (ii) a charge packet is introduced into the loop, as shown in Fig 5, which shows a lumped model equivalent circuit. Results (14) from a test structure (Fig 6), illustrate that careful choice of epitaxial layer thickness and well depth and doping can give latchup-free circuits but this is a difficult series of compromises, especially for n-well CMOS (Fig 7).

Fig 5

A CMOS latch-up model showing parasitic transistors and the equivalent circuit for both triggering latch-up and sustaining it.

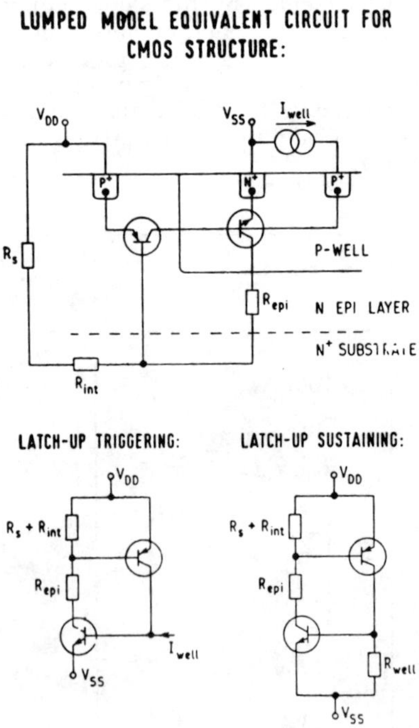

Fig 6

Latch-up test structure (denominations in micrometers) used to obtain the results of figure 7 for p-wells of various depths.

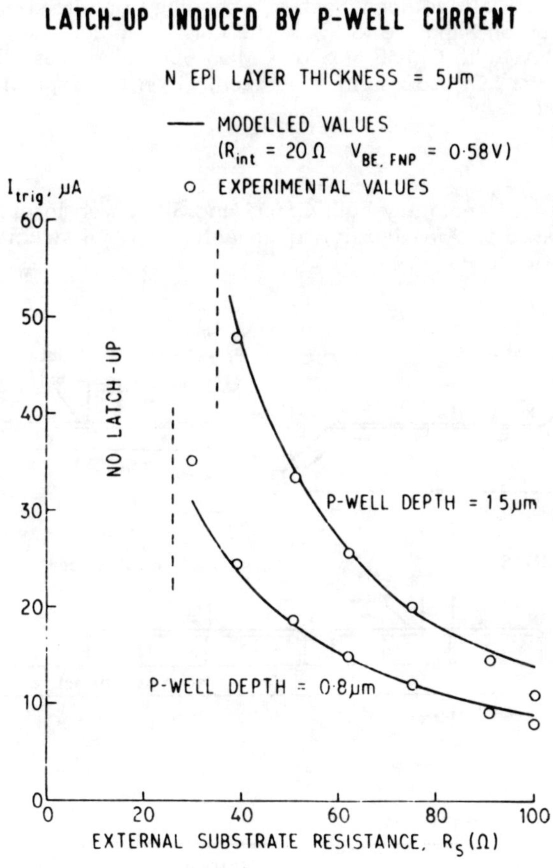

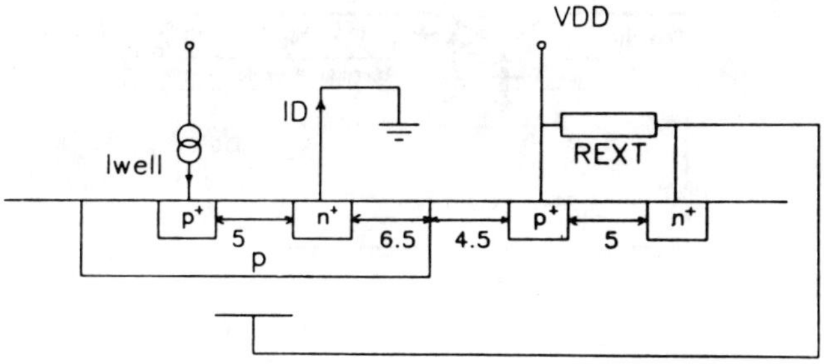

Fig 7

The latch-up effects in the test structure of figure 6 for the Marconi Electronic Devices 1.5μm CMOS process showing that by careful choice of substrate and well doping and control of external contact resistances, latch-up free structures can be built.

Silicon on Insulator

Many of these considerations lead to a reconsideration of the ideal MOS structure. This would avoid latch-up and other bulk material problems by avoiding the substrate altogether. Such a technology is Silicon-on-Insulator or SOI of which Silicon on Sapphire (SOS) is the current major example. Figure 8 compares bulk CMOS with CMOS-SOS or CMOS-SOI structures. The inherently simpler nature of the SOI process may be seen, given the capability of making the starting material.

Fig 8

A comparison of small geometry bulk CMOS and Silicon on Insulator (e.g. SOS) showing the increased packing density achievable with SOI structures of the same nominal denominations.

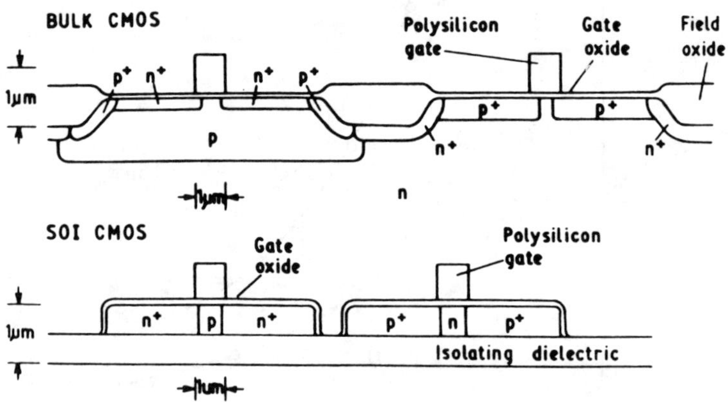

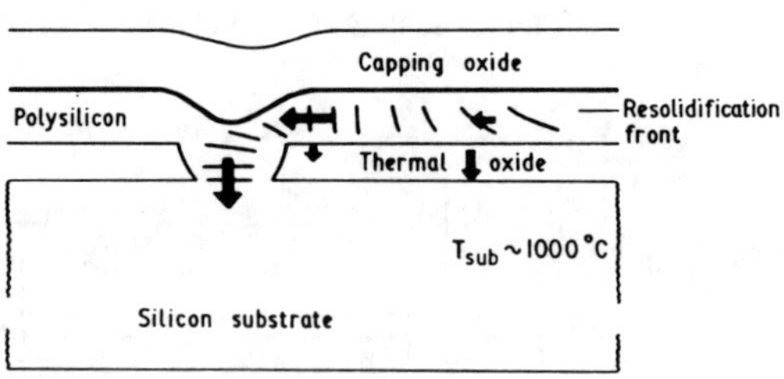

Fig 9

This is an example of a technique using dual electron beams to (simultaneously) heat the substrate and melt and recrystalize the silicon layer on top of the insulator structure. This gives a single crystal silicon layer on top of an insulating thermal oxide substrate. Devices can be constructed in both the substrate and the top layer of single crystal silicon leading to both two dimensional and three dimensional circuits.

Currently CMOS-SOS material with 0.3 to 0.6μm epitaxial Si is freely available; new technologies for the formation of silicon layers on insulating substrates are well advanced (e.g. SIMOX or Separation by Implanted Oxygen, wherein an iinsulating layer is formed beneath the Silicon surface by implanting up to 10^{18} Oxygen atoms per cm^2. Alternatively layers of silicon can be formed by deposition of polysilicon followed by annealing by laser or, preferably electron beam systems, to produce good silicon layers as shown in Fig 9.

The table below shows the leakage currents obtained by each technique, with commercial silicon-on-sapphire as a comparison :

	n-channel A/μm	p-channel A/μm
SOS	7×10^{-12}	1×10^{-11}
e-beam	$<10^{-13}$	$<10^{-13}$
SIMOX	3×10^{-12}	8×10^{-10}
Bulk Si	$<10^{-13}$	$<10^{-13}$

all these techniques produce acceptable results.

The insulating substrate technologies, whether produced by SIMOX, oxidation of porous silicon, recrystallization of polysilicon, or further development of silicon on sapphire, may well be the choice for the next generation of integrated circuits in both 2 - dimensions and possibly 3 - dimensions. They offer significant advantages in freedom from latch-up problems, increased packing density for the same nominal lithographic rules, lower dynamic power and higher speeds, and a vastly increased tolerance to radiation effects, with reduced parasitic capacitance effects and simple dielectric isolation. This materials technology, still under development, is extremely promising at the time of writing. A recent article summarises many of the developments (15).

This technology has no significant advantage over bulk CMOS from the point of view of electromigration. Current densities of $10^6 A/cm^2$ or greater should be avoided, and will be a considerable limitation as dimensions are reduced in all technologies.

Architectures

Considerations from earlier sections, particularly those associated with the problems of driving large capacitances associated with large track lengths, lead us to several conclusions. Firstly, SOI technologies have an advantage because of lower track capacitance per cm^2. Secondly - we should favour architectures which give minimum interconnect distances. In practice we may distinguish a number of generic types of device architecture : unstructured (such as early microprocessors etc); regular (such as e.g. gate arrays, Fig 10, or cell arrays Fig 11) and structured arrays shown diagrammatically in Fig 12.

Fig 10

A schematic gate array showing lines of identical cells with spaces for wiring. The latest concept in gate arrays include total coverage. Silicon with gates (sea of gates) which use multilayer metal for the interaction and simply sacrifice cells where they are not needed.

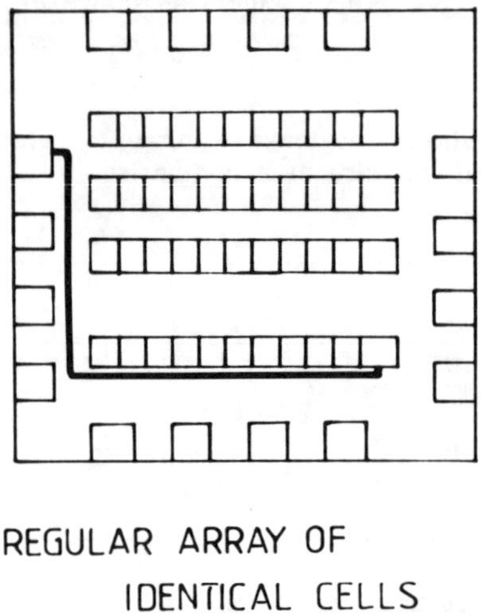

Fig 11

An example of a Marconi Electronic Devices CELLSOS chip which uses a structured area of custom built cells from a cell library to perform the required total function. Latest concepts include the use of macrocells which might be as much as a whole microprocessor in an individual cell.

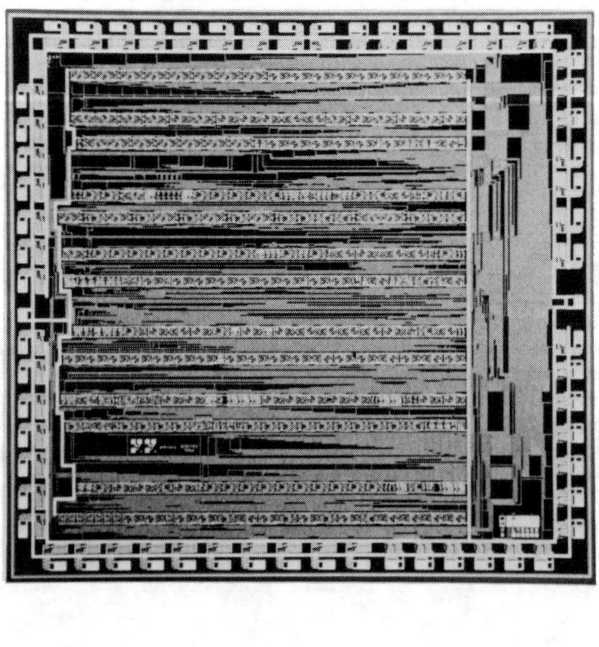

Fig 12

A schematic structured array indicating the short distances of nearest neighbour connections.

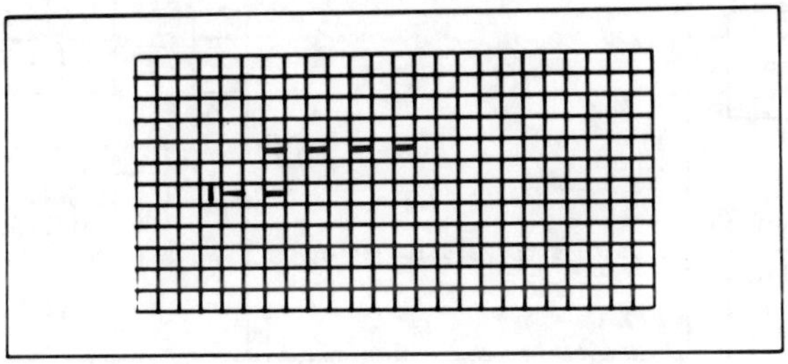

ARRAY LAYOUT

Structured arrays are particularly well suited to the exploitation of VLSI technology. An example of such an array is a bit-level systolic array correlator (MEDL device MA717) wherein a series of cells perform multiplication of data streams by co-efficients, for purposes of 64 point correlation. These architectures lead to well-structured designs which are densely packed and therefore efficient in use of silicon, and in addition reduce the speed constraints associated with long wiring distances, since each device drives only a nearest neighbour.

Further factors which have made a major impact at chip level and should also do so at the system level are redundancy and fault-tolerance. Redundancy techniques have been the key to successful large scale memory fabrication. Surplus extra rows and columns are fabricated and used, after test, to replace non-working segments of memory : their use has given the same yields in production on 256K RAMS that were previously obtained on 16K devices using the same technology but without redundancy. Fault tolerance is a further factor in VLSI design and Architecture.

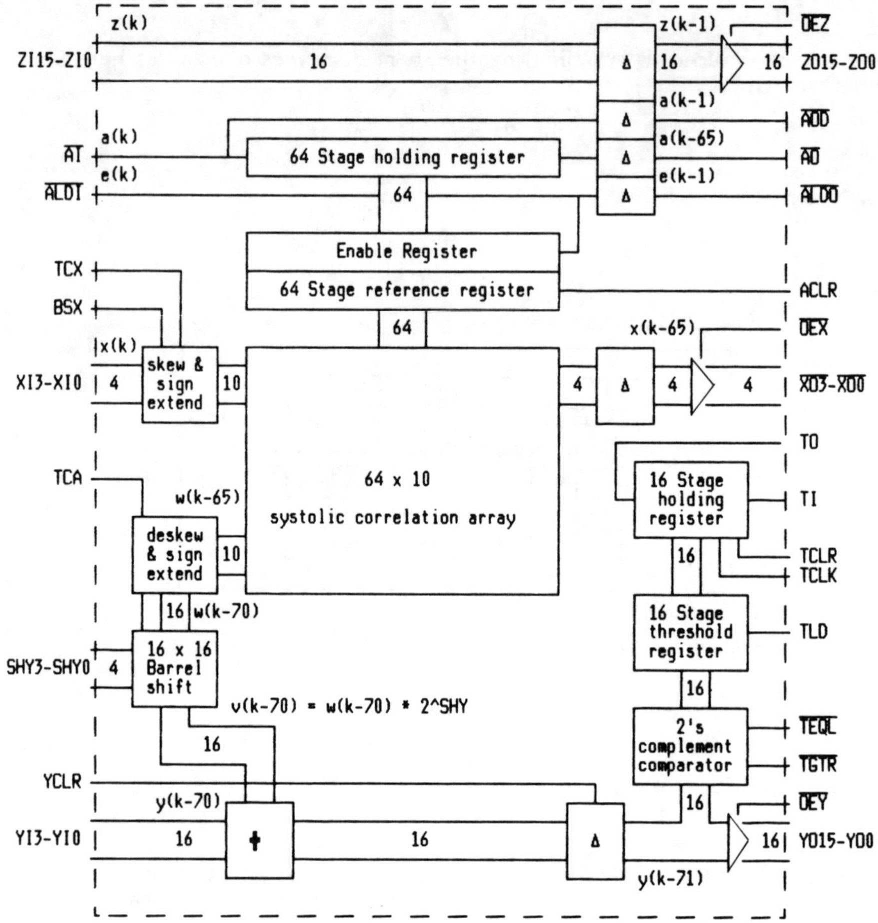

Functional Block Diagram

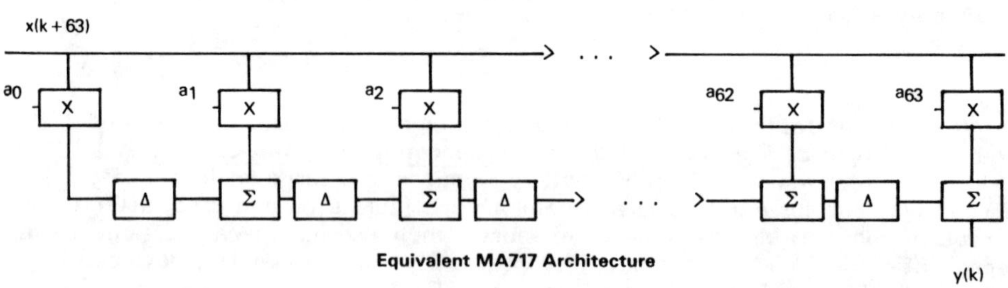

Equivalent MA717 Architecture

Fig 13

A bit level systolic correlator the Marconi MA717 showing the architecture

In this case appropriate algorithms are built into the system structure such that, in service, continuous error detection and correction takes place, with, as necessary, hardware replacement either on - or off - chip as necessary, under software control or by EPROM/EEPROM techniques. Current work is focussed on fault tolerance in systolic array systems; the incorporation of built in self test (BIST) is also potentially very important for overall system integrity.

Practical Considerations

The practical acquisition of LSI/VLSI components by a non-manufacturing customer from competitive and competing vendors is a difficult task.

Performance in delivery, speed, packing density etc seldom compares with verbal promises and technology availability is often at the whim of manufacturing tolerances or product crashes. There is usually up to a four year delay between the reporting at learned conferences of a technology breakthrough and its emergence as routine in foundries. Multilayer metal was reported frequently in the 1960's for example;widespread use had to wait till 1983;in the early 1980's devices were designed on hypothetical 6μm pitch design rules, which had to wait till 1985 for acquisition from practical vendors of gate arrays or ASIC's (Application Specific ICs). The technological break-throughs of trench isolation, ultra thin oxides etc, currently being described are unlikely to be available in production quantities till later in the decade for custom circuits though they will appear sooner in memory products. One emerging capability which will have a major impact on technology availability will be the "one month silicon system". Here it is planned to go from design concept to full custom silicon in less than four weeks, including processing and testing time. This will require integration of design tools such as MACROMOS® i.e. suites of programmes incorporating Macro cell libraries including parameterizable and technology independent cells, with electron-beam or fast optical lithography and possibly e-beam related testing procedures and rapid packaging and assembly. Prototype delivery in a few weeks will be the goal of such systems which should revolutionize the capability to build prototype large systems in VLSI. The key to success is avoiding self-deception, accepting practical reality in design tools and technology, and building in redundancy and fault tolerance as far as possible.

MACROMOS® is a trade-mark of Marconi Electronic Devices Limited

1. B S Landman and R L Russo, IEEE Trans. Comput. C-20, 1469 (1971)

2. R W Keyes IEEE Trans Ele Dev ED-26 271 (1976)

3. R W Keyes, Chap 5 in Vol 1 of VLSI Electronics ed by N G Einspruch Academic Press (1981)

4. Takahashi et al, NEC Tech J.36, 85 (1983)

5. Marconi Electronic Devices, 1553B System (available from MEDL, Lincoln (1986)

6. G E Moore, IEEE Spectrum 16, 30, (1979)

7. V L Rideout, Chap 5, VLSI Electronics Vol 7 ed by N Einspruch Academic Press 1983

8. R H Dennard et al, IEEE J Sol. St. Ccts. SC-9, 256 (1974)

9. Y A El-Mansy, Proc IEEE Ind Conf. Ccts and Computers ICCC80 1, 457 (1980)

10. P L Shah and R H Havemam, Chap 2 "VLSI Electronics" Vol 7 ed by N G Einspruch

11. P K Chatterjee et al, IEEE Trans Ele. Dev. Lett. EDC-1, 220 (1980)

12. C G Sodini and J L Moll, 1982 IEDM Tech Dig, 103 (1982)

13. D H Roberts, Electronics Letters 19, 1, (1983)

14. A G Lewis, IEEE Trans Ele. Dev. ED31, 1472 (1984)

15. H W Lam et al, Chap 1 of Vol 4, VLSI Electronics Academic Press (1982)

16. D V McCaughan and J C White, Handbook of Semiconductors, Vol 4, ed by C Hilsum, North Holland, 1980

ON ULTIMATE THERMODYNAMIC LIMITATIONS IN COMMUNICATION AND COMPUTATION

JEROME ROTHSTEIN

DEPARTMENT OF COMPUTER AND INFORMATION SCIENCE
THE OHIO STATE UNIVERSITY
2036 NEIL AVENUE MALL
COLUMBUS, OHIO 43210-1277=

0. SUMMARY

We view reversible computation as follows. A physical system (computer), prepared in a desired initial state (i.e., given a program and data and turned on), will undergo a time evolution to a subsequent state which can be (at least partially) measured, the result of the measurement being the result of the computation (recorded on a tape, for example). Alternatively, the computer prepares the output tape, just as the programmer prepares the input tape. If the computer is a reversible dynamical system, irreversibilty occurs only in such tape preparation processes. Similarly in a communication system (transmitter plus channel plus receiver), preparation of the "input end" of the system (sending the message), after time evolution of the whole system, permits (partial) measurement at the "output end" of a new state, the result of the measurement now being the received message. Reliable (noiseless) communication is approached over noisy channels via use of redundancy and coding which is computation. As real systems, with tapes counted as part of system, must be used for successions of computations and communications, entropy production connected with preparation (including erasure) and reading of tapes is the fundamental thermodynamic limit even for perfect machines and noiseless channels. In general purpose computers there is also internal communication (between memory and CPU for example), status flags, measurements ancillary to conditional branching, etc. With refrigeration and use of devices with closely spaced quantum levels energy threshold limitations appear to have no definite positive lower limit in principle. The entropy limititations, in contrtast, are absolute (k log m for selection from m possibilities, either for measurement or preparation) and related to the

J. K. Skwirzynski (ed.), Performance Limits in Communication Theory and Practice, 43–58.

third law of thermodynamics. Analysis of the complex systems involved, in terms of subsystems brought into temporary interaction and then decoupled, implies a chain of wave-packet reductions and concomitant entropy increases. Reversible time evolution occurs in intervals between such wave-packet reductions. The number of unavoidable wave-packet reductions is algorithm and problem dependent for a general purpose computer, always being at least two. This minimum is approachable in principle for a special purpose computer. Reversible computers discussed in the literature are essentially special purpose; those which appear to be general purpose require preparation procedures whose complexity is algorithm and problem dependent, or whose size grows exponentially with the size of the problem, or both.

1. INTRODUCTION

Engineering devices and systems eventually demand as close an approach to perfect operation as is economically achievable. This inspires studies of ultimate theoretical limitations on performance. In the case of heat engines they led to discovery of the laws of thermodynamics. The first law of thermodynamics, often phrased as the conservation of energy, bars construction of perpetual motion machines of the first kind, namely those whose work output exceeds their total energy input. This now so familiar that it is hardly even viewed as a limitation. Designs for machines which violate it are simply dismissed as naive. We therefore concentrate on what the second and third laws of thermodynamics imply for the ultimate performance of the systems of interest here, namely communication systems and computers.

The history of both laws abounds with controversies, echoes of which still resound, sometimes loudly. They are often stated negatively, as assertions that certain reasonable sounding procedures can not be carried out. The many equivalent ways of stating the second law include several such "principles of impotence". One says that a quantity of heat energy drawn from a hot reservoir can not be converted completely into mechanical energy. Another says it is impossible for a quantity of heat to flow from a reservoir at one temperature to another at higher temperature without doing work which is dissipated as additional heat. Consider a heat engine operating between high (T_2) and low (T_1) temperature heat reservoirs, taking in heat Q_2 at T_2, rejecting Q_1 at T_1, and doing work $W = Q_2 - Q_1$. If the second statement were false we could then put Q_1 back in the T_2 reservoir without doing work. The net result is that heat $Q_2 - Q_1$ has been drawn from it and converted completely into work W, falsifying the first statement. It can also be shown that if the first is false so is the second. As shown in texts on thermodynamics any of these statement imply both the existence of a thermodynamical state function called entropy, which increases monotonically in all physical processes, remaining unchanged only in idealized (reversible) ones, and the existence of an absolute temperature scale with a uniquely defined absolute zero. One of the ways of stating the third law of thermodynamics is that the temperature of no physical system can be reduced to absolute zero in a finite number of operations. Stating physical laws as principles of impotence is perfectly general, as shown by Rothstein [1], "demons" able to violate them are outlawed by those very laws.

There are two other famous cases where fundamental physical laws are frequently formulated as principles of impotence, namely relativity and quantum mechanics. One might think, a priori, that their roles in establishing ultimate limits on elementary acts of communication or computation would be more fundamental than any possible thermodynamic limitations, particularly those which might be ascribed to quantum mechanics. In our opinion, however, this is not the case; we believe that essentially all limitations ascribed to them, on closer analysis, are more appropriately viewed as basically thermodynamic in nature. As this view may be thought heretical by many, we discuss it further below. Thermodynamical limitations, using a generalized entropy concept in which information and organization are handled consistently in an operationally sound manner will be our chief concern thereafter.

Relativity modifies classical physical laws appreciably at high velocities (special relativity) or for phenomena on a very large scale (general relativity). Communication systems or computers which do not have subsystems in motion relative to each other at an appreciable fraction of the velocity of light or which are not large compared to the Milky Way will thus be described by classical laws to an accuracy far beyond what engineering requires. Even if distributed systems were built in which relativistic effects were no longer negligible, it seems hard to imagine limits on computation or communication not already present when the interacting subsystems are relatively at rest. Doppler shifts due to relative motion or gravitational red shifts of quanta emitted by a source, do not change the detection problem at the receiver from the classical case. Also, despite the controversies over relativistic thermodynamics, the laws themselves emerge unscathed. Changes in the first law affect only the form of the energy function (mass-energy equivalence), entropy is invariant, and absolute zero is still unattainable. For the time being it thus seems reasonable to assume that ultimate limits of computation and communication are to be sought in quantum mechanics and thermodynamics. Should the need arise to study complex exotic systems whose particles or subsystems are moving with relativistic velocities, or whose size is super-galactic, that assumption may have to be re-examined. For example, relativity of simultaneity may impose new limitations on distributed computation in a very large system which do not apply to distributed systems in which a common clock is possible in principle.

Limitations of quantum origin, in sharp contrast, seem natural, immediate and inherent. To communicate we send signals, but the minimum energy needed is a whole quantum. But this could be a low frequency, i.e. a low energy quantum. It is the need to overcome thermal noise, of the order of k T, where T is the temperature of the detector, that sets a lower limit on the energy $h\nu$ of a useful, signal quantum. The dissipation of this quantum yields the well-known Szilard ΔS value of k T ln 2. We can lower T (refrigerate the detector) and thus admit the use of lower frequency photons, but by the third law we can only approach, but never attain, T=0. There is also the problem of the noisy channel. If the transmission medium has an effective temperature T, it may radiate spurious photons to the detector. Reliable communication in the presence of noise can be obtained by the use of error-correcting codes (Shannon [2]) or statistical filtering (Wiener [3]). We can view both techniques as using computation on a more complexly prepared signal to substitute for unavailable or impractical means to refrigerate the channel. This technique therefore trades off important practical limitations on communication for limitations of computation. However, computation itself requires communication to and from the computer (I/O) as well as internal measurement (e.g. reading of status flags, reading operations on stored data) and preparation procedures (writing operations), as well as communication between subunits. Despite several ingenious attempts to achieve reversible computation, including conceptual designs for quantum mechanical computers, we remain convinced that an entropy price for unavoidable selection, measurement, or preparation acts must be paid for every such act in physical communication or computation. Our discussion is thus, in a sense, a further development of Szilard's fundamental result on the entropic cost of obtaining physical information from measurement.

The plan of this paper is then the following. First we review the interplay between physical information, measurement and quantum mechanics, irreversibility of measurement or preparation procedures, and their essentiality for communication of information. This gives us both the Szilard lower entropy limit of k ln2 per bit communicated and the conclusion that although there is no lower limit on the energy required to transmit a bit, the entropy requirement is absolute. Its persistence, even as absolute zero is approached also sheds an interesting light on the third law of thermodynamics. We then turn to computation, finding the minimal communication acts required, and examine so-called reversible computing. We conclude that analogous to the reversible time evolution between measurement acts described by Schroedinger's equation one can have special-purpose idealized computers which are piece-wise reversible in the intervals between irreversible acts of measurement, preparation, or communication. A general purpose computer requires a number of such acts which is algorithm and problem dependent, that number being non-computable, in general. We accept results in the literative on reversible computers as correct, when corrected for omission of the entropic price of initial preparation and read-out, for a class of idealized special-purpose computers. We believe them to be seriously misleading for general purpose computers in that not even bounds on the number of dissipative acts needed in the course of the computation can be computed in principle (partial recursive functions). In some important problems, for example those in which procedures are iterated until given conditions are satisfied, irreversibility accompanying each decision or conditional branch instruction can be avoided, but at the cost of increasing the size of the computer. This blow-up increases exponentially with the number of such decisions, resulting in a computer bigger than the universe for reasonable size problems. For example, 10^6 iterations is not out of the ordinary, but a number of chips equal to two raised to that power would use a mass vastly greater than that of the universe, and the universe would be too small to hold it. It would be interesting to calculate properties of the largest computer the universe could hold, but this paper will not attempt even to set bounds on them. Relativity would surely be needed!

Our attitude to reversible computing is thus much like that to Maxwell's demon - it is amusing, instructive, challenging and conceptually useful in helping to mark the boundary between what we would like to achieve and what can actually be achieved. We stress its value here to avoid misconstruing the intent this paper. Though it obviously denies that reversible computers hold any hope of being "practical" (in addition to all the foregoing they are too slow), it affirms the value and necessity of scrutinizing scientific foundations carefully. Thought experiments are not idle amusements. Good ones confront accepted concepts and techniques with paradoxes and new challleges, clarifying new advances and pointing out the limits of the old. Reversible computers are valuable additions to the library of thought experiments, and we hope they help in pushing computers to their ultimate thermodynamic limits.

2. COMMUNICATION, MEASUREMENT AND PHYSICS

Information is conveyed by choices from a set of alternatives; prior uncertainty about the choices to be made is eliminated by their actually being made. In discrete communication theory and computer science, messages (words, strings) consist of a succession of choices from a set of symbols called an alphabet (usually the binary alphabet {0,1}, whereby arbitrarily large sets of messages are easily constructed over a finite alphabet. Following Shannon [2] and Wiener [3], uncertainty about a message ensemble is measured, (up to a multiplicative constant k called Boltzmann's constant by Planck), by the entropy function introduced into physics by Boltzmann. The information conveyed by a message is precisely the prior entropy of the message ensemble (at the receiver) diminished by the posteriori entropy of the ensemble compatible with the received message.

Consider n possible alternatives, with p_i the prior probability of the i' th one. The usual normalization condition is

$$\sum_{i=1}^{n} p_i = 1 \tag{2.1}$$

Boltzmann's entropy is given by

$$S = -k \, \Sigma \, p_i \, \ln p_i \tag{2.2}$$

while the entropy used in information theory is

$$H = - \Sigma \, p_i \, \log_2 p_i \tag{2.3}$$

The constant, which reflects the choice of base 2 for the logarithms, has been chosen to make H = 1 for a choice between two equally probable alternatives, and the unit of information is called one "bit".

Confusion has frequently arisen about the sign of the entropy, clearly positive above, particulary when information and physical entropy are combined in one discussion. In statistical mechanics it is clear that the entropy corresponding to a given probability distribution in phase space, where the phase space is divided into cells and the probability that the representative point for the system has probability p_i of being found in the $i^{\underline{th}}$ cell, is exactly given by (2.2). The maximally precise description, i.e. the sharpest possible distribution, is when the point is in a single cell, say the $k^{\underline{th}}$. In this case $p_k = 1$, all other p_i are zero, and S=0. So the entropy is proportional to the "missing information" needed to convert the statistically smeared description into the sharply defined case, and the higher the initial entropy, the greater the information conveyed when the distribution is maximally sharpened. The maximum entropy case (equilibrium) is the one where maximal information is missing for the physicist, no message is sent. But for information theorists the maximum entropy source corresponds to maximal information conveyed-when the message is sent. To make matters worse, Boltzmann's famous H-theorem uses a quantity H which is -S. Brillouin suggested the term negentropy [4] for the reduction in entropy, compared to equilibrium, of physical systems in which physical information is stored, and in information transfer in physical systems. Schroedinger, considering the problem of basing biology on physics, talks of life feeding on negative entropy [5]. Characterizations of entropy as a measure of randomness, disorder, complexity, or ignorance have cropped up in the physical literature for many decades long before the birth of contemporary information theory. However, physicists were reluctant, in the decades after 1948, to admit any fundamental role for an informational generalization of entropy, fearing it would introduce undesirable subjective considerations into physics. Information theory even became rather disreputable, for a while, because of the excesses of naive enthusiasts who thought information theory would solve many, if not most, problems in many, if not most, fields. A sarcastic editorial [6] reflected the feelings of many "hard-headed" people in those days.

However Rothstein [7] realized that any measuring instrument presented a set of alternatives, namely the set of possible indications or results, from which a selection is made by carrying out the measuring procedure. Measurement thus produces information in the precise sense understood in information theory. This fact inheres in the operational viewpoint independent of what theory one is testing or what physical phenomenon one is investigating. The meaning or interpretation of the alternatives is independent of information theory, just as properties of real countable objects like cows and pebbles are independent of number theory. The exceptions are arithmetic properties, as exemplified in statements like "two pebbles in a bottle with two more thrown in gives four pebbles in the bottle"; pebbles were no doubt actually used in prehistoric (and later) computers, say to tally flocks, as reflected in words like calculus and calculate. Computers, like piles of pebbles, simulate systems of interest by exploiting non-specific properties common to both, like logical or numerical ones.

It is obvious that the above informational concept per se, as long as popular meanings of information do not contaminate it, introduces no subjectivity beyond that implied by the use of measuring equipment, i.e. none at all, as far as science is concerned. Its quantitative measure, probability, is sometimes asserted to introduce subjectivity, principally by those consciously or unconsciously using a subjective concept of probability. We believe this to be a red herring also, for the physicist's use of probability is essentially measure-theoretical, with the choice of measure reflecting accumulated objective experience (instrumental and multi-observer), rather than passing personal idiosyncrasy. The calculus of probability, like logic, is pure mathematics. The assignments of probability distributions, like assignments of "true" or "false" as truth values of statements about experience, are justified empirically. They are the data on which formal theory operates. In [8] this is carried a step further; information and the related notion of organization earlier introduced in [9] are presented as the language of the operational viewpoint. The informational interpretation of the wave-function of quantum mechanics proposed in [7] is thus no more subjective than any possible interpretation, as opposed to the informational interpretations discussed by Heisenberg [10] and Wigner [11], which really are subjective. The strictures applied by Ballentine [12] to the subjective interpretation do not appply to [7], which has all of the advantages of the statistical interpretation he advocates.

Entropy has a purely thermodynamic definition which antedates that of statistical mechanics. Indeed, Gibbs regarded his statistical entropies as mere analogs of thermodynamic entropy. But the latter can be defined operationally, so if there really is a fundamental connection between it and information it should be possible to establish it by operational analysis. This was done in [13], which also, in a sense, derives the second law from the first as a necessary condition that the first law be more than a theorem of mechanics (in a generalized sense including electromagnetics etc.). It also gave a novel form of the second law: the existence of modes of energy transfer not admitting macroscopic mechanical description. Another consequence of [7] and [13] is to provide a methodological foundation for Szilard's entropy equivalent of one bit of physical information [14]. His discussion, aimed at exorcising Maxwell's demon, used a thought experiment which Popper later attempted to refute by another one. Rothstein refuted the latter with a third, in which the same apparatus could be used in two mutually exclusive ways to simulate either of the two [15], thereby also illustrating the entropy equivalent of imposing a constraint (organization), which had been earlier introduced in [9]. For a general procedure to create and outlaw demons at will, and to reduce most special-purpose demons to the three thermodynamic ones see [1].

The informational interpretation of entropy, which could now be relied on as microscopically, macroscopically, and epistemologically sound, also led to a simple natural resolution of Loschmidt's and Zermelo's paradoxes in statistical mechanics [16]. In addition it was shown that in spin-echo experiments dynamic information storage (in the coherent phases of the precessing spins), by leading to an apparent entropy decrease when the field was reversed, actually realized a Loschmidt reflection. A later paper [17], examining statements of the paradoxes from an operational viewpoint, showed that they resulted from using entropy with two semantically distinct meanings (operational and macroscopic versus non-operational and microscopic, the former increases, the latter doesn't) and confusing the two. There are no paradoxes when one talks consistently.

The discussion of information storage in precessing spins discussed in [16] applies, with essentially no change, to any kind of dynamic information storage, like mercury or quartz delay lines. Information stored in static memories, like punched cards or magnetic tapes, exploits stable, or better, metastable configurations of mater. These are, of course, not equilibrium configurations, and thermodynamic entropy has traditionally been viewed as an equilibrium concept. In [9] the generalization of entropy and temperature to metastable equilibrium is considered (in actual experiments it is generally the case that what is called equilibrium is actually metastable equilibrium, c.f. glass, steel, etc.). The effect of the constraints maintaining the metastable equilibrium is to prevent the ergodic wandering of the representative point for the system into certain regions of phase space. This conveys information I about the whereabouts of the representative point. We can write for the entropy S of the representative point, in terms of the equilibrium entropy S_{equ} and I

$$S = S_{equ} - k\ I \tag{2.4}$$

The corresponding defintion of temperature T is then given by

$$(\partial S/\partial E)_{V,I} = I/T \tag{2.5}$$

It differs from the usual thermodynamic relation only in requiring that the constraints not change when d E is added to the system. This "freezing" of constraints is crucial in applying the third law of the Nernst heat theorem [18]. Such freezing is clearly tantamount to stabilizing the information (or organization) stored in the material by applying the constraints.

The ultimate limitation implied in (2.4) is often assumed to be at the atomic level, as in the case of flipping a single spin. But it takes rather special conditions to approach it. Even in spin echo experiments one deals with a population of spins and observes their summed effects, in complete analogy to when one observes gas pressure as the summed effect of molecular impacts on a container wall. In magnetic information storage one deals with flips of a domain rather than with those of individual spins. Thought experiments in which one prepares or observes the orientation of a single spin may be requiring an unjustifiable classical accuracy at an atomic level or assuming an ability to intervene simultaneously in a large number of microscopic cases. If one could intervene with mechanical reflectors simultaneously for all molecules of an expanding gas one could reverse an irreversible process, realize Loschmidt's reflection and be a Maxwell's demon. One would be multiplying up the kind of situation described in Lamb's analysis of quantum measurement [19] to the level of macroscopic manipulation. Care is needed to avoid coming to a conclusion similar to undoing Slzilard's argument. His thought experiment, after all, derived the k T log 2 entropy price of a binary observation as a requirement that the second law not be invalidated. The thought experiment permitted the confrontation of mechanics and the second law and the setting of limitation on the former by the latter.

This primacy of thermodynamics was well stated by Einstein [20]. On page 33 he says,

> "It is the only physical theory of universal content concerning which I am convinced that, within the framework of the applicability of its basic concepts, it will never be overthrown (for the special attention of those who are

skeptics on principle)".

It motivated his search for universal principles (p.53), and on p.57, speaking of the Lorentz transformation he says,

"This is a restricting principle for natural laws, comparable to the restricting principle of the non-existence of the perpetuum mobile which underlies thermodynmics".

This singular methodological importance of thermodynamics, so profoundly intuited by Einstein, stems from the information-measurement-entropy-organization-operational nexus. Already in [9] it was realized that this extended to making meaningful the concept of the information contained in a physical law with respect to operationally defined situations. This is more fully developed in [21], where generalized entropy is turned into a quantitative measure of how well the theory organizes observation or performs other functions we demand of theory. It gives a measure for preferring one theory over another, as well as measures of simplicity or complexity of a theory. Applied to a computer program embodying the computations of the theory it goes over to Kolmogoroff complexity [22] or algorithmic complexity [23] in computer science. Operational-informational analysis of primitive sense-impressions gave a basis for both logic as part of physical theory [24] and a (3+1)-dimensional topological space-time [25]. Paradoxes of quantum mechanics and statistical mechanics are tamed or eliminated in [7],[16],[17],[26]. Novel forms of the second law, applicable to the behavior of "well-informed heat engines" emerge as undecidable questions [27], and those devices suggest themselves as physical models for biology, with the genesis of new undecidable questions [15],[26],[28],[29],[30],[31],[32],[33]. Also, the Einstein-Podolsky-Rosen paradox, viewed informationally in [7], reappears with a new twist in [33]. It is shown there that incompleteness of quantum mechanics is needed to preserve it from contradiction, and that irreversibility of measurement/preparation performs that function. Furthermore this incompleteness is not to be viewed as a defect of quantum theory, for Goedel's theorem demands this choice between completeness and consistency for every formal system complex enough to include arithmetic.

We conclude this section by showing that communication necessarily involves irreversible acts. We communicate either by manufacturing an object which serves as the bearer of the message (magnetic tape, punched card, printed page, hand-written letter, etc.) which the recipient reads (makes a measurement), or we generate signals which the recipient detects. To say, as has been said by some we shall, in kindness, not name, that because a magnetic tape can be carried from the preparer (source) to the destination (receiver) by mechanical means which can approach the perfectly reversible, and that therefore there is no lower limit to dissipation in communicating one bit, is to commit an elementary blunder. Part is confused with whole. The complete communication process necessarily involves preparation and reading of the tape. Indeed, communication via a bulletin board needs no spatial transmission at all, only preparation and reading of a note. The ultimate limits are the energy and entropy costs of the preparation and measurement acts by source and receiver, and as we have seen earlier, by refrigeration and use of "soft" materials we can reduce the energy requirement. There is, of course, a minimum energy characteristic of any system, but nothing prevents us from getting a new system with a lower energy requirement. The entropy cost, on the other hand, is absolute, being set by the selectivity of the measurement/preparation acts. As we have seen, this is connected with the third, as well as the second, law of thermodynamics and irreversibility of measurement/preparation processes.

The situation is similar if signals are used, say photons, where a shutter is used to provide a binary choice of light or dark, say. The sender can't avoid choosing one of two ways to prepare his transmitter, namely shutter open or shutter closed. Similarly the receiver must, in effect, measure whether the shutter is open or closed, as indicated by whether or not photons have been detected. Again the selectivity required sets the minimun entropy price, the minimum energy price being characteristic of the system and its temperature.

3. THERMODYNAMIC LIMITS ON COMPUTATION

In recent years interest has developed in achieving as close to thermodynamically reversible computation as possible, starting with Bennett [34]. The ballistic computer of Fredkin and Toffoli [35] is purely classical, but quantum mechanical computers have been examined by Benioff [36],[37], Deutsch [38], Feynman [39] and Peres [40]. Devices holding promise of achieving closer approaches to computational reversibility in practise than current ones are the magnetic bubbles of Chang [41] and the parametric quantrons of Likharev [42]. The bubbles have properties suggesting their use to realize a ballistic computer, while the quantron is suggested to realize Bennett's Turing machine.

The writer finds it hard to close the gap between what practical computation is generally understood to be and what the computational capabilities of quantum computers whose Hamiltonians are actually realizable as physical systems might be. Should it be possible to do so, it would appear that the need for initial system preparation and final measurement would still remain. But it is not clear how real computations, with their myriads of decision points, not able to be predicted in advance, could ever be modeled with such Hamiltonians.

The nature of general computation itself poses more problems. It is far more than evaluating Boolean functions or doing addition. In both those cases one can set tight bounds on how many operations , and of what kinds, need to be done. With a classical ballistic computer one can then set up a set of reflectors, the number depending on the size of the problem, with a large number of different evaluations (for various truth value assignments) possible for each set-up. But when the size of the problem goes up the needed precision of setting the reflectors goes up exponentially. The operational situation soon demands an entropy price which can not be paid. Similarly with Turing machine calculations, it is impossible to predict in advance, in general, when or how often the direction of reading must be reversed. This must be determined in the course of the computation, and the proper direction selected. Either might be selected, so they must both be possible. A selective act must therefore occur at least on each reversal. Fig. 7 of Bennett's review article [43] shows a mechanical Turing machine whose parts lift, lock, rotate, and the head moves in either direction. Its behavior must be selective, and done in response to tape information. We therefore demand an entropy price for each selective act,though, as emphasized, we set no lower limit energy-wise, for the act. Bennett asserts (abstract and text) that the making of a measurement can be done reversibly in principle. For reasons earlier set forth we can not accept that statement as valid.

We are willing to grant that for limited kinds of computation physical systems can be set up in principle whose dynamical equations will generate a succession of states isomorphic to the computation, and, as an idealization, such systems can be reversible. We deny that possibility for a true general purpose computer. Information must be generated and stored until it needs to be consulted. The storage is writing, i.e. a preparation of some subsystem. The consultation is reading, i.e. a measurement on some subsystem. Both kinds of operation are selective and thus demand their entropy costs.

Between such selective acts, from a quantum viewpoint, the system evolves reversibly according to Schroedinger's equation. But on each occasion where information has to be stored or consulted subsystem preparation or measurement must be carried out. The quantum description of the process, reduction of the wave-packet, involves a discontinuity not described by Schroedinger's equation, and a newly specified system. It can not be stressed too much that one specifies initial and boundary conditions and then finds the solution of Schroedinger's (or other) equations. The setting of boundary or initial conditions doesn't come out of the equations of motion! The critical importance of this point for biology is discussed in detail in [15], and for the behavior of all physical systems capable of exhibiting slective behavior, including computers, in [33].

Decisions on program branching can be reduced to the foregoing. As emphasized before, it is in the nature of computation in general for the occasions where branching will occur to be unpredictable. One must either build a monster machine with branching hardware at every possible branch point (a "many-worlds" machine, a quantum version of which is examined in [38]) or one must store the information relevant to the branch condition, read it at each possible branch point, and select which way to go on accordingly. This technique, of course, is ubiquitous in computer architecture and in living things. As long as computers of unbounded size are not available, such chains of selective, and thus irreversible, acts must punctuate the general computation. This remains the case even for an idealized finite computer capable of performing some class of computations reversibly.

REFERENCES

1. J. Rothstein: Physical Demonology, Methodos 11 94 (1959).
2. C.E. Shannon: Mathematical Theory of Communication, Bell Syst. Tech. Jour. 27, 379, 623 (1948).
3. N. Wiener: The Extrapolation, Interpolation and Smoothing of Stationary Time Series, MIT Press and Wiley (Cambridge and New York, 1948) and Cybernetics, same publishers (1948).
4. L. Brillouin: Science and Information Theory Academic Press (New York 1956, second ed. 1962).
5. E. Schroedinger: What is Life, Cambridge U.P. and Macmillan (Cambridge, New York, 1945).
6. P. Elias: Two Famous Papers, IRE Trans. Info. Theory IT-4, 99 (1958).
7. J. Rothstein: Information, Measurement, and Quantum Mechanics, Science 114 171(195').
8. J. Rothstein: Information and Organization as the Language of the Operational Viewpoint, Philosophy of Science 29, 406(1962).
9. J. Rothstein: Organization and Entropy, Jour. Applied Physics 23, 1281(1952).
10. W. Heisenberg: Daedalus 87.3, 95, p. 99 (1958).
11. E.P. Wigner: Remarks on the Mind-Body Question, in the Scientist Speculates, J. Good, ed. Capricorn Books (New York, 1965), pp.

284-302.

12. L.E. Ballentine: The Statistical Interportaation of Quantum Mechanics, Rev. Mod. Physics, 42, 358(1970).
13. J. Rothstein: information & Thermodynamics, Phys. Rev. 85 1135(1952).
14. L. Szilard: Zeit. Physik 53 840(1929).
15. J. Rothstein: Generalized Entropy, Boundary Conditions and Biology, pp. 423-468 in The Maximum Entropy Formalism, R.D. Levine and M. Tribus, eds. MIT Press (Cambridge, Mass. 1979).
16. J. Rothstein, Spin Echo Experiments and the Foundations of Statistical Mechanics, Amer. Jour. of Physics 25 510(1957).
17. J. Rothstein: Loschmidt's and Zermelo's Paradoxes Do Not Exist, Foundations of Physics 4 83(1974).
18. R.H. Fowler and E.A. Guggenheim: Statistical Thermodynamics, Cambridge U.P. (Cambridge, 1949).
19. W.E. Lamb Jr.: An Operational Interpretation of Nonrelativistic Quantum Mechanics Physics Today, 22 23 (April 1969).
20. Albert Einstein: Philosopher-Scientist, P.A. Schilpp, ed. Library of Living Philosophers, (Evanston Ill. 1949, reprinted by Dover Publicatons, New York. 1957).
21. J. Rothstein: A Physicist's Thoughts on the Formal Structure and Psychological Motivation Internaationale de Philosophic, no. 40, 211(1957).
22. A.N. Kolmogorov: Three Approaches to the Quantitative Definition of Information, Information Transmission 1 3(1965) and IEEE Trans. Info. Th, IT-14 662(1968).
23. G.J. Chaaitin: Algorithmic Information Theory, IBM J. Res. Dev. 21 350(1977).
24. J. Rothstein: Information, Logic and Physics, Philos. of Science 23 31(1956).
25. J. Rothstein: Wiggleworm Physics, Physics Today 15 28(Sept. 1962).
26. J. Rothstein: Informational Generalization of Entropy in Physics, pp. 291-305 in Quantum Theory and Beyond, T. Bastin, ed. Cambridge U.P. Cambridge 1971).
27. J. Rothstein: Thermodynamics and Some Undecidaable Physical Questions, Philos. of Science 31 40(1964).
28. J. Rothstein: Heuristic Application of Solid State Concepts to Molecular Phenomena of Possible Biological Interest pp. 77-85 in Proceedings of the First National Biophysics Conference, Yale University Press (New Haven, 1959).
29. J. Rothstein: On Fundamental Limitations of Chemical and Bionic

Information Storage Systems, IEEE Trans. Mil. Electronics, MIL-7, Bionics Issue, 205 (April-July 1963).

30. J. Rothstein: Excluded Volume Effects as the Basis for A Molecular Cybernetics, pp. 229-245 in Cybernetic Problems in Bionics, H.E. Oestreicher and D.R. Moore, eds., Gordon and Breach (New York 1968).
31. J. Rothstein and P. James: Families of Chain Configurations on the Quadratic Lattice and on Narrow Lattice Channels, Jour. Appl. Physics 38 170(1967).
32. J. Rothstein: Generalized Life, Cosmic Search 1 35(1979).
33. J. Rothstein: Physics of Selective Systems: Computation and Biology Internat. Jour. Theor. Physics 21 327(1982).
34. C.H. Bennett: Logical Reversibility of Computation, IBM Jour. Res. Dev. 17 525(1973).
35. E. Fredkin and T. Toffoli: Conservative Logic, Internat. Jour. Theor. Physics 21 219 (1982).
36. P.A. Benioff: The Computer as a Physical System, Jour. Stat. Physics 22 563(1980).
37. P.A. Benioff: Quantum Mechanical Hamiltonian Models of Discrete Processes that Erase Their Own Histories, Internat. Journ. Theor. Physics 21 177(1982).
38. D. Deutsch: Quantum theory, The Church-Turing principle and the universal quantum computer Proc. Roy. Soc. Lond. A 400 97(1985).
39. R.P. Feynman, Quantum Mechanical Computers, Optics News 11 11(1985).
40. A. Peres: Reversible Logic and Quantum Computers, Phys. Rev. A 32 3266(1985).
41. H. Chang: Magnetic Bubble Conservative Logic, Internat. Jour. Theor. Physics 21 955(1982).
42. K.K. Likharev: Classical and Quantum Limitations on Energy Consumption in Computation, Internat. Jour. Theor. Physics 21 311(1982).
43. C.H. Bennett: The Thermodynamics of Computation-A Review, Internat. Jour. Theor. Physics 21 905(1982).

Part 2.
Statistical, Informational, Computational and Cryptographic Limits

ON THE CAPACITY OF PEAK POWER CONSTRAINED GAUSSIAN CHANNELS

I. BAR-DAVID

DEPARTMENT OF ELECTRICAL ENGINEERING
TECHNION - ISRAEL INSTITUTE OF TECHNOLOGY
HAIFA 32000, ISRAEL.

1. INTRODUCTION

The most interesting and important results on the performance of communication systems have been obtained when a peak power constraint has been imposed on the transmitted signals. While such a constraint afforded analytical tractability, in many cases, real-life systems are naturally limited in their peak excursion. This talk is devoted to the influence of a peak power limitation on the information rates of various models of practical communication systems.

In his epochal work [1], Shannon did calculate, inter alia, the mutual information for a bounded input of a time-discrete, additive Gaussian channel but under the assumption of a uniform distribution; the capacity-achieving distribution had to wait for its discovery by Smith [2], who proved that, in one-dimension, it is discrete. In an earlier publication, Farber [2] assumed that the distribution is discrete, and proceeded to calculate the optimum probability mass points and their weights. Smith's result thus established digital communications as being preferable in a practical <u>1-dimensional</u> time -discrete setting. I emphasize "practical" since it is well known that under an average-power constraint the optimum input is the impractical Gaussian one. The penalty for the peak limitation vs. the Gaussian laissez-faire is about 1.53db, at very large SNR and decreases to zero at vanishing SNR. This last result is one facet of the folk theorem claiming that input quantization does not decrease capacity at low SNR [3].

Of course we are interested in higher dimensional signals, two being required for narrow-band signalling with piece-wise (per symbol interval) constant parameters. Furthermore, interest lies also in time varying modulation formats, involving random transition instants, shaped pulses, partial response signals and in bandwidth-conserving continuous phase modulations (CPM).

Figure 1 offers an up-to-date classification of various envelope-constrained communication formats over the additive white Gaussian noise (AWGN) channel and indicates the state of knowledge on the achievable capacity or, mutual information rates for specific input distributions, when the optimizing one is not known.* The two diagonal lines separate continuous- from discrete-time and modulated-envelope from constant-envelope formats, respectively. Outside the circle, the signalling spectrum is unrestricted, whereas some kind of restriction is imposed within. The qualifier "discrete time" here means that the signal parameters, amplitude and phase, are piecewise constant (over symbol duration). We

*In Figure 1 no distinction is made between capacity and mutual information, for the sake of brevity

J. K. Skwirzynski (ed.), Performance Limits in Communication Theory and Practice, 61–73.

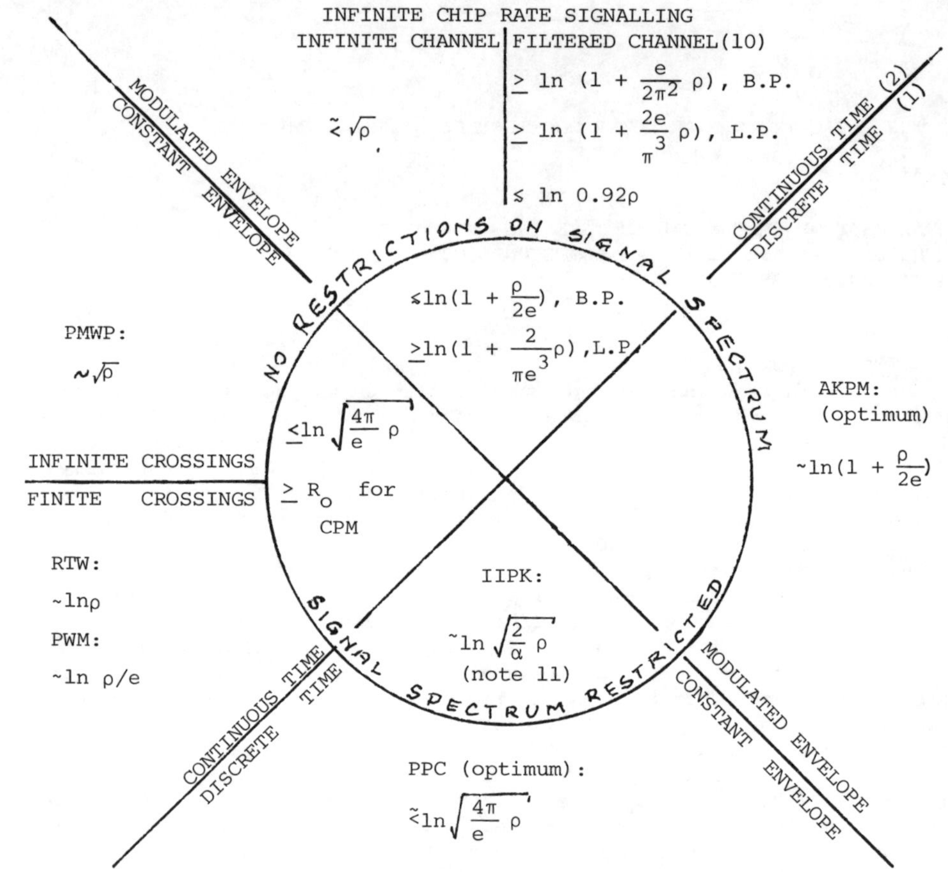

FIGURE 1. CAPACITY IN PEAK POWER LIMITED COMMUNICATIONS

NOTES: Capacity values are in nats per dimension. ρ is signal to noise ratio.

1. "DISCRETE TIME" includes sinusoids with piecewise constant parameters.
2. "CONTINUOUS TIME" includes signals with infinite rate of parameter changes.
3. PMWP: Phase Modulation by Wiener Process
4. AKPM: Amplitude Keying and Phase Modulation
5. CPM : Continuous Phase Modulation
6. IIPK: Independent Increment Keying
7. RTW : Random Telegraph Wave
8. PWM : Pulse Width Modulation
9. PPC : Polyphase Coding
10. Lower bounds are for signals not necessarily bounded at the filter output. Filters are strictly bandlimiting.
11. Here α is approximately also the factor of spectral sidelobe suppression.

have included within the "continuous-time" class binary signals, the rate of alterations of which is not restricted to be finite.

The purpose of this talk is to review the results on capacity and mutual information for the various classes, to relate them to results on the computational cut-off rate, R_o, and to indicate the main ideas beyond some of the derivations. For details the references should be consulted. The results presented in Fig. 1 are for asymptotically large SNR.

As a yardstick we use the capacity of the average-power (P_a) constrained channel which, in 1-dimension, is

$$C_1 = \frac{1}{2}\ell n(1+\rho) \;;\quad \rho \triangleq \frac{P_a}{\sigma_n^2} \tag{1}$$

and, in a strictly limited band of width W, with $\sigma_n^2 = N_o W$, and with N_o the one-sided power density of the assumedly white noise, the capacity density γ (per unit bandwidth) is

$$\gamma \triangleq \frac{C}{W} = \ell n(1+\rho) \to \rho, \tag{2}$$

as either $W \to \infty$, or $P_a \to o$ [3], [4]. We refer to ρ as the SNR. These results are achieved when the input is a Gaussian random variable in (1) and a Gaussian bandlimited white process in (2). Since $\ell n(1+\rho)<\rho$, the conclusion is that the capacity density is at most linear in SNR for given P_a constraint. For arbitrary P_a, this holds only in a strictly infinite bandwidth. In all further discussion the peak power will be constrained to P_p and, where applicable, the envelope to $(P_p/2)^{1/2} = A_p$.

2. MODULATED ENVELOPE

2.1. <u>Modulated envelope-continuous time</u>

The most general case is that of an arbitrary waveform which, not being Gaussian because of the peak-constraint, cannot be sufficiently well characterized for derivation of analytical results.

One may visualize a specific case, composed of narrow chips that change at a large rate. In the limit of "infinite chip rate signalling" (ICRS) the capacity density achievable is

$$\gamma \to \rho \,, \tag{3}$$

exactly as in (2), the difference being that in (2) the strict band limit, $W \to \infty$, whereas in (3) the chip rate $\to \infty$, while the spectrum is of infinite support, for any rate. This equality is achieved, for example, with the signal values being exactly $\pm A_p$ [3], and is just a restatement of the folk theorem alluded to in the Introduction and is a direct consequence of the presence of overwhelming noise at <u>infinite</u> bandwidth, under which practically all signal distributions fare equally. Other well known examples of constant envelope, capacity-achieving signals in infinite bandwidth are the orthogonal sine-cosine and Hadamarc sets.

Recently, Ozarow, Wyner and Ziv [5] investigated the penalty incurred in capacity when originally peak-constrained signals are passed through brickwall filters, of the band-pass and low-pass varieties, of width W. Their lower bound is the mutual information achieved with random signalling at a chip rate of $2W$, with the chip amplitudes uniformly distributed in $(-\sqrt{P}, +\sqrt{P})$. It is, for the bandspass case,

$$\gamma = \frac{C}{W} \geq \log(1 + \frac{e}{2\pi^2}\rho) \quad , \quad \text{B.P.F.} \tag{4}$$

With the LPF case the multiplier of ρ, above, need be changed to $2e/\pi^3$. It should be noted that this model does not guarantee the boundedness of the filter output, because of the sin t/t-nature of the filter response. Therefore, if a peak constraint is required at all stages of the communication link, eq. (4) cannot be considered as a lower bound.

An upper bound for a channel that includes a strictly band-limiting filter is, of course, given by (2), since the latter is the capacity for the optimum case when the output of the filter is Gaussian. In a more recent work [6] we have derived a tighter upper bound for the case of peak limited binary wave forms passed through a strictly band limiting low pass filter. The asymptotic expression for it is

$$\gamma \leq \ell n\ 0.92\rho \quad , \qquad \rho \to \infty \tag{5}$$

We consider this to be an important result since it indicates that, with otherwise unconstrained signals, peak-limiting cascaded with band-limiting does decrease capacity, compared to that attainable with Gaussian signals.

This bound is based on a theorem by McMillan [7] that restricts the allowable spectra of binary waveforms (unit processes): Using a signal that has a particular spectrum, say $S(f)$ limited to W, the mutual information is overbounded by

$$\gamma \leq \frac{1}{W}\int_0^W \log(1 + \frac{S(f)}{N_o})df \quad , \tag{6}$$

from which (2) also results. Equality would be achieved with a Gaussian input. Therefore, among all spectra satisfying a McMillan-type restriction, the one that maximizes the right hand side of (6) yields an upper bound to capacity. Eq. (5) is not the lowest upper bound; for it is obtained for the simplest McMillan-type restriction, arrived at by an "educated" guess. [It is also easy to show that this technique of upper bounding cannot achieve a power-degradation factor below 0.63 (as compared with 0.92 in (5)), since the latter is obtained if the Random Telegraph Wave (RTW), which is a particular binary waveform, is used as the input and its Lorenzian spectrum is used in the integral in (6). We emphasize that $\log(0.63\rho)$ is not necessarily an upper bound to capacity since it is obtained for the RTW input and some other binary wave-form might be superior after passage through the brickwall filter. It is not a lower bound either, since (6) was used for its calculation and the filtered RTW is, also, not Gaussian.]

Turning to spectrum-constrained inputs to the channel (with or without filter) we recall the lower bound due to Shannon, who considered strictly band-limited, $\text{sinc}^2 t$-type signals, in order to ensure a peak-constraint for all t. The thus-obtained lower bound to capacity is

$$\gamma \geq \log\ (1 + \frac{2}{\pi e^3}\rho) \tag{6a}$$

which is a rather loose bound but still the only generally known one.

An upper bound for the strict low pass case can be easily obtained by a thought experiment using Smith's result, mentioned in the Introduction, to the effect that with 1-dimensional signalling the asymptotic degradation is 1.53db, as compared to Gaussian imputs. Suppose samples are attached to sinc t functions. The time-continuous signal will certainly be band limited but, guaranteed- peak-limited <u>only</u> at the sampling instants. A signal, peak-limited <u>for all time</u> is more restricted and hence the penalty of 1.53db compared to $\ell n(1+\rho)$ gives an upper bound. We shall return to this after extending Smith's result to the band-pass case.

2.2. <u>Modulated envelope - discrete time</u>

This class of signals is the generalization to two dimensions of the 1-dim channel discussed in the Introduction and is widely known as the (time-discrete) Quadrature Gaussian Channel (QGC). The input is a vector and to it a Gaussian random vector with i.i.d. components is added. The QGC is a well established model for quadrature amplitude modulation (QAM), physically implemented by a sinusoid of selected amplitude and phase, constant over the symbol duration and passed through the AWGN channel. Wyner [8] has shown the sufficiency of the QGC model in representing this time-continuous situation for capacity calculations. Here, peak power constraint means that the domain of the distribution of the input vector is within a circle of radius A_p.

We have recently shown [9] that the capacity-achieving distribution, within the circle, is uniform in angle (phase) and discrete in radius (amplitude), i.e. its loci are distinct concentric circles. This result in a sense re-establishes analog communications as the leader, capacity-wise. We refer to this optimum input format as Amplitude Keyed Phase Modulation (AKPM).

The proof of the optimality of AKPM hinges on the known fact [9] that the entropy for each input symbol, in the optimum distribution, has to equal the average entropy. It is then shown that this condition cannot be met by a <u>bounded</u> continuous amplitude distribution. The same condition is also used to show that, for SNR's less than 7.8db, a single circle, i.e. constant-envelope PM (CEPM) is optimum; at slightly higher SNR's zero-power (signal OFF) symbols need be used and as SNR increases the hub-point develops into a second, inner circle and eventually a new hub-point is required, etc. The following, later sections are devoted to the discussion of the practical constant envelope inputs; therefore this condition for their optimality should be recalled. It is also interesting that a very low SNR is the phase distribution on the single circular circumference need not be uniform. Any phase density that has a vanishing first Fourier coefficient, such as for example, QPSK, is also optimum. This is another facet of the folk-theorem that input quantization does no harm at low SNR.

Even though the title of the talk is capacity, it is compelling to digress slightly into the discussion of the various 2-dim. constellations in the quadrature phase plane, which have been shown to be optimum (see the survey by Forney et al. [10]), vs. the optimality of the concentric circles. The word "optimum" has been been abused; rather, we are reminded again that it has meaning only if qualified by the underlying criterion. The circles are capacity-wise optimum, i.e. they should be used with coded communications. Constellations, on the other hand, have been optimized for minimum expected energy per symbol for given lattice structures, typically square or hexagonal; a particular structure ensuring some given probability of correct decision per symbol. In practically all analyses, equal probabilities have been assigned to all signal points. Ref. [10] suggests a way to distribute the probability weights to the symbols according to a close-to-Gaussian law. The rationale for this choice is that the Gaussian law is capacity-achieving. This is indeed true under an average power constraint, but the use of a constellation tacitly implies a peak-power constraint. If such a constraint is explicitly imposed, the optimizing distribution becomes, as discussed above, essentially different: At low SNR's, the hub-point should have zero weight, compared to largest weight under the Gaussian law. At very large SNR, the optimum law tends to become uniform over the area of the circle.

This last, important fact, can be seen from the asymptotic result, previously reported by Blachman [11],

$$\lim_{\rho_p \to \infty} [\rho_p F_r(r)] = r^2 \; ; \quad \rho_p \triangleq \frac{P_p}{\sigma_n^2} \; . \tag{7}$$

Here $F_r(r)$ is the probability distribution function of the amplitude; then, the density is linear in r and so is the number of equidistant points on the circumference of a circle, creating a uniform density over the disc.

Bounds on the capacity for the 2-dim QAM, denoted by C_2 have also been derived [9]:

$$\ln(1 + \frac{\rho_p}{2e}) \leq C_2 \leq \ln(1 + \frac{4}{\pi}\frac{\rho_p}{2e}) \; . \tag{8}$$

The lower bound, that depends on the convulutional equality of entropies [11], is asymptotically tight. Comparing C_2 with the capacity for Gaussian input having the same average power, and using the lower bound for comparison, one finds an asymptotic penalty of 1.34db, caused by the peak constraint - somewhat smaller than in the 1-dim case. The performance of efficient original codes for QAM constellations and their comparison to mutual information has been presented by Ungerboeck [12].

Returning temporarily, to continuous-time, spectrum-constrained inputs, a thought-experiment, similar to the one made in conjunction with the low pass case, yields an upper bound to the capacity in the band pass case as well. Interpolation of quadrature sinc t function between sample values yield a band pass signal and the penalty of 1.34db carries over to the upper bound on time-continuous transmissions.

The concentric circles have been shown by Saleh and Salz [13] to be optimal for the R_o-criterion as well (see also [14]). A comparison of results shows that under a capacity-criterion less circles are preferable than under the R_o-criterion.

Finally, it should be noted that the spectrum in the above formats decreases, naturally, as sinc-squared. It is an interesting challenge to devise practical modulation methods over several concentric circles that also obey some spectral restriction. Meanwhile, the inner region in this quadrant of Fig. 1 is left vacant.

3. CONSTANT ENVELOPE, DISCRETE TIME

The constant-envelope (CE) discrete-time case has been treated by Wyner [8], who referred to it as polyphase coding (PPC). He derived upper and lower bounds, to both capacity and R_o. We present here only the asymptotic expression for capacity, for comparison (see also [11]):

$$C_{ce} \sim \begin{cases} \rho \ , & \rho << 1 \ , \\ \frac{1}{2} \ln \left(\frac{4\pi}{e} \rho \right) \ , & \rho >> 1 \ . \end{cases} \tag{9}$$

It appears that the penalty for the constant envelope constraint as compared to the peak power constraint (see (8)) is considerable at large SNR: ρ enters only as its square root into the argument of the logarithm. This is by far a stronger penalty than the one incurred by the peak-constraint. At low SNR, as also seen before, there is no penalty.

With PPC, the information-carrying phase changes randomly and independently from symbol to symbol, resulting in the unrestricted sinc-squared power spectral density (psd). In an attempt to restrict the spectrum, we have introduced [15] dependence between successive symbols by defining the Independent Increment Phase Keying (IIPK) format in which the phase sequence is an independent increment process. The probability distribution function f_ψ of the phase increments controls to some extent the psd $S(\omega)$ of the transmitted signal. Analysis shows that $S(\omega)$ depends only on the first Fourier coefficient F of f_ψ and becomes narrower as F increases from 0 to 1. If f_ψ is uniform $(0,2\pi)$ then $F = 0$ and IIPK reduces to PPC. The mutual information with IIPK depends, however, not only on F but on the details of f_ψ as well. The capacity $C(F)$, for the IIPK class under the spectral constraint defined by F, is then the supremum of the mutual information over all pdf's that have a given F. At large SNR, the optimizing density turns out to be Tikhonov, i.e.

$$f_\psi(\psi) = (2\pi I_o(\alpha))^{-1} \exp(\alpha \cos\psi) \ , \quad |\psi| \leq \pi \ , \tag{10}$$

where α is related to F by the modified Bessel functions,

$$F = I_1(\alpha)/I_o(\alpha) \ ; \ 0 \leq \alpha \leq \infty \ . \tag{11}$$

In [15], upper and lower bounds on the mutual information and on R_o, as well as graphs of $S(\omega)$ as a function of F are derived and graphed. Here we present results for capacity at very large SNR and large α:

$$C_{ce}(F) \sim \frac{1}{2} \ln \frac{2\rho}{\alpha} \ , \quad \rho >> 1 \ , \tag{12}$$

which, using (11), indicates that an attempt to increase F, in order to achieve a sizable restriction in spectral width, is very detrimental to capacity. A more detailed investigation reveals the sobering result that, if spectral excess is the criterion of bandwidth then, at large SNR, IIPK is not superior to PPC. This is because the reduction of the spectral sidelobes is also by a factor of α and if capacity is to be restored to its original value (see (12)) by increasing signal power, the power in the sidelobes also increases to its original value.

This observation should not be interpreted as a case against introduction of Markovian dependence in the phase process. It only shows that the one-step memory of the IIPK, in the presence of the large discontinuity at the symbol transition instants, is not sufficient. A finite state Markov sequence of more complicated structure is needed to generate phase variations with smaller steps. Such an approach has been taken by Maseng [16], but he did not present capacity calculations. In principle, Markov chains are good candidates for source models since they have maximum entropy for given covariance [17], [18]. Of course, the most promising constant envelope signals for spectral reduction have continuous phases and belong to the next group in our classification.

A last point of interest concerns the class of receivers that observe only the phase of the channel output of IIPK transmission. The penalty in mutual information is nil at high SNR and exactly a factor of $\pi/4$ at low SNR [15], which is equal to the degradation in SNR due to hard-limiting. These results are of interest because they serve as a lower bound to the performance of continuous-time, hard-limiting receivers that observe the entire phase path of IIPK transmissions.

The main step in the derivation of the results for the IIPK is the reduction of dimensionality of the problem, based on the Markov property of IIPK. This reduction was possible only by bounding the mutual information from above and from below, respectively. The asymptotic results are obtained by first proving the independence of the noisy phase and amplitude at the channel output and then maximizing the conditional entropy of the output phase and output amplitude given the previous input phase. To derive some of the lower bounds, the convexity cap of $\ln(I_o(\sqrt{x}))$ was proven, in contrast to the convexity cup of $\ln(I_o(x))$.

4. CONSTANT ENVELOPE, CONTINUOUS TIME

We consider first the subclass that has no explicit restriction on bandwidth. As we shall see, even within this subclass there is an essential difference in information transfer between signals, depending on the rate of their level crossings (refer to the horizontal line in Fig. 1).

Among the waveforms that have a finite rate of zero crossings which we consider first, we concentrate on the Random Telegraph Wave (RTW). It has the distinction that it is a Markov process and that the intervals between its points of transition are independent and exponentially distributed. This distribution has largest entropy for given mean value so that it is a good candidate for the capacity-achieving input. We have shown [19] that the Information Transfer (IT), which is the term we use for the mutual information per unit time, is given by

$$\gamma_R \triangleq \frac{1}{\lambda}(IT_R) = \lambda\rho^{\frac{1}{2}} \frac{W_{o,o}(R^{-1})}{W_{\frac{1}{2},\frac{1}{2}}(R^{-1})} \rightarrow \begin{cases} \rho \quad , & \rho \rightarrow o \\ \ln\rho \quad , & \ell \rightarrow \infty \end{cases} , \tag{13}$$

where λ is the mean transition rate of the RTW and is, of course, also the 3-db decrease point of its power spectral density, which is Lorentzian. $W_{i,i}(\cdot,\cdot)$ are Whittacker's functions and are well tabulated. The subscript R refers to the RTW. We observe the asymptotic increase in capacity as $\ln\rho$ is considerably superior to the $\ln\sqrt{\rho}$ behavior for the time discrete formats (9). The inevitable conclusion is that with constant envelope, at high SNR, the transition instants are an important element in information transfer. In the time-discrete formats, the transition instants convey no information. This facet of the results appears again in the next subclass. Even though it has not been proven that RTW is the optimum input and that (13) is the capacity for binary waveforms with finite rate, a heuristic argument based on results from estimation theory is presented in [19], in support.

The similarity of the result (13) with that for the Gaussian B.L. case (2) might be misleading. We recall that the spectrum of RTW is Lorentzian, $2P\lambda(\omega^2+\lambda^2)^{-1}$ and use of this spectrum in (6) yields (with subscript G for Gaussian)

$$\gamma_G = \frac{1}{\lambda} C_G = \frac{2\rho}{1+\sqrt{1+4\rho}} \rightarrow \begin{cases} \rho, & \rho \rightarrow o \\ \sqrt{\rho}, & \rho \rightarrow \infty \end{cases} , \tag{14}$$

from which we conclude that a Gaussian model for the input with the same (Lorentzian) spectrum would have an essentially superior performance: $\sqrt{\rho} >> \ell n\rho$, at large ρ.

It appears that such superior performance of the Gaussian model, with Lorentzian spectrum, depends on its infinite effective bandwidth which implies an infinite expected rate of level crossings; each level crossing conveying some information. Since no practical system can possibly match such an idealized model, we conclude that the $\sqrt{\rho}$-type behavior is not to be expected. The transition from the $\sqrt{\rho}$- to the $\ell n\rho$-law with a realistic cut-off of the Lorentzian spectrum is discussed in [19].

In spite of the finite expected rate of transitions in an RTW, rapid local transitions do occur. To allow for the requirement of practical circuits, a "guard interval" RTW is proposed, in which a fixed amount Δ is added to each random intertransition interval. It is shown that the logarithmic increase with ρ of the information transfer is not changed, with just a slight degradation in SNR. Nor is the logarithmic behavior changed if the information-conveying random transitions are organized in a synchronous pulse width modulation (PWM) format, which has a practical appeal. The loss in IT compared to RTW is but 1 nat per transition, at large SNR.

We now turn to signals with unbounded rate of level crossings. A theoretically interesting one is the limiting case of PPC, obtained when the random phase variations are modelled by a Wiener process, $w(t)$. We define the PMWP signal to be a sinusoid, phase modulated by such a process:

$$s(t) = \sqrt{2P} \cos (\omega_c t + kw(t)) \tag{15}$$

where k is the modulation index. It is well-known that its psd is also Lorentzian about ω_c with the 3db corner frequency, ω_o, given by

$$\omega_o = k^2/2 . \tag{16}$$

A lower bound and asymptotic expressions for the IT are obtained in [19]. The asymptotic result is

$$\frac{1}{\lambda} IT_P \rightarrow \sqrt{\rho} , \quad \rho \rightarrow \infty , \tag{17}$$

where $\omega_o = \lambda$ has been set for comparison with the RTW result (13), and of course with the Gaussian (14), all three with the same Lorentzian spectrum.

It is remarkable that asymptotically, the constant-amplitude PMWP input achieves capacity, as does the GM input, which is only average-power limited. This may again be attributed to the fact that the PMWP has infinite effective bandwidth, as has the Gaussian process. Intuitively, about every level (of course limited to $(-\sqrt{2P}, +\sqrt{2P})$ for the PMWP) there are an infinity of wiggles that potentially contribute to information transfer. This level crossing behavior is well-known for the Gaussian process; the PMWP behaves in a similar way. We contrast these infinitely dense wiggles of the PMWP to the finite density of information-carrying transitions of the RTW, that has the same psd. That this finiteness is responsible for the loss of the R-type behavior of IR_R, with R, is supported by the observation that the band-limited Gaussian process, that has a finite rate of zero-crossing has also a $\ln\rho$ behavior (2). Again, there is no direct proof that the PMWP is the optimum constant-envelope input with Lorenzian spectral restriction. However, the fact that asymptotically its IT approaches capacity for the same spectral constraint as optimum Gaussian input puts in on solid ground.

The proofs for both the RTW and the PMWP signals are based on the following version for stationary processes of a theorem by Duncan [20],

$$\lim_{T\to\infty} \frac{1}{T} I(s(t);\, r(t)) = \frac{1}{N_o}\, \bar{\varepsilon}^2$$

where $I(\cdot,\cdot)$ is the mutual information between the input s and the output r of an AWGN channel, and $\bar{\varepsilon}^2$ is the causal steady state (minimum) mean square estimation error. The expression for $\bar{\varepsilon}^2$ is available in [21] for RTW. For the PMWP it is bounded in [19] using techniques based on [22].

Finally we turn to **maybe** the practically most important area in fig. 1: The class of constant envelope, bandwidth efficient signals. Technology has made rapid advances using continuous phase modulation, (CPM) and thorough analyses investigated minimum distance properties and R_o-type measures for a variety of signal shapes, full- and partial-response. The most recent and comprehensive publication is by Anderson et al. [23]The paper by Omura and Jackson [24] is also noteworthy. However, the optimum signal distribution under some given spectral restriction for time-continuous inputs has not been derived.

We can offer only an upper bound derived by the same thought experiment as conducted in conjunction with the peak-constrained signals. Quadrature sinc t-functions interpolated between sampling points yields a band-limited signal, not necessarily of constant envelope; only the sample points have this property. Therefore eq. (9) is also an upper bound for strictly band-limited continuous-time waveform inputs. Of course, CPM signals cannot be strictly bandlimited. The conclusion is that the performance limits with simultaneously defined spectral and amplitude constraints are still an open question.

5. SUMMARY AND CONCLUSIONS

5.1.

Most of the conclusions appear in concise form on Fig. 1. In infinite-bandwidth channels with unrestricted signals,peak-limitation does not reduce capacity. This appears, for example, from the comparison of the PMWP signal with the Gaussian one that has the same Lorenzian spectrum; for both the $\sim\sqrt{\rho}$ law holds.

5.2.

Peak limiting combined with filtering <u>does</u> reduce capacity. One example is the reduced upper bound, $\ell n(1+0.92\rho)$, obtained for the strict band-limited low-pass case.

5.3.

Restriction to constant envelope coupled with bandwidth restrictions (or in time-discrete cases) drastically reduces capacity; from a $\ell n\rho$- to a $\ell n\sqrt{\rho}$- law. However, in time-continuous, unrestricted spectrum cases, such as the RTW, or PWM the $\ell n\rho$- law is maintained.

5.4.

There is strong evidence that the transition instants, or more generally, the level crossings of peak limited signal, are important information carrier. This appears from the $\sqrt{\rho}$- law for the PMWP vs. the $\ell n\rho$- law for the RTW with both signals sharing the same spectrum. But, in the same vein RTW is much superior to PPC ($\ell n\rho$ vs. $\ell n\sqrt{\rho}$), even though they have very similar spectra. PWM, which may be considered a slotted version of the RTW has also a $\ell n\rho$- law performance. An important contribution would be the evaluation of the performance of these waveforms over a filtered channel.

5.5.

The attempt to reduce bandwidth by using the IIPK format is disappointing: The reduction in bandwidth is counterbalanced by the required increase in transmitted power. A more complex inter-symbol dependence (possibly Markov) is needed for a sizable bandwidth reduction.

Finally, we have to realize that, notwithstanding the variety of more or less interesting capacity results presented, the most important problem has not been directly addressed. One can say that we have been beating most of the time around the inner circle where spectrum is free, whereas nowadays bandwidth is becoming maybe the most important resource, even in satellite communications. The upper bounds that we have suggested for capacity in the strictly band-limited cases cannot serve to evaluate the ultimate performance achievable with the most promising signals: the CPM class. The latter are not strictly band-limited and it is exactly in the influence of spectral restriction that we are interested.

The final conclusion then is that the performance limits on "practical" communications with both spectral and amplitude constraints are still an open problem.

6. ACKNOWLEDGEMENT

Thanks are due to Shlomo Shitz who contributed to the lucidity of this compendium.
This work was supported by the Technion Fund for Research and Development.

REFERENCES

1. Shannon CE: A mathematical theory of communication, (BSTJ, 379-423, 1948; A mathematical theory of communication, BSTJ, 623-656, October 1948; Communication in the presence of noise, Proc. IRE, Jan. 1949: all reprinted in: Key Papers in the Development of Information Theory, D. Slepian (ed.) pp 5-46, N.Y., N.Y. IEEE Press, 1974.

2a. Smith JG: The information capacity of amplitude and variance-constrained scalar Gaussian channels, Inf. and Contr., vol. 18, pp 203-219, 1971.

2b. Smith JG: On the information capacity of peak and average power constrained Gaussian Channels, Ph.D. Dissertation, Dept. of Electrical Engineering, University of California, Berkeley, California.

2c. Farber G: Die Kanalkapazitat Allgemeiner Ubertragunskanale Bei Begrenztem Signalwertbereich, Beliebigen Signalubertragungszeiten Sowie Beliebiger Sturung, Arch. Elektr. Ubertr. 21, H.11, S. 565-574, and H.12, S. 653-660, 1967.

3. McEliece RJ: The theory of information and coding, Encyclopedia of Mathematics and its Application, vol. 3, Addison-Wesley, 1977.

4. Gallager RG: Information theory and reliable communication, Wiley, New York, 1968.

5. Ozarow L, Wyner AD and Ziv J: Achievable rates for a constrained Gaussian channel, submitted for publication in IEEE Trans. on Information Theory, 1986.

6. Shitz S and Bar-David I: Peak power constraint and bandlimiting reduces capacity, EE Pub No. 593, Technion, Israel, June 1986.

7. McMillan B: A history of a problem, J. Soc. Indust. Appl. Math. vol. 3, 114-128, 1955.

8. Wyner AD: Bounds on communication with polyphase coding, BSTJ, 523-559, April 1966.

9. Shitz S and Bar-David I: Capacity of peak and average-power limited-quadrature Gaussian channel, EE Pub No. 533, Technion, Israel, December 1985.

10. Forney Jr GD, Gallager RG, Lang GR, Longstaff FM and Qureshi SU: Efficient modulation for band-limited channels, IEEE J. SAC-2, No. 5, 632-647, September 1984.

11. Blachman NM: Capacities of amplitude- and phase-modulation, Proc. IRE, vol. 41, 748-759, June 1953.

12. Ungerboeck G: Channel coding with multilevel/phase signals, IEEE Trans on IT, vol. IT-28, 55-67, January 1982.

13. Saleh ARM and Salz J: On the computational cut-off rate, R_o, for the Peak-Power-Limited Gaussian Channel, Internal Report AT & T Bell Labs, Holmdel, New Jersey, submitted for publication in the IEEE Trans. on Communications.

14. Einarsson G: Signal design for the amplitude limited Gaussian channel by error bound optimization, IEEE Trans on Comm, vol. COM-27, No. 1, 152-158, January 1979.

15. Shitz S and Bar-David I: Capacity bandwidth trade-off for a class of constant envelope modulations, EE Pub. No. 532, Technion, Israel, December 1985.

16. Masseng T: Digitally phase modulated (DPM) signals, IEEE Trans. on Comm. Vol. COM-33, No. 9, 911-918, September 1985.

17. Slepian D: On Maxentropic discrete stationary processes, BSTJ, 629-653 March 1972.

18. Mcliece, RJ, Rotemich ER and Swanson L: An entropy maximization problem to optical communications, submitted for publication in the IEEE Trans. on Inf. Theory.

19. Bar-David I and Shitz S: Information transfer by random telegraph vs. phase modulation with relevance to magnetic recording, EE Pub. No. 568 Technion, Israel, January 1986.

20. Duncan TE: On the calculation of mutual information, SIAM J. Appl. Math. vol. 19, no. 1, 215-220, July 1970.

21. Wonham WM: Some applications of stochastic differential equations to optimal nonlinear filtering, J. SIAM. Contr. Ser. A, vol. 2, 347-369, 1965.

22. Bobrovsky BZ and Zakai M: Asymptotic a Priori estimates for the error in nonlinear filtering problem, IEEE Trans. on Inf. Th., vol. IT-28, 371-376, March 1982.

23. Anderson JB, Aulin T and Sundberg C-E: Digital phase modulation, textbook to be published 1986.
24. Omura JK and Jackson D: Cut-off rates for channels using bandwidth efficient modulations, pp. 14.1.1-14.1.11, NTC-1980.

Complexity Issues for Public Key Cryptography

Ian F. Blake, Paul C. van Oorschot and Scott A. Vanstone*,
University of Waterloo, Waterloo, Ontario, Canada

1. Introduction.

The proliferation of large computer networks, of both a public and private variety and either local, national or international, has intensified the need for the security of data transmission. One component of the solution to the security problem is the use of a cryptosystem which, together with the appropriate protocols, might provide authentication, digital signatures as well as data integrity and security. The Data Encryption Standard has been in use over the past decade as a reliable and secure data encipherment algorithm and thus an important component in many such systems.

Assuming that a high speed cryptosystem is available to each user of a network, the question arises as to how the keys required in its operation can be distributed to the users. Ideally, every pair of potential users in the network should possess a unique key for each session. A solution to this problem, which avoids prior physical distribution of keys and the associated security threats, is the elegant notion of public key cryptosystems proposed by Diffie and Hellman [11]. In such a system two users are able to establish a common key in a secure manner by means of a communication protocol using only the public network. In theory many of the public key systems can also be used to encipher/decipher data but the known examples of such systems tend to be relatively slow and it seems the main practical use of them at present is for key exchange, with the common key thus established used for a high speed system.

The two systems that are currently receiving the most attention for adoption as an international public key exchange standard are apparently the RSA and the discrete logarithm. These systems are briefly described in the next section. RSA depends for its security, as far as is known, on the inability to efficiently factor large integers. This problem has received considerable attention from mathematicians over the centuries and, although the steady progress made over the past twenty five years has been impressive, many doubt that a dramatic breakthrough in the problem of proportions required to compromise the security of a well designed RSA system with currently suggested parameters, is likely. Progress has been made, however, from the ability to factor twenty five digit numbers in the late 1960's to about eighty digit numbers in the present.

The discrete logarithm problem has received less attention. Over GF(p), as will be described later, the security is similar to that of RSA for parameters of the same magnitude [10]. Over fields of characteristic two, recent work [9] has yielded quite dramatic improvements to find a discrete logarithm, although the system is by no means broken. The fact remains that the structure available for exploration in finding an efficient algorithm for discrete logarithms is richer in GF(2^n) than in GF(p).

By the same token, the structure in GF(2^n) can be exploited in a variety of ways to

*This work was supported by NSERC Grant No. G 1588.

J. K. Skwirzynski (ed.), Performance Limits in Communication Theory and Practice, 75–97.

make exponentiation far faster than in GF(p) or modulo an integer, as required for RSA. Thus, if the question of likelihood of further progress in algorithm design for GF(2^n) is put aside, the question of tradeoffs between system speed, complexity and security for the systems might be asked. The thought here is that if a discrete logarithm in GF(2^n) can be implemented more easily and made to operate faster for key exchange than either GF(p) or RSA, *for the same level of security*, then it might still be the preferred choice. It is acknowledged that psychological factors play no small role in such decisions and, further, that speed might not be a very important parameter for key exchange.

This paper considers these questions. The complexity of implementing the three systems is considered in section 3, where complexity is measured by the number of clock cycles required to perform the various operations. The best known (by the authors) algorithms to break the systems and the amount of work required is given in the sections 4,5 and 6 respectively for RSA, GF(p) and GF(2^n). A comparison of the results follows in section 7.

It is noted that many of the estimates used for the various algorithms can be questioned and often depend on a variety of issues which are difficult to resolve. It is hoped the choices made here are reasonable and that the alternatives available would not dramatically affect the outcome. The recent paper by Odlyzko [26] is the definitive work on the discrete logarithm problem, particularly for fields of characteristic two, and was invaluable in preparing this work, containing as it does a great many interesting and useful results previously unknown to the authors. The present paper is a review and comparison of known results for the purpose of addressing the particular questions stated.

2. A Brief Description of the Public Key Systems.

This section gives a brief description of the RSA and discrete logarithm algorithms for key distribution. It does not consider other issues such as parameter selection or other uses of the systems such as authentication or signatures.

In the RSA system, user A in the network chooses two large primes p_A and q_A and forms the integer $n_A = p_A q_A$. An integer e_A is then chosen so that $(e_A, \phi(n_A)) = 1$, $\phi(.)$ the Euler phi function, and by using the Euclidean algorithm, its inverse d_A in the group of units of Z_{n_A} determined. The integers (e_A, n_A) are stored in a public file available to each user in the network. Any user wishing to communicate a message or key M to user A may transmit

$$C \equiv M^{e_A} \pmod{n_A}$$

User A is able to recover M by forming

$$C^{d_A} \equiv M \pmod{n_A}$$

The message M in the present context is the key the two users will use with the high speed encipherment algorithm for subsequent transmissions. Here and in the sequel we ignore such considerations as the relative size of the required key and n_A, such matters being easy to resolve.

The security of this system depends on the difficulty of determining d_A from e_A and n_A. Certainly if $\phi(n_A)$ can be determined then n_A can be factored [32]. Under the extended Riemann hypothesis Miller [24] has shown that determining $\phi(n)$ is polynomially equivalent to factoring n and Williams [38] has described a version of RSA whose security is essentially equivalent to factoring n. It is generally believed that the security of the above version is equivalent to factoring and it will be assumed so here. The problem of choosing large primes for this system is not difficult. The probabilistic test of Solovay and Strassen [36] and the more efficient one of Rabin [30] are regarded as sufficient for most purposes. For the more cautious, the deterministic test of Cohen and Lenstra [7] will also readily provide primes of the magnitude required. The work of Goldwasser and Killian [14] is also interesting in this regard, although at the moment their algorithm does not seem as fast as that of Cohen-Lenstra.

For the discrete logarithm system let GF(q) be the finite field with q elements, where q is either p, a large prime or 2^n, and let α be a primitive element. If users A and B wish to communicate, A chooses a random integer a, $1 \leq a \leq q-2$ and transmits α^a to B. Similarly B chooses a random integer b and transmits α^b. The common key is then α^{ab} and requires four exponentiations to determine, as opposed to two for RSA. As far as is known the only way to compute α^{ab} is to compute a or b i.e. to compute discrete logarithms.

In all three systems the parameters should be chosen carefully as certain classes of values have simple breaking algorithms. For example if the prime factors of $p-1$ are all small, it is a simple matter to find logarithms [27]. The determination of a primitive element in GF(p) is an interesting problem with some information available on the maximum value of the smallest such element [1]. The only technique known to the authors to determine a primitive element in GF(p) is by trial and error which is quite efficient in the sense that the probability of choosing a primitive element for a trial is, at $\phi(p-1)/(p-1)$, relatively high. However determining the period of an element in prime fields is a computationally expensive task. In fact, an efficient algorithm to determine periods in prime fields would lead to an efficient factoring algorithm [39]. Similarly, it has been observed [23] that an efficient algorithm to find logarithms modulo an integer would lead to an efficient factoring algorithm.

For GF(2^n) it is straightforward to determine an irreducible polynomial of degree n over GF(2). To determine a primitive element, given an irreducible polynomial as field generator, appears to be a problem similar to that for GF(p).

3. Complexity of Implementation.

The complexity of modular exponentiation, which will be used in both RSA and GF(p), will be discussed first. We assume that an n-bit integer is to be raised to an n-bit integer and note that $n = \lceil N \log_2 10 \rceil$ is the number of bits required to represent an N-digit integer. In the following sections N will denote the product of two primes for RSA, p will be used for the discussion of logarithms in GF(p) and 2^n for logarithms in GF(2^n).

Ordinary integer multiplication ranges from being an $O(n^2)$ bit operations for a conventional algorithm to an $O(n^{\ln 3})$ (ln denotes natural logarithms) operation by recursive subdivision to $O(n \ln(n) \ln\ln(n))$ for the Schönhage-Strassen method [34], based on the fast Fourier transform.

A variety of modular multiplication techniques have been proposed in hardware. One by Brickell [5] operates in $(n+7)$ clock cycles and others (eg Norris [25], Kochanski [17]) have similar performance. Raising an n-bit integer to an n-bit exponent, where the binary expansion of the exponent is of weight m requires n squarings, assumed to be the same as multiplies, and m multiplications, for a total of $n+m$ modular multiplies. On average, if the exponents are unrestricted, m will be $n/2$ and in the worst case, n. Thus a modular exponentiation will take at most $2n^2$ clock cycles and on average about $1.5n^2$.

With the RSA system the hardware would presumably be required to raise an arbitrary integer to an arbitrary exponent modulo an integer. The situation is similar with GF(p) since the prime p and primitive element α might be changed from time to time. With GF(p) one might limit the size of the exponent to reduce the time for exponentiation. Reducing the size too far however will make the system vulnerable to an attack by Shanks (see [16]). Alternatively, the weight of the exponent might be limited to some integer t, giving a total of

$$\sum_{i=0}^{t} \binom{n}{i} = 2^{H_2(\lambda)}, \quad H_2(\lambda) = -\lambda\log_2\lambda - (1-\lambda)\log_2(1-\lambda)$$

possible exponents where $\lambda = t/n$. So far as is known, if t is chosen sufficiently large to make a brute force attack infeasible, bounding the weight of the exponent does not simplify the problem of taking discrete logarithms. However, n squarings are still required making the savings modest. If squaring were a significantly more efficient operation than multiplication then real savings may be realized by bounding the exponent weight. This is the case with GF(2^n). Much of the point of this paper revolves around the fact that squaring is a linear operation in fields of characteristic two.

Bounding the enciphering exponent size in RSA might lead to certain types of attack and will lead to a full size deciphering exponent. In general it does not appear to be a useful idea for RSA. To increase the efficiency of enciphering and deciphering in RSA it might be possible to exponentiate modulo p and q individually and combine the results using the Chinese Remainder Theorem [29].

Exponentiation and multiplication in GF(2^n) can use quite a different approach than modular arithmetic. We assume the elements are represented as binary n-tuples with respect to a normal basis $\{\alpha,\alpha^2,\alpha^{2^2},\alpha^{2^2}, \ldots ,\alpha^{2^{n-1}}\}$ of GF(2^n) over GF(2). It is known [20] that such bases always exist and in such a basis squaring an element corresponds to a cyclic shift. Furthermore it can be shown [22] that if circuitry is found to compute one bit of a product of two elements, the same circuitry operating on cyclic shifts of the operands will compute the remaining bits. To see this let $a,b \epsilon \mathrm{GF}(2^n)$

$$a = \sum_{i=0}^{n-1} a_i\alpha^{2^i} \qquad b = \sum_{i=0}^{n-1} b_i\alpha^{2^i}$$

and let

$$c = ab = \sum_{i=0}^{n-1} c_i\alpha^{2^i}, \quad c_k = \sum_{i,j} \lambda_{ij}{}^{(k)} a_i b_j, \quad \lambda_{ij}{}^{(k)} \epsilon GF(2)$$

Then

$$c^2 = (ab)^2 = \sum_{i=0}^{n-1} c_i \alpha^{2^{i+1}}$$

showing that c_{k-1} is the coefficient of α^{2^k} in a^2b^2. Thus the bits of the representation of c can be computed independently of each other and, in particular, in parallel with the circuitry used to compute the others. This is in contrast to modular arithmetic where the process is essentially sequential in nature, although special purpose hardware can reduce the time for some of the computation required.

Since squaring corresponds to a cyclic shift with a normal basis representation, bounding the weight of an exponent significantly reduces the complexity of the operation. If the exponent is bounded by weight t then t multiplies are required, each multiply taking either n clock cycles (sequential) or 1 clock cycle (parallel). It appears to be an open question as to whether bounding the weight of the exponent to some reasonable value weakens the system but there is no attack known to the authors based on this premise.

4. The Quadratic Sieve Algorithm for Integer Factorization.

Many modern integer factoring algorithms designed for general use (i.e. not for integers of a special form) are based on Fermat's observation that if two integers x and y can be found such that $x^2 \equiv y^2 \pmod{n}$, $x \not\equiv \pm y \pmod{n}$, then $(x+y,n)$ and $(x-y,n)$ are proper factors of n. The various techniques differ only in how the integers x and y are arrived at.

While the more recent elliptic curve method of Lenstra [19], which is not of the above type, has promise of being a fast general purpose algorithm, requires less storage and is amenable to parallelism, it appears to still be in a development stage and will not be considered. It has a running time dependent on the least prime factor. At the moment the best general purpose algorithm appears to be the quadratic sieve (QS) [28] and its variations such as the multiple polynomial QS [35]. For RSA type numbers (the product of two large primes of approximately equal magnitude) the two techniques will have a similar operation count but the QS will be faster due to the simpler operations involved. This section will describe and analyze the performance of the basic QS as representative of the amount of work required to break RSA. It should be noted that both the elliptic curve and QS factoring techniques depend on certain unproved hypotheses for their operation.

Description of the Algorithm: For all of the methods in this class, to factor the integer N, one starts with a factor base of primes $B = \{p_1, p_2, \ldots, p_b\}$, $|B| = b$. For a reason that will become apparent we will require that the Jacobi symbol $(N/p) = 1$ and also that $-1, 2 \epsilon B$. The general procedure then requires a search for pairs of integers (a_i, b_i), $i=1,2,...,b+1$, such that

$i)$ $a_i^2 \equiv b_i \pmod{N}$

$ii)$ b_i factors completely in the factor base, $b_i = \prod_j p_j^{c_{ij}}$.

For each i, associate to b_i the vector $v_i = (v_{i1}, v_{i2}, ..., v_{ib})$ where $v_{ij} \equiv c_{ij} \pmod{2}$. With $(b+1)$ such vectors in the vector space of dimension b over Z_2, a linear dependency relationship must exist. Specifically there must exist a set S such that

$$\sum_{i \epsilon S} v_i = 0$$

Using the Wiedemann algorithm [37], such a solution can be determined in $O(b^2)$ operations. Thus

$$\prod_{i \epsilon S} b_i = y^2$$

is a perfect square and elements x and y are found such that

$$x^2 \equiv \prod_{i \epsilon S} a_i{}^2 \equiv \prod_{i \epsilon S} b_i \equiv y^2 \pmod N$$

Notice there are four solutions to the congruence $x^2 \equiv a^2 \pmod{pq}$ which can be found by solving the congruences modulo p and q individually and combining using the Chinese Remainder Theorem. Of these four solutions, two will correspond to $x \equiv \pm a \pmod{pq}$ and thus there is a probability of one half that $x \not\equiv \pm a \pmod{pq}$. If n is the product of more than three primes, the probability of noncongruent solutions is greater than one half.

This procedure is common to many general purpose factoring techniques (such as continued fraction, QS, Dixon, etc.). The difference in the algorithms lies in how the pairs (a_i, b_i) are generated and the properties they have. The two most significant properties appear to be the magnitudes of the b_i (smaller is better) and the difficulty in factoring the b_i's (as the name implies, the sieve methods avoid factoring by sieving).

As a first step in the QS algorithm the residues

$$f(z) = ([\sqrt{N}] + z)^2 - N$$

are computed for all integers z such that $|z| \leq M$ for some integer M. Here $[\cdot]$ is the integer part of function. The maximum size of such a residue is on the order of $2M[\sqrt{N}]$ for $M \ll [\sqrt{N}]$. By the choice of the factor base, the quadratic congruence

$$f(z) \equiv 0 \pmod{p^\alpha} \ , \ p \epsilon B$$

will have solutions for any integer power α. If $z(p^\alpha)$ is a solution then $z(p^\alpha) + kp^\alpha$ is also a solution. This indicates that the $f(z)$ may be factored by sieving and the algorithm may be described in the following way.

The residues $f(z)$ are stored in a linear array, indexed by z. For a given prime $p \epsilon B$ the solutions $z_1(p)$, $z_2(p)$ to

$$f(z) \equiv 0 \pmod p \ , \ p \epsilon B$$

are determined and all residues corresponding to $z_i(p) \pm kp$, i=1,2 are divided by p. The procedure is repeated for each prime $p \epsilon B$. It is also done for powers of small primes since the probability of repeated prime factors is not negligible. Judicious choices are made as to the magnitudes of the powers and the primes. In this procedure it is noted that the two solutions of the quadratic congruences modulo p^α can be built up from those modulo p (see also Appendix 2) and that a solution modulo p^α is also a solution modulo $p^{\alpha-1}$. Thus when sieving from a solution (modp^α) the array elements are divided by p (not p^α) since the same element has already been divided for each lower power. As a matter of practice, the logarithms are stored, using a suitable fixed precision arithmetic, and the logarithms of the primes subtracted. The prime 2 requires a special sieving procedure since quadratic congruences modulo a power α of 2 have four solutions for $\alpha \geq 3$.

At the conclusion of the sieving procedure for each prime in B, the array is scanned for elements close to zero in magnitude. These correspond to residues which completely factor over B, including the small prime powers. Another refinement of the procedure which turns out to be important in practice, is to further consider those residues corresponding to array elements greater than p_b and less than $p_b{}^2$ in magnitude, where p_b is the largest prime in B. Such elements correspond to prime factors of the residue greater than p_b or prime powers of primes in the base but not sieved , or some combination. If another residue leaves the same cofactor the two can be combined to yield another completely factored residue. This appears [13] to be an essential step in practice for efficient implementation.

Analysis of the Algorithm: Following the discussion in Pomerance [28] we will choose the factor base to have size

$$b = |B| = L(a)$$

for some value of a, where $L(a)$ is defined in Appendix 1. Recall that the largest residue is on the order of $2M[\sqrt{N}]$. Certain assumptions are required to proceed further. It is assumed that for large N:

(i) For any a, $1/10 < a < 1$, the number of primes $p \leq L(a)$ for which p does not divide N and $(N/p) = 1$ is at least $\pi(L(a))/3$, where $\pi(\cdot)$ is the prime counting function.

(ii) For any $c > 1/10$, the fraction of the numbers $|f(z)|$ with $|z| \leq L(c)$ that are smooth with respect to $L(a)$ is the same as $\psi(2\sqrt{N}L(c),L(a))/(2\sqrt{N}L(c))$, i.e. $1/(L(1/4a))$, (Appendix 1), where M is chosen as $L(c)$.

(iii) There is a constant n_1 such that for $n \geq n_1$, if $1 + [\log^2 n]$ pairs (x,y) are found such that $x^2 \equiv y^2 \pmod{n}$, then at least one pair will satisfy $x \not\equiv y \pmod{n}$.

The first assumption assures an adequate supply of primes with the required properties, the second allows estimates of the probability of smoothness to be used, as discussed in Appendix 1, and the third assures us of a factor when enough of the appropriate pairs are found. In particular the probability a randomly chosen residue from the interval $[1,2M[\sqrt{N}]]$ is smooth with respect to $L(a)$ is $L(-1/4a)$. From the $L(c)$ residues approximately $L(c)L(-1/4a) = L(c-(1/4a))$ will factor into B. Since $L(a)$ such residues are required, c is chosen so that $c-(1/4a) \geq a$ or $c = a+(1/4a)$. The number of operations required to sieve the $L(c)$ values is

$$\sum_{p \leq L(a)\,(n/p)=1} L(c)/p = L(c)$$

assuming at least a few small primes in B. Thus the number of operations for sieving is on the order of $L(a+(1/4a))$. To solve the equations requires on the order of (Appendix 2) $L^2(a) = L(2a)$ operations. To solve the quadratic congruences for each prime in the factor base is at most an $L^2(a) = L(2a)$ operation. Thus the overall operation count is of the form $L(c)$ where $c = \max(2a, a+(1/4a))$ and c may be minimized by choosing $a=1/2$, giving an operation count of $L(1)$. The storage requirement for the equation matrix is $L(2a) = L(1)$ and for the sieve is $L(c) = L(a + 1/4a) = L(1)$, although in fact both of these can be reduced to $L(1/2)$ by using a sparse encoding for the equations and by segmenting the sieve.

5. Discrete Logarithms in GF(p).

For certain types of primes, for example those with the property that $p-1$ has all its prime factors small, it is a relatively easy matter to find logarithms using the Pohlig-Hellman algorithm, as has been mentioned. The algorithm discussed in this section is the best general one known to the authors and is due to Coppersmith, Odlyzko and Schroeppel [10]. It has very much the flavor of the linear sieve factoring algorithm of Schroeppel.

Description of the Algorithm. Using the terminology of [10], it is assumed that the logarithm to some base γ, a primitive element in GF(p), is required. The algorithm is in two stages, requiring a precomputation stage which involves finding the logarithms of all elements in a factor base, followed by a computation to find the logarithm of the actual element. Let

$$H = [\sqrt{p}] + 1 , \qquad J = H^2 - p$$

and let S be a factor base consisting of two parts:

$$A = \{ q \mid q \text{ prime} , q < L(1/2)\} \quad \text{and} \quad B = \{ H + c \mid 0 < c < L(1/2 + \epsilon)\}$$

and $S = A \cup B$. Notice that in the definition of $L(\cdot)$ in Appendix 1, p is used for this discussion rather than N. Thus A consists of small primes and B a set of consecutive integers beginning at about $\sqrt{p}$ and S has size about $L(1/2 + \epsilon)$. It is desired to find the logarithms of all elements in S by constructing a sufficient number of linear equations over Z_{p-1} such that the unknown logarithms may be solved for. Assuming that the base γ is in A one equation is obtained immediately, namely $\log_\gamma(\gamma)=1$ and the resulting set of equations will be nonhomogeneous. To find the other equations consider the residue of the product of two elements in B:

$$(H + c_1)(H + c_2) \equiv J + (c_1 + c_2)H + c_1c_2 \pmod{p}$$

For p large enough $c_1 < L(½ + \epsilon) < p^{\epsilon/2}$ i.e. $(½ + \epsilon)\ln(p)^{½}\ln^{½}(\ln(p)) < \epsilon\ln(p)/2$ and $H,J < 2\sqrt{p}$. Thus the residue is $O(p^{½ + \epsilon/2})$. The probability that a residue is smooth with respect to $L(½)$ i.e. factors into A is $L(-(1 + \epsilon)/2)$ using a result of Appendix 1. Suppose that all pairs (c_1,c_2), $c_1 < c_2$, are used to generate residues $(H + c_1)(H + c_2) \pmod{p}$ and checked for smoothness with respect to $L(½)$. Of the $½L^2(½ + \epsilon) \approx L(1 + 2\epsilon)$ possible pairs approximately

$$L(1 + 2\epsilon)L(-½ - \epsilon/2) \approx L(½ + 3\epsilon/2)$$

equations should be obtained. There should thus be more equations than unknowns allowing solution for the logarithms of all elements of S. Each such equation is of the form

$$(H + c_1)(H + c_2) \equiv \prod_{q_i \in S} q_i^{\,f_i} \pmod{p}$$

or

$$\log(H + c_1) + \log(H + c_2) \equiv \sum_i f_i \log q_i \pmod{p-1}$$

and these equations are sparse with each containing two terms from the set B and, at most, $\log(p)$ terms from the set A.

As with RSA, the Wiedemann algorithm outlined an Appendix 2 might be used to solve the equations over the ring Z_{p-1} and this is an $L^2(½ + \epsilon) \approx L(1 + 2\epsilon)$ operation. As with the quadratic sieve algorithm, factorization of the residues is not necessary - the values can be sieved as follows. For a fixed value of c_1 set up a linear array of size $L(½ + \epsilon/2)$ with the ith position containing the real logarithm to an appropriate fixed precision of the residue corresponding to c_1 and $c_2 = c_1 + i$. For a prime $q \epsilon A$ compute

$$d \equiv -(J + c_1 H)(H + c_1)^{-1} \pmod{q}.$$

Consider the residue

$$r \equiv J + (c_1 + c_2)H + c_1 c_2 \pmod{p}$$

and note that for any value of $c_2 \equiv d \pmod{q}$ i.e. $c_2 = aq + d$

$$r \equiv (J + c_1 H) + (aq + d)(H + c_1) \equiv aq(H + c_1) \pmod{p}$$

and is divisible by the prime q. Thus from an appropriate starting place in the array, $\log(q)$ is subtracted from every d-th element. The procedure is repeated for each prime in A and also for prime powers less than $L(1/2)$ as long as a solution for d exists. As with the QS algorithm, for each small prime q and for $c_2 \equiv d \pmod{q^j}$, $\log(q)$ is subtracted from every d-th element since the sieving is done sequentially for increasing powers of q each time $\log(q)$ being subtracted. At the end of the sieving procedure for all primes in A those elements in the array close to zero correspond to pairs (c_1,c_2) for which the residue is smooth and yields an equation.

The procedure for finding an individual logarithm is quite complicated and involves expressing the given element whose logarithm is required first in terms of medium sized primes (about $L(14)$). This is achieved using a randomization of the element, use of the extended Euclidean algorithm and sieving about a carefully chosen interval. These medium sized primes are then, by a sieving procedure, expressed as products of elements in S from which the logarithm of the original element can be determined.

Analysis of the Algorithm: For each of the $L(½ + \epsilon)$ elements c_1, a sieve is required of size $L(½ + \epsilon)$. Each array is initialized and sieved requiring a use of the extended Euclidean algorithm to compute the value d. Both the initializing and updating are $L(½ + \epsilon)$ operations and since there are $L(½ + \epsilon)$ values of c_1 to consider, the equation gathering stage requires on the order of $L^2(½ + \epsilon) = L(1 + 2\epsilon)$ operations. The equations require on the order of $L(½ + \epsilon)$ storage and $L^2(½ + \epsilon) = L(1 + 2\epsilon)$ operations to solve. It is shown [10] that to find an individual logarithm requires $L(½)$ space and $L(½)$ operations.

6. Discrete Logarithms in Fields of Characteristic Two.

Description of the Algorithm: This section will give a brief description of the Coppersmith algorithm [9],[26]. Assume that the field GF(2^n) is generated by the polynomial

$f(x) = x^n + f_1(x)$ and assume that $\deg(f_1(x)) \approx \log_2(n)$ (Appendix 3). Only the precomputation stage of the algorithm is considered in detail here as the second stage to compute a logarithm of a given element turns out to be much faster and is thus not a limiting factor.

Choose an integer b approximately $c_1 n^{1/3} \ln^{2/3}(n)$ for some small constant c_1. The database consisting of the logarithms of all irreducible polynomials of degree less than or equal to b is to be constructed. There are approximately $B{=}2^{b+1}/b$ of them (Appendix 3). The idea will be to generate at least B equations in these B unknowns and solve them over the ring of integers modulo N, $= Z_N$, where $N = 2^n - 1$. The following parameters are chosen for reasons that will become apparent later:

$$d \text{ near } b$$

$$k \text{ such that } 2^k \text{ is near } \sqrt{n/d}$$

$$h = \left\lfloor n/2^k \right\rfloor + 1$$

Now choose polynomials $a(x)$,$b(x)$ such that $(a(x),b(x)) = 1$, to avoid redundant equations, and $\deg a(x)$, $\deg b(x) \leq d$. There are precisely 2^{2d+1} such ordered pairs (Appendix 3) and let

$$d(x) \equiv c(x)^{2^k} \pmod{f(x)}$$

where

$$c(x) = x^h a(x) + b(x)$$

Thus

$$d(x) = x^{h2^k} a(x^{2^k}) + b(x^{2^k})$$

and if

$$x^{h2^k} \equiv r(x) \pmod{f(x)} \quad \text{then} \quad d(x) \equiv r(x)a(x^{2^k}) + b(x^{2^k}) \pmod{f(x)}$$

The degree of $c(x)$ is at most $h{+}d$ and the degree of $d(x)$ is at most $r{+}d2^k$, where $r{=}\deg(r(x))$ and is small. From the above choice of parameters it follows that $d2^k \approx \sqrt{nd}$, $h \approx \sqrt{nd}$, $\sqrt{nd} \approx n^{2/3}\ln^{1/3}(n)$ and $d \approx n^{1/3}\ln^{2/3}(n)$. Thus $c(x)$ and $d(x)$ both have degrees on the order of $\sqrt{nd} \approx n^{2/3}$. Now

$$\log(d(x)) \equiv 2^k \log(c(x)) \pmod{2^n - 1}$$

and if both $c(x)$ and $d(x)$ have all their irreducible factors of degree less than or equal to b, an equation among the B unknowns is obtained. We require B such equations. For a fixed b, d must be chosen to provide a sufficient number of such equations.

Analysis of the Algorithm: To obtain an estimate of the amount of work which this will involve, proceed as follows. The probability that a randomly chosen polynomial of degree exactly n has all of its irreducible factors of degree less than or equal to m is (Appendix 1)

$$p(n,m) = \exp((1+o(1))\frac{n}{m}\ln(\frac{m}{n}))$$

$$\approx\left(\frac{m}{n}\right)^{\frac{n}{m}} \approx \left(\frac{n}{m}\right)^{-\frac{n}{m}}$$

The approximation will be quite good for $n^{\frac{1}{100}} \leq m \leq n^{\frac{99}{100}}$ and is sufficiently accurate for use in deriving estimates.

The probability that both $c(x)$ and $d(x)$ are smooth with respect to b is then approximately

$$p(d+h,b)p(2^k d+r,b) \approx p^2(\sqrt{nd},b)$$

$$\approx\left(\frac{\sqrt{nd}}{b}\right)^{-2\sqrt{nd}/b} = \exp\left(-\frac{2\sqrt{nd}}{b}\ln(\frac{\sqrt{nd}}{b})\right)$$

The analysis given here differs slightly from that in [26] due to our approximation that $d+h$ and $2^k d+r \approx \sqrt{nd}$. Following the technique of Odlyzko [26] let

$$2^k \approx \alpha n^{1/3}(\ln(n))^{-1/3}$$

$$b \approx \beta n^{1/3}\ln^{2/3}(n)$$

$$d \approx \gamma n^{1/3}\ln^{2/3}(n)$$

and

$$K(a) = \exp(a n^{1/3}\ln^{2/3}(n))$$

On average, the amount of work to obtain one equation is

$$\exp\left(\frac{2\sqrt{nd}}{b}\ln(\frac{\sqrt{nd}}{b})\right)$$

and the average amount of work to generate the equations is

$$\exp\left(\frac{2\sqrt{nd}}{b}\ln(\frac{\sqrt{nd}}{b})\right)\cdot\frac{2^{b+1}}{b} \approx \exp\left(\frac{\sqrt{nd}}{b}\ln(\frac{nd}{b^2}) + (b+1)\ln(2)\right)$$

For large n it is clear that, relatively, d will be only slightly greater than b and the simplifying assumption that $d=b$ is made to yield

$$\exp\left(\left(\frac{n}{b}\right)^{1/2}\ln(\frac{n}{b}) + (b+1)\ln(2)\right)$$

as the amount of work to find the equations. To minimize this, differentiate the exponent with respect to b and equate to zero:

$$-\frac{\sqrt{n}}{2b\sqrt{b}}\ln(\frac{n}{b}) + \ln(2) = 0$$

or

$$b^{2/3} = \frac{\sqrt{n}}{2}\ln(n) - \frac{\sqrt{n}}{2}\ln(b) + \sqrt{n}$$
$$\approx \frac{\sqrt{n}}{2}\ln(n)$$

or

$$b \approx \left(\frac{1}{2\ln(2)}\right)^{2/3} n^{1/3}\ln^{2/3}(n)$$
$$\approx c_1 n^{1/3}\ln^{2/3}(n)\,, \quad c_1 \approx .8043.$$

The more refined analysis in [26] yields a constant of .9743. Thus the total number of trials that must be performed to generate the $2^{b+1}/b$ equations is approximately

$$\exp\left(\frac{n^{1/2}}{c_1^{1/2} n^{1/6}\ln(n)^{1/3}}(\ln(n) - \ln(b)) + c_1 n^{1/3}\ln^{2/3}(n)\ln(2)\right)$$
$$\approx \exp\left(\frac{2}{3c_1^{1/2}} n^{1/3}\ln^{2/3}(n) + c_1\ln(2)n^{1/3}\ln^{2/3}(n)\right)$$
$$\approx \exp\left((\frac{2}{3c_1^{1/2}} + c_1\ln(2))n^{1/3}\ln^{2/3}(n)\right) \approx \exp(1.3009n^{1/3}\ln^{2/3}(n)) = K(1.3009)$$

(where the corresponding constant in [26] is 1.3507). The amount of storage required for these equations is proportional to $2^{b+1}/b$ which is on the order of

$$\exp(b\ln(2)) \approx \exp(.5575n^{1/3}\ln^{2/3}(n)) = K(.5575).$$

Each storage location for an equation contains the coefficient of each log term (typically unity) and a designation of which log terms are involved. The equations are quite sparse since there is a low probability that a randomly chosen polynomial will have a large number of factors.

In order to find a sufficient number of equations , d must be chosen large enough that 2^{2d+1} is greater than the expected number of trials to obtain the $B = 2^{b+1}/b$ equations. Asymptotically it is required that

$$2^{(2d+1)} \approx \exp(2\ln(2)d) \geq \exp(1.3009n^{1/3}\ln^{2/3}(n)) \approx \exp(\frac{1.3009}{.8043}b)$$

or d should be about $1.1667b$.

To conclude the analysis the following estimates are required:

(i) The number of operations to test a chosen pair of polynomials $a(x)$ and $b(x)$ to determine an equation. This includes polynomial generation, exponentiation, factoring etc.

(ii) The number of operations to solve the resulting $2^{b+1}/b$ equations.

(iii) The amount of work to find an individual logarithm.

The generation of the pairs of polynomials is straightforward. As noted, exactly one half of all possible pairs will be relatively prime (Appendix 3). There is a fairly low probability that a given pair of polynomials will yield an equation and thus it is desirable to find techniques that lead to an early abort for pairs that will not give an equation. This topic is briefly discussed in the Appendix 3. The number of operations to solve a BxB system of linear equations over a ring is on the order of B^2 (Appendix 2), on the order of $K(1.115)$. To find an individual logarithm, assuming the database has been constructed, Coppersmith [9] gives an algorithm that, given a polynomial of "medium" degree, proceeds to iteratively reduce the degree of the polynomial whose logarithm is to be found, at each stage producing more polynomials whose logarithms are required. As the degrees of the polynomials are reduced there is a higher probability they will factor into the database. It can be shown that this stage of finding a particular logarithm is much faster than for the precomputation stage of constructing the database. We will assume that the difficulty of finding discrete logarithms in fields of characteristic two is that of finding the database, determined as $K(1.3507)$, using the value of Odlyzko [26].

7. A Comparison of the Public Key Systems.

In this section an attempt is made to compare the systems in terms of speed of operation and level of security afforded and to begin, the requirements to break the systems are reviewed. In the QS algorithm, $L(1)$ residues are formed. For each prime in the factor base of size $L(½)$ a congruence is solved and the residues sieved. The resulting $L(½)$ equations are solved over Z_2 in $L(1)$ operations, for a total number of operations on the order of $L(1)$. The algorithm requires $L(1)$ storage for both the equations and the sieving.

For the precomputation stage to find logarithms in $GF(p)$, a factor base of size on the order of $L(½+\epsilon)$ is formed and $½L^2(½+\epsilon) \approx L(1+2\epsilon)$ residues found and sieved. Each residue requires modular multiplications and the computation of an inverse element modulo a prime for $L(½+\epsilon)$ of them. The storage required during the equation gathering and solving stages is $L(½+\epsilon)$ and the number of operations during each of these stages is $L(1+2\epsilon)$. To find an individual logarithm requires on the order of $L(½)$ for both space and operations.

For logarithms in $GF(2^n)$ the factor base consists of all irreducible polynomials up to a given degree on the order of $\ln K(.8043)$. Approximately $K(.5575)$ equations are to be found, requiring the generation of $2^{2d} \approx K(1.3009)$ pairs of polynomials, each resulting polynomial being tested for factoring in the database. Those that pass the early abort test are factored using the Berlekamp algorithm. The resulting equations are solved in about $K(1.115)$ operations, over the ring Z_N, $N=2^n-1$. As in $GF(p)$, the work involved in finding an individual logarithm is much less than that of the equation gathering stage and will be ignored in the comparison.

A more precise description of what is meant by an operation in each case would be desirable. Measuring such things as the number of clock cycles or logic gates involved, however, is difficult and the significance may be doubtful.

A few characteristics of the systems are worth mentioning. In RSA when the integer is factored, the system is broken in that all enciphered messages are easily deciphered. With the discrete logarithm systems, after the logarithms of the database are found, there is still a considerable amount of work to find an individual logarithm, although less than that of the precomputation stage. In the discrete logarithm problems the size of the database was chosen to minimize the amount of work in finding the logarithms of the database. Another approach might be to choose the size of the database to make the amount of work in finding it approximately the same as that of finding an individual logarithm. This approach was not pursued.

As far as implementation is concerned from a practical point of view, it appears that RSA chips for n=664, (N about 200 digits), can be made to run at about 25kb/s [33]. For discrete exponentiation in $GF(2^n)$, a chip for n=23–27 was built and cascaded to give n=593. With this experience a chip with n over 1000 was designed, but not yet built, with a speed estimated at approximately 1Mb/s when the exponent weight is bounded by 20. More professional industrial interests have considered the design and estimate that with 1-2 micron technology a chip with n=1500–2000 is quite feasible at comparable speeds. Once again, the speed required of a key passing scheme will be application dependent, with relatively slow speeds tolerated in many such systems.

It is an interesting mathematical exercise to compare the speed/security of RSA versus discrete logarithms in $GF(2^n)$, to illustrate the issues involved. It has already been commented upon that modular exponentiation for RSA will take, on average, about $1.5n^2$ clock cycles. For $GF(2^n)$, bounding the exponent to weight 20, an exponentiation will require $20n$ clock cycles. Thus in the comparison advantage will be taken of the ability of discrete logarithms to exploit weight bounding but not of the parallelism opportunities it affords.

For purely mathematical purposes, it will be assumed that the measure of security for RSA is $\ln L(1)$ and for discrete logarithms in $GF(2^n)$ is $\ln K(1.35)$ (since the exponent obtained by Odlyzko [26] using a more refined analysis is 1.3507 as opposed to 1.3009 obtained here). It is assumed the speed and security of logarithms in $GF(p)$ is comparable to that of RSA and it is not considered further. Using the arguments for these functions developed in the previous sections will change the results of the analysis but not the fundamental nature of it.

With these given speed and complexity measures, two questions can be asked. First, for the same encoding speed, which system is more secure? Second, for the same level of security, which system is faster?

For the first question, choose n for RSA ($N \sim 2^n$), and to equate the speeds of the systems, choose n_1 for $GF(2^n)$ where $1.5n^2 = 20n_1$ or $n_1 = (1.5n^2/20)$. The security of the two systems compares as

$$n^{1/2}\ln^{1/2}(n) \quad \text{versus} \quad 1.35(1.5n^2/20)^{1/3}\ln^{2/3}(1.5n^2/20) \approx n^{2/3}\ln^{2/3}(n)$$

and the discrete logarithm is the more secure. This comparison however is not practical since for $n \approx 600$, $n_1 \approx 27000$ which could not be realized.

The second question appears to be the more interesting since, in practice, a given level of security of competing systems would be desired. For this comparison, resort is made to computation. For n=664 for RSA the level of security is $\ln^{1/2}(N)\ln^{1/2}(\ln(N)) = 53.12$ for $N=2^{664}$. To achieve the same exponent for the discrete logarithm case a value of n_1 slightly less than 1250 would be required. For n=512 for RSA the exponent is 45.65 and for discrete logarithms to achieve the same exponent n_1 is approximately 850. To compare the resulting speeds, for n=664 for RSA, one exponentiation would require $1.5n^2 = 6.61\text{x}10^5$ clock cycles and for n_1=1250 for discrete logarithms $20n_1 = 2.5\text{x}10^4$. For n=512 for RSA, $3.93\text{x}10^5$ cycles are required and for $n_1 = 850$ for discrete logarithm, $1.7\text{x}10^4$. From very limited knowledge the authors would suggest that the architecture and design of the chip for $GF(2^n)$ would be simpler than that for RSA and would operate much faster for the same level of security.

For comparison purposes the exponents of $K(1)$, $K(1.35)$ and $L(1)$ are shown in figure 1. It is noted that the multiplicative factor of 1.35 in the exponent of the $K(\cdot)$ function has a significant effect on the results. As it is unlikely that RSA will be realized in the forseeable future with more than 200 digits, this limits the range of interest of the graph. In this range of interest it appears that discrete logarithms in $GF(2^n)$ using bounded weight exponents offers significant advantages.

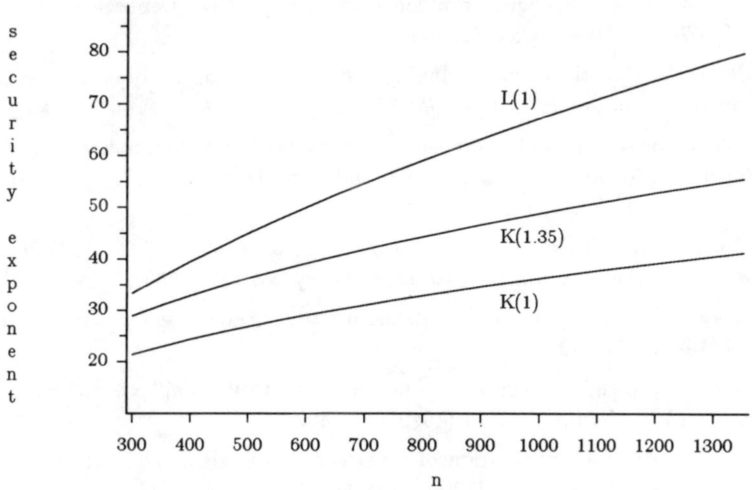

Figure 1. Comparison of Security Exponents.

The above results are mainly of theoretical interest but nonetheless serve to illustrate the issues. If they are combined with the possibilities for parallelism in discrete logarithms, the results appear favorable to that system. In practice other aspects, harder to quantify, may be more important.

It is not fruitful to speculate on the likelihood of further progress on the factoring and logarithm algorithms. Determining the exact complexity of these problems seems beyond present abilities. This work has reviewed approaches to analyzing three particular problems in the setting of public key cryptography to focus attention on the speed, complexity and security tradeoffs between them.

Acknowledgement

The authors are grateful for the careful reading of the original version of this manuscript and the many useful comments of Carl Pomerance.

References

[1] E. Bach, *Analytic Methods in the Design and Analysis of Number-Theoretic Algorithms,* The MIT Press: Cambridge, Mass., 1985.

[2] E.R. Berlekamp, *Algebraic Coding Theory,* McGraw-Hill: New York, 1968

[3] I.F. Blake, M. Jimbo, R.C. Mullin and S.A. Vanstone, Computational Algorithms for certain shift register sequence problems, Final Report for Dept. Supply and Services, Project 307-16, Ottawa, Canada, 1984.

[4] I.F. Blake, R. Fuji-Hara, R.C. Mullin and S.A. Vanstone, Computing logarithms in finite fields of characteristic two, *SIAM J. Alg. Disc. Meth.,* Vol. 5 (1985), 276-285.

[5] E. Brickell, A fast modular multiplication algorithm with applications to two key cryptography, 51-60, in *Advances in Cryptology,* D. Chaum, R. Rivest and A. Sherman eds., Plenum Press, 1983.

[6] E.R. Canfield, P. Erdös and C Pomerance, On a problem of Oppenheim concerning "Factorisation Numerorum", *J. Number Theory,* Vol. 17 (1983), 1-28.

[7] H. Cohen and H.W. Lenstra, Jr., Primality testing and Jacobi sums, *Math. Comp.,* Vol. 42 (1984), 297-330.

[8] D. Coppersmith and S. Winograd, On the asymptotic complexity of matrix multiplication, *SIAM J. Comp.,* Vol. 11 (1982), 472-492.

[9] D. Coppersmith, Fast evaluation of logarithms in fields of characteristic two, *IEEE Trans. Inform. Theory,* Vol IT-30 (1984), 587-594.

[10] D. Coppersmith, A.M. Odlyzko and R. Schroeppel, Discrete logarithms in GF(p), preprint.

[11] W. Diffie and M. Hellman, New directions in cryptography, *IEEE Trans. Inform. Theory,* Vol. IT-22 (1976), 644-654.

[12] J.D. Dixon, Factorization and primality tests, *Am. Math. Monthly,* Vol. 91 (1984), 333-352.

[13] J.L. Gerver, Factoring large numbers with a quadratic sieve, *Math. Comp.,* Vol. 41 (1983), 287-294.

[14] S. Goldwasser and J. Killian, Almost all primes can be quickly certified, preprint.

[15] G.H. Hardy and E.M. Wright, *An Introduction to the Theory of Numbers,* Oxford University Press: Oxford, 1962.

[16] D.E. Knuth, *The Art of Computer Programming, Vol. 3,* Addison Wesley: Reading, Mass., 1973.

[17] M. Kochanski, Developing an RSA chip, presented at *Crypto '85,* Santa Barbara, CA,

August, 1985.

[18] D.H. Lehmer, Computer technology applied to the theory of numbers, 117-151, in *Studies in Number Theory,* W.J. LeVeque, ed., MAA Studies in Math., Vol. 6, 1969.

[19] H.W. Lenstra, Jr., Factoring integers using elliptic curves over finite fields, to appear.

[20] R. Lidl and H. Niederreiter, *Finite Fields,* Addison-Wesley: Reading, Mass., 1983.

[21] J.L. Massey, Shift register synthesis and BCH decoding, *IEEE Trans. Inform. Theory,* Vol. IT-15 (1969), 122-127.

[22] J.L. Massey and J.K. Omura, patent application, Computational method and apparatus for finite field arithmetic, submitted 1981.

[23] J.L. Massey, Logarithms in finite cyclic groups - cryptographic issues, *4th Symposium on Inform. Theory,* Belgium, 1983.

[24] G.L. Miller, Riemann's hypothesis and tests for primality, *J. Comput. System Sci.,* Vol. 13 (1976), 300-317.

[25] M.J. Norris and G.J. Simmons, Algorithms for high speed modular arithmetic, *Congressus Numerantium,* Vol. 31 (1981), 153-163.

[26] A. Odlyzko, Discrete logarithms in finite fields and their cryptographic significance, in *Advances in Cryptology,* 224-314, T. Beth, N. Cot and I. Ingemarsson editors, Vol. 209, Lecture Notes in Computer Science, Springer-Verlag: Berlin, 1984.

[27] S. Pohlig and M. Hellman, An improved algorithm for computing logarithms over GF(p) and its cryptographic significance, *IEEE Trans. Inform. Theory,* Vol. IT-24 (1978), 106-110.

[28] C. Pomerance, Analysis and comparison of some integer factoring algorithms, 89-139 in *Computational Methods in Number Theory: Part 1,* H.W. Lenstra and R. Tijdeman, eds., Math. Centr. Tract 154, Math. Centr.: Amsterdam, 1982.

[29] J.J. Quisquater and C. Couvreuer, Fast decipherment algorithm for RSA public key cryptosystem, *Electronic Letters,* Vol. 18 (1982), 905-907.

[30] M. Rabin, Probabilistic algorithm for primality testing, *J. Number Theory,* Vol. 12 (1980), 128-138.

[31] J.A. Reeds and N.J.A. Sloane, Shift-register synthesis (modulo m), *SIAM J. Comput.,* Vol. 14 (1985), 505-513.

[32] R. Rivest, A. Shamir and L. Adleman, A method for obtaining digital signatures and public key cryptosystems, *Comm. ACM,* Vol. 21 (1978), 120-128.

[33] R. Rivest, *Advances in Cryptology,* Proceedings of Eurocrypt '84, Berlin: Springer-Verlag, 1985.

[34] A. Schönhage and V. Strassen, Schnelle multiplikation grosser Zahlen, *Computing,* Vol. 7 (1971), 282-292.

[35] R.D. Silverman, The multiple polynomial quadratic sieve, preprint.

[36] R. Solovay and V. Strassen, A fast Monte-Carlo test for primality, *SIAM J. Comput.,* Vol. 6 (1977), 84-85, (Vol. 7 (1978), 118, erratum).

[37] D. Wiedemann, Solving sparse linear equations over finite fields, *IEEE Trans.*

Inform. Theory, Vol. IT-32 (1986), 54-62.

[38] H.C. Williams, A modification of the RSA public-key encryption procedure, *IEEE Trans. Inform. Theory,* Vol. IT-26 (1980), 726-729.

[39] H.C. Williams, personal communication.

Appendix 1. Smooth Numbers and Polynomials.

A positive integer x is said to be smooth with respect to y if all the prime divisors of x are of magnitude at most y. The function $\psi(x,y)$ is the number of integers not exceeding x that are smooth with respect to y. The probability that a randomly chosen integer in the interval $[1,x]$ is smooth with respect to y will be approximated by $\psi(x,y)/x$. This function has been extensively investigated and good bounds and approximations are available for certain ranges of parameters. The following theorem will be useful:

Theorem [6, corollary to theorem 3.1, p.15].

If ϵ is arbitrary and $3 \leq u \leq (1-\epsilon)\ln(x)/\ln\ln(x)$, then

$$\psi(x,x^{1/u}) = x\cdot\exp\Big(-u(\ln(u) + \ln\ln(u) - 1 + ((\ln\ln(u)-1)/\ln(u)) + E(x,u))\Big)$$

$$= x\cdot\exp\Big(-u(1+o(1))\ln(u)\Big)$$

where

$$|E(x,u)| \leq c_\epsilon \frac{\ln^2(\ln(u))}{\ln^2(u)}$$

where c_ϵ is a constant that depends only on the choice of ϵ.

It is sufficient for the theorem that $x \geq 10$ and $y \geq \ln(x)^{(1+\epsilon)}$.

It is convenient to define the quantity

$$L(\alpha) = \exp\Big(\alpha(\ln(N)\ln\ln(N))^{1/2}\Big)$$

In the analysis of the QS algorithm the following smoothness estimate is needed:

$$\ln\Big(\psi(2\sqrt{N}L(c),L(a))/(2\sqrt{N}L(c))\Big) = -u\ln(u)$$

where

$$u = \ln(2\sqrt{N}L(c))/\ln(L(a)) = \frac{\ln(2) + \frac{1}{2}\ln(N) + c(\ln(N)\ln\ln(N))^{1/2}}{a(\ln(N)\ln\ln(N))^{1/2}}$$

$$\approx \frac{1}{2a}\left(\frac{\ln(N)}{\ln\ln(N)}\right)^{1/2}$$

and

$$-u\ln(u) \approx -\frac{1}{2a}\left(\frac{\ln(N)}{\ln\ln(N)}\right)^{1/2}\cdot\frac{1}{2}\ln\ln(N)$$

$$\approx -\frac{1}{4a}(\ln(N)\ln\ln(N))^{1/2}$$

In the discussion of the algorithm for discrete logarithms in GF(p) we require approximations for $\psi(x,y)$ where x is of the form $A\cdot N^b$, A a small constant, and $y = L(a)$. It follows from the above theorem that

$$\ln\left[\psi(AN^b,L(a))/(AN^b)\right] = -u\ln(u), \quad u = \ln(AN^b)/\ln(L(a))$$

$$= -\left[\frac{b}{a}\left(\frac{\ln(N)}{\ln\ln(N)}\right)^{1/2}\right]\ln\left[\frac{b}{a}\left(\frac{\ln(N)}{\ln\ln(N)}\right)^{1/2}\right]$$

$$\approx -\frac{b}{2a}\left(\frac{\ln(N)}{\ln\ln(N)}\right)^{1/2}\left[\ln\ln(N) - \ln\ln\ln(N)\right] \approx -\frac{b}{2a}$$

It should be noted that expressions such as $L(a)$ should be $L(a + o(1))$ throughout the paper but it is convenient, and sufficient for our purpose, to use the shorter, slighlty inaccurate version.

A similar development is available for irreducible polynomials over a finite field. The number of such polynomials of degree k over GF(2) is (Appendix 3)

$$I(k) = 2^k/k + O(2^{k/2}/k)$$

Analogous to integers, a polynomial of degree n will be called smooth with respect to m if all of its irreducible factors are of degree at most m. To estimate the probability that a randomly chosen polynomial of degree n is smooth with respect to m let $N(n,m)$ be the number of polynomials of degree exactly n that are smooth with respect to m. It is readily seen [4],[26] that this number satisfies the recursion

$$N(n,m) = \sum_{k=1}^{m}\sum_{r\geq 1} N(n-rk,k-1)\binom{r+I(k)-1}{r}$$

where the binomial coefficient gives the number of ways of choosing r irreducible polynomials, with repetition, of degree exactly k. The generating function of this quantity is given by

$$f_m(z) = \prod_{k=1}^{m}(1-z^k)^{-I(k)} = \sum_{n=0}^{\infty} N(n,m)z^n$$

and using the saddle point method of asymptotic analysis, Odlyzko [26] derives the following approximations when $n^{1/100} \leq m \leq n^{99/100}$:

$$(i)\ \ N(n,m) \approx \left[2\pi b(r_o)\right]^{-1/2} f_m(r_o)r_o^{\,-n} \quad ,n \to \infty$$

where r_o is the unique positive solution to

$$r\frac{f'_m(r)}{f_m(r)} = n$$

$$(ii) \quad N(n,m) = 2^n \left(\frac{m}{n}\right)^{(1+o(1))n/m} \qquad \text{as } n \to \infty$$

Thus the probability that a random polynomial of degree exactly n is smooth with respect to m is well approximated by

$$p(n,m) = \left(\frac{n}{m}\right)^{-(1+o(1))n/m} \quad , n^{1/100} \le m \le n^{99/100}.$$

The probability that a randomly chosen polynomial of degree at most n is smooth with respect to m is approximately

$$\sum_{k=1}^{n} 2^{-k} \, p(n-k,m) \approx p(n,m)\frac{(ne/m)^{1/m}}{2-(ne/m)^{1/m}}$$

which is close to $p(n,m)$ for large n.

Appendix 2. Solving Systems of Linear Equations over Z_N.

Solving a txt system of equations of the form $Ax = y$ by Gaussian elimination requires $O(t^3)$ operations and $O(t^2)$ storage. There have been many attempts to improve on this situation and perhaps the most efficient is the Coppersmith-Winograd [8] algorithm which operates on general matrices to find a linear dependency in $O(t^{2.495548})$ operations. It seems that this technique is difficult to implement. When the coefficient matrix is sparse however, the recent paper by Wiedemann [37] reduces the number of operations to $O(t\omega)$ where ω is the total number of nonzero elements in the matrix. If this number is ct for some constant c the number of operations is $O(t^2)$ and the algorithm seems quite practical for implementation. It is briefly described here.

It is first noted that the equations of interest here will, intuitively, be sparse, a consequence of the fact that a randomly chosen integer or polynomial will have, on average, relatively few prime or irreducible factors. Suppose that the matrix A is nonsingular and assume that the characteristic polynomial of A is

$$f(x) = \sum_{i=0}^{t} c_i x^i.$$

Techniques for finding this polynomial will be mentioned later. Since A is nonsingular, $c_0 \neq 0$ and

$$\sum_{i=0}^{t} c_i A^i = 0 \quad or \quad c_0 I = -\sum_{i=1}^{t} c_i A^i$$

and

$$A^{-1} = -\,c_0^{-1}\left(\sum_{i=1}^{t} c_i A^{i-1}\right)$$

Thus to compute $x = A^{-1}y$ compute

$$-\,c_0^{-1}\left(\sum_{i=1}^{t} c_i A^{i-1} y\right)$$

and this can be done in about $t(\omega + t) \approx 2t\omega$ field operations (the scalar multiplications by c_i are ignored). Notice that the saving arises from the only ω multiplies to compute $A^i y$ from $A^{i-1}y$.

For both integer factoring and the discrete logarithm problem in the precomputation stages, a few more equations than are required are generated. An arbitrary selection of these equations may be made to obtain a square matrix which is then tested for nonsingularity by attempting to solve the equation. If the attempt is unsuccessful one of the unused equations is substituted and the attempt to solve is repeated until success is achieved.

There are two problems to be solved in using this method. First it is necessary to know the characteristic polynomial of the coefficient matrix and secondly it is necessary, for the discrete logarithm problem, to solve the equations over the ring Z_N. Both problems are relatively easy to overcome, assuming the factorization of N is known.

To determine the characteristic polynomial of A, use of the Berlekamp-Massey algorithm is suggested ([2],[21]). Note that any $2t+1$ consecutive symbols from a linear feedback shift register of length t over a field uniquely determine the shift register connections. For some randomly chosen t-vector y compute the first K components of the vectors $A^j y = u_j$, $j=0,1,2,...,2t$ and let the resulting K-vectors be v_j. The amount of storage for the matrix is proportional to ω, for the vectors $A^j y$ (only one stored at a time) is t and for the results v_j is $(2t+1)K$. Since the characteristic polynomial of A is of degree at most t

$$\sum_{i=0}^{t} c_i A^i = 0 \quad \text{and} \quad \sum_{i=0}^{t} c_i A^{i+k} = 0$$

or

$$\sum_{i=0}^{t} c_i v_{i+k} = 0$$

and so each component of the K-vectors v_i satisfies a linear recurrence relationship of degree at most t. For each component the minimal such recurrence can be determined by the Berlekamp-Massey algorithm as the feedback connection polynomial. This must be determined over the ring Z_N and either the technique to be described next or that reported by Reeds and Sloane [31] could be used. The polynomial that satisfies all the recurrence relationships is the least common multiple of the polynomials and, for K sufficiently large, with high probability this will be the characteristic polynomial of A. The number of operations to determine this is $O(t^2)$.

To solve the equations over a ring Z_N, where $N = p-1$ for discrete logarithms in

$GF(p)$ and $N = 2^n - 1$ for discrete logarithms in $GF(2^n)$, the prime factorization of N is first determined as $\prod p_i^{e_i}$. This may be a computationally difficult task for N very large. The equations can first be solved mod $p_i^{e_i}$, individually and the solution mod N obtained by combining these with the Chinese Remainder Theorem. The solutions mod $p_i^{e_i}$ are related to the solutions mod p_i, in an easy to determine manner ([15],[18]). The solutions mod p_i are found using the technique described since it is a field. Thus the fact that the equation must be solved in Z_N, while taking more storage and time than for a field, is of the same complexity as that for a field and is not otherwise felt to be a factor in the calculations of feasibility. For RSA the equations are solved over Z_2, which is a computationally much easier task than for the discrete logarithms, but of the same order of complexity. Recently adaptations of the Conjugate Gradient and Lanczos methods for solving sets of equations, standard techniques for equations over the reals, have been given for finite fields [10],[26].

Appendix 3. Polynomials over GF(2).

Several results on polynomials over $GF(2)$ were required in the text and while these are easy to derive, they are shown here for reference. The number of irreducible polynomials of degree m over $GF(2)$ is

$$I_2(m) = \frac{1}{m} \sum_{d \mid m} \mu(d) 2^{m/d}$$

Since the smallest possible nonunit divisor is 2, $I_2(m) = 2^m/m + O(2^{m/2}/m) \approx 2^m/m$. The number of irreducible polynomials of degree at most m can then be approximated by

$$\begin{aligned} I_2(m) + I_2(m-1) + \cdots &\approx \frac{2^m}{m} + \frac{2^{m-1}}{m-1} + \cdots \\ &\approx \frac{1}{m}\left(2^m + 2^{m-1} + \cdots\right) \approx 2^{m+1}/m \end{aligned}$$

To consider the possibility of irreducible polynomials of the form $x^m + f_1(x)$ where the degree of $f_1(x)$ is as small as possible the following random and heuristic argument is suggested. Suppose the irreducible polynomials of degree m are distributed at random among the 2^{m-1} possible polynomials (we assume the coefficients of x^m and x^0 are unity). The probability that a randomly chosen polynomial has all powers (except for m) less than b is $2^{b-1}/2^{m-1}$. If X is the number of such polynomials then the expected number of them is $E(X) = (2^m/m)(2^{b-1}/2^{m-1})$ and in order to have $E(X) = 1$, b must be at least $\log_2(m)$.

Let P_d be the number of unordered pairs of relatively prime polynomials $a(x)$, $b(x)$, $\deg(a(x))$, $\deg(b(x) \leq d$. To determine P_d the following argument is used [3]. There are $t = 2^{k+1} - 1$ nonzero polynomials of degree at most k and thus

$$\binom{t}{2} + t = 2^k(2^{k+1} - 1)$$

distinct unordered pairs of nonzero polynomials. There are exactly $2^i P_{k-i}$ pairs of polynomials having a greatest common divisor of degree exactly i and so

$$P_k = 2^k(2^{k+1} - 1) - \sum_{i=1}^{k} 2^i P_{k-i}\ ,\quad k \geq 1\ ,\quad P_0 = 1$$

or

$$\sum_{i=0}^{k} 2^i P_{k-i} = 2^k(2^{k+1} - 1)$$

For $k+1$ the equation reads

$$\sum_{i=0}^{k+1} 2^i P_{k+1-i} = 2^{k+1}(2^{k+2} - 1)$$

and multiplying the previous equation by 2 and subtracting from this last equation gives the result $P_{k+1} = 2^{2(k+1)}$ as required. The number of ordered such pairs of polynomials, as can be used in the Coppersmith algorithm, is twice this number.

Consider the problem of testing whether the polynomial $a(x)$ has all of its irreducible factors of degree at most b without actually factoring the polynomial. Clearly such a test would greatly reduce the amount of computation in the Coppersmith algorithm and he [9] suggests the following. Consider $(a'(x),a(x))$ where $a'(x)$ is the derivative of $a(x)$, which consists of

$$(a'(x),a(x)) = \prod_i p_i^{2\lfloor e_i/2 \rfloor}(x) \quad \text{where} \quad a(x) = \prod_i p_i^{e_i}(x)$$

i.e. all even powers of irreducible factors survive and odd powers are reduced by one in the gcd. Consider the polynomial

$$a'(x) \prod_{i=\lceil m/2 \rceil}^{m} (x^{2^i} + x) \pmod{a(x)}.$$

If this polynomial is zero then $a(x)$ has all of its irreducible factors of degree at most m with the possible exception that it could, although is unlikely to, have irreducible factors of degree greater than m to an even power. Thus with a single polynomial division (of a high degree polynomial) the more expensive factoring procedure is avoided.

COLLABORATIVE CODING FOR OPTICAL FIBRE MULTI-USER CHANNELS

P. BRIDGE

ELECTRICAL ENGINEERING LABORATORIES, UNIVERSITY OF MANCHESTER· MANCHESTER.UK*

1. INTRODUCTION

Optical fibre communications is now a mature technology, and is becoming the dominant medium for line communications. The advantages of optical fibre are well known, but are repeated here for completeness:-

(a) High bandwidth
(b) Low loss
(c) Low cost, size and weight
(d) Immune to eavesdropping and external pickup

In the past the main application of optical fibres has been high speed point to point links for telecommunications, but as the technology becomes more robust and reliable it is becoming attractive for use in applications such as Local Area Networks (LANs) and Subscriber Services Networks (SSNs). So far, as in trunk telecommunications, LAN or SSN optical systems have simply been used to provide point to point transmission capacity in systems originally designed around wire transmission. This results in reduced costs and improved reliability, but the basic network structure is unaffected.

However it is in a multi-user context that the differences between optical systems and other media become most apparent. Hence there is a need to develop new multi-user network topologies and protocols based around the characteristics of optical fibres.

2. PROPERTIES OF OPTICAL FIBRE SYSTEMS

2.1 Segmented bandwidth

The optical passband of each low loss window in an optical fibre is of the order of 10^{15}Hz, but access to the fibre is via electro-optics with bandwidths of the order of 10^{10}Hz. Hence, communication between two users is normally restricted to 'subchannels' of at most 0.1% of the potential bandwidth. This may not be sufficient if the fibre is to be shared between, say, several hundred users. One way to allow a multi-user system access to a greater fraction of the potential bandwidth is to use several subchannels in parallel, for example by the technique of wavelength multiplexing. With present methods only a few tens of subchannels are possible, but with the development of coherent detection and integrated optics several hundred subchannels may be possible.

2.2 Variation in path loss

In principle, the paths between the users of an optical fibre network should remain within the optical domain as much as possible to make full use of the fibre properties, so it can be argued that an 'good' optical network should use passive branching to distribute a signal. The problem here is that at each branching the signal is split, resulting in progressive reduction of signal level along the path, in contrast with a wire conductor. Practical topologies for more than a very few users have different numbers of branchings, connectors and lengths of fibre on paths between different user pairs, resulting in loss variations of several orders of magnitude.

J. K. Skwirzynski (ed.), Performance Limits in Communication Theory and Practice, 99–111.

2.3 Shot noise

The more sensitive optical receivers are limited by shot noise on photon arrivals rather than thermal noise in the channel. The important consequence is that the noise level is dependent on the optical power level incident on the detector. This can cause problems if power from many sources is simultaneously incident on a detector whilst the receiver is trying to make decisions on the signal from only one of these.

2.4 Signal format

Practical limitations on source linearity together with the dominance of shot noise result in a binary format being the optimum for optical signals. Also, the signal is unipolar since there is no such thing as a negative optical signal.

2.5 Lack of synchronisation

The low loss of optical fibres makes long transmission distances possible. Also, the subchannels accessed via conventional electro-optics have a high absolute bandwidth despite being only a small fraction of the potential fibre bandwidth. These two facts means that if the medium is utilised to the full in terms of bit rate and distance then the propogation delay variation between paths can be several thousand bits.

3. OPTICAL MULTIPLE ACCESS NETWORK

A multi-user network can be divided into a Multiple Access Channel (MAC) representing messages from many sources to a single given receiver, and a Broadcast Channel representing messages for diverse receivers from a single given source. The rest of this discussion concentrates on the MAC aspect since this has more impact on the performance of SSNs and LANs. The MAC model and a possible optical fibre realisation are shown in Figures 2.1 and 2.2.

The nature of the optical channel leads us to make the following assumptions:-

(a) There is a wide variation in path loss.
(b) The network is entirely passive.
(c) The source signal format is unipolar binary.

The following assumptions are consistent with making the network as flexible and simple as possible:-

(d) Sources are independent.
(e) There is no feedback channel.
(f) The network is asynchronous.

Finally, we make the following assumptions to simplify the analysis:-

(g) The channel is noise free, linear and memoryless.
(h) Each source has an inexhaustible supply of information.

Next we consider the protocols which control access to the MAC.

4. MULTIPLE ACCESS PROTOCOLS AND CODING

Multiple access protocols can be classifed as stochastic or deterministic. Stochastic protocols for optical fibre networks have received a great deal of attention in recent years, and their performance is fairly well characterised. Also, stochastic protocols are generally inferior to deterministic ones under conditions of inexhaustible sources. This directs our attention towards the potential offered by some deterministic protocols, and it is these that are the concern of the remainder of this text.

Wolf has suggested a useful classification of deterministic protocols into the following three cases [1]. We suppose that there are N sources

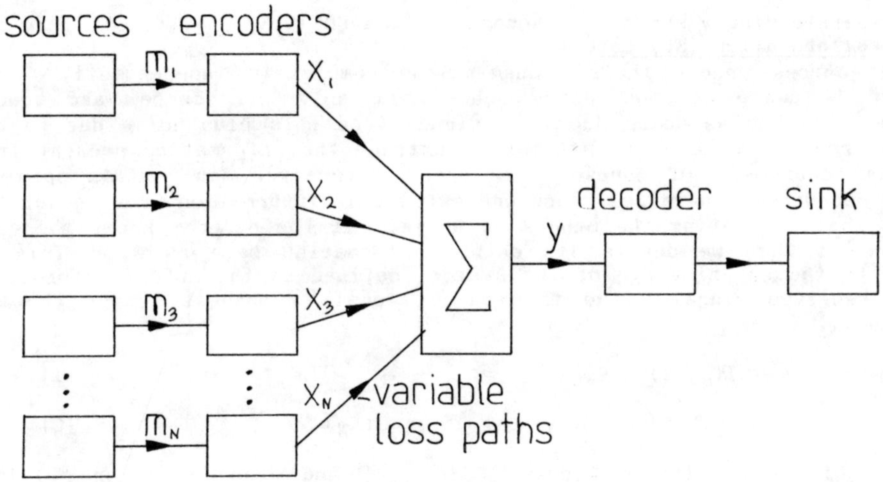

Figure 2·1 Multiple Access Channel

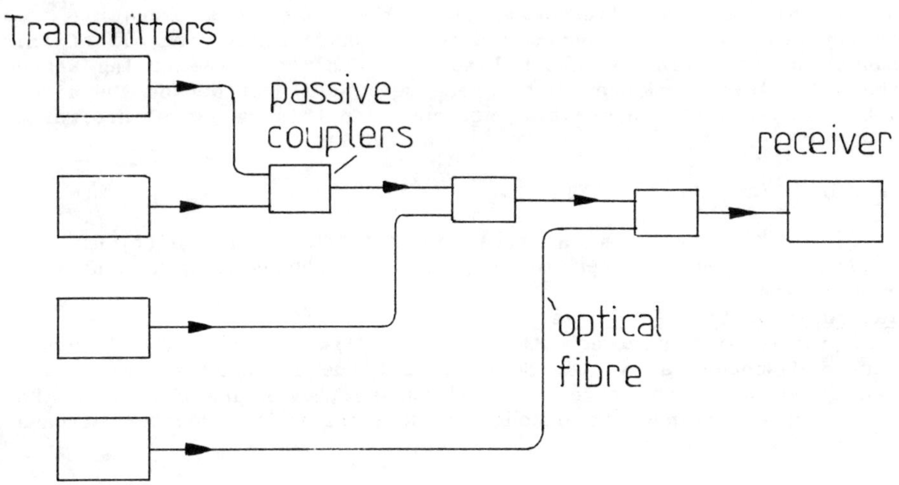

Figure 2·2 Optical fibre MAC

which generate binary symbols x_i according to a PDF $Q(X_1, X_2, \ldots X_N)$.

4.1 No collaboration (NC) case

The sources encode their transmissions completely independently. The receiver is aware of the coding scheme for only one source, and treats received signal y as being due to a signal from this plus noise due to the other sources. It makes no attempt to extract the information present from the other sources. Of course in practice a receiver may contain up to N similar decoders in parallel, each one extracting information from one of the sources whilst treating the others as noise. If the decoder knows the code of source i then we denote its output information rate as R_i. This is bounded by the capacity C_{NC} of the system, defined as the mutual information of the received signal y and the source signal x_i maximised over the PDF $Q(X_1, X_2, \ldots, X_N)$:-

$$R_i \leqslant \max I(X_i;Y) = C_{NC} \tag{1}$$

where $$Q(X_1, X_2, \ldots X_N) = Q(X_1)\, Q(X_2) \ldots Q(X_N) \tag{2}$$

Code Division Multiple Access (CDMA) [2] and Pulse Position Multiple Access (PPMA) [3,4,5] are examples of NC coding schemes. These are characterised by relatively low channel efficiency. In essence this is because the duty ratio, the ratio of the time that sources are sending information to the time that they are idle, must be low if the number of errors induced in the information from a particular source is to be within the capability of available EDC codes to correct.

4.2 Total collaboration (TC) case

In this case the N sources can interact directly. In effect there exists a multiplexer which can simultaneously observe the messages emanating from all of the sources, and then produce a single message containing all of this information to be sent via the channel to the receiver. The coding scheme used by the multiplexer is known to the receiver, so it can decode the signal and extract the N separate information streams. In this case the information rate is givne by:-

$$R \leqslant \max I(X_1, X_2, \ldots X_N;Y) = C_{TC} \tag{3}$$

Where $Q(X_1, X_2, \ldots X_N)$ is a true joint PDF. The difficulty in implementing such a multiplexer means that TC schemes are not normally considered practical.

4.3 Partial collaboration (PC) case

This is intermediate between the previous two cases. The N source messages are independent as in the NC case, but the encoding scheme used by the sources is known to the receiver, and these schemes are chosen so that the receiver can simultaneously decode the N separate information streams. The information rate is given by:-

$$R \leqslant \max I(X_1, X_2, \ldots X_N;Y) = C_{PC} \tag{4}$$

where $$Q(X_1, X_2, \ldots X_N) = Q(X_1)\, Q(X_2) \ldots Q(X_N) \tag{5}$$

Time Division Multiple Access (TDMA) may be considered a form of PC However, the results of previous work reviewed in the next section indicate that there are <u>collaborative coding</u> (CC) techniques for implementing PC on an optical fibre based MAC which have potentially better channel utilisation than either TDMA or NC schemes, though relatively little work has been published in this area.

5. COLLABORATIVE CODING

The sources encode their transmissions independently, but the encoding methods are known to the receiver, which can recover the separate message streams simultaneously. The rate of information flow out of the receiver for the ith source is R_i, and the performance of a particular coding scheme can be described as a point in an N- dimensional rate space denoted by the coordinates $(R_1, R_2, \ldots R_N)$. The set of achievable coordinates for a particular channel defines a closed volume K about the origin in this space, known as the capacity region. All points on the boundary and in the interior of the K are in principle achievable with zero error probability. This is the multi-user version of Shannon's theorem.

The boundaries of K are set by the rate sum equations over all possible subsets of sources, for example:-

$$R_1 + R_2 + \ldots + R_N \leqslant I(X_1, X_2, \ldots X_N; Y) = H(Y) \tag{6}$$

$$R_1 + R_2 \leqslant I(X_1, X_2; Y) \tag{7}$$

$$R_1 \leqslant I(X_1; Y) \tag{8}$$

Here H(Y) is the entropy of the received waveform y. As an example we consider the adder channel with two users:-

Source 1 sends symbol x_1 = 0 or 1 according to PDF $Q(X_1)$
Source 2 sends symbol x_2 = 0 or 1 according to PDF $Q(x_2)$
Received symbol $y = x_1 + x_2$ = 0, 1 or 2.

The capacity region K is defined by:-

$$R_1 + R_2 \leqslant I(X_1, X_2; Y) = H(Y); \quad R_1 \leqslant I(X_1; Y); \quad R_2 \leqslant I(X_2; Y)$$

It can be shown that in this case there are points at which:-

$$I(X_1, X_2; Y) = I(X_2; X_1, Y) + I(X_1; Y) = H(Y) \tag{9}$$

or

$$I(X_1, X_2; Y) = I(X_1; X_2, Y) + I(X_2; Y) = H(Y) \tag{10}$$

Equation (9) corresponds to the receiver decoding method where first message x_1 is decoded from y, treating x_2 as noise, and then the estimate of x_1 used to decode x_2, yielding capacity region K_1 shown shaded in Figure 5.1. Equation (10) corresponds to the case where the roles of x_1 and x_2 are reversed, yielding region K_2. Hence the union $K_1 \cup K_2$ is a subset of K. It is at this point that recourse is usually made to the assumption that it is possible to timeshare between these two cases so that K also contains rate points in the convex hull of $K_1 \cup K_2$, as shown in Figure 5.2. The ability to timeshare implies that it must be possible to achieve network synchronisation.

Collaborative coding schemes are often compared with TDMA, since of all conventional protocols this offers the best channel utilisation under high loading. Consider the application of TDMA to the 2 source example. We allow user 1 exclusive use of the channel for a fraction α of the time, and allocate the remaining $1-\alpha$ to user 2. The result is the unit capacity time sharing line shown on Figure 5.2, the specific operating point depending on α. Thus collaborative coding offers significantly improved channel utilisation over TDMA.

Most work on collaborative coding has concentrated on getting as close to

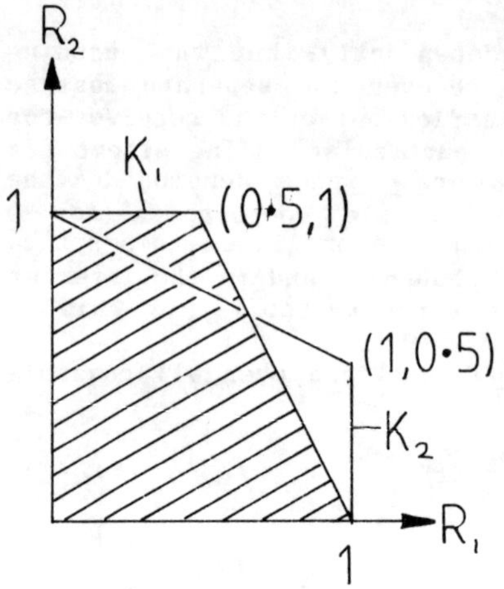

Figure 5·1

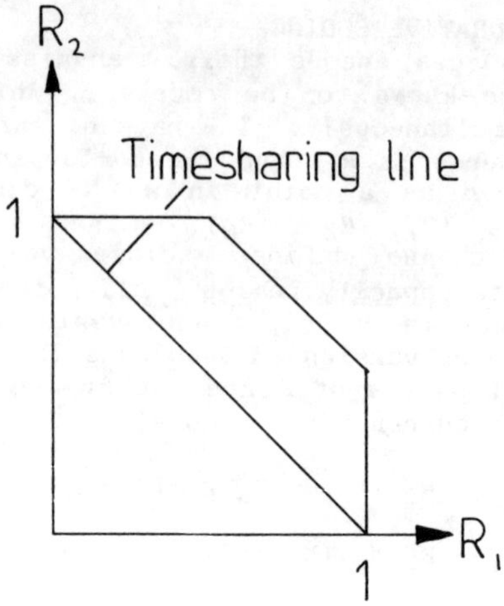

Figure 5·2

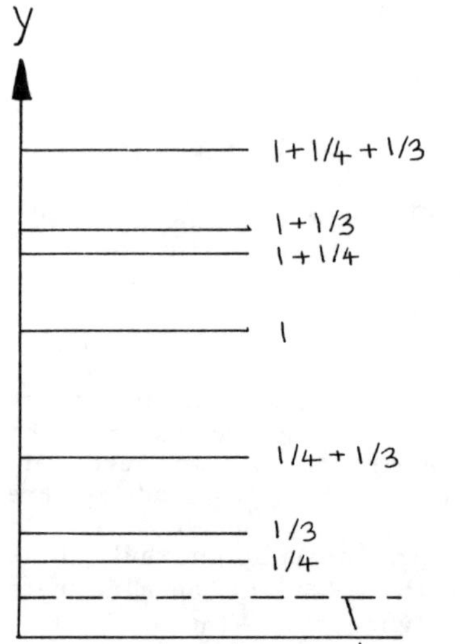

Figure 5·3

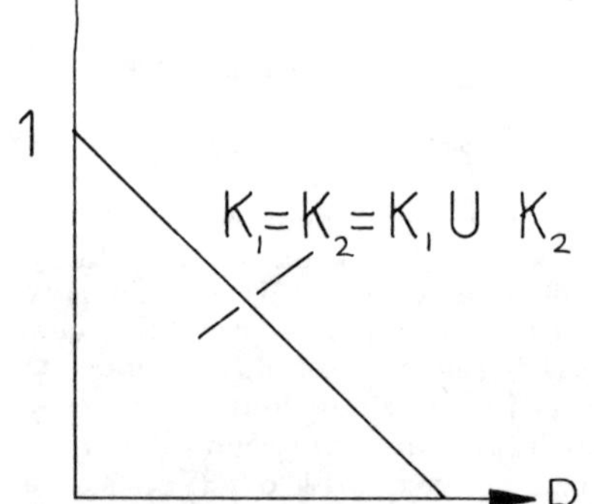

Figure 5·4

the maximum channel capacity as possible, since this is of importance when bandwidth is a scarce resource, as in radio systems for example.

5.1 The optical channel model

The adder channel is acceptable for modelling media where there is not a large variation in path loss between users. In this case y has identical intervals between all of its discrete levels, so that placing the N decision thresholds for a hard-decision decoding scheme is a relatively simple task.

The situation is much more complex in the optical case because the large variation in path loss means that the intervals between the signal values are not uniform. For example, suppose N=3, with path losses from the three sources to the receiver being factors of 1, 1/3 and 1/4, so the received signal y has the discrete levels shown in Figure 5.3. Clearly the situation will become intractable for more than very few users. In this analysis we assume that the only feasible solution is to have a single threshold to decide whether one or more sources sent a 1 symbol.

A second reason for adopting the OR channel model arises because of the dominance of shot noise in many optical receivers. Suppose we have two sources, one with a large path loss to the receiver under consideration and one with negligible path loss. If '1' symbols, represented by pulses of optical power, from both sources are coincident at the receiver then the shot noise generated by the low loss pulse can easily dominate the signal from the high loss source. The simplest answer is to assume that if the receiver observes a pulse then all it can say is that <u>at least</u> one source sent a 1.

5.2 Collaborative coding for the OR channel

The concept of a closed capacity region around the origin of an N-dimensional rate space described by equations (6) to (8) also applies to the OR channel.

Since in this case y is a two level signal H(Y) and hence the rate sum over all sources is upper bounded by unity. For two users the system is again described by (9) and (10), yielding capacity regions K_1 and K_2, which in this case are identical and equivalent to the region enclosed by the timesharing line. Hence taking the convex hull of $K_1 \cup K_2$ does not add any extra rate points to K, which is shown in Figure 5.4.

5.3 Collaborative coding without synchronisation

For the sake of simplicity much previous work on collaborative codes has assumed that network synchronisation is possible, since:-

(a) If block codes are used by the sources, then practical collaborative codes are much simpler to implement and analyse if we can assume that all of these codes are block synchronised at the receiver.

(b) A time sharing argument can be used to prove that the capacity region of a scheme is the convex hull of $K_1 \cup K_2$.

McEliece et al., have demonstrated that for the two source case K_1 and K_2 are theoreticaly achievable without assuming block synchronisation of source codewords [6,7]. This is proven by postulating that source 1 uses a block length of L symbols whilst source 2 uses a length of L+1 symbols such that L and L+1 are relatively prime. In this way 'quasi' block synchronisation can be achieved over a block of L(L+1) symbols irrespective of the relative phases of the blocks from each source. Hui and Humblett have studied the two source case with <u>complete asynchronism</u>, defined as the relative phase difference d between source codewords as a fraction of block length L being unbounded such that d/L remains finite as L is made larger [8]. It can be shown that K is equivalent to $K_1 \cup K_2$ with points in the convex hull being unattainable.

However, complete asynchronism may be an over-stringent restriction. Cover, McEliece and Posner have considered the case of <u>limited asynchronism</u>, where an upper bound can be placed on d such that d/L now tends to zero as L is made larger [9]. By employing a source timesharing argument it can be shown that points in the convex hull of $K_1 \cup K_2$ are attainable. An unfortunate trade-off is that the coding delay must tend to infinity if points on the outer boundary of the convex hull are to be achieved. Nevertheless, this result may be of importance for adder channel collaborative coding schemes where the aim is to maximise channel utilisation by operating as near as possible to the outer boundary of the convex hull, since it means that the stringent condition of network synchronisation is not required to implement collaborative coding.

The ability to achieve points in the convex hull is of less consequence for the optical OR Channel since $K_1 \cup K_2$ is not convex so the hull contains no extra rate points. However, the result that the capacity regions K_1 and K_2 can be achieved under complete asynchronism is crucial, since this gives collaborative coding a theoretical advantage over TDMA in terms of channel utilisation. Under asynchronous conditions TDMA cannot achieve the unit capacity represented by the timesharing line because a 'guard time' is required adjacent to each user slot, whereas the asynchronous capacity region of OR channel collaborative coding has an outer boundary equivalent to the timesharing line. Furthermore, long coding delays are not implicit in this result, which can be extended to more than two users.

5.4 Practical collabrative coding

The synchronous adder channel has received much attention, and several good coding schemes have been proposed [10,11]. Coding schemes for the asynchronous adder channel are less numerous. Wolf has suggested a scheme which is independent of symbol as well as block synchronisation for the two source case [12,13]. Source 1 is assigned the all 1's and all 0's codewords of length L. Source 2 is assigned variable length codewords such that observing y through an L length window only yields L consecutive nonzero symbols if source 1 sent the all 1's word. Thus the decoder can synchronise to source 1 by looking for L consecutive nonzero symbols and can then decode message x_1 by looking for a 0 in the window. Source 2 is then decoded by subtracting x from y.

Wolf has also suggested a coding method for the synchronous OR channel [1] which uses a form of pulse position modulation (PPM), with each of N>2 users choosing one slot out of a block of M in which to send a symbol. A similar scheme is proposed by Chang and Wolf [14], where the sources choose one out of M frequencies to send a symbol. In fact any set of orthogonal waveforms could be used. These slots or frequencies can be treated as independent subchannels at the receiver. If N>M then one way to use these subchannels is to divide the sources into N/M groups and then assign each group an exclusive subchannel to be used collaboratively. This results in a total channel capacity bounded by M, since each subchannel has a potential capacity of unity. In Wolf's analysis each source can choose from among all subchannels, utilising a PDF which puts most of the weight onto the jth frequency such that $H(Y_{i\neq j})$ is unity and $H(Y_j)$ tends to zero. This results in a total capacity of M-1. However, if we admit the possibility of a source sending no frequenices then we can choose a PDF which makes all $H(Y_i)$ unity. Thus, as far as theoretical synchronous OR channel capacity is concerned, it makes no difference whether a source is restricted to sharing a single subchannel or whether it can spread its messages across subchannels. In [14] codes are suggested which achieve the asymptotic synchronous capacity of M-1 for N>>M. both PPM and multi-frequency (MF) subchannels can be considered appropraite for the optical case. PPM is often mooted as a good choice of line code for optical links because it minimises the effect of shot noise. On the other hand the MF case can be straightforwardly implemented using WDM.

Another scheme suggested by Wolf is concerned with the asynchronous 2-source OR channel [1]. Essentially, source 1 uses codewords 1010 and 0101 and source 2 uses 10 and 1001. Synchronisation to source 1 is via a preamble sequence, and subsequent determination of the relative phase of source 2 is achieved statistically by observing that this source sends 2/3 of its 1's in the first slot of its codewords. The nature of the codewords is such that once the relative phases of the sources is known then x_1 and x_2 can be decoded. This code has a rate of 0.583.

In [6] McEliece and Rubin suggest another code for the asynchronous 2-source OR channel. Source 1 sends m interleaved (n,k) cyclic codewords, the last j of which are all 0's. Source 2 sends n concatenated (m,j) cyclic codewords, the last k of which are all 0's. Whatever the relative phases of the L=m.m blocks, the pattern of errors in x_1 caused by x_2, and vice versa, is such that the cyclic codes can correct them. If R_1 that R_2 are the rates of the cyclic codes, then it can be shown that:-

$$R_{sum} \leqslant \surd R_1 + \surd R_2 \qquad (11)$$

with equality when $R_1 + R_2 = 1$. McEliece cites the example of (m,j) = (n,k) = (2,1), a simple repetition code, yielding the rate point (1/4,1/4) and $R_{sum} = 1/2$.

The remainder of this text describes how a scheme proposed by Massey and Mathys for the collision channel [15] can be used to extend the basic idea of using interleaved cyclic codes to the case where N>2.

Massey describes a general case where time is divided up into equal slots in which sources send either Q-ary packets x_i or idle characters I. If any part of one packet overlaps another at the receiver then a collision occurs and both packets are replaced with a collision symbol E. Packets coincident with I are unchanged. for N=2, Q=2 the channel can be represented by the following table:-

Source 1	Source 2	y
I	x_2	x_2
x_1	I	x_1
x_1	x_2	E
I	I	I

We shall consider the restricted case where 'packets' are equivalent to binary symbols x_i. It can be shown that the rate of the ith source is given by:-

$$R_i \leqslant d_i \prod_{j=i} (1-d_j) \qquad (12)$$

where d_i is the duty ratio of symbols to I's from source i. The set of rate points $(R_1, R_2, \ldots R_N)$ defines a capacity region K in N-dimensional rate space. Equality is achieved if the sum of d_i's over all sources is unity. If all the sources have the same duty ratio N^{-1} then each will have the same rate, and the rate sum is given by:-

$$R_{sum} = (1-N^{-1})^{N-1} \qquad (13)$$

so the capacity tends to e^{-1} as N becomes large.

Massey describes a coding technique which can achieve points on the boundary of this capacity region. Like McEliece's 2-user OR channel scheme described previously, the essence of this technique is to encode the message packets and then interleave them with idle symbols so as to restrict the location and number of collisions in such a way that the code can correct for them and extract the original message. Massey proposes the vehicle of a <u>protocol matrix</u> S for implementing these ideas. The ith row of S, s_i, corresponds to a block of symbols and idle characters from source i. S has elements 1 and 0, and where a row contains a 1 then the source transmits a symbol, where there is a 0 it transmits I. By carefully choosing the elements of S the location and number of collisions can be controlled such that points on the outer boundary of K can be achieved.

The elements are chosen as follows: the duty factors d_i are written as the ratio of integers q_i and q, with the latter common to all sources and chosen to be as small as possible. Each row is q^N elements long, and S is a N X q^N matrix. Row s_i contains q^{N-i} repeated sequences of 1's and 0's. Within each sequence are $q_i.q^{i-1}$ 1's in consecutive locations from the beginning of the sequence, and 0's in the remaining locations. For N=2, $q_1=q_2=1$, q=2 then S is:-

$$S = \begin{bmatrix} 1 & 0 & 1 & 0 \\ 1 & 1 & 0 & 0 \end{bmatrix} \qquad (14)$$

which is essentially that used in McEliece's scheme. For N=3, $q_1=q_2=q_3=1$, q=3 then S is:-

$$S = \begin{bmatrix} 100100100100100100100100100 \\ 111000000111000000111000000 \\ 111111111000000000000000000 \end{bmatrix} \qquad (15)$$

When a column contains more than a single 1 then a collision results in y, otherwise x_i's or I's occur. For the example above y takes the form:-

$$y = EEEEx_3x_3Ex_3x_3Ex_2x_2x_1IIx_1IIEx_2x_2x_1IIx_1II \qquad (16)$$

Protocol matrices constructed in this way have the following properties:-

(a) There are $q_i \prod (q-q_j)$ columns of S where a 1 occurs in the ith row only. This means that y always contains this many symbols from source i out of a block of q^N.
(b) It is theoretically possible to determine the sources of all the correct symbols in y.
(c) Cyclic shifts of each row of S by arbitrary amounts yields a matrix which is a column permutation of S. This means that properties (a) and (b) remain true even if the sources are not block synchronised. For the moment we retain the assumption of symbol synchronism.

In practice separation of the symbols from different sources in y can be accomplished by the technique of <u>decimation decoding</u>. The pth phase of the dth decimation of a sequence z of length L is the sequence obtained by taking every dth element of z starting with the pth, d being a divisor of L. The pth phase of the q^jth decimation of s_i has the following properties:-

(a) for $i \leqslant j$ the sequence obtained is either all 1's or all 0's.
(b) for $j<1$ at least one element of the sequence is 0.

Denote the construction of those phases of the qth decimation of y that contain <u>no</u> idle symbols as D[y]. Denote the converse operation of taking those phases that contain I's as B[y]. The above properties allow the receiver to identify sources of symbols appearing in y by use of the following algorithm:-

(1) Take D[y]. This yields source 1 symbols and E's only.
(2) Take D[B[y]]. This yields x_2's and E's only. Etc.

In this way the ith step generates $D[B^{i-1}[y]]$, comprising in total $k_i=q_i \prod (q-q_j)$ x_i's, the remaining elements being E, corresponding to the theoretical rate in (12).

The next step is to find suitable ways of encoding the sources. Suppose that during the time interval corresponding to block s_i source i generates k_i message symbols m_i. These must be encoded into $n_i=q_i.q^N$ transmitted symbols x_i in such a way that the decoder can determine the original k_i symbols from some choice of k_i out of n_i transmitted symbols.

It can be shown that the collisions in each phase of D[y] due to source 2 alone form closed loop bursts of length q_2 and period q, as do the collisions due to source i alone in $D^{i-1}[y]$. This provides the key to encoding the sources. Massey points out that systematic linear (n,k) codes over ring Z_Q exist which can correct all closed loop erasure bursts of length n-k. Furthermore, cyclic codes belong to this group, known as Maximum Erasure Burst Correcting (MEBC) codes. Massey describes how 'nesting' of MEBC codes at the sources, and a complementary 'de-nesting' at each stage of the decimation decoding algorithm, can be used to apply MEBC codes to the collision patterns that occur in y. For protocol matrix (14) this scheme is similar to McEliece's use of (2,1) repetition codes, and equation (12) results in R_{sum} given by equation (11).

The most significant differences between the collision channel described in Massey's work and the optical OR channel is that there are no explicit collision symbols or idle symbols in the OR channel. Lack of E symbol implies that collisions in the OR channel constitute errors rather than erasures, and MEBC codes only work in this scheme with erasures. Furthermore, decimation decoding hinges on identification of idle symbols.

However, these problems can be overcome by identifying idle and collision slots statistically rather than explicity as in Massey's scheme. This can be done because use of the S matrix means that every block of y will have idles and collisions in the same slots. Idle slots can be identified as those locations where a 1 is never observed. Correct symbol locations are those where the ratio of 1's to 0's over a long time is determined to be 0.5, and collision slots are those with more 1's than 0's observed over a long period.

Massey also describes how this scheme can be extended to the case where symbol as well as block synchronisation cannot be assumed. The technique involves replacing every occurrence of a 1 in S with the sequence $1^{m-1}0$ and every occurrence of 0 with 0^m to yield the new matrix S^m If the rate of source i for the slot synchronised case is R_i, then it can be shown that the new matrix S^m yields a rate $R_i(m-1)/m$ in the completely asynchronous case. Thus the rate tends to the slot synchronised value for large m. We again observe a trade-off between coding delay and rate sum.

There are some disadvantages with this scheme. Firstly, there is a long initialisation sequence while the decoder performs its statistical analysis in order to determine the locations of idle and collision slots. Secondly, the lowest value of q is N, the number of sources, so the block length and coding delay will increase as N^N at least. The rate of increase will be faster if a large value of m is used under symbol asynchronous conditions.

Since these protocol matrix methods decode each source independently rather than jointly it could be argued that they are not strictly PC methods. However, the statistical analysis to determine the location of message, idle and collision symbols made possible by the sources 'collaborating' according to the protocol matrix can be considered a form of joint decoding. Furthermore, the method is not true NC since the decoder uses a knowledge of the encoding of each source in the 'independent' decoding process.

6. CONCLUSIONS

The OR channel has been presented as the most appropriate model for an optical fibre multiple access channel. For the asynchronous case a review of past work reveals that partial collaboration between sources permits a potentially better channel utilisation than Time Division Multiple Access and cases where no collaboration between sources exists. However, the following questions remain unanswered:-

(a) How to TDMA, PC and NC compare with different numbers of sources and degrees of asynchronism?

(b) What is the role of MF schemes in an asynchronous system? Can the idea of a protocol matrix be extended to these schemes?

7. REFERENCES

1. J. K. Wolf, "Coding techniques for multiple access communication channels", in "New concepts in multi-user communcation", Ed. J. K. Skwirzynski, Sijthoff and Noordhoff, 1981, p. 83-103.
2. S. Tamura, S. Nakano and K. Okazuki, "Optical code-multiplex transmission by Gold sequences", IEEE Journal of Lightwave Tech., Vol. LT-3, No. 1, Feb. 1985.
3. A. R. Cohen, J. A. Heller and a. J. Viterbi, "A new coding technique for asynchronous multiple access communication", IEEE Trans., Vol. COM-19, pp. 849-855, 1971.
4. I. F. Blake, "Performance of non-orthogonal signalling on an asynchronous multiple access On-Off channel", IEEE Trans., Vol. COM-30, No. 1, pp. 293-298, Jan. 1982.
5. L. Gyorfi and I. Kerekes, "A block code for noiseless asynchronous multiple access OR channel", IEEE Trans., Vol. IT-27, No. 6, pp. 788-791, Nov. 1981.
6. R. J. McEliece and A. L. Rubin, 'Timesharing without synchronisation", Proc. ITC, Los Angeles, pp. 16-20, Sept. 1977.
7. R. J. McEliece and E. C. Posner, "Multiple access channels without synchronisation", Proc. ICC, Chicago, pp. 29.5-246 to 248, June 1977.
8. J. Y. N. Hui and P. A. Humblett, "The capacity region of a totally asynchronous multiple access channel", IEEE Trans., Vol. IT-31, No. 2, pp. 207-216, March 1985.
9. T. M. Cover, R. M. McEliece and E. C. Posner, "Asynchronous multiple access channel capacity", IEEE Trans., Vol. IT-27, No. 4, pp;. 409-413, July 1981.
10. T. Kasami and S. Lin, "Coding for a multiple access channel", IEEE Trans., Vol. IT-22, No. 2, pp. 129-137, March 1976.
11. S. C. Chang and E. J. Weldon, "Coding for T-user multiple access channels", IEEE Trans., Vol. IT-25, No. 6, pp. 684-691, Nov. 1979.
12. M. A. Deatt and J. K. Wolf, "Some very simple codes for the nonsynchronised two-user multiple access adder channel with binary inputs", IEEE Trans., Vol. IT-24, No. 5, pp. 635-636, Sept. 1978.
13. J. K. Wolf, "Multi-user communication networks", in "Communication systems and random process theory", Ed. J. K. Skwirzynski, Sijthoff and Noordhoff, pp. 37-53, 1978.

14. S. Chang and J. K. Wolf, "On the T-user M-frequency multiple access channel with and without intensity information", IEEE Trans., Vol. IT-27, No. 1, pp. 41-48, Jan. 1981.
15. J. L. Massey and P. Mathys, "The collision channel without feedback", IEEE Trans., Vol. IT-31, No. 2, pp. 192-206, March 1985.

* This work has in part been funded by the Science and Engineering Research Council of the United Kingdom.

WHAT HAPPENED WITH KNAPSACK CRYPTOGRAPHIC SCHEMES?

Y.G. Desmedt

aangesteld navorser NFWO, Katholieke Universiteit Leuven, ESAT
Kardinaal Mercierlaan, 94, B–3030 Heverlee, Belgium

1 INTRODUCTION

The knapsack problem originates from the economic world. Suppose one wants to transport some goods which have a given economical value and a given size (e.g. volume). The transport medium, e.g. a car, is however limited in size. The question then is to maximize the total economical value to transport, given the size limitations of the car.

The above mentioned knapsack problem is not used (today) in cryptography, but only a special case: namely if *the economical value of each good is equal to its size. This special problem is also known as the subset sum problem* [28]. This problem was first used by Merkle and Hellman to make a public key system (an introduction to the concept of a public key scheme can be found in [56,26,22]). In the subset sum problem n integers a_i are given (the size of n goods). Given a certain integer S the problem is then to determine a subset of the n numbers such that by adding them together one obtains S. Remark that it is possible that for some S and n given integers a_i no such subset exists. The decision problem wondering if such a subset exists is an NP–complete problem [28]. The same is true if the a_i and S are positive integers. In other terms solving the subset sum problem in its generality would solve a lot of problems, as the travelling salesman problem. It is *expected* (not proven) that for some *worst case* large inputs such problems are unfeasible, in other words that NP is different from P [28].

If we mention in the next sections "the knapsack problem" we use this as a synonym for "the subset sum problem". Remark that there also exists a subset product problem, which was used in the so–called "multiplicative public key knapsack cryptographic systems [48]". The multiplicative knapsack and its security will be shortly discussed in Section 3.4. Remark that most cryptographic knapsack schemes protect only the privacy of the message. Cryptographic knapsack schemes which protect the authenticity are shortly discussed in Section 3.5. Sometimes knapsack problems (subset sum problems) are also used completely differently in cryptographic systems (see Section 3.6).

We will now introduce the Merkle–Hellman knapsack scheme. This introduction will be given in such a way that some terms can be used later on when other knapsack public key schemes are discussed. This allows to first give an overview of the history of the cryptographic knapsack systems. Later more mathematical aspects of the breaking techniques are given.

Except for x which is binary and except when explicitly mentioned, all numbers in this text are natural numbers or integers (depending of the context).

J. K. Skwirzynski (ed.), Performance Limits in Communication Theory and Practice, 113–134.

We remember here that the terms decryption and deciphering in English are synonymous (what not always the case is in other languages). The terms corresponding with breaking techniques are: breaking, attacking, cryptanalysing, and so on.

2 THE CRYPTOGRAPHIC KNAPSACK SCHEME: AN INTRODUCTION

2.1 The enciphering

We will now explain the enciphering operation in the Merkle–Hellman scheme to protect privacy protection. Suppose we want to send a message to Alice, and Alice's public key is $\mathbf{a} = (a_1, a_2, \ldots, a_n)$. To encipher a message $\mathbf{x} = (x_1, x_2, \ldots, x_n)$ of n bits, we make the sum:

$$S = \sum_{i=1}^{n} x_i \cdot a_i \tag{1}$$

This S is sent to Alice. If the message is long it can be split up into blocks of n bits. More secure methods can be used as e.g. the CBC mode [50].

2.2 Its apparent security

Because the enciphering key is public and S can be eavesdropped, it must be (at least) "difficult" to determine $\mathbf{x}$ out of S and $\mathbf{a}$. If $\mathbf{a}$ is chosen as any sequence of integers Alice can *mostly* not find $\mathbf{x}$ in a *reasonable* amount of CPU time, otherwise NP is equal to P (because the subset sum problem is an NP–complete problem). The only method which is known today is to try for all 2^n possible $\mathbf{x}$ if Eq. 1 is satisfied. This is unreasonable (or unfeasible) if e.g. $n \geq 100$. So it seems that eavesdropping is indeed a hard problem if the knapsack enciphering scheme is used. So if the above enciphering is used, it is hard to find $\mathbf{x}$. It is important to notice the term *mostly.* Indeed if $\mathbf{a} = (1, 2, 4, 8, \ldots, 2^{n-1})$ it is trivial to find for all S the corresponding x, by writing S in binary form. Sequences for which it is easy to find for all S the corresponding $\mathbf{x}$ will be called *easy* sequences.

2.3 The deciphering

If $\mathbf{a}$ is chosen randomly by Alice, (in most cases) also for her it will be unfeasible to decipher S and find the plaintext $\mathbf{x}$. To allow this, Merkle and Hellman [48] introduced some trapdoor. The trapdoor allows Alice to overcome the unfeasibility of finding $\mathbf{x}$ given S and some secret information. It is important to remark that this trapdoor information was used when Alice made her public key. The secret information is called the *deciphering key. It is now exactly the use of such a trapdoor technique which turned out later to be* (mostly) *insecure.*

Remark that in a cryptographic knapsack scheme (used to protect the privacy) the enciphering function defined in Eq. 1 which maps $\mathbf{x}$ into S, has to be one–to–one. Indeed if two different plaintexts $\mathbf{x}$ and $\mathbf{y}$ give the same ciphertext S, the receiver cannot uniquely recover the plaintext. A sequence $(a_1, a_2, \ldots, a_n)$ that leads to such a one–to–one function was called by Shamir [62] a *one–to–one system.* E.g. (4,2,2) is not a one–to–one system, because both cleartexts (1,0,0) and (0,1,1) generate 4 as ciphertext. One can wonder how one can generate one–to–one enciphering keys. In general, it is co–NP–complete to determine if a given sequence is a one–to–one system [62]. One believes (not proven) that problems which are co–NP–complete are even much harder to solve than NP–complete problems [28]. Here also the trapdoor will find a way around the problem.

Almost all knapsack schemes differ only in the use of other trapdoor techniques. Some knapsack schemes allow the x_i to have more values than only one and zero. Others add some kind of noise to the plaintext.

2.4 The Merkle–Hellman trapdoor

When Alice constructed her public enciphering key $\mathbf{a}$ (in the Merkle–Hellman case), she first generated a superincreasing sequence of natural numbers $(a_1^1, a_2^1, \ldots, a_n^1)$. The vector $\mathbf{a}^1 = (a_1^1, a_2^1, \ldots, a_n^1)$ is said to be a ***superincreasing*** sequence if:

$$\text{for each } i, \text{ with } 2 \le i \le n, \qquad a_i^1 > \sum_{j=1}^{i-1} a_j^1$$

E.g. $(1, 2, 4, 8, \ldots, 2^{n-1})$ is a superincreasing sequence. Remark that a superincreasing sequence is an "easy" sequence, as will be explained further on.

The second part of the construction of the public key consists of applying one or several modulo transformations in order to hide the superincreasing structure, such that eavesdroppers cannot use it. These transformations are of the following type:

$$\sum_{i=1}^{n} a_i^j \quad < \quad m_j \tag{2}$$

$$\gcd(w_j, m_j) \quad = \quad 1 \tag{3}$$

$$\begin{array}{rrcllcl} & a_i^{j+1} & \equiv & a_i^j \cdot w_j \quad \bmod m_j & \text{and} & 0 < a_i^{j+1} < m_j \\ \text{or} & a_i^j & \equiv & a_i^{j+1} \cdot w_j^{-1} \quad \bmod m_j & \text{and} & 0 < a_i^j < m_j \end{array} \tag{4}$$

When k transformations are used, the public key $\mathbf{a}$ is equal to$(a_1^{k+1}, a_2^{k+1}, \ldots, a_n^{k+1})$. We will refer to this transformation defined in Eq. 2 –4 as the *Merkle–Hellman transformation*. We call the condition in 2 the *Merkle–Hellman dominance*, or the MH–dominance condition. In the case one uses this transformation in the direction from $\mathbf{a}^{j+1}$ to $\mathbf{a}^j$, we call it the *reverse Merkle–Hellman transformation*. Remark that in this case it is not trivial to satisfy the MH–dominance condition. When only one transformation is discussed we will drop the indices j, $j+1$, k and $k+1$.

The case for which $\mathbf{a}^1$ is superincreasing and only one transformation is used, is called the *basic Merkle–Hellman scheme*, or sometimes the *single iterated Merkle–Hellman scheme*. The case that two transformations are used instead of one is called *double iterated*.

Let us now explain the deciphering. The legitimate receiver receives S. The idea is to calculate $S^1 = \sum_{i=1}^n x_i \cdot a_i^1$ starting from S and the knowledge of the secret parameters $(w_1, m_1), \ldots, (w_k, m_k)$. Because $\mathbf{a}^1$ is easy it is possible to find $\mathbf{x}$ easily. Hereto first $S^{k+1} = S$ and iteratively for all j $(1 \le j \le k)$:

$$S^j = S^{j+1} \cdot w_j^{-1} \bmod m \qquad \text{and} \qquad 0 \le S^j < m_j \tag{5}$$

It is trivial to understand that $S^j \equiv \sum_{i=1}^n x_i a_i^j \bmod m$ and, as a consequence of the inequality condition in Eq. 5, $S^j = \sum_{i=1}^n x_i a_i^j$.

We now explain that if S^1 and the superincreasing sequence $(a_1^1, a_2^1, \ldots, a_n^1)$ are given, it is "easy" [48] to find the message $\mathbf{x}$. Hereto start with $h = n$. If $S^1 > \sum_{i=1}^{h-1} a_i^1$ then x_h has to be one, else zero. Continue iteratively by subtracting $x_h a_h$ from S^1, with h decrementing from n to 1 during the iterations. In fact a rather equivalent process is used to write numbers in binary notation. Indeed the sequence $(1, 2, 4, 8, \ldots, 2^{n-1})$ is (a) superincreasing (sequence).

In Section 2.3 we have seen that an important condition for the public key is that it has to form a one-to-one system. This is the case for the Merkle–Hellman knapsack scheme by applying Lemma 1 as many times as transformations were used, and by observing that a superincreasing sequence forms a one-to-one system.

Lemma 1 *Suppose that* $(a_1^1, a_2^1, \ldots, a_n^1)$ *is a one-to-one knapsack. If* $m > \sum a_i^1$ *and* $\gcd(w, m) = 1$, *then any set* $(a_1, a_2, \ldots, a_n)$, *such that* $a_i \equiv a_i^1 * w \bmod m$, *is a one-to-one system.*

Proof: By the contrary, suppose that $(a_1, a_2, \ldots, a_n)$ does not form a one-to-one system, then there exist $\mathbf{x}$ and $\mathbf{y}$ such that $\mathbf{x} \neq \mathbf{y}$ and $\sum x_i a_i = \sum y_i a_i$. Thus evidently, $\sum x_i a_i \equiv \sum y_i a_i \bmod m$, and also $(\sum x_i a_i) * w^{-1} \equiv (\sum y_i a_i) * w^{-1} \bmod m$, because w^{-1} exists $(\gcd(w, m) = 1)$. So $\sum x_i a_i^1 \equiv \sum y_i a_i^1 \bmod m$. Since $0 \leq \sum x_i a_i^1 \leq \sum a_i^1 < m$ and analogously $0 \leq \sum y_i a_i^1 \leq \sum a_i^1 < m$ we have $\sum x_i a_i^1 = \sum y_i a_i^1$. Contradiction.

3 AN OVERVIEW OF THE HISTORY OF THE CRYPTOGRAPHIC KNAPSACK

Some of the results which will be mentioned here were found independently by many authors. On the other hand some results strongly influenced others. It is sometimes very hard to figure these things out. We will present the research on the cryptographic knapsack (subset sum problem used to protect privacy) as much as possible in chronological order.

We will mainly discuss (see Section 3.2) the additive knapsack public key systems protecting privacy and using the same encryption function as the Merkle–Hellman one. We will abbreviate this as ***the class of usual knapsacks***. Then we will shortly discuss similar schemes but using different encryption functions (see Section 3.3). Before giving the state of the art (see Section 3.7), we very briefly discuss the history of: the multiplicative knapsack schemes (see Section 3.4), the use of trapdoor knapsacks in signatures (see Section 3.5), and other uses of knapsacks in cryptography (see Section 3.6). Let us now start by discussing the driving force of the research on knapsacks.

3.1 Hardware and practical aspects

Almost all research on the cryptographic knapsack systems was focused on ***analysing and/or breaking knapsack systems*** and on ***trying to find a secure version.*** The main reason that so much research has been done on the knapsack scheme is that the encryption is very fast, because it consists mainly of an addition of natural numbers (see Eq. 1). In the case that $n = 100$, and that the sizes of the a_i are 200 digits, an encryption speed of 10 Mbits/sec. is easily achievable. In the deciphering operation a modulo multiplication is necessary for each transformation (which was used during the construction of the public key). At first look this modulo transformation slows down the decryption speed enormously. Henry found a nice idea to speed up the decryption, such that for the basic Merkle–Hellman scheme a decryption speed of 10 Mbits/sec. is obtainable [32]. This idea started a VLSI chip integration of the knapsacks system. So, from the point of view of speed, cryptographic knapsacks algorithms are much better than RSA [57]. We now overview other research on the knapsack system.

3.2 The trials to avoid weaknesses and attacks for the class of usual knapsacks

About immediately after the publication [48] of the Merkle–Hellman scheme, T. Herlestam found in September 1978 some weakness for simulated results [33]. Mainly he found that (for his simulations) he was mostly able to recover at least *one* bit of the plaintext. Hereto he defined some *"partially easy"* sequence. Indeed if $a_r^{k+2} > \sum_{i \neq r} a_i^{k+2}$ for all $S = \sum x_i a_i^{k+2}$ it is easy to recover x_r. Because he did not use the reverse Merkle–Hellman transformation, but the Merkle–Hellman one, he had to add another condition (see [33]).

At the end of 1978 and in the begin of 1979 Shamir found several results [62,63] related to the security of cryptographic knapsack systems. First of all he remarked that the theory of NP–completeness is a bad method to analyse the security of a cryptosystem [62]. Indeed NP–completeness and similarly the theory of NP only discuss *worst case* inputs! In cryptography, problems have to be hard *almost always* for a cryptanalyst. New measures were proposed to overcome this problem. However until today no deeper results have been found related to these new measures. The second weakness that Shamir found was related to what he called the *density* of a knapsack system. The density of a knapsack system with public key $(a_1, a_2, \ldots, a_n)$ is equal to the cardinality of the image of the encryption function (see Eq.1) divided by $\sum a_i$. Knapsack systems which have a very high density can (probabilisticly) easily be cryptanalysed as Shamir found [62]. This result is independent of the trapdoor used to construct the public key. Finally Shamir and Zippel figured out some weakness related to a remark in the paper of Merkle and Hellman. They considered the case that the public key was constructed using the *basic Merkle-Hellman scheme* and using parameters proposed by Merkle and Hellman [48] and that *m (a parameter of the secret Merkle–Hellman transformation) would be revealed.* For that special case the knapsack system can almost always be broken [62,63]. We will refer to this case as the *Shamir–Zippel weakness.*

Graham and Shamir [63] and Shamir and Zippel proposed to use other easy sequences than the superincreasing ones and then to apply Merkle–Hellman transformations to obtain the public key. The case that only one transformation is used is called the *basic Graham–Shamir* and *basic Shamir–Zippel* scheme. The basic Graham–Shamir and basic Shamir–Zippel scheme do not suffer from the Shamir–Zippel weakness. E.g. in the Graham–Shamir scheme $\mathbf{a}^1$ is not superincreasing but can be written as:

$$\mathbf{a}^1 = \mathbf{a}' + 2^q \mathbf{a}'', \qquad \text{with} \quad a'_n < 2^{q-1} \qquad \text{and} \quad \mathbf{a}' \text{ superincreasing.}$$

It is trivial to understand that such a sequence is easy, using Section [48].

At the end of 1980 Ingemarsson [34] found sequences **a** which, when used as a public key, could not be broken by the Herlestam attack. He called these sequences *NIPS.* Almost all of his theorems related to NIPS, however, are not useful in the discussion of knapsack systems used for encryption, because the sequences discussed correspond with non–one–to–one systems.

In the beginning of 1981 Lenstra [44] found a polynomial time (practical) algorithm to solve the integer linear programming problem, when the number of unknowns is fixed. The complexity of the algorithm grows exponentially if the size of the number of unknowns increases. A part of Lenstra's algorithm uses a lattice reduction algorithm (more details are given in 4.2). The importance for the security of knapsack cryptosystems will be explained later.

In 1981 Desmedt, Vandewalle and Govaerts found several results [16,17] related to the security of cryptographic knapsack systems, which are obtained using Merkle–Hellman transformations. First they proved that any public key which is obtained from a superincreasing sequence using the Merkle–Hellman transformation, has infinitely many deciphering keys. In general, if some public key is obtained using a Merkle–Hellman transformation, then there exist infinitely many other parameters, which would result in the same public key when used to construct it. This allowed to generalize the Shamir–Zippel weakness. It was no longer necessary to know m in order to be able to apply their ideas (infinitely many other m's allow to break). Secondly it has been shown by examples that iterative transformations do not necessarily increase the security. Thirdly a new type of partially easy sequences has been proposed. This one, together with the idea of Herlestam, led mainly to new versions of knapsack systems. Remember that in the Merkle–Hellman case, n is *fixed* during the construction of the public key. Remember also that for deciphering S it was transformed k times using the (w_j^{-1}, m_j) to obtain S^1, which allows to find *all* x_i' about at *once* (using the superincreasiong sequence). Here n *grows* during the construction of the public key. In the deciphering process here, transformations with (w_j^{-1}, m_j) are applied *mixed up with retrieval of some* bit(s) x_i. Let us briefly explain the other type of partially easy sequence, called *ED*. If d divides all a_i^j, except a_r^j, then if $S^j = \sum_{i=1}^n x_i a_i^j$, it is easy to find x_r, by checking if d divides S^j or not. The here discussed method to construct the public key, together with the discussed partially easy sequence will be called the *Desmedt-Vandewalle-Govaerts* knapsack. They also proved that some sequences which correspond to one–to–one knapsack systems cannot be used when the public key would be build up using Merkle–Hellman transformations. In fact these sequences are either easy, or unobtainable (even using infinitely many transformations) from other sequences (e.g. easy or partially easy sequences). They called these sequences together with the non–one–to–one systems *useless*, and called the set of these sequences U, and the intersection with the one–to–one sequences UB. Finally they proved that the security of Merkle–Hellman transformations is reduced to a problem of simultaneous diophantine approximations. All these results were obtained by regarding the problem of reversing the Merkle–Hellman transformation as an integer linear programming problem.

The same year Karnin and Hellman discussed a special case of the Herlestam partially easy sequence, and its consequences on the security of knapsack cryptographic systems. Their main idea was to look to the probability that some subsequence of the public key is a superincreasing sequence. Their main conclusion was that the security was not affected by their algorithm [37].

Ingemarsson analysed the security of the knapsack algorithm by applying several Merkle–Hellman transformations on the public key and on the ciphertext S [35]. Because he did not used the reverse transformation, he obtained only congruences. In order to turn them over to equations, he had to add extra unknowns. Not enough information is available today to estimate the performance of this method.

In the beginning of 1982 Lenstra, Lenstra and Lovasz found some algorithm for factoring polynomials with rational coefficients [45]. A part of this algorithm is an improvement of the lattice reduction algorithm (described in [44]). This improvement is known in the cryptographic world as the *LLL algorithm* (nevertheless that it was mainly Lovasz who found it). Remark that the LLL algorithm speeds up the algorithm to solve the integer linear programming (with the number of variables fixed) [46]. Another application of it is that it allows to find some simultaneous diophantine approximations [45].

In April 1982 Shamir broke in almost all cases the basic Merkle–Hellman scheme [64,66]. His attack uses the same observation as above, related to the integer linear programming problem. Shamir was able to reduce dramatically the number of unknowns (in almost all cases) in the integer linear programming problem. In fact the cryptanalyst first guesses the correct subsequence of the public key corresponding with the smallest superincreasing elements. The number of elements in the subsequence is small. Because the Lenstra algorithm (to solve the integer linear programming problem) is feasible if the number of unknowns is small, Shamir was able to break (in almost all cases) the basic Merkle–Hellman scheme.

A few months later Brickell, Davis and Simmons [5] found that by a careful construction of the public key (using the basic Merkle–Hellman scheme) the designer can avoid Shamir's attack. This work demonstrated clearly that one has to be careful with attacks, which break systems in *almost all cases.* Indeed this kind of attacks seems only very dangerous, if no method is found to overcome the weakness. (This last remark is less important today, as a consequence of further research, which will now be overviewed.)

About the same time Davio came with a new easy sequence [15]. This easy sequence is based on ED, but it allows to find all x_i at once. The construction is similar to the proof of the Chinese Remainder Theorem. It is used then instead of the superincreasing sequence.

Lagarias started a deeper research on the computational complexity of simultaneous diophantine approximation [39] in a more general sense.

Adleman broke the basic Graham–Shamir scheme [1]. He also claimed that he could break the iterated Merkle–Hellman scheme [1]. For the case of the basic Graham–Shamir scheme, Adleman demonstrated with a home computer that his breaking method works on small (n small) examples. The main idea of Adleman was to treat the cryptanalytic method as a lattice reduction problem and not as a linear integer programming problem. *This idea was one of the most influential in the area of breaking cryptographic knapsack algorithms.* To solve the lattice problem he used the LLL algorithm [45]. The choice of a good lattice and avoiding undesired results play a key role in his paper. Remark that in a lot of breaking algorithms a lot of heuristics is used, however mostly some deep arguments can be found that the breaking technique works almost always. This was e.g. the case for the basic Graham–Shamir case.

In August 1982 Shamir presented a new knapsack scheme, known as *Shamir's ultimate knapsack scheme* [65]. The main idea is that instead of applying k Merkle–Hellman transformations, one uses "exactly" $n-1$ of such transformations to construct a public key. "Exactly" means here, that after each transformation (e.g. j^{th}) one checks if $\mathbf{a}^j$ is linearly independent of $(\mathbf{a}^1, \ldots, \mathbf{a}^{j-1})$, if not, one drops $\mathbf{a}^j$, makes a new one and tries again. The final result $\mathbf{a}^n$ is the public key. To decipher S, the legitimate receiver applies his $n-1$ reverse secret transformations. He starts by putting $S^n = S$ and by calculating the other S^j, similar as in the Merkle–Hellman case (see Section 2.3). So he obtains a set of linear equations:

$$\begin{pmatrix} S^1 \\ \vdots \\ S^{n-1} \\ S^n \end{pmatrix} = \begin{pmatrix} a_1^1 & \dots & a_n^1 \\ \vdots & \ddots & \vdots \\ a_1^{n-1} & \dots & a_n^{n-1} \\ a_1^n & \dots & a_n^n \end{pmatrix} \cdot \begin{pmatrix} x_1 \\ \vdots \\ x_{n-1} \\ x_n \end{pmatrix} \tag{6}$$

After the discussed transformations to find $\mathbf{x}$, the legitimate receiver then only has to solve the set of linear equations. It is important to observe that the obtained public key is one–to–one, even if $\mathbf{a}^1$ is *not* an easy sequence, or even if no partially easy sequences are used. This follows from the nonsingularity of the matrix in Eq. 6. In order to speed up the deciphering the receiver can do all calculations modulo p, with p a small (or if posible the smallest) prime such that the matrix in Eq. 6 is nonsingular in $\mathbf{Z_p}$ [7, pp. 29]. This works because $\mathbf{x}$ is binary.

Other research went on, trying to obtain other easy (or partially easy) knapsack sequences. Petit [54] for example used what she called ***lexicographic knapsacks*** as easy sequence. Roughly speaking $\mathbf{a}$ is lexicographic, if the binary words $\mathbf{x}$ are arranged by the ordering of the integers $\mathbf{a}^T\mathbf{x}$, as in a dictionary. The exception is that if the Hamming weight $w(\mathbf{x})$ of $\mathbf{x}$ is smaller than that of $\mathbf{y}$, with $\mathbf{x}$ and $\mathbf{y}$ binary, then $\mathbf{a}^T\mathbf{x} < \mathbf{a}^T\mathbf{y}$. More formally $\mathbf{a}$ is lexicographic, if and only if, $\mathbf{a}^T\mathbf{x} < \mathbf{a}^T\mathbf{y}$ for all binary $\mathbf{x}$ and $\mathbf{y}$, with $\mathbf{x} \neq \mathbf{y}$ and one of the two cases (i) $w(\mathbf{x}) < w(\mathbf{y})$ or (ii) $w(\mathbf{x}) = w(\mathbf{y})$ and $\mathbf{x}$ and $\mathbf{y}$ satisfy together $x_k\overline{y}_k = 1$ and $x_i \oplus y_i = 0$ for all $i < k$, with $\oplus$ the exclusive or. The construction the public key is as in the Merkle–Hellman case, using Merkle–Hellman transformations.

In August 1982 Desmedt, Vandewalle and Govaerts [18] found that for very special public key, the weaknesses found for the basic Merkle–Hellman scheme carry over to similar ones for the special public keys. A similar attack as Shamir can be used to break such knapsack systems. These special ones, were e.g. obtained by more than one transformation. The main criticisms on this paper is that such special knapsacks are very rare.

Willett [70] also came with another easy sequence and a partially easy sequence, which are then used similar as in the Merkle–Hellman and in the Desmedt–Vandewalle–Govaerts knapsack. We only discuss the easy sequence. It is not to difficult to figure out how it works in the case of the partially easy sequence. The ith row of the matrix in Eq. 7 corresponds with the binary representation of a_i^1.

$$(\; T_n \;\; C_n \;\; O_{n-1} \;\; T_{n-1} \;\; C_{n-1} \;\; \dots \;\; O_2 \;\; T_2 \;\; C_2 \;\; O_1 \;\; T_1 \;\; C_1 \;), \tag{7}$$

In Eq. 7 the T_i are randomly chosen binary matrices, the C_i are $n \times 1$ binary column vectors such that they (C_i) are linearly independent modulo 2, and the O_i are $n \times l_i$ zero binary matrices, where $l_i \geq \log_2 n$. Let us call the locations of the C_i t_i. To find $\mathbf{x}$ out of S_1, we first represent S_1 binary, and we call these bits s_h. As a consequence of the choice of l_i, the bits s_{t_i} are not influenced by T_{i-1} and C_{i-1}. To find $\mathbf{x}$ we have to solve modulo 2:

$$\begin{pmatrix} s_{t_1} \\ \vdots \\ s_{t_{n-1}} \\ s_{t_n} \end{pmatrix} = \begin{pmatrix} c_1^1 & \dots & c_n^1 \\ \vdots & \ddots & \vdots \\ c_1^{n-1} & \dots & c_n^{n-1} \\ c_1^n & \dots & c_n^n \end{pmatrix} \cdot \begin{pmatrix} x_1 \\ \vdots \\ x_{n-1} \\ x_n \end{pmatrix} \bmod 2,$$

where the c_i^j are coefficients of C_i.

McAuley and Goodman [47] proposed in December 1982 a very similar knapsack scheme as the one proposed by Davio (see higher). The differences are that no Merkle–Hellman transformations are used and that the $\mathbf{x}$ can have more values than binary (they have to be smaller than a given value and larger or equal than zero). The trapdoor information consists only in secrecy of the primes which were used in the construction.

Another method to construct public keys in a knapsack system was found between the end of 1982 and the beginning of 1983 by Desmedt, Vandewalle and Govaerts [20]. They called their scheme *the general knapsack scheme.* The main purpose of this paper was to stop looking for new easy and partially easy sequence in a heuristic way, and to generalize as well as the construction of public keys, as well as the deciphering. Let us briefly explain it from a point of view of deciphering. Because the algorithm to construct public keys is quite involved we refer to [24]. The basic idea is similar to the deciphering method of Shamir's ultimate knapsack scheme. Let us explain the differences. Instead of going in the deciphering from $\mathbf{a}^{j+1}$ to $\mathbf{a}^j$ with a reverse Merkle–Hellman transformation, an intermediate vector $\mathbf{b}^j$ and some integer R^j are used and the reverse Merkle–Hellman transformation is generalized. Let us first explain the generalized reverse transformation idea which was called *extended map.* For a vector $\mathbf{a}^j$ a mapping g_{j-1} is an extended map of a subset of $\mathbf{Z}$ into $\mathbf{Z}$, if and only if, for each binary $\mathbf{x}$:

$$g_{j-1}(\sum x_i a_i^j) = \sum x_i g_{j-1}(a_i^j)$$

Remark that the reversed Merkle–Hellman transformation is a special case of an extended map. Desmedt, Vandewalle and Govaerts found another extended map [20], they later improved their idea after a suggestion of Coppersmith. In order to be practical the calculation of the transformation and its reverse corresponding with the extended map have to be fast. Let us now explain The rows in the matrix of Eq. 6 corresponding with the vectors $\mathbf{a}^1, \mathbf{a}^2, \ldots, \mathbf{a}^{n-1}$ are replaced by the vectors $\mathbf{b}^1, \mathbf{b}^2, \ldots, \mathbf{b}^{n-1}$ (remark that $\mathbf{a}^n$ remains in the matrix). The numbers $S^1, S^2, \ldots, S^{n-1}$ in Eq. 6 are replaced by $R^1, R^1, \ldots, R^1$. These new vectors $\mathbf{b}^j$ and new numbers R^j were obtained as from $\mathbf{a}^{j+1}$ and S^{j+1} by the extended map g_j, or $b_i^j = g_j\left(a_i^{j+1}\right)$ and $R^j = g_j\left(S^{j+1}\right)$. $\mathbf{a}^{j+1}$ and S^j are nothing else than linear combination of previous vectors and corresponding sums, or:

$$\begin{aligned} S^{j+1} &= e_n^{j+1} S^n + e_{n-1}^{j+1} R^{n-1} + e_{n-2}^{j+1} R^{n-2} + \cdots + e_{j+1}^{j+1} R^{j+1} \\ \mathbf{a}^{j+1} &= e_n^{j+1} \mathbf{a}^n + e_{n-1}^{j+1} \mathbf{b}^{n-1} + e_{n-2}^{j+1} \mathbf{b}^{n-2} + \cdots + e_{j+1}^{j+1} \mathbf{b}^{j+1} \end{aligned},$$

where the e_l^j are rationals (such that S^{j+1} and $\mathbf{a}^{j+1}$ are integers and integer vectors). These e_l^j, and many other parameters used in the construction of the public key are kept secret. This method of using linear combinations of previous results in the deciphering allows easily to prove that all previously discussed knapsack systems are special cases of the general one [24]. At first sight the difference with the ultimate knapsack scheme of Shamir seems to be small. However, details of the construction method of the public key show the converse. In Shamir's scheme one can only choose one vector and start the transformation, while here n choices of vectors are necessary (or are done implicitly). The idea of extended map allows also to generalize the idea of useless knapsacks, and may be this explains the failure of so many knapsack systems [21,24]

Around the same period Brickell found some method to cryptanalyse low density knapsacks [6,7] (the density of a knapsack was informally discussed at the begin of this section). A similar attack was independently found by Lagarias and Odlyzko [40]. To perform his attack, Brickell first generalized the Merkle–Hellman dominance condition. The integers he used may also be negative. Brickell called a ***modular mapping*** $*w \bmod m$ from $\mathbf{a}$ into $\mathbf{c}$ to have the ***small sum*** property if $c_i \equiv a_i w \bmod m$, and $m > \sum |c_i|$. He called mappings satisfying this property *SSMM*. (Remark that the condition $\gcd(w, m) = 1$ is not necessary here because the reverse transformation is only used to break systems, such that a w^{-1} is not necessary). Given $\sum x_i a_i$ one can easily calculate $\sum x_i c_i$. This is done exactly as in the reverse Merkle–Hellman case. If the result is greater than $\sum_{c_i>0} c_i$ M is substracted from it. He tries to find $n-1$ such transformations all starting from the public key $\mathbf{a}$. He can then solve a set of equations similar as in the ultimate scheme of Shamir (remark the difference in obtaining the matrix). To obtain such transformations in order to break, he uses the LLL algorithm choosing a special lattice. If all the reduced lattice basis vectors are short enough, he will succeed. This happens probably when the density is less than $1/\log_2 n$. In the other cases he uses some trick to transform the problem into one satisfying the previous condition. Arguments were given that this will succeed almost always when the density is less than 0.39. The low dense attack proposed by Lagarias and Odlyzko is expected to work when the density of the knapsack is less than 0.645. These attacks break the ultimate scheme of Shamir, because the density of the public key is small as a consequence of construction method of the public key.

Lagarias found some nice foundation for the attacks on the knapsack system, by discussing what he called ***unusually good*** simultaneous diophantine approximations [41]. The term unusually good is motivated by the fact that such approximations do not exist for almost all randomly generated sequences of rational numbers. His theory underlies the low dense attack of Brickell and of Lagarias and Odlyzko as well as the Adleman's attack on the basic Graham–Shamir scheme. Lagarias used similar ideas [42] to analyse Shamir's attack on the basic Merkle–Hellman scheme. In this context it is worth to mention that an improved algorithm for integer linear programming was found earlier by Kannan [36]. The main result is that Shamir overlooked some problems, but nevertheless his attack works almost always.

Brickell, Lagarias and Odlyzko performed an evaluation [8] of the Adleman's attack on multiply iterated Merkle–Hellman and Graham–Shamir schemes. They concluded that his attack on the basic Graham–Shamir scheme works, but that the version to break iterated Merkle–Hellman or Graham–Shamir scheme failed. The main reason for it was that the LLL algorithm found so called undesired vectors, which could not be used to cryptanalyse the cited systems. Even in the case that only two transformations were applied (to construct the public key) his attack fails.

Karnin propsosed in 1983 an improved time–memory–processor tradeoff [38] for the knapsack problem. The idea is related to exhaustive machines [25] and time–memory tradeoffs [31], in which an exhaustive search is used to break the system using straightforward or more advanced ideas. This paper is completely theoretical if the dimension of the knapsack system n is large enough (e.g. $n \geq 100$).

In 1984 Goodman and McAuley proposed a small modification [30] to their previous system [47]. In the new version some modulo transformation is applied.

Brickell proposed the same year how to cryptanalyse [10] the iterated Merkle–Hellman and Graham–Shamir scheme. As usual no proof is provided that the breaking algorithm works, arguments for the heuristics are described in [10]. Several public keys were generated by the Merkle–Hellman and Graham–Shamir scheme and turned out to be breakable by Brickell's attack. Again the LLL algorithm is the driving part of the attack. First the cryptanalyst picks out a subset of the sequence corresponding with the public key. He enters these elements in a special way in the LLL algorithm. He obtains a reduced basis for that lattice. He then calculates what the linear relation is between the old and new basis for the lattice. This last information will allow him to determine if he picked out some "good" subset of the sequence. If not he restarts at the beginning. If it was a good set, he will be able to calculate the number of iterations that were used by the designer during the construction of the public key. Some calculation of determinants will then give him an almost superincreasing sequence. Proceeding with almost superincreasing sequences was yet discussed by Karnin and Hellman [37] (remarkable is the contradiction in the conclusion of their paper and its use by Brickell!).

In October 1984, Odlyzko found an effective method to cryptanalyse the McAuley–Goodman and the Goodman–McAuley scheme, using mainly gcd's [53].

Later on Brickell [11] was able to break with a similar idea as in [10] a lot of other knapsack schemes, e.g., the Desmedt–Vandewalle–Govaerts, the Davio, the Willett, the Petit and the Goodman–McAuley. The attack affects also the security of the so called general knapsack scheme.

At Eurocrypt 85 Di Porto [27] presented two new knapsack schemes, which are very close to the Goodman–McAuley one. However they were broken during the same conference by Odlyzko.

3.3 The case of usual knapsacks with other encryption functions

Arazi proposed in 1979 a new knapsack based additive knapsack algorithm to protect the privacy of the message [2]. It's main difference with the Merkle–Hellman encryption is that random noise is used in the enciphering function. The parameters which are chosen during the construction of the public key have to satisfy some properties (see [2]).

In 1983 Brickell also presented a new knapsack system [9], which is similar to the Arazi one.

Brickell declared one year later his new own scheme insecure, as a consequence of his attack on iterated knapsacks [10].

Chor and Rivest proposed in 1984 another knapsack based system [13]. The encryption process is very close to the one in the Merkle–Hellman scheme. The main difference in the enciphering is that the $\sum x_i \leq h$ for some given h. The trapdoor technique does not use a modular multiplication (as do almost all other knapsack schemes). The trapdoor uses the discrete log problem [4,52,55] (see also Section 3.4). A study of possible attacks was done, but it turned out that by a good choice of parameters all attacks known at that moment could be avoided. New attacks were set up by the authors [13] but this did not change the above conclusion.

In 1985 Brickell broke the Arazi knapsack system [11].

Cooper and Patterson [14] proposed also in 1985 some new trapdoor knapsack algorithm, which can however be cryptanalysed by Brickell [11]. The same attack of Brickell can break this knapsack as well as the Lagger knapsack [43].

3.4 The multiplicative knapsack scheme and its history

It is in fact completely wrong to discuss the so called multiplicative *knapsack here, because it uses exactly the same enciphering function as the Merkle–Hellman additive knapsack scheme.* However the trapdoor is completely different in nature, because it is mainly based on a transformation from an additive knapsack problem into a ***multiplicative*** one.

Up till now the only multiplicative knapsack scheme was presented by Merkle and Hellman in their original paper [48].

Let us first explain the construction of the public key. One chooses n relative prime positive numbers $(p_1, p_2, \ldots, p_n)$, some prime q, such that $q-1$ has only small primes and such that

$$q > \prod_{i=1}^{n} p_i \tag{8}$$

and some primitive root b modulo q. The designer then finds integers a_i, where $1 \leq a_i \leq q-1$, such that $p_i \equiv b^{a_i} \bmod q$. Or the a_i are the discrete logarithms of the p_i to base b modulo q. This last formulation explains why $q-1$ was chosen as the product of small primes, because an algorithm exists to calculate easily these discrete logarithms in that case [55] (remark that a lot of research in that area was done recently (see [4,52]).

To decipher the message S one calculates $S' = b^S \bmod q$, because $b^S = b^{\sum x_i \cdot a_i} = \prod b^{x_i \cdot a_i} = \prod p_i^{x_i} \bmod q$. The last equality is a consequence of the condition in Eq. 8. One can easily find the corresponding $\mathbf{x}$ starting from S', using the fact that the numbers p_i are relative prime. This last point is important, because in the general case the subset product problem is NP–complete [28].

This scheme can be cryptanalysed by a low dense attack [7,40]. However the disadvantage is that it requires a separate run of the lattice reduction algorithm (which takes at least on the order of n^4 operations) to attack each n bit message. To overcome that problem, Odlyzko tried another attack [51]. Herein he starts from the assumption that some of the p_i are known. He then tries to find q and b. He also assumes that b, q and the a_i consist of approximately m bits. His attack will take a polynomial time if $m = O(n \log n)$. Also in this attack the LLL algorithm is the driving force. A special choice [51] of the lattice is used to attack the system. Once the b and q were found the cryptanalysts can as easily cryptanalyse ciphertexts as the receiver can decipher them.

3.5 The trapdoor knapsack schemes to protect signatures and authenticity

In modern communication the protection of the authenticity of messages and signatures is important [56,26,22], sometimes more important than the privacy of the message. The discussion here assumes that no privacy protection is necessary (otherwise both protections are used in cascade). In a public key system a secret key is used by the receiver, the public key is used by the receiver. This can be considered that the sender applies the decryption function on the plaintext. From this point of view it is easy to understand that the higher discussed knapsack schemes are not well suited for this purpose. Indeed if the deciphering function is not "enough" (pseudo) invertible the sender has to perform other trials in order to generate a signature. Such a scheme was presented in the original Merkle–Hellman paper [48]. Shamir suggested a more practical one [61] in 1978.

In 1982 Schöbi and Massey proposed another version of [60] a fast signature scheme, more related to the Merkle–Hellman knapsack.

In 1982 –1983 Odlyzko broke [51] the Shamir's fast signature and the Schöbi–Massey one. Here also the LLL algorithm plays an important role. Shamir and Tulpan [67] found recently also an attack on the Shamir's signature scheme. Unlike the attack of Odlyzko it can be proved to succeed with high probability, but its running time is exponential in n. Nevertheless the attack is still realistic for the case $n = 100$.

Remark that in 1983 Desmedt, Vandewalle and Govaerts invented a protocol [23] to use a usual knapsack scheme to protect authenticity of messages (not signature). Its security is less or equal than the security of the used knapsack system.

3.6 Other uses of knapsacks in cryptography

In 1981 Schaumuller [58] and in 1982 Desmedt, Vandewalle and Govaerts [19], proposed several ideas for using the knapsack (subset sum) problem as a one–way function instead of the S–boxes in DES [49]. The use of the knapsack here is completely different compared with the previous ones. Here no condition as one–to–one is necessary and no trapdoor is used. So its use is there based on the NP–completeness of the knapsack problem.

In a completely different application very recently Shamir used the knapsack problem in order to come up with a "provably" secure protocol to protect passports [69]. Here again the knapsack is used as a hard to invert problem. So the security is not based on the security of a trapdoor. Here also the use of knapsacks in cryptography is completely different from that in the schemes presented in other sections. This idea is another version of his protocol for the same purpose using an RSA like function [68].

3.7 State of the art of the use of knapsacks in cryptography

Almost all *trapdoor* knapsack public key schemes have been broken. All modern breaking algorithms use the LLL algorithm. The best breaking techniques (today) are the low dense attacks [7,40] and the attack on iterated knapsacks by Brickell [10] and its improvement [11]. These attacks break almost all knapsack systems, which protect privacy. In the case of the protection of signatures by a trapdoor public key knapsack system, the attacks of Odlyzko [51] break the proposed schemes.

Today only two trapdoor knapsack systems are *not* broken, the Chor–Rivest scheme [13] and the general knapsack [20,24]. This *does not mean at all* that they are secure. As a consequence of the history of other trapdoor knapsack systems (almost) *everybody* is very sceptical about their security. So for the first one it is not excluded that other lattice basis reduction algorithms (than LLL) could help to break the scheme. It is not the purpose of this paper to overview all other lattice reduction algorithms. The reader interested in it can find some in [51, pp. 598] and in [59]. For the general knapsack, first remark that if the construction of the public key is done without care a low dense one is obtained which can be cryptanalysed. In [24] (see also [29]) the authors have given some method to avoid the last problem. Unfortunately they did not yet construct a public key of acceptable dimension to be analysed on its security.

3.8 Is there a future for knapsacks?

Another formulation of this question is: *Is there life after death?*.

The answer to this question depends on the case which we discuss. In the case of trapdoor public key knapsack schemes one has to be very sceptical. We strongly advise *not* to use such schemes, even new unbroken ones. In general the history of trapdoor knapsack schemes demonstrates clearly enough that schemes which were not enough analysed are dangerous to be used (the iterated Merkle–Hellman scheme was only broken after 6 years, or after about 2 years of intensive research). Everybody who comes up today with a new trapdoor knapsack scheme has first to investigate possible attacks. But even if nobody can break them today, what will happen after intensive research during two years? Probably the academic and scientific world will no longer be interested in such research turning around in circles, what does not exclude that others are well interested in attacks, but for other than scientific purposes! However the use of knapsack in non-trapdoor application, as e.g. protocols may have some future. However the research will be completely different and will focus on other aspects as for example speed and ease of implementation.

4 MORE DETAILS OF THE WEAKNESSES AND OF THE CRYPTANALYTIC ATTACKS ON TRAPDOOR KNAPSACK SCHEMES

Only a complete book can describe enough details about weaknesses and attacks. To overcome that problem we have to restrict ourself. We will prove that Merkle–Hellman transformations lead to the possibility of more than one deciphering key to break. After explaining the LLL algorithm in Section 4.2 we will give an example of its use in the low density attack of Brickell.

4.1 The existence of infinitely many deciphering keys

To explain this we focus on the basic Merkle–Hellman scheme. Suppose w^{-1} and m correspond with the reverse Merkle–Hellman transformation and that a'_i was the used superincreasing sequence. We will demonstrate that other values allow to break (call these V, M, and a''_i). In order to analyze for which V and M Eq. 2 -4 holds let us reformulate the Merkle–Hellman transformation in terms of *linear inequalities*. $a''_i \equiv a_i \cdot V \bmod M$ and $0 < a''_i < M$ can be reformulated into:

$$0 < a''_i = (a_i \cdot V - s_i \cdot M) < M, \quad s_i \text{ integer} \tag{9}$$

Remark that s_i is equal to $\lfloor (a_i \cdot V)/M \rfloor$ with $\lfloor \gamma \rfloor$ the floor or the largest integer less than or equal to γ.

With the aid of Eq. 9 we can reformulate the conditions in Eq. 2 -4 and the condition of superincreasing of $\mathbf{a}''$ as *linear inequalities* on V/M :

$$\text{Eq. 9 gives :} \quad \frac{s_i}{a_i} < \frac{V}{M} < \frac{1+s_i}{a_i} \leq 1 \tag{10}$$

$$\text{Eq. 2 gives :} \quad \frac{V}{M} < \frac{1+\sum_{i=1}^{n} s_i}{\sum_{i=1}^{n} a_i} \tag{11}$$

the condition on superincreasing of $\mathbf{a}''$ gives for all j, with $2 \leq j \leq n$:

$$\text{if } a_j > \sum_{i=1}^{j-1} a_i : \quad \frac{V}{M} > \frac{s_j - \sum_{i=1}^{j-1} s_i}{a_j - \sum_{i=1}^{j-1} a_i} \tag{12}$$

$$\text{if } a_j < \sum_{i=1}^{j-1} a_i : \frac{V}{M} < \frac{s_j - \sum_{i=1}^{j-1} s_i}{a_j - \sum_{i=1}^{j-1} a_i} \tag{13}$$

Observe that the condition in Eq. 3 does not impose an extra condition on the ratio V/M. Indeed, for any V/M which satisfies the conditions in Eq. 10–13 one can take coprime V, M in order to satisfy Eq. 3.

Theorem 1 *For each enciphering key $(a_1, a_2, \ldots, a_n)$ constructed using Eq. 2-4 from a superincreasing sequence $(a'_1, a'_2, \ldots, a'_n)$, there exist infinitely many superincreasing sequences satisfying the conditions in Eq. 2-4.*

Proof: The conditions Eq. 2, Eq. 4 and superincreasing reformulated as Eq. 10, Eq. 11, Eq. 12, Eq. 13 can be summarized as:

$$L < \frac{V}{M} < U \tag{14}$$

where L and U are rational numbers. Now since there exists a superincreasing deciphering key, which satisfies Eq. Eq. 2–4 there exist L and U such that $L < U$. Since $L \neq U$ there exist infinitely many (V, M) for which the condition in Eq. 14 holds and $gcd(V, M) = 1$.

One can generalizes Theorem 1. The above is true if **a** *is obtained by Merkle-Hellman transformations,* [16,17,24].

4.2 The LLL algorithm

We will first define a lattice (in the geometrical sense of the word) which should not be confounded with the set theoretic concept [3]. A main theorem is then shortly surveyed. The LLL algorithm is shortly described.

Definition 1 *Let $(\mathbf{v}^1, \ldots, \mathbf{v}^n)$ be a linearly independent set of real vectors in a n-dimensional real Euclidean space. The set of all points*

$$u_1\mathbf{v}^1 + \cdots + u_n\mathbf{v}^n$$

with integral $u_1, \ldots, u_n$, is called the lattice with basis $(\mathbf{v}_1, \ldots, \mathbf{v}_n)$.

An example of a lattice is the set L_0 of all vectors with integral coordinates. A basis for this lattice is clearly the set of vectors:

$$\mathbf{e}^j = (0, \ldots, 0, 1, 0, \ldots, 0) \qquad (1 \leq j \leq n), \quad \text{with the 1 in location } j.$$

Theorem 2 *Let $(\mathbf{v}^1, \ldots, \mathbf{v}^n)$ be a basis of a lattice L and let $\mathbf{v}'^i$ be the points*

$$\mathbf{v}'^i = \sum_j z^i_j \mathbf{v}^j \qquad \text{for} \qquad (1 \leq i \leq n) \quad \text{and} \quad (1 \leq j \leq n)$$

where z^i_j are integers, then the set $(\mathbf{v}'^1, \ldots, \mathbf{v}'^n)$ is also a base for the same lattice L, if and only if $\det(z^i_j) = \pm 1$. We call an integer matrix Z with $\det(z^i_j) = \pm 1$ an unimodular matrix.

Proof: See [12].

As a consequence of Theorem 2 it is clear that $|\det(\mathbf{v}^1, \ldots, \mathbf{v}^n)|$ is independent of a particular basis for a lattice.

In the geometric theory of numbers it is well know that in general in a lattice L there does not exist a set of n vectors such that they form an orthogonal set. As a consequence of Theorem 2, this raises the problem that basises for lattices sometimes have large coefficients. The Lenstra Lenstra Lovasz (LLL [45, pp. 515–525]) algorithm finds in polynomial time a basis for a lattice L, which is nearly orthogonal with respect to a certain measure of non–orthogonality. The LLL algorithm does however not find in general the most orthogonal set of n independent vectors. As a consequence of Theorem 2 it finds short (probably not the shortest) vectors. A basis is called ***reduced*** if it contains relatively short vectors.

Let us briefly describe LLL. Let $\mathbf{v}_1, \mathbf{v}_2, \ldots, \mathbf{v}_n$ belong to the n-dimensional real vector space. To initialize the algorithm a orthogonal real basis $\mathbf{v}'_i$ is calculated, together with μ^i_j $(1 \leq j < i \leq n)$, such that

$$\mathbf{v}'_i = \mathbf{v}_i - \sum_{j=1}^{i-1} \mu^i_j \mathbf{v}'_j \tag{15}$$

$$\mu^i_j = (\mathbf{v}_i, \mathbf{v}'_j)/(\mathbf{v}'_j, \mathbf{v}'_j), \tag{16}$$

where (,) denotes the ordinary inner (scalar) product. In the course of the algorithm the vectors $\mathbf{v}_1, \mathbf{v}_2, \ldots, \mathbf{v}_n$ will be changed several times, but will always remain a basis for L. After every ***change*** the $\mathbf{v}'_i$ and μ^i_j are ***updated*** using Eq. 15 and Eq. 16. A current subscript k is used during the algorithm. LLL starts with $k = 2$. If $k = n + 1$ it terminates. Suppose now $k \leq n$, then we first achieve that $|\mu^k_{k-1}| \leq 1/2$ if $k > 1$. If this does not hold, let r be the integer nearest to μ^k_{k-1}, and ***replace*** $\mathbf{v}_k$ by $\mathbf{v}_k - r\mathbf{v}_{k-1}$, (don't forget the update). Next we distinguish ***two cases***. Suppose that $k \geq 2$ and $|\mathbf{v}'_k + \mu^k_{k-1}\mathbf{v}'_{k-1}|^2 < (3/4)|\mathbf{v}'_{k-1}|^2$, then we ***interchange*** $\mathbf{v}_{k-1}$ and $\mathbf{v}_k$, (don't forget the update), afterwards replace k by $k - 1$ and restart. In the ***other case*** we want to achieve that

$$|\mu^k_j| \leq \frac{1}{2} \qquad \text{for} \qquad 1 \leq j \leq k-1 \tag{17}$$

If the condition in Eq. 17 does not hold, then let l be the largest index $< k$ with $\mu^k_l > 1/2$, let r be the nearest to μ^k_l and ***replace*** $\mathbf{b}_k$ by $\mathbf{b}_k - r\mathbf{b}_l$, (don't forget the update), repeat until the conditions Eq. 17 hold, afterwards replace k by $k + 1$ and restart. Remark that if the case $k = 1$ appears one replaces it by $k = 2$.

4.3 The use of the LLL algorithm in Brickell's low dense attack

In Section 3.2 we yet briefly discussed Brickell's low dense attack. We introduced the concept of *SSMM* and have given a sketch of Brickell's low dense attack. Remember also that if the density is not low enough ($> 1/\log_2 n$) it has to be artificially lowered. We will only discuss the case that it is indeed low enough. This last part is always used as the main technique of the breaking algorithm.

The breaking is based on Theorem 3. Hereto we first have to define ***short enough vector***.

Definition 2 *A vector* $\mathbf{c}$ *in a lattice* L *is called short enough related to* a_1 *if*

$$\sum_{i=2}^{n} |c'_i| < a_1 \quad ,$$

where $c_1' = 0$ *and* $c_i' = c_i/n$ *for* $2 \leq i \leq n$.

Theorem 3 *If all vectors in the reduced basis for the lattice, with basis vectors* $\mathbf{t}^i$ *defined in Eq. 18, are short enough related to* a_1*, then we can find* $n-1$ *independent SSMM for* $a_1, \ldots, a_n$.

$$\begin{array}{lcl} \mathbf{t}^1 & = & (\begin{array}{ccccccc} 1 & na_2 & na_3 & na_4 & \ldots & na_n \end{array}) \\ \mathbf{t}^2 & = & (\begin{array}{ccccccc} 0 & na_1 & 0 & 0 & \ldots & 0 \end{array}) \\ \mathbf{t}^3 & = & (\begin{array}{ccccccc} 0 & 0 & na_1 & 0 & \ldots & 0 \end{array}) \\ \mathbf{t}^4 & = & (\begin{array}{ccccccc} 0 & 0 & 0 & na_1 & \ldots & 0 \end{array}) \\ \vdots & & \begin{array}{ccccccc} \vdots & \vdots & \vdots & \vdots & \ddots & \vdots \end{array} \\ \mathbf{t}^n & = & (\begin{array}{ccccccc} 0 & 0 & 0 & 0 & \ldots & na_1 \end{array}) \end{array} \tag{18}$$

Proof: Call the vectors of the reduced basis $\mathbf{v}^1, \mathbf{v}^2, \ldots, \mathbf{v}^n$. We will first prove that a modular mapping by $v_1^j \bmod a_1$ has the small sum property (see Section 3.2). Since $\mathbf{v}^j$ is an integral linear combination of the vectors in Eq. 18, there exist integers $(y_1^j, \ldots, y_n^j)$ such that $v_1^j = y_1^j$ and $v_i^j = y_i^j na_1 + y_1^j na_i$ for $2 \leq i \leq n$. Since n divides v_i^j let $u_i^j = v_i^j/n$ for $2 \leq i \leq n$. This implies evidently that $0 \equiv a_1 y_1^j$ and $u_i^j \equiv a_i y_1^j$ for $2 \leq i \leq n$. As a consequence of the short enough property we have indeed the small sum property. The independence of the $n-1$ vectors so obtained with SSMM, is then easy to prove.

Arguments are given in [7] that the condition in Theorem 3 are almost satisfied if the density is low enough.

5 CONCLUSION

Let us conclude from a point of view of ***limits in security performances*** of cryptography. While the enciphering in the Merkle–Hellman knapsack is based on NP–completeness, its trapdoor was not and opens the door for attacks. In secure public key cryptosystems the enciphering process must be hard to invert but it must also be hard to find the original trapdoor *or another trapdoor*. So the security *performance* of the trapdoor knapsack schemes is so *limited* that they are (presently) useless!

Remark finally that in VLSI and in communication NP-completeness causes a lot of troubles and *limits performances*. Cryptography now just tries to use these ***limits in performances*** to be used as ***limits in performances*** of cryptanalysis. One may however not forget another limit: the theory of NP–completeness is based on ***unproven*** assumptions.

Acknowledgements

The author wants first to thank E. Brickell, from Bell Communication Research for the received information on several results on knapsack systems. He also is grateful to J.Vandewalle from the Kath. Univ. Leuven for suggestions by reading this text and to J.–J. Quisquater from Philips Research Brussels for discussions about the subject.

References

1. L. M. Adleman, "On Breaking the Iterated Merkle–Hellman Public–Key Cryptosystem," *Advances in Cryptology, Proc. Crypto 82*, Santa Barbara, California, U. S. A,

August 23 – 25, 1982, Plenum Press, New York, 1983, pp. 303 – 308, more details appeared in "On Breaking Generalized Knapsack Public Key Cryptosystems," TR–83–207, Computer Science Dept., University of Southern California, Los Angeles, U. S. A., March 1983.

2. B. Arazi, "A Trapdoor Multiple Mapping," *IEEE Trans. Inform. Theory*, vol. 26, no. 1, pp. 100 – 102, January 1980.

3. Birkhoff and MacLane, *"A Survey of Modern Algebra,"* MacMillan Company, 1965.

4. I. F. Blake, "Complexity Issues for Public Key Cryptography," *Proc. of this Nato Advanced Study Institute.*

5. E. F. Brickell, J. A. Davis, and G. J. Simmons, "A Preliminary Report on the Cryptanalysis of the Merkle–Hellman Knapsack Cryptosystems", *Advances in Cryptology, Proc. Crypto 82*, Santa Barbara, California, U. S. A, August 23 – 25, 1982, Plenum Press, New York, 1983, pp. 289 – 301.

6. E. F. Brickell, "Solving Low Density Knapsacks in Polynomial Time," *IEEE Intern. Symp. Inform. Theory*, St. Jovite, Quebec, Canada, September 26 – 30, 1983, Abstract of papers, pp. 129 – 130.

7. E. F. Brickell, "Solving low density knapsacks," *Advances in Cryptology, Proc. Crypto 83*, Santa Barbara, California, U. S. A, August 21 – 24, 1983, Plenum Press, New York, 1984, pp. 25 – 37.

8. E. F. Brickell, J. C. Lagarias and A. M. Odlyzko, "Evaluation of the Adleman Attack on Multiple Iterated Knapsack Cryptosystems," *Advances in Cryptology, Proc. Crypto 83*, Santa Barbara, California, U. S. A, August 21 – 24, 1983, Plenum Press, New York, 1984, pp. 39 – 42.

9. E. F. Brickell, "A New Knapsack Based Cryptosystem," presented at Crypto 83, Santa Barbara, California, U. S. A, August 21 – 24, 1983.

10. E. F. Brickell, "Breaking Iterated Knapsacks," *Advances in Cryptology, Proc. Crypto 84*, Santa Barbara, August 19 – 22, 1984, Lecture Notes in Computer Science, vol. 196, Springer–Verlag, Berlin, 1985, pp. 342 – 358.

11. E. F. Brickell, "Attacks on Generalized Knapsack Schemes," presented at Eurocrypt 85, Linz, Austria, April 9 – 11, 1985.

12. J. W. S. Cassels, *"An Introduction to the Geometry of Numbers,"* Springer–Verlag, Berlin, New York, 1971.

13. B. Chor and R. L. Rivest, "A Knapsack Type Public Key Cryptosystem Based on Arithmetic in Finite Fields," *Advances in Cryptology, Proc. Crypto 84*, Santa Barbara, August 19 – 22, 1984, Lecture Notes in Computer Science, vol. 196, Springer–Verlag, Berlin, 1985, pp. 54 – 65.

14. R. H. Cooper and W. Patterson, "Eliminating Data Expansion in the Chor–Rivest Algorithm," presented at Eurocrypt 85, Linz, Austria, April 9 – 11, 1985.

15. M. Davio, "Knapsack trapdoor functions: an introduction", Proceedings of *CISM Summer School on: Secure Digital Communications*, CISM Udine, Italy, June 7 – 11 1982, ed. J. P. Longo, Springer Verlag, 1983, pp. 41 – 51.

16. Y. Desmedt, J. Vandewalle and R. Govaerts, "The Use of Knapsacks in Cryptography public key systems (Critical Analysis of the Security of Knapsack Public Key Algorithms)," presented at: Groupe de Contact Recherche Operationelle du F. N. R. S., Mons, Belgium, February 26, 1982, appeared in *Fonds National de la Rechereche Scientifique, Groupes de Contact, Sciences Mathématiques*, 1982.

17. Y. G. Desmedt, J. P. Vandewalle and R. J. M. Govaerts, "A Critical Analysis of the Security of Knapsack Public Key Algorithms," *IEEE Trans. Inform. Theory*, vol. IT-30, no. 4, July 1984, pp. 601 – 611, also presented at *IEEE Intern. Symp. Inform. Theory*, Les Arcs, France, June 1982, Abstract of papers, pp. 115 – 116.

18. Y. Desmedt, J. Vandewalle and R. Govaerts, "How Iterative Transformations can help to crack the Merkle–Hellman Cryptographic Scheme," *Electronics Letters*, vol. 18, 14 October 1982, pp. 910 – 911.

19. Y. Desmedt, J. Vandewalle and R. Govaerts, "A Highly Secure Cryptographic Algorithm for High Speed Transmission," *Globecom '82*, IEEE, Miami, Florida, U. S. A. , 29 November–2 December 1982, pp. 180 – 184.

20. Y. Desmedt, J. Vandewalle and R. Govaerts, "Linear Algebra and Extended Mappings Generalise Public Key Cryptographic Knapsack Algorithms," *Electronics Letters*, 12 May 1983, vol. 19, no. 10, pp. 379 – 381.

21. Y. Desmedt, J. Vandewalle and R. Govaerts, "A General Public Key cryptographic Knapsack Algorithm based on linear Algebra," *IEEE Intern. Symp. Inform. Theory*, St. Jovite, Quebec, Canada, September 26 – 30, 1983, Abstract of papers, pp. 129 – 130.

22. Y. Desmedt, J. Vandewalle and R. Govaerts, "Can Public Key Cryptography provide Fast, Practical and Secure Schemes against Eavesdropping and Fraud in Modern Communincation Networks?," *Proc. 4th World Telecommunication Forum 83*, Geneva, Switzerland, October 29 – November 1, 1983, Part 2, Vol. 1, pp. 1. 2. 6. 1 – 1. 2. 6. 7.

23. Y. Desmedt, J. Vandewalle and R. Govaerts, "Fast Authentication using Public Key Schemes," *Proc. International Zurich Seminar on Digital Communications 1984*, IEEE Catalog No. 84CH1998–4, March 6 – 8 1984, pp. 191 – 197, Zurich, Switzerland.

24. Y. Desmedt, "*Analysis of the Security and New Algorithms for Modern Industrial Cryptography*," Doctoral Dissertation, Katholieke Universiteit Leuven, Belgium, October 1984.

25. W. Diffie and M. E. Hellman, "Exhaustive cryptanalysis of the NBS Data Encryption Standard," *Computer*, vol. 10, no. 6, pp. 74 – 84, June 1977.

26. W. Diffie and M. E. Hellman, "Privacy and Authentication: An Introduction to Cryptography," *Proc. IEEE*, vol. 67, pp. 397 – 427, March 1979.

27. A. Di Porto, "A Public Key Cryptosystem Based on a Generalization of the Knapsack Problem," presented at Eurocrypt 85, Linz, Austria, April 9 – 11, 1985.

28. M. R. Garey and D. S. Johnson, "*Computers and Intractability: A Guide to the Theory of NP - Completeness*," W. H. Freeman and Company, San Francisco, 1979.

29. P. Goetschalckx and L. Hoogsteijns, "*Constructie van veilige publieke sleutels voor het veralgemeend knapzak geheimschriftvormend algoritme: Theoretische studie en voorbereidingen tot een computerprogramma,*" (Construction of Secure Public Keys for the General Cryptographic Knapsack Algorithm: Theoretical Study and Preparations for a Computerprogram, in Dutch), final work, Kath. Univ. Leuven, May 1984.

30. R. M. Goodman and A. J. McAuley, "A New Trapdoor Knapsack Public Key Cryptosystem," *Advances in Cryptology. Proc. Eurocrypt 84*, Paris, France, April 9 – 11, 1984, Lecture Notes in Computer Science, vol. 209, Springer–Verlag, Berlin, 1985, pp. 150 – 158.

31. M. E. Hellman, "A cryptanalytic time–memory trade–off," *IEEE Trans. Inform. Theory*, vol. IT–26, no. 4, July 1980, pp. 401 – 406.

32. P. S. Henry, "Fast Decryption Algorithm for the Knapsack Cryptographic System," *Bell Syst. Tech. Journ.*, vol. 60, no. 5, May – June 1981, pp. 767 – 773

33. T. Herlestam, "Critical Remarks on Some Public Key Cryptosystems," *BIT*, vol. 18, 1978, pp. 493 – 496.

34. I. Ingemarsson, "Knapsacks which are Not Partly Solvable after Multiplication modulo q," *IBM Research Report* TC 8515, 10/10/80, Thomas J. Watson Research Center, see also *IEEE International Symposium on Information Theory, Abstract of papers*, Santa Monica, California, 9–12 February 1981, pp. 45.

35. I. Ingemarsson, "A New Algorithm for the Solution of the Knapsack Problem," *IEEE Intern. Symp. Inform. Theory*, Les Arcs, France, June 1982, Abstract of papers, pp. 113 – 114.

36. R. Kannan, "Improved Algorithms for Integer Programming and Related Lattice Problems," *Proc.* 15th *Annual ACM Symposium on theory of Computing*, 1983, pp. 193 – 206.

37. E. D. Karnin and M. E. Hellman, "The largest Super–Increasing Subset of a Random Set," *IEEE Trans. Inform. Theory*, vol. IT-29, no. 1, January 1983, pp. 146 – 148, also presented at *IEEE Intern. Symp. Inform. Theory*, Les Arcs, France, June 1982, Abstract of papers, pp. 113.

38. E. D. Karnin, "A Parallel Algorithm for the Knapsack Problem," *IEEE Trans. on Computers*, vol. C-33, no. 5, May 1984, pp. 404 – 408, also presented at *IEEE Intern. Symp. Inform. Theory*, St. Jovite, Quebec, Canada, September 26 – 30, 1983, Abstract of papers, pp. 130 – 131.

39. J. C. Lagarias, "The Computational Complexity of Simultaneous Diophantine Approximation Problems," *Proc. Symp. on Foundations of Computer Science*, November 1982, pp. 32 – 39.

40. J. C. Lagarias and A. M. Odlyzko, "Solving Low Density Subset Sum Problems," *Proc.* 24th *Annual IEEE Symposium on Foundations of Computer Science*, 1983, pp. 1 – 10.

41. J. C. Lagarias, "Knapsack Public Key Cryptosystems and Diophantine Approximation," *Advances in Cryptology, Proc. Crypto 83*, Santa Barbara, California, U. S. A, August 21 – 24, 1983, Plenum Press, New York, 1984, pp. 3 – 23.

42. J. C. Lagarias, "Performance Analysis of Shamir's Attack on the Basic Merkle–Hellman Knapsack Cryptosystem," *Proc. 11th Intern. Colloquium on Automata, Languages and Programming (ICALP)*, Antwerp, Belgium, July 16 – 20, 1984, Lecture Notes in Computer Science, vol. 172, Springer Verlag, Berlin, 1984.

43. H. Lagger, "*Public Key Algorithm based on Knapsack Systems* (in German)," dissertation, Technical University Vienna, Austria.

44. H. W. Lenstra, Jr., "Integer Programming with a Fixed Number of Variables", University of Amsterdam, Dept. of Mathematics, Technical Report, 81–03, April, 1981.

45. A. K. Lenstra, H. W. Lenstra, Jr., and L. Lovasz, "Factoring Polynomials with Rational Coefficients," *Mathematische Annalen 261*, pp. 515 – 534, 1982.

46. H. W. Lenstra, Jr., "Integer Programming with a Fixed Number of Variables", *Math. of Operations Research*, Vol. 8, No. 4, November 1983, pp. 538 – 548.

47. A. J. McAuley and R. M. Goodman, "Modifications to the Trapdoor–Knapsack Public Key Cryptosystem," *IEEE Intern. Symp. Inform. Theory*, St. Jovite, Quebec, Canada, September 26 – 30, 1983, Abstract of papers, pp. 130.

48. R. C. Merkle and M. E. Hellman, "Hiding Information and Signatures in Trapdoor Knapsacks," *IEEE Trans. Inform. Theory*, vol. 24, no. 5, pp. 525 – 530, September 1978.

49. National Bureau of Standards (NBS), "Data Encryption Standard," FIPS publication 46, *Federal Information Processing Standards Publ.*, U. S. Department of Commerce, Washington D. C. , U. S. A. , January 1977.

50. National Bureau of Standards (NBS), "DES Modes of Operation," FIPS publication 81, *Federal Information Processing Standard*, U. S. Department of Commerce, Washington D. C. , U. S. A. , 1980.

51. A. M. Odlyzko, "Cryptanalytic Attacks on the Multiplicative Knapsack Cryptosystem and on Shamir's Fast Signature System," *IEEE Trans. Inform. Theory*, vol. IT-30, no. 4, July 1984, pp. 594 – 601, also presented at *IEEE Intern. Symp. Inform. Theory*, St. Jovite, Quebec, Canada, September 26 – 30, 1983, Abstract of papers, pp. 129.

52. A. M. Odlyzko, "Discrete Logarithms in Finite Fields and their Cryptographic Significance," *Advances in Cryptology, Proc. Eurocrypt 84*, Paris, France, April 9 – 11, 1984, Lecture Notes in Computer Science, vol. 209, Springer–Verlag, Berlin, 1985, pp. 225 – 314.

53. A. M. Odlyzko, personal communication.

54. M. Petit, "*Etude mathématique de certains systèmes de chiffrement: les sacs à dos*," (Mathematical study of some enciphering systems: the knapsack, in French), doctor's thesis, Université de Rennes, France.

55. S. C. Pohlig and M. E. Hellman, "An Improved Algorithm for Computing Logarithms over GF(p) and its Cryptographic Significance," *IEEE Trans. Inform. Theory*, vol. 24, no. 1, pp. 106 – 110, January 1978.

56. F. C. Piper, "Recent Developments in Cryptography," *Proc. of this Nato Advanced Study Institute.*

57. R. L. Rivest, A. Shamir and L. Adleman, "A Method for Obtaining Digital Signatures and Public Key Cryptosystems," *Commun. ACM*, vol. 21, pp. 294 – 299, April 1978.

58. I. Schaumuller–Bichl, "On the Design and Analysis of New Cipher Systems Related to the DES," *IEEE Intern. Symp. Inform. Theory 1982*, Les Arcs, France, pp. 115.

59. C. P. Schnorr, "A More Efficient Algorithm for a Lattice Basis Reduction," October 1985, preprint.

60. P. Schöbi, and J. L. Massey, "Fast Authentication in a Trapdoor Knapsack Public Key Cryptosystem," *Cryptography*, Proc. Burg Feuerstein 1982, Lecture Notes in Computer Science, vol. 149, Springer–Verlag, Berlin, 1983, pp. 289 – 306, see also *Proc. Int. Symp. Inform. Theory*, Les Arcs, June 1982, pp. 116.

61. A. Shamir, "A Fast Signature Scheme," *Internal Report*, MIT, Laboratory for Computer Science Report RM – 107, Cambridge, Mass. , July 1978.

62. A. Shamir, "On the Cryptocomplexity of Knapsack Systems," *Proc. Stoc 11 ACM*, pp. 118–129, 1979.

63. A. Shamir and R. Zippel, "On the Security of the Merkle–Hellman Cryptographic Scheme," *IEEE Trans. Inform. Theory*, vol. 26, no. 3, pp. 339 – 340, May 1980.

64. A. Shamir, "A Polynomial Time Algorithm for Breaking the Basic Merkle–Hellman Cryptosystem," *Advances in Cryptology, Proc. Crypto 82*, Santa Barbara, California, U. S. A, August 23 – 25, 1982, Plenum Press, New York, 1983, pp. 279 – 288.

65. A. Shamir, "The strongest knapsack–based cryptosystem," presented at CRYPTO'82, Santa Barbara, California, U. S. A, August 23 – 25, 1982.

66. A. Shamir, "A Polynomial Time Algorithm for Breaking the Basic Merkle–Hellman Cryptosystem," *IEEE Trans. Inform. Theory*, vol. IT-30, no. 5, September 1984, pp. 699 – 704.

67. A. Shamir and Y. Tulpan, paper in preparation.

68. A. Shamir, "Unforgeable passports," presented at Workshop: Algorithms, Randomness and Complexity, CIRM, Marseille, France, March 23 – 28, 1986.

69. A. Shamir, personal communication.

70. M. Willett, "Trapdoor knapsacks without superincreasing structure," *Inform. Process. Letters*, vol. 17, pp. 7 – 11, July 1983.

OPTICAL LOGIC FOR COMPUTERS

Robert W. Keyes
IBM T.J. Watson Research Center
P.O. Box 218, Yorktown Heights, NY 10598

1. Introduction

The year 1960 marked the advent of the integrated circuit and solidification of the conviction that silicon microelectronics contained enormous potential. Transistors rapidly became the dominant device in the logic circuitry of computers. Large research and development efforts devoted to miniaturization and increasing integration of transistors were launched and met with great success, making a procession of ever larger and faster machines available. Within a decade, silicon transistors had evolved to dominate memory technology and extended computing to a new regime of cost and availability in the form of the microprocessor chip.

1960 was also the year of the invention of the laser, which made great advances in optical science and technology possible. Highly coherent and very intense light sources suddenly became available. Metrology, laser ranging, frequency multiplication, and optical information storage were quickly demonstrated, for example. The development of the semiconductor laser and, subsequently, the continuously operating, room temperature, semiconductor laser, made the advantages of the laser available in a small, low power form and greatly expanded the scope of applications of lasers. The semiconductor laser also stimulated developments in junction luminescence and led to efficient low cost light-emitting diodes. These advances, in particular, have led to an important role for optical devices and techniques in information processing hardware, in such aspects as communication, storage, displays, printers, and input devices.

2. Information processing

Coherent light also allowed a new form of optical information processing. Complicated operations could be performed on images, such as transformation into a spatial frequency domain and holography. These techniques are useful and have become known as

J. K. Skwirzynski (ed.), Performance Limits in Communication Theory and Practice, 135–141.

"Optical Computing"[1]. The systems that perform this kind of computing are quite distinct from the general purpose computers that are familiar in business and industry.

The general purpose computer is capable of performing very long series of operations, such as iterative solutions of equations and simulations of the evolution of complex physical systems with time. The course of the calculation is controlled by logical decisions based on results already obtained; for example, a calculation may be terminated when a certain accuracy is attained. These functions are carried out by an assembly of logic gates that accept information represented as binary digits and produce a binary digit as output. Binary representation of information is the method of choice because it is easy to establish two standard values to which a digit can be set at each step in a calculation. The deterioration of the representation of information as a physical quantity in the course of hundreds or thousands of operations is thereby prevented. Even if the representation of a digit is not perfect when received at a logic gate it can be restored by reference to the standard.

3. Logic with Transistors

Binary digital reference values are established by the power supply and ground voltages distributed throughout the system in electrical computation. The FET NOR shown in Fig. 1 illustrates the way in which this can be accomplished. The transistors that receive the inputs on their gates are of enhancement type, that is, they are non-conductive when their gates are connected to ground. A positive input voltage turns them on, thereby establishing a connection between the source and drain. The load transistor is of depletion type, it is on when its gate is connected to its source. It acts as a non-linear resistor. When all inputs V_i are zero (ground potential), the output, V_o, is connected to the power supply through the load transistor and is close to V_B. If at least one of the inputs is positive it connects the output to ground potential through the active FET and V_o is nearly zero. The operation of the circuit is shown in the form of a load line on the FET characteristics in Fig. 1. Fig. 1 also presents the result as a curve of output as a function of input.

The high gain of the input-output characteristic makes the standardization of the output possible. High gain means that a small change in the input near the threshold at which the FET becomes conductive effects a large change in the output. The change in input voltage needed to cause the output to change from the high output state to the low output state is approximately the current through the circuit divided by the sum of the load conductance and the drain conductance of the FET. The signal amplitudes used in digital processing are substantially larger than this minimal value; the excess signal swings above and below the transition region are called noise margins. These noise margins allow standardization of the output values over a range of inputs; even if a signal has been degraded by attenuation or noise it still produces the standard output value. Further, the threshold can vary, as shown by the dashed line, without affecting the operability of the circuit. Therefore, the necessity for high precision in the fabrication of the devices is relieved. The low cost that is necessary to allow many thousands, even millions, of logic gates to be assembled into systems can be attained.

4. Bistability

Bistability is a common phenomenon in physical systems. A bistable response to an input signal is illustrated in Fig. 2. The meaning of the curves is that, starting from zero and increasing the input, the output switches to the high state at *Y*; on decreasing the input again it returns to the low state at *Z*. When a characteristic of this type is observed in a physical system, researchers, hoping to find some application for their discoveries, note that the two states can represent binary information: one branch of the curve is a one, the other is a zero. It is not a large further step to think of using a system with such a characteristic to perform logic operations.

However, most kinds of gates cannot be formed from the bistable response function. Only ANDs and ORs can be implemented by a method known as threshold logic. Simply stated, the idea is that M ONE inputs to a logic stage are summed and if the sum is equal to or greater than some value N the stage switches. M and N are small integers. If $N = 1$ an OR is created, if $N = M$ the function is an AND. In other words, if the bias is such that a single input causes the device to switch, then it acts as a logical OR. If all of the inputs are needed to cause switching, then an AND function is formed.

The method is illustrated in Fig. 2. A constant bias signal S maintains the system in the bistable range in the absence of inputs. Initially the system is on the low branch of the characteristic. Inputs are added to the bias; when the sum of the bias and the inputs exceed the threshold Y the device switches to the high branch. Then the high output is retained even after the removal of the inputs. Obvious difficulties in using this method are that inputs cannot cause switching in the opposite direction, from the high output branch to the low output branch (the operation of negation or NOT is not implemented), and that the bias must be removed to restore the system to the low state to prepare it for the next operation.

Although working electrical circuits have been built with threshold logic and used for computation, the method places great demands on the accuracy of components and signal levels and is not suitable for integrated circuitry. To illustrate the point, consider that a three-input AND is to be implemented by adding three inputs with a nominal value of 1 in appropriate units. The threshold for switching is then arranged to be 2.5, so that two nominal inputs do no excite a response but three do. The inputs I must satisfy

$$2I < 2.5, \ 3I > 2.5 \tag{1}$$

to insure that two inputs do not cause switching and that three inputs do. That is, $0.83 < I < 1.25$. There is little room for the noise margins that are needed for reliable operation with signals that have been transmitted from place to place in a large system.

Poor tolerance of component variability is another difficulty of threshold logic. If, for example, the threshold varies by ±10%, then, instead of (1),

$$2I < 2.25, \ 3I > 2.75 \tag{2}$$

In no case must two inputs cause switching, while three inputs must do so in every case. Thus $0.92 < I < 1.12$. The ability to tolerate signal distortion is greatly diminished. If the threshold may vary by ±20%, i.e., lie between 2 and 3, the interval in which the signal must lie vanishes! The use of a bias to maintain the device in the bistable region, S in Fig. 2, exacerbates the demands upon the precision of signal amplitudes and device parameters.

5. Optical logic

Optical bistability arises from non-linear effects that cause the index of refraction or the absorption constant of a substance to depend on light intensity. Bistability can be produced by providing feedback through the use of an interferometer, an optical cavity with reflecting end faces. The transmitted intensity depends on the history of the incident intensity and not just on its current value. In other words, the optical device can be in either of two states, as in Fig. 2, depending on its previous exposure to light.

Optical logic is the attempt to perform the threshold logic described in the preceding section with optically bistable devices. In addition to the basic problems of threshold logic, discussed in the preceding section, several aspects of the physics of optically bistable devices are not favorable to their use as the logic elements of a general purpose computer. The number of wavelengths in the cavity is a critical parameter, and careful tuning is necessary to demonstrate switching action [2,3]. For the same reason, precision manufacturing of the devices would be needed. The critical tuning requires precise temperature control because of the temperature dependence of energy levels in solids. Thus, although it is possible to demonstrate transitions in single devices, the assembly of such devices into systems of many thousands would encounter difficulties. The devices dissipate power which must be carried away by the driving force of temperature gradients, making variability of temperature inevitable. And the high cost of insuring very accurately controlled components would limit the availability of any such assembly of devices.

The size of optical devices is limited by diffraction effects. Thus the attainment of the high packing densities needed to achieve short signal transit times through a large system seems unlikely.

Efficient transmission of signals from one device to another requires that their optical cavities by closely coupled. Modes involving reflections in more than one cavity will exist, as will some mode-mode coupling by the optically nonlinear media. A change in the index of refraction of a receiving cavity will be reflected in the sending cavity. Isolation of inputs from outputs will be poor, in other words. Extra componentry can be introduced to improve isolation [2, 4], again at the expense of system size and cost.

In spite of these limitations and a long and not very encouraging history [e.g., 5-8], interest in logic based on optical and electro-optical bistability persists.

6. What makes a good computing device?

The great power of modern information processing machines derives from the ability of silicon microelectronics to provide low cost, low power, very reliable switching devices. Low cost is made possible by mass fabrication methods, the large-scale integrated circuit. The same methods have also proved to be the key to high reliability. One of the key ingredients of integration is miniaturization, reducing the dimensions of all components: devices, insulators, wires, and connectors. Miniaturization leads to low power because all capacitances are reduced and less change must be drawn from the power supply to charge and discharge them. Miniaturization and the reduction of capacitance have also led to high speed of operation.

Large electrical computers have been constructed from three kinds of devices, electromagnetic relays, vacuum tubes, and transistors. These devices have common qualities that have made it possible to use them in large systems. These same qualities are are not found in various devices which have been the focus of large but unsuccessful development efforts during the last quarter-century, notably tunnel diodes and Josephson cryotrons.

Three terminals allow separate terminals for input and output, and permit the devices to be used in a wide variety of circuit configurations. In contrast, two terminal bistable devices can be used to perform logic operations only in the configuration of Fig. 2 and are thereby limited to using threshold logic to perform AND and OR operations.

Three terminals also allow good input-output isolation; the state of the output is not reflected at the input terminal. The switching of successful logic components is determined by their inputs and not affected by the state of the following devices.

With high gain only part of the signal amplitude is needed to effect the change of state of the logic gate, much of it is available as the noise margins throughout which the output has one of the standard values that represent the binary digits. Circuits can provide both current gain and voltage gain, and both are needed to insure signal standardization and permit fan-out. Other proposed devices do not share the high gain.

Logic circuits built from relays, vacuum tubes, and transistors switch in each direction in comparable amounts of time. It is ordinarily easy to switch bistable devices in one direction, but difficult and time consuming to switch them in the other direction.

Certain properties seem too obvious to deserve comment if it were not for the fact that they are sometimes overlooked or dismissed as unimportant in proposals for logic technologies: Outputs are suitable as inputs to another logic stage. And materials from which good relays and vacuum tubes and transistors can be fabricated exist.

The circuit flexibility of transistors is enhanced by the availability of two polarities, that is, *npn* and *pnp* in the case of the bipolar transistor and *p* channel and *n* channel in the FET. Thus CMOS circuitry is possible for example. The possibility of adjusting the threshold voltages of FETs to create both normally on and normally off devices lends still more versatility to FETs in circuits. It is difficult to conceive of another device with equally favorable properties.

1. Society of Photo-Optical Instrumentation Engineers, 10th International Optical Computing Conference, April 6-8, 1983, SPIE Volume 422.
2. M. Dagenais and W.F. Sharfin, Optical Engineering **25**, 219-224 (1986).
3. A. Migus, et. al., Appl. Phys. Letters **46**, 70-72 (1985).
4. S.S. Tarng, et al., Appl. Phys. Letters **44**, 360-61 (1984).
5. M. Nathan, et. al., J. Appl. Physics **36**, 473-480 (1965).
6. G.J. Lasher and A.B. Fowler, IBM J. Res. Devel. **8**, 471-76 (1964).
7. W.F. Kosonocky, IEEE Spectrum **2** (3), 183-195 (1965).
8. R. Landauer in "Optical Information Processing", eds. Nesterkin, Stroke, and Kock, New York: Plenum, 1976, pp. 219-254.

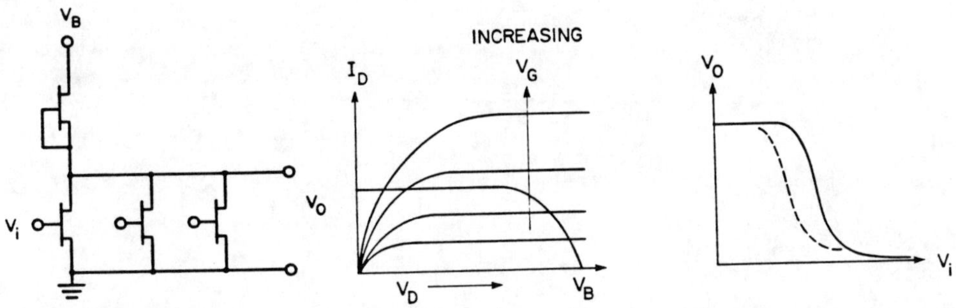

Fig. 1 (a) A NOR circuit constructed from field-effect transistors. (b) The operation of the NOR as a load line on the FET characteristics. (c) The output voltage of the NOR as a function of an input.

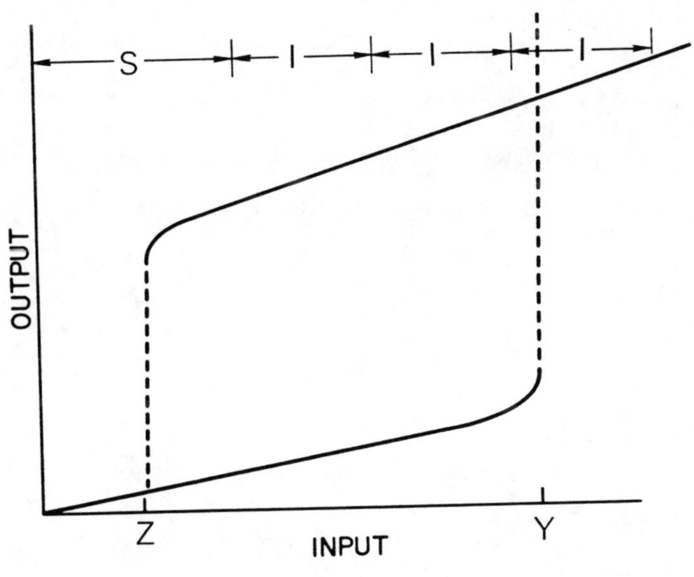

Fig. 2. The dependence of output on input in a bistable device.

LIMITATIONS OF QUEUEING MODELS IN COMMUNICATION NETWORKS

Anthony Ephremides

University of Maryland
College Park, MD 20742 USA

ABSTRACT

The use of queueing models in analytical studies of store-and-forward networks is natural and often fruitful. The physical nature of communication networks, however, imposes restrictions on the models that quickly push them to the limits of their usefulness. Deviations from the traditional approaches of queueing theory can sometimes provide solutions to otherwise intractable problems. The use of stochastic techniques that concentrate on sample function properties and dominance concepts is illustrated here by focusing on two basic problems of interest in communication networks: 1) an optimization problem in capacity allocation that arises from flow control considerations, and 2) an analysis problem of delay and stability in the collision channel.

1. INTRODUCTION

In a communication network it is natural to model each link as a queueing system. It is also consistent with physical considerations to ignore processing and propagation delays in most cases. Thus the cause of queueing delays is essentially the finite transmission capacity of the link (the "speed" of the line) and the length of the message unit. Thus if a link has capacity of C bits/s and the message length is a random variable of mean value $1/\mu$ bits, the service time for each message is random with mean $1/\mu C$ secs. The messages to be transmitted over the link arrive at one end of it according to a random process with average rate λ messages/sec. Thus we have all the makings of a queueing system for the link model.

Complications arise as soon as links are interconnected. A key assumption in queueing theory is that the system under consideration be <u>standard</u>, namely that it satisfy the following two properties:

1) arrival and service processes are statistically independent
2) successive service times are also statistically independent.

Only under these fundamental assumptions are the traditional methods of queueing theory capable of analyzing queueing systems. Even then the traditional methodology, that relies heavily on "brute-force" probability distribution manipulations, leads to extremely complicated formulations and difficult solutions. A ray of hope is allowed to shine when it is realized that often there are simple fundamental properties of wide applicability in queueing systems that can be easily established, leading, thus, one to suspect that perhaps the complexities of queueing problems are only a facade, responsible for which are some unimportant technicalities in the arrival and departure processes, and that, fundamentally, queueing problems may not be that complex.

In networks of interconnected links these assumptions are violated. Successive messages that are transmitted over a link constitute part of the arrival process at the next link. It is clear, however, that two suc-

J. K. Skwirzynski (ed.), Performance Limits in Communication Theory and Practice, 143–153.

cessive messages will have arrival instants at the next link separated by exactly the transmission time of the first message over the first link. That transmission time is a function of the message length which also determines the transmission (service) time over the next link. Thus, arrivals and service at the next link are correlated. This correlation between arrivals and service causes also statistical dependence between successive service times.

Another instance in which the basic assumptions of a standard queueing system are violated is the case of sharing resources. In communication networks buffer-space at a node is shared typically between several message streams traveling through that node. In radio networks the shared resource can be the transmission medium itself for the use of which different terminals contend. In both these cases the queueing delays in one traffic stream (or at one radio terminal) depend on the service and arrival processes of the other streams (or other radio terminals). This dependence translates into statistical correlation between successive service times of <u>each</u> stream (or terminal) and, consequently, between its arrival and service processes.

The difficulties in handling networks of interconnected links were soon realized to be almost insurmountable. This realization led to the two most famous assumptions that, in the model of "poetic license", were accepted by the community because they served the purpose of penetrating the deceptively complicated armor of queueing problem analysis. They are nevertheless wrong or, more precisely, physically unjustified.

The first is <u>Kleinrock's independent assumption</u> which, in effect, states that, despite the obvious dependence between successive links as explained above, the queueing models for different links will be considered independent queueing systems. The theory of Jacksonian networks becomes, then, applicable and useful insights and results can be obtained for several network operational issues such as routing, flow control, and capacity assignment.

The second is <u>Abramson's Poisson assumption</u> for the channel traffic in the ALOHA version of the collision channels. This one suppresses the obvious fact that retransmissions of messages occur only in response to previously attempted unsuccessful first transmissions and treats the entire process as one of independent increments. Again, one is able then to perform elementary, albeit erroneous, analysis that happens to predict physical system behavior reasonably well under a proper interpretation.

Still, the work of the last decade has extracted every ounce of usefulness from these two assumptions without being able to resolve several, rather basic, problems of interest. Therefore, there is a need for new way to approach these problems in order to ascertain whether the difficulties we are encountering are truly fundamental or, as there is evidence to suspect, only a consequence of the ways we've been looking at these problems.

In recent years there has been a cautious shift in methodology towards probabilistic or stochastic techniques, that is techniques that rely on "path" or "sample function" arguments rather than on distributions and moments. This shift has yielded some promising results that are cause for optimism.

In this paper we review two problems that are rather simple in their statement and quite basic for their significance and implications in communication networks. In both cases we rely on non-traditional solution approaches.

2. A PROBLEM IN CAPACITY ALLOCATION

There are stituations in multi-processor systems as well as in virtual circuit-switched communication networks in which a certain amount of processing or transmission power is made available to service a stream of customers (jobs or messages, as the case may be). As congestion builds up, because of a period of slow service or accelerated arrivals, it may become necessary to provide additional service resources (more processors or more virtual circuits).

The simplest version of this problem (which has proven highly non-trivial to solve) can be modeled as follows: Consider a single queue with an arbitrary (for the moment) arrival process and two servers whose service times are (for the moment) arbitrary independent random variables but such that one (say, the first one) is, on the average, faster than the other. It is of interest to keep the total weighting time as small as possible. Clearly the first customer should be assigned to the faster server. If a second customer arrives before the first one exits, the question arises whether it is better to send that customer to the (idle) slower server, or keep him waiting in order to eventually send him to the faster one as soon as the latter becomes available. In other words, the problem is when to "activate" the extra "capacity" represented by the slower server.

This problem, unlike many that have been studied in queueing theory or in communication network modeling, is not one of analysis, but of control. An interesting body of literature has been developing over the last decade on queueing-theoretic control problems. Again, in these problems breakthroughs started appearing when sample-path-based approaches were applied to them as opposed to standard minimum principle or dynamic programming optimization methods. The most widely studied problem of queue-control is the problem of elementary routing or flow control. A good survey of work on this problem can be found in [1]. The problem we propose here is in some ways similar to the one studied in [1], but is quite different in formulation. Now, something that a novice to the field must learn to contend with fast, is the fact that a small perturbation in the assumptions of the problem may seriously affect the validity of its solution or even the usefulness of its solution approach.

The problem of capacity activation was first studied as a multi-processor system by R. Larsen [2] in his thesis. Under the assumption of Poisson arrivals and exponentiality of service times he <u>conjectured</u> that the optimal solution is of the threshold type, namely that the slower server should be activated only when the queue-size exceeds a certain value that depends on the parameters of the problem (arrival rate λ and service rates μ_1 and μ_2). This conjecture was proven by W. Lin and P.R. Kumar [3]. J. Walrand in [4] gave a simpler proof of the conjecture by using probabilistic rather than dynamic programming methods. Finally, in [5], the result was extended for certain types of non-Poisson arrivals and non-exponential service times. It is interesting to note that with rather elementary reasoning the open loop version of the problem (that is, for th case in which the queue size is not observable) results in a threshold-type policy also, but on the value of the arrival rate λ (see [6]).

Let us, briefly, consider the details of the problem. The state of the system is represented by $\bar{x} = (x_0, x_1, x_2)$, where x_0 denotes the queue size, and x_i is the indicator of whether server i is busy or not, i.e.

$$x_i = \begin{cases} 1 & , \quad \text{server i busy} \\ 0 & , \quad \text{else} \end{cases} \qquad i = 1,2$$

The events of interest are arrivals and departures; when these events occur a decision must be made where to dispatch the head-of-the-line customer.

Let D_i, i = 1,2, denote the operator that acts on the state vector upon a departure from server i; thus

$$D_1(x_0,x_1,x_2) = (x_0,(x_1 - 1)^+,x_2)$$
$$D_2(x_0,x_1,x_2) = (x_0,x_1,(x_2 - 1)^+)$$

Similarly, A denotes the operator of an arrival, i.e.

$$A(x_0,x_1,x_2) = (x_0 + 1, x_1,x_2)$$

and P_i, i = 1,2,12, the control action operator of dispatching customers to server 1, 2, or both respectively, i.e.

$$P_1(x_0,x_1,x_2) = (x_0-1,1,x_2) \quad , \text{ assuming } x_1 = 0$$
$$P_2(x_0,x_1,x_2) = (x_0-1,1,x_1,1) \quad , \text{ assuming } x_2 = 0$$
$$P_{12}(x_0,x_1,x_2) = (x_0-2,1,1) \quad , \text{ assuming } x_1 = x_2 = 0$$

and P_0 represents the operator of "holding" the customer, i.e.

$$P_0(x_0,x_1,x_2) = (x_0,x_1,x_2)$$

The state of the system evolves as a continuous or discrete time Markov chain depending on whether we assume slotted or unslotted time. We must characterize the optimum control sequence of actions $u_1,u_2,\ldots$ at the decision points, where, in effect, $u_i \varepsilon \{0,1,2,12\}$, so as to minimize either the discounted average cost or the long-term average cost, i.e. either

$$E \int_{t=0}^{\infty} |x(t)| \beta^t dt$$

where $|x(t)| = x_0 + x_1 + x_2$ at time t and β is a discount factor in (0,1), or

$$\lim_{t\to\infty} \frac{1}{t} \int_0^t E|x(t)|dt$$

We consider the case of the discounted average cost. Let $J^\beta(x)$ denote the value of the cost function at the optimum as a function of the initial value of the state x. For any stationary control policy π^* we define the dynamic programming operator T on a Banach function space F by

$$(T_\pi f)(x) = |x| + \beta E[f(\pi(x))]$$

It is then a standard result that

$$J^\beta(x) = \lim_{n\to\infty} T^{(n)} f(x) \qquad \forall f \varepsilon F$$

where

---

* a control policy π is stationary if it does not depend explicitly on time.

$$(Tf)(x) = \min_{\pi} (T_{\pi}f)(x)$$

With this characterization of the optimum it is possible to show the optimality of the threshold policy in several steps. First one shows that under the optimum policy the following properties of the cost functions hold:

(i) $J^{\beta}(P_1x) \leq J^{\beta}(P_0x)$

(ii) $J^{\beta}(P_1x) \leq J^{\beta}(P_2x)$

that is, it is better to use the fast server rather than the slow one, if both are available, and it is better to use the fast server, if available, rather than hold the customer. Next, one proves that whenever the fast server is available he must be utilized if a customer is waiting. Thus the admissible controls are restricted to simply deciding whether to utilize the slower server or not. Following that, one proves that the policy iteration algorithm of dynamic propgramming converges to the optimal policy. Finally it is shown that the policy iteration algorithm applied to a threshold policy yields a threshold policy again and that there exists some policy of the threshold type which achieves the minimum overall policies at some step of the algorithm.

The actual computation of the threshold is difficult because it requires the computation of the average cost at each step of the policy iteration process. However, if $\beta \to 1$, it is possible to numerically compute the threshold.

The extension of the results to non-exponential services follows the same formulation, provided a Markovian evolution of the state can be preserved. This can be ensured if the service is Erlangian with an arbitrary number of stages r.

However, the extension to non-Poissonian arrivals is not possible via the approach just described. Instead, a probabilistic approach is needed that can extend the results to the case of an arrival process with almost <u>arbitrary</u> independent interarrival times (certainly not <u>only</u> Erlangian). The main idea of the approach is to show that if the optimal policy does not satisfy the properties of the threshold policy, performance can be improved via a modified policy. The improvement is established by a sample-function-based comparison between the two policies.

In fact, one can revisit the case of Poisson arrivals and totally unrestricted service times by using this probabilistic approach. The problem that remains unsolved in that case is whether an optimal policy exists. If it does the preceding approach can show that it must be a threshold-based one.

As a closing remark, we note that a generalization toward the direction of increasing the number of servers has proven to be very resisting. It looks like the optimal policy may have several thresholds depending on the status (idle or not) of the different servers. Unfortunately the weakness of the outlined approach is that the nature of the optimal policy must be <u>guessed</u> before it can be shown to be optimal.

3. A PROBLEM OF STABILITY IN A MULTI-ACCESS CHANNEL

A generic problem of interfering queues in a resource-shaping environment is provided by the collision channel operating under the ALOHA protocol and with a finite number of buffered terminals.

Consider M discrete-time queueing systems each of which accepts messages that arrive according to a Bernoulli process. Let λ_i be the probability that a message arrives at the queue of system i at a given time slot. The length of the message is not important and may be considered fixed. Message arrivals at different queues are statistically independent. All queues have infinite capacity (this assumption is not strictly necessary, but it corresponds to the interesting case). The server in the ith queueing systems attempts transmission of the head-of-the-line message at a given slot with constant probability p_i. Transmission attempts by the servers of the different queues are statistically independent. Service is completed when a message is successfully transmitted. If two or more servers attempt transmission at the same time slot, none are successful. Thus the probability of successful transmission by the ith user at a given time slot is given by

$$p_i \prod_{j=i_1}^{i_k} (1 - p_j)$$

where i_1, $i_2,\ldots,i_k$ are the identitites of those queues that have non-empty buffers at that time slot. Hence the queues are coupled and interact.

This problem has been considered by several researchers in recent years. Fayolle and Iasnogorodski [7] considered two symmetric queues in continuous time in the context of coupled processors. Ephremides and Sadawi [8] have considered an approximate model for the case of M symmetric queues. Sidi and Segall [9,10] have considered the case of two queues (non-symmetric) and obtained precise solutions in certain special cases, while proposing another approximate model for other cases. Szpankowski [11,12] has looked at the determination of the ergodic (stable) region of such systems and has obtained upper and lower bounds. He has also proposed approximate models for the steady-state analysis, and so have other researchers as well [13-15]

Interestingly enough, information theorists have looked at the same problem but not from the queueing-theoretic point of view. Their concern has been the determination of the maximum stable throughput (capacity) of the collision model just described. Massey and Mathys [16] have obtained the capacity for the no-feedback case of M users. Here we derive this region for M=2 in a new and much simpler way and show that it is identical to that obtained by Tsybakov in [18]. We can also obtain bounds for the case of M>2 that are consistent with a conjecture that the stability region of the system we study is the same as the capacity region of the no-feedback channel.

Let us return to the original system of interacting queues that we shall call system A. The service rate for the ith terminal is μ_i, which takes different values of the form

$$\mu_i = p_i \prod_j (1 - p_j)$$

where the product ranges over the values of j corresponding to the non-empty queues. Thus the Markov chain corresponding to A is an M-dimensional random walk that permits "diagonal" transitions and such that the service rates for each queue increase when a boundary is reached (i.e. when some queue empties.

The key idea is to introduce a "dominant" system B in which all queue sizes are, on a sample-function-basis, dominating those of A. Thus, let B consist of a set of M queues, as in A, each of which accepts identical arrival processes as its counterpart in system A, but for which there is no difference in service rates when queues empty; that is, we may consider that each terminal continues to attempt transmission of "dummy" messages with the same probability p_i after its own queue of real messages empties. This system is clearly strongly dominant of A; that is, if A and B start with the same initial conditions, then at any subsequent time slot the size of each queue of B will be no less than the size of the corresponding queue of A. Furthermore it is easy to analyze system B because it decomposes and we can study each of its queues separately. In fact we can obtain analytical, closed form expressions for the marginal distributions of each queue of system B. Specifically if q_i is the size of queue i then

$$\Pr[q_i = k] = \frac{1-\rho_i}{1-\mu_i} \cdot [\rho_i \frac{1-\mu_i}{1-\lambda_i}]^k$$

where

$$\mu_i = p_i \prod_{j\neq i} (1 - p_j) \text{ and } \rho_i = \frac{\lambda_i}{\mu_i}$$

We should note that, even though system B decomposes, its queues are not independent since simultaneous departures are not permitted.

The conditions for stability for each individual queue of system B is given by

$$\lambda_i < p_i \prod_{j\neq i} (1 - p_j)$$

However, it is possible that system A is stable for values of the λ_i's that do not satisfy the above condition. In fact it is known from [17] that for M=2 the region of stability is given by

$$\lambda_1 = (1 - \frac{\lambda_2}{p_2\bar{p}_1}) p_1 + \frac{\lambda_2}{p_2\bar{p}_1} p_1\bar{p}_2, \text{ for } \lambda_2 \leq \bar{p}_1 p_2$$

and by

$$\lambda_2 = (1 - \frac{\lambda_1}{p_1\bar{p}_2}) p_2 + \frac{\lambda_1}{p_1\bar{p}_2} p_2\bar{p}_1 \text{ for } \lambda_2 > \bar{p}_1 p_2.$$

It is also an easy exercise to show that for any pair of input rates λ_1 and λ_2 in the region bounded by the curve

$$\sqrt{\lambda_1} + \sqrt{\lambda_2} = 1$$

there exist values of p_1 and p_2 that make the system ergodic. We would like first to present a simple proof of this result that doesn't require use of Malyshev's theorem [18], which was used in [17]. In proving the result we will establish an appropriate dominant system with the same ergodic region as A, which, at least for the case M=2, can be partly analyzed.

Proposition: For a system of two queues as described above the stability region is given by Eq. (1).

Proof: For a given pair of λ_1 and λ_2, if both queues are stable, then at least one of them is strongly stable in the sense that it satisfies the condition $\lambda_i < p_i(1-p_j)$. If this was not the case, that is if both queues satisfied

$$\lambda_1 > p_1(1-p_2) \text{ and } \lambda_2 > p_2(1 - p_1),$$

then we might consider another system B (dominant) as before, in which when a queue empties it doesn't stop interfering with the other queues, that is it continues transmitting dummy packets. Such a system B would then be unstable. Therefore if B started from an initial condition with both queues non-empty, than with a finite (non-zero) probability it would diverge without visiting the zero state first*. However system A is indistinguishable from B if it starts with the same initial condition, so long as its queues don't empty. Thus with the same (finite) probability, system A would also diverge without visiting the zero start first. Thus it couldn't have been stable in the first place.

Consequently if A is stable, at least one of its queues, say the 2nd, must be strongly stable, i.e. satisfy

$$\lambda_2 < p_2(1 - p_1) = p_2\bar{p}_1$$

Given that, we now consider the 1st queue. We construct another dominant system B', in which if queue 2 empties it doesn't continue transmitting dummy packets, that is it continues behaving as in system A, but if queue 1 empties it goes on as in system B, that is it continues transmitting dummy messages. Clearly this system dominates A. Since the second queue sees always a service rate of $p_2\bar{p}_1$, it remains stable, but with generally longer queue sizes. Notice that 2 can be independently analyzed. In fact we have

$$\Pr[q_2 = 0] = 1 - \frac{\lambda_2}{p_2\bar{p}_1}$$

and

$$\Pr[q_2 > 0] = \frac{\lambda_2}{p_2\bar{p}_1}$$

Therefore in order for the 1st queue to be stable we must have

$$\lambda_1 \leqslant (1 - \frac{\lambda_2}{p_2\bar{p}_1})p_1 + \frac{\lambda_2}{p_2\bar{p}_1} \cdot p_1\bar{p}_2$$

The above inequality defines the region for $\lambda_2 < \bar{p}_1 p_2$. We claim that the stability condition for the 1st queue is the same in both A and B'. The proof uses the same argument as earlier. If both A and B' get started with non-empty queues, they behave identically so long as queue 1 does not empty. Thus if queue 1 in system B is unstable, it will diverge without first emptying with some non-zero probability. Since A, in that case, is indistinguishable from B', the same divergence must be experienced by queue 1 in A. Therefore if queue 1 is unstable in B' it is also unstable in A. On the other hand B' dominates A. therefore if queue 1 is stable in B' it is also stable in A. Thus the stability condition for system 1 is the same.

We can obtain the remaining part of the ergodic region by reversing our assumption about which of the queues is strongly stable. Thus, if we assume

---

* there is a subtlety in this argument that is not being addressed here.

$$\lambda_1 < p_1\bar{p}_2$$

we can show that

$$\lambda_2 < (1 - \frac{\lambda_1}{p_1\bar{p}_2})p_2 + \frac{\lambda_1}{p_1\bar{p}_2} p_2\bar{p}_1$$

It is now a straightforward matter to establish that the envelope of these stable regions, as p_1 and p_2 vary, becomes indeed, the curve $\sqrt{\lambda_1} + \sqrt{\lambda_2} = 1$.

Unfortunately, the situation becomes more complicated if $M > 2$. On the basis of known results it is reasonable to make the following conjecture: The stability region for M queues is defined by the condition

$$\lambda_i \leq p_i \prod_{j \neq i} (1 - p_j)$$

under the constraint that $\sum_{i=1}^{N} p_i = 1$. Recent results in [19] support this conjecture by showing that inner bounds to the ergodicity region are consistent with the above expression.

Note that for the symmetric case this condition becomes

$$\lambda_i = \lambda \leq \frac{1}{M} (1 - \frac{1}{M})^{M-1}$$

obtained for $p_i = p = \frac{1}{M}$. This result can be independently established for symmetric systems. Note that as $M \to \infty$ the above expression approaches the famous quantity e^{-1}.

4. CONCLUDING REMARKS

There are two major clases of problems in communication networks in which queueing models come into play. the first class consists of what one might call "hard-wired", point-to-point networks in which the concept of a link is well defined. A link in such networks provides dedicated service between the two nodes at its ends. As mentioned in the introduction these links can be modeled reasonably well by queueing systems. The validity of the models becomes strained when these links are considered interconnected in the form of a network. Kleinrock's independence assumption can extend the use of these models somewhat by allowing use of the theory of Jacksonian networks and of the reversibility theory of Markov chains. Even that extension is curtailed considerably when fixed length packets (rather than messages of random (exponential) length) are considered. Various attempts for performance analysis of such systems have been made. A good survey remains the article by Reiser in [20]. Recently a small breakthrough was achieved in [21] in that it becomes possible to analyze a <u>tandem</u> connection of queues in which the independence assumption is <u>not</u> made.

In this class of problems elementary control questions can be asked that prove formidable to answer. In Section 2 a simple such control problem was addressed that shows the flavor of the difficulties encountered and the techniques that have, or can, be used for its solution.

The second class of problems consists of what we may call "multiple-access" or "shared-resource" problems. They reflect the case of radio channels in which the concept of a "link" is not well-defined in the presence of several users. Abramson's assumption was crucial in opening up this class of problems to careful investigations. In addition to queueing considerations there are stability and protocol analysis problems that have been the object of study for over ten years, always with moderate-to-limited success. The queueing aspects of these problems have only very recently been looked at.

As explained in the introduction, the interaction between the queues is very complex even in the simplest of such systems. A brief expose of an approach to such a simple problem was presented in section 3.

In conclusion, it can be stated that either the use of queueing models in communication networks has stretched their capabilities to their fundamental limits or there is a need for new analysis methodologies that can by-pass the "apparent" limitations of these models that we seem to be confronted with today. Recent attempts to use new points of view (based on sample-path properties) have met with some success and lend evidence for the validity of the latter point of view.

REFERENCES

1. S. Stidham, Jr., "Optimal Control of Admission to a Queueing System", IEEE Trans. AC, Vol. 30, No. 8, pp. 705-713, August 1985.
2. R.L. Larsen, "Control of Multiple Exponential Servers with Application to Computer Systems", Ph.D. Dissertation, University of Maryland, 1981.
3. W. Lin, P.R. Kumar, "Optimal Control of a Queueing system with two Heterogeneous Servers", IEEE Trans. AC, Vol. 29, No. 8, pp. 696-703, 1984.
4. J. Walrand, "A Note on 'Optimal Control of a Queueing System with two Heterogeneous Servers'", Systems and Control Letters, Vol. 4, pp. 131-134, 1984.
5. I. Viniotis, A. Ephremides, "Extension of the Optimality of the Threshold Policy in Heterogeneous Multiserver Queueing Systems", submitted to IEEE Trans. AC.
6. R. Gallager, D. Bertsekas, Data Networks, manuscript under preparation.
7. G. Fayolle, R. Iasnogorodski, "Two Coupled Processors: The Reduction to a Riemann-Hilbert Problem", Wahrscheinlichkeits theorie, pp. 1-27, 1979
8. T. Saadawi, A. Ephremides, "Analysis, stability, and Optimization of Slotted Aloha", IEEE Trans. AC, June 1981.
9. M. Sidi, A. Segall, "Two Interferring Queues in Packet-Radio Networks", IEEE Transactions on Communications, Vol. 31, No. 1, pp. 123-129, January 1983.
10 A. Segall, M. Sidi, "Priority Queueing Systems with Applications to Packet-Radio Networks", Proc. International Seminar on Modeling and Performance Evaluation, pp. 159-177, Paris 1983.
11. W. Szpankowski, "A Multiqueue Problem: Bounds and Approximations", Performance of Computer-Communication Systems, (H. Rudin, W. Bux, editors), IFIP, 1984.
12. W. Szpankowski, "Performance Evaluation of a Protocol for Multiaccess Systems," Proc. Performance '83, pp. 337-395, College Park, 1983.
13. S.S. Kamal, S.A. Mahmond, "A Study of Users' Buffer Variations in Random Access Satellite Channels", IEEE Trans. on Communications, Vol. 27, pp. 857-868, June 1979.
14. K.K. Mittal, A.N. Venetsanopoulos, "On the Dynamic Control of the Urn Scheme for Multiple Access Broadcast Communication Systems", IEEE Trans. Communications, Vol. 29, No. 7, pp. 962-970, July 1981.

15. H. Kobayashi, "Application of the Diffusion Approximation to Queueing Networks I: Equilibrium Queue Distributions", Journal ACM, Vol. 21, No. 2, pp. 316-328, April 1974.
16. J. Massey, P. Mathys, "The Collision Channel Without Feedback", IEEE Trans. IT, Vol. 31, No. 2, pp. 192-204, March 1985.
17. B. Tsybakov, W. Mikhailov, "Ergodicity of Slotted ALOHA System", Problems of Information Transmission, Vol. 15, No. 4, pp. 73-87, 1979.
18. W. Malyshev, "A Classification of Two-dimensional Markov Chains and Piece-Linear Martingales", Doklady Akademii Nauk USSR, Vol. 3, pp. 526-528, 1972.
19. R. Rao, A. Ephremides, "On the Stability of Interacting Queues in a Multiple Access System", submitted to the IEEE Trans. IT.
20. M. Reiser, "Performance Evaluation of Data Communiucation Systems", Proceedings IEEE, Vol. 70, No. 2, pp. 171-196, February 1982.
21. M. Shalmon, M.A. Kaplan, "A Tandem Network of Queues with Deterministic Service and Intermediate Arrivals", Operations Research, Vol. 32, pp. 753-773, 1984.

LIMITS TO NETWORK RELIABILITY

Günter G. Weber
Kernforschungszentrum Karlsruhe
Institut für Datenverarbeitung in der Technik
Postfach 3640
D-7500 Karlsruhe

INTRODUCTION

Frequently, for computer systems and communication networks a reliability analysis is carried out. However, for "degradable" computer systems a unified measure for performance and reliability is preferable. By degradable we mean that, depending on the history of the computer and on the environment, the system can show various levels of performance. The interplay of reliability and performance is significant for these systems.

First, some methodological developments of system reliability analysis will be discussed. Here special emphasis is on fault tree techniques. It is possible to obtain unavailability and reliability of these systems. Then the use of these techniques for certain networks is mentioned. If we have, however, a system with a phased mission, its relevant configurations may change during consecutive periods (called phases).

Next, systems are discussed which have - in contrast to the models mentioned above - a performance related dependence. Here the state of a subsystem at one time depends on at least one state at an other time.

Finally, suitable concepts for functional dependence are introduced, leading also to criteria whether a system is functionally dependent or not. Such considerations are clearly related to the decomposition theory of systems and also to combinatorial theory (especially matroid theory). Based on system analytic and stochastic considerations it is possible to evaluate the performability of such a degradable system.

1. FAULT TREE ANALYSIS

1.1 FUNDAMENTAL CONCEPTS

With a fault tree analysis it is possible to get for a previously specified event (e.g. system failure):

J. K. Skwirzynski (ed.), Performance Limits in Communication Theory and Practice, 155–172.

- a systematic identification of possible failure combinations, and
- an evaluation of safety relevant characteristics (e.g. unavailability, expected number of failures).

Now we define a fault tree: A fault tree is a finite directed graph without (directed) circuits. Each vertex may be in one of several states. For each vertex a function is given which specifies its state in terms of the states of its predecessors. The states of those vertices without predecessors are considered the independent variables of the fault tree /1 /.

Note: We assume two states for each vertex, thus we obtain Boolean expressions. This definition of a fault tree corresponds to a combinational circuit.

1.2 STRUCTURE FUNCTION

A Boolean function is introduced which describes the fault tree. Evidently this function is closely related to a switching function. This Boolean function specifies the state of each vertex in terms of its predecessors. The structure function may be used for all fault trees e.g. consisting of AND-, OR, NOT-gates. However, for sequential systems the structure function cannot be used.

Frequently, a system is coherent, i.e. the following conditions (a) and (b) hold:

(a) If S is functioning, no transition of a component from failure to the non failed state can cause system failure (positivity of structure function).

(b) If all components are functioning the system is functioning and if all components are failed, the system is failed.

If the system may be represented by AND and OR, but without complements, then the structure is coherent. For coherent systems exactly one irredundant polynomial exists which is also minimal (min cut representation).

We use the following notation for structure function:

$x_i = 1$ if element i is intact

$x_i = 0$ if element i is defect 1-1

and, similarly,

$$\phi(x_1, x_2, ..., x_n) = 1 \qquad \text{if system S is intact}$$

$$\phi(x_2, x_2, ..., x_n) = 0 \qquad \text{if system S is defect} \qquad 1\text{-}2$$

(See also / 1 / for more details).

1.3 PROBABILISTIC EVALUATION

Based on (1-1), (1-2) it is possible to evaluate the reliability of a coherent structure. We have for component i the state X_i which is random (and may be also time dependent). Thus we get

$$P(X_i = 1) = p_i = E(X_i)\,, \quad i = 1,2,...,n \qquad 1\text{-}3$$

For the system S represented by ϕ we get

$$P(\phi(\underline{X}) = 1) = E(\phi(\underline{X})) \qquad 1\text{-}4$$

where, $\underline{X} = (X_1, X_2, ..., X_n)$.

For repairable systems it is interesting to evaluate the system availability A_s (t),unavailability A_s (t), and the expected number of failures $W_s(t)$ /1 /.

1.4 COMMUNICATION NETWORKS

As has been shown by Murchland /3 /, the fundamental concepts of fault tree analysis can be also applied to communication networks. Here average state probabilities and transition rates are useful either as time functions or in the long run asymptotic form. For networks the number of paths and cuts may be very high. Thus recently methods have been developed, which do not require all minimal paths (or cuts) for reliability evaluation of a network. This is a considerable improvement saving much computer time / 4/. Also note that there are many different methods related to network reliability .

It may be sometimes useful to introduce notations and methods from multistate system analysis /3 /.

2. SYSTEMS WITH PHASED MISSION

Untll now we discussed systems which have the same configuration during the whole life time. If we have, however, a system with a phased mission, its configuration may change during consecutive periods (called phases). Reliability and performance analysis requires the use of a (generalized) multistate structure function and the concept of association.

It is possible to give bounds for unavailability. It is interesting to note that there is also a criterion showing the admissibility of phased structure functions for these systems. This can be based on some algebraic properties of the so called functional dependence (see Meyer /5/).

It will be sufficient to consider here systems having two states for each component. Fore more general information see Esary and Ziehms /6/, A. Pedar and V. Sarma /7/.

2.1 DISCUSSION OF A PHASED MISSION

We consider the system of Fig. 2.1 given as block diagram. It has different structures in the three phases of its mission (see /6/).

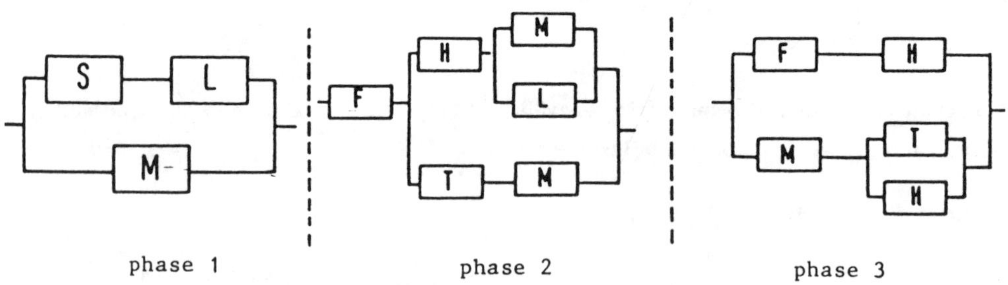

Fig. 2.1 System with phased mission

For this system we obtain as minimal cuts:

Phase	Minimal cuts
1	{M, L}, {M, S}
2	{F}, {H, M}, {H, T}, {M, L}
3	{F, M}, {H, M}, {H, T}

2.2 SIMPLIFICATION OF A SYSTEM

A minimal cut in a phase can be deleted (without loss of information) if it contains a minimal cut of a later phase. This is similar to absorption. But it would not refer to deleting of a minimal cut regardless of time ordering. Thus we obtain the following reduced list ("after cancellation") of cut sets:

Phase	Cuts	cancelled cuts
1	{M, S},	{M, L}
2	{F}, {M, L},	{H, M}, {H, T}
3	{F, M}, {H, M}, {H, T}	no cancellation possible

This can be also given as a simplified block diagram:

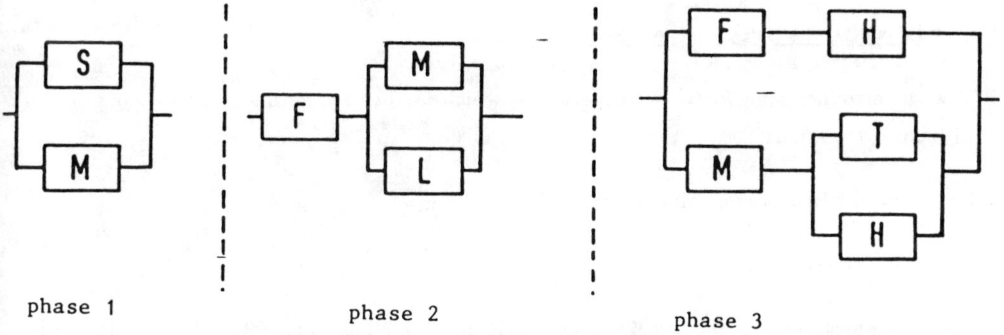

Fig. 2.2 System after cancellation

An equivalent representation is by a structure funktion ϕ_i referring to phase i.

$x_{Mi}(i=1,2,3)$ refers to the success of component M in phase i. If for a phase $j < i$, M would be failed, it could not be successful in phase i.

We obtain:

$$\phi_1(\underline{x}_1) = x_{M1} + x_{S1} \qquad 2\text{-}1$$

$$\phi_2(\underline{x}_2) = x_{F2}(x_{M2} + x_{L2})$$

$$\phi_3(\underline{x}_3) = x_{F3}\,x_{H3} + x_{M3}(x_{T3} + x_{H3})$$

We obtainas probability that this system is operative for the whole mission

$$P_{system} = P\left(\prod_{j=1}^{n} \phi_j(\underline{X}_j) = 1\right) = E\left(\prod_{j=1}^{n} \phi_j(\underline{X}_j)\right) \qquad 2\text{-}2$$

and

$$E\left(\prod_{j=1}^{n} \phi_j(\underline{X}_j)\right) \leq \prod_{j=1}^{n} E(\phi_j(\underline{X}_j)) \qquad 2\text{-}3$$

(see Esary /6/).

This is an example for a "structure based" capability function, i.e. a function which can be related to structure functions ϕ_i (see sect. 4.5, 4.6).

2.3 KERNEL OF A BOOLEAN MAPPING

Now we introduce some further considerations which can be used for a methodology of systems with phased missions.

Let ϕ_j be a Boolean mapping, from B to A 2-4

$$\phi_j : B \rightarrow A$$

Then the kernel of ϕ_j is the set M_j of elements in B which ϕ_j maps onto 1 in A. This can be written

$$M_j = \{p \mid \phi_j(\underline{p}) = 1\} \qquad 2\text{-}5$$

Example: The kernel M_1 of ϕ_1 is

$$M_1 = \{(x_{M1}, x_{S2}) \mid \phi_1(x_{M1}, x_{S1}) = 1\}$$
$$= \{x_{M1}, x_{S1}\}$$

Note: p refers to the variables of ϕ_j.

2.4 APPLICATION TO OUR SYSTEM

We obtain as kernels:

$M_1 \quad = \{x_{M1}, x_{B1}\}$ 2-6

$M_2 \quad = \{x_{F2} x_{M2}, x_{F2}\, x_{L2}\}$

$M_2 \quad = \{x_{F3}\, x_{H3}, x_{M3}\, x_{T3}, x_{M3}\, x_{H3}\}$

By a Cartesian product of these kernels

$M_1 \times M_2 \times M_3$

we obtain all success trajectories of our system. This can be rewritten:

$M_1 \times M_2 \times M_3$
$= \{x_{M1}, x_{S1}\} \times \{x_{F2}\, x_{M2}, x_{F2}\, x_{L2}\}$ 2-7
$\times \{x_{F3}\, x_{H3}, x_{M3}\, x_{T3}, x_{M3}\, x_{H3}\}$
$= \{x_{M1}\, x_{F2}\, x_{M2}\, x_{F3}\, x_{H3}, x_{M1}\, x_{F2}\, x_{M2}\, x_{H3}\, x_{M3}, x_{M1}\, x_{F2}\, x_{M2}\, x_{H3}\ x_{T3}, \ldots\ldots, x_{S1}\, x_{F2}\, x_{L2}\, x_{H3}\, x_{T3}\}$

This Cartesian product can be also given as a tree. In this tree each path from left to right is a single term of the Cartesian product (Fig.. 2.3).

Each term is a success trajectory /1/.

For example

M1 · F2 M2 · F3 H3

is a success trajectory. But failure of M1 <u>and</u> S1 would lead to system failure.

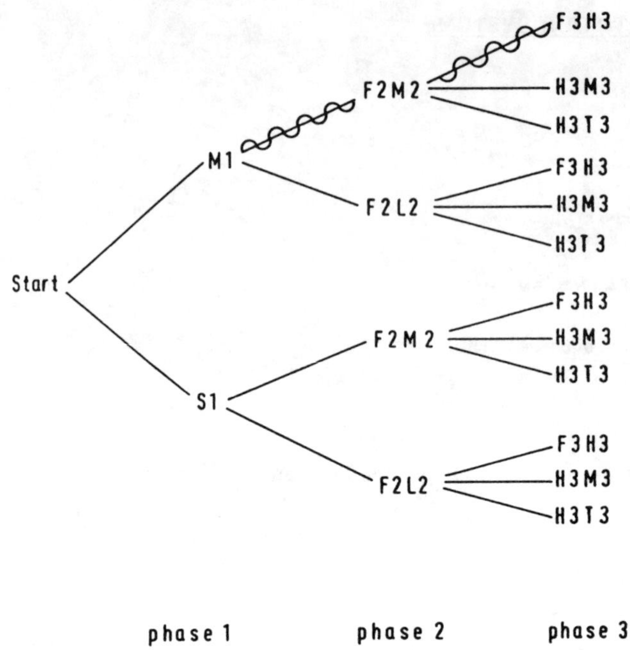

Fig. 2.3 Tree for a system with phased missions

(∿∿∿ success trajectory M1 · F2 M2 · F3 H3)

Note:

We may also use a success tree or a fault tree for representation (see also /1/).

3. PERFORMABILITY

Now we introduce a quantity which can be used for systems with degradable performance. It is called "performability". This concept has been introduced and developed by J.F. Meyer /5/. One special case of performability is the usual reliability (based on intact and defect-states without degradation). We need various concepts for our consideration. Note that a computing system may be viewed at several levels. At a low level we have a detailed picture of the computer's behaviour. At a higher level we have a view what the computer accomplishes for the user during a given period T.

3.1. BASE MODEL X_S

Let (Ω, E, P) be a probability space for the system S where Ω is the description space and E a set of measurable events. S can be viewed by a stochastic process

$$X_S = \{X_t \mid t \varepsilon T\} \qquad \text{3-1}$$

with T utilization period. For all $t \varepsilon T$, X_t is a r.v.

$$X_t : \Omega \rightarrow Q \qquad \text{3-2}$$

where Q is the state space of the system. X_S is also called the base model of S. For fixed $\omega \varepsilon \Omega$ the following time function is called a state trajectory $u_\omega : T \rightarrow Q$, with 3-3

$u_\omega(t) = X_t(\omega) \quad \forall t \varepsilon T$,

where ω is an elementary event, and

$U = \{u_\omega \mid \omega \varepsilon \Omega\}$ state trajectory space of S.

3.2 SYSTEM PERFORMANCE MODEL Y_S

Here we have an accomplishment set A with $A = \{a_0, a_1, a_2, ...\}$
(accomplishment levels).
Example: $a_k = k$ ($k = 0, 1, 2, ...$) where k is the number of system failures during time T.

In terms of the aacomplishment set, system performance can be regarded as a r.v.

$$Y_S : \Omega \rightarrow A \qquad \text{3-4}$$

where $Y_S(\omega)$ is the accomplishment level corresponding to an outcome ω in the underlying description space Ω.

It is also possible, to introduce a "worth measure" for such a system which is related to the economic benefit (see /5/).

3.3 PERFORMABILITY AND CAPABILITY

For the performance variable Y_S a measure which quantifies both

- system performance and
- system reliability (ability to perform)

can be introduced.

Definition 1: If S is a system and A is the accomplishment set associated with system performance Y_S, then the performability of S is the probability function $p_S : A \to [0, 1]$ where $p_S(a)$ = the probability that S performs at level a, that is

$$p_S(a) = P(\{\omega \mid Y_S(\omega) = a\}) \qquad \text{3-5}$$

We also note that system effectiveness can be evaluated:

$$Eff(S) = \sum_{a \varepsilon A} w(a)\, p_S(a) \qquad \text{3-6}$$

Here w(a) is the worth of performance level a. System effectiveness is a measure of the expected benefit for the user (see also /5/). Similarly, reliability can be evaluated:

$$R_S = p_S(B) = \sum_{a \varepsilon B} p_S(a) \qquad \text{3-7}$$

Here B is the subset of all accomplishment levels related with system success.
To evaluate performability we need a relation between the base model X_S and the user oriented performance model Y_S. Let us assume that the base model is refined enough to distinguish the levels of accomplishment observed by the user, i.e. for all

$$Y_S(\omega) \neq Y_S(\omega') \text{ implies } u_\omega \neq u_{\omega'} \qquad \text{3-8}$$

where u_ω and $u_{\omega'}$ are state trajectories associated with outcomes ω. ω'. This means that each trajectory $u \varepsilon U$ is related to a unique accomplishment level $a \varepsilon A$. Thus the capability of S can be defined as follows:

Definition 2: If S is a system with trajectory space U and accomplishment set A then the capability function of S is the function
$\gamma_S : U \to A$ where $\gamma_S(u)$ is the level of accomplishment resulting from state trajectory u, that is,

$$\gamma_S(u) = a \text{ for some } \omega \varepsilon \Omega,\ u_\omega = u \text{ and } Y_S(\omega) = a. \qquad \text{3-9}$$

Relations between capability and performability:
It is useful to know properties of the base model, e.g. that X_S is Markovian (see also /5/). If a is an accomplishment level then (by definition 1) we have:

$p_S(a) = P(\{\omega \mid Y_S(\omega) = a\}) = P(\{u_\omega \mid Y_S(\omega) = a\})$ 3-10

Moreover (by definition 2) wehave:

$p_S(a) = P(\{u \mid \gamma_S(u) = a\}) = P(\gamma^{-1}{}_s(a))$. 3-11

The inverse $\gamma^{-1}{}_s(a)$ is referred to the trajectory set of a.

The capability function has certain relations to the structure function. To see this more precisely, we have to introduce the concept of functional dependence.

4. FUNCTIONAL DEPENDENCE

The concept of functional dependence (FD) is very useful for performability evaluation. Before we enter this field, let me give a few remarks on the origin and application of this concept:

The concept of functional dependence is very useful for logical consideration of relational data bases. For relational data bases, certain first order sentences, called "dependencies" about relations have been defined and studied /8/. It is also interesting, to note that for the decomposition theory of systems these dependencies are very important /9/. Also our problems referring to performability are closely related to decomposition theory. Also more general dependencies are used for various purposes. We will not discuss these here.

It has been noted that the theory of matroids can be used with some difficulties to obtain results analogous to decomposition theory /9/. Especially for systems which are not finite sets, matroid theory seems not to be of much use for decomposition theory /9/.

Example

We give an example which is known from relational data bases /8/.

Consider a relation with three collumns with headings EMP (employee), DEPT. (department), MGR (manager).

EMP	DEPT	MGR
Pythagoras	Geometry	Euclid
Pappus	Geometry	Euclid
J.F. Meyer	Computer Science	von Neumann
B. Littlewood	Computer Science	von Neumann

Fig. 4.1 Relation

The relation in Fig. 4.1 obeys the functional dependence DEPT → MGR. This means that whenever two tuples (i.e. rows) agree in the DEPT-collumn, then they necessarily also agree in the MGR-collumn. We note that the FD can also be related to formal logic.

4.1 DEFINITION OF FUNCTIONAL DEPENDENCE

Here we need the following notations (see also /10/):

Let S be a system with subsystems $S_1, \ldots, S_n$. Each subsystem is observed at k different times $t_1, \ldots, t_k$.

Then $D = \{(i,j) \mid 1 \leq i \leq n,\ 1 \leq j \leq k\}$ where (i, j) represents S_i observed at time t_j.

$\underline{Q} = \{Q_d \mid d \in D\}$ is a family of sets indexed by D and Q = state of the base model (see sect. 3.1).

Definition 3 : A structured set R (relative to D and $\underline{Q}$) is a subset of the Cartesian product of the sets in Q (family of state sets), that is

$$R \subseteq \underset{d \in D}{\times} Q_d \qquad \text{4-1}$$

where the product is taken according to the ordering of D.

Definition 4: If $C \subseteq D$ is a coordinate set where $C = \{c_1, c_2, \ldots\}$ (c_1 is the first element of C according to the ordering of D etc.) the projection of R on D is the function

$$\xi_C : R \to \underset{c \in C}{\times} Q_c \qquad \text{4-2}$$

where $\xi_C(r) = (\xi_{c1}(r), \xi_{c2}(r), \ldots)$.

Note: If $C = \emptyset$ (empty set), $\xi_\emptyset : R \to 1_\emptyset$, where $1_\emptyset$ is an arbitrary constant.

Example:

Suppose $D = \{1, 2, 3\}$ and $Q_i = \{0, 1\}$, $i \in D$. Then $C_1 = \{1, 2\} \subseteq D$

$C_2 = \{3\} \subseteq D$ with $\xi_{\{1,2\}}(0, 1, 1) = (0, 1)$, $\xi_{\{3\}}(0, 1, 1) = (1)$.

Definition 5: If R is a structured set indexed by D, and $A, B \subseteq D$, then A R-depends on B ($A \Delta_R B$) if

$\exists\, v \in \xi_A(R)$, $\exists\, w \in \xi_B(R)$ such that

$\forall r \in R\, [\xi_B(r) = w$ implies $\xi_A(r) \neq v]$ 4-3

If $A \Delta_R B$, we are saying that, relative to sequences $r \in R$, knowledge of the coordinates in B (e.g. when $\xi_B(r) = w$) increases our knowledge of the values of coordinates in A (i.e. $\xi_A(r) \neq v$).

Examples: It can be seen from Fig. 4.1 , that knowledge regarding the DEPT increases the knowledge regarding the MGR! See also /1/, /10/ for more detailed examples.

Note:
A is R-independent of B ($A \not\Delta_R B$) if A does not R-depend on B.

4.2 FUNDAMENTAL PROPERTIES OF FUNCTIONAL DEPENDENCE

Now we introduce some fundamental properties of FD (R-dependence) which contribute to a better understanding of this concept and are used for performability of systems.
The formal definition (4-3) of FD can be related to partitions as follows:
Theorem 1: Let R be a structured set indexed by D, and let $A, B \subseteq D$. Then A R-depends on B if and only if $\exists\, v \,\varepsilon\, \xi_A(R)$, $\exists\, w \,\varepsilon\, \xi_B(R)$ such that $Bl_A(v) \cap Bl_B(w) = \emptyset$. 4-4

(see also /10 /.) Here $Bl_A(v)$, $Bl_B(w)$ are blocks of partitions π_A, π_B.

Theorem 1 gives an algebraic criterion of FD, which, given partitions π_A and π_B, says that there is a FD between A and B if and only if there is a block in π_A and a block in π_B, which have no elements in common. For similar considerations see also the "decomposition theory" /9 /.
Based on our definition (4-3) and on theorem 1 we can derive additional properties of FD. We show first that FD is preserved by supersets:

Theorem 2:
Let $A, B \subseteq D$. If $A \Delta_R B$, then $\forall A' \supset A$ and $\forall B' \supseteq B$ such that $A', B' \subseteq D$, $A' \Delta_R B'$. 4-5

Theorem 2 which says that dependence is preserved by supersets, has the following "dual" which says that independence is preserved by subsets, that is:

Theorem 3:
Let $A, B \subseteq D$. If $A \not\Delta_R B$, then, $\forall A' \subseteq A$, $\forall B' \subseteq B$, $A' \not\Delta_R B'$. 4-6

4.3 REMARKS ON LINEAR DEPENDENCE

For Theorems 2, 3, 4 and for Definition 7 it is important to note the following facts from linear algebra:

An arbitrary, not necessarily finite system X of elements of a vector space V over a field K is called linearly dependent if K contains at least one finite linearly dependent subsystem of elements, and linearly independent if all finite subsystems of X are linearly independent. Clearly, every subsystem of a linearly independent system is itself linearly independent.

Remarks on Matroids

Some rather deep considerations which are related to the theory of decomposition (see e.g. Naylor /9/) and to the theory of matroids (see e.g. Whitney /11/ and Welsh /12/) also come from properties of linear dependence. The field of matroid theory grew from its beginning into a considerable discipline which is now closely related to combinatorial theory, graph theory and modern algebra.

Let me give a definition and an example which illustrates this discipline. Matroid theory has a similar relationship to linear algebra as does point set topology to the theory of real variables. Thus, it postulates certain sets to be "independent" (=linearly independent) and develops a theory from certain axioms which it demands hold for this collection of independent sets /12 /.

Definition 6: A matroid is a finite set S and a collection F of subsets of S (called independent sets) such that (I, II, II) are satisfied.

(I) $\emptyset \varepsilon F$

(II) If $X \varepsilon F$ and $Y \subseteq X$ then $Y \varepsilon F$

(III) If U, V are members of F with $|U| = |V| + 1$ there exists $x \varepsilon U \setminus V$ such that $V \cup x \varepsilon F$.

A subset of S not belonging to F is called dependent.

Example: Let V be a finite vector space and let F be the collection of linearly independent subsets of vectors of V, Then {V, F} is a matroid. As has been shown by Naylor /9/, for FD and decomposition theory of sets which are not necessarily finite, matroid theory presents considerable difficulties. We will not discuss these questions here, but see Naylor /9 /.

4.4 MORE PROPERTIES OF FUNCTIONAL DEPENDENCE

We skip a dew details regarding R-dependence /10/ and go over to a further definition which is related to the use of FD in decomposition theory.

Definition 7: If R is a structured set indexed by D and $C \subseteq D$, then C is R-dependent if there exist finite sets $A, B \subseteq C$ with $A \cap B = \emptyset$ such that $A \Delta_R B$.

C is R-independent if C is not R-dependent.

Note: Here we require that A and B are finite sets (compare also our remarks on linear dependence). Moreover, the requirement $A \cap B = \emptyset$ ensures that C is not regarded as R-dependent simply because some subset of C depends on itself.

Now we come to a theorem which allows to characterize R-independence by means of the structure of ξ_C (R) and a Cartesian product:

Theorem 4: If R is a structured set indexed by D, and $C \subseteq D$, then C is R-independent if and only if

$$\xi_C(R) = \times_{d \varepsilon D} \xi_d(R) \qquad 4\text{-}7$$

Corollary 4': If R is indexed by D then D is R-independent if and only if R is a Cartesian product:

$$R = \times_{d \varepsilon D} \xi_d(R) \qquad 4\text{-}8$$

4.5 CAPABILITY AND STRUCTURE FUNCTION

Corollary 4' says that if R is Cartesian, the index set D must be free of R-dependence and that the absence of R-dependencies among finite subsets of D guarantees that D will have the structure of a Cartesian product. Based on this criterion we can see if the structure function is admissible for certain problems. We first define a concept which relates certain capability functions and structure functions /10/.

Definition 8: If Q is the state set of the base model X_S and A = {0, 1} (where 1 denotes "success" and 0 "failure"), then a capability function is structure based if there exists a decomposition of T into k consecutive time periods $T_1, T_2, ..., T_k$, and there exist functions $\phi_1, \phi_2, ..., \phi_k$ with
$\phi_k: Q \to \{0, 1\}$ such that for all u ε U

$$\gamma_S(u) = 1 \quad \text{if } \phi_i(u(t)) = 1 \text{ for all } i \varepsilon \{1, 2, ..., k\} \qquad 4\text{-}9$$

and for all t ε T. Here the $\phi_i(u(t))$ are structure functions.

Based on this definition we make this consideration: Assuming that $\phi_i(u(t)) = 1$ for all t ε T_i whenever the structure function is = 1 at the end of a phase i. Then due to our Corollary 4', the trajectory space U can be represented as a Cartesian product.

$$U = Q_1 \times Q_2 \times .... \times Q_k \qquad 4\text{-}10$$

Now if $u = (q_1, q_2, ..., q_K)$, then

$$\gamma_S(u) = 1 \text{ iff } \phi_i(q_i) = 1 \text{ for all } i \varepsilon \{1, 2, ..., k\}. \qquad 4\text{-}11$$

Thus, u ε $\gamma_S^{-1}(1)$ iff $\xi_i(u) \,\varepsilon\, \phi_i^{-1}(1)$ for all i, and we conclude that the set of success trajectories $R = \gamma_S^{-1}(1)$ is also Cartesian.

On the other hand, when a capability function is such that $R = \gamma_S^{-1}(1)$ it admits a structure based representation using the ϕ_i. Thus we get the follwing theorem:

Theorem5: Let S be a phased system with trajectory space $U = Q_1 \times Q_2 \times \ldots \times Q_k$ and capability function $\gamma_S : U \rightarrow \{0, 1\}$. Then γ_S is structure based iff the set of all phases $D = \{1, 2, \ldots k\}$ is R-independent, where

$$R = \gamma_S^{-1}(1). \qquad \text{4-12}$$

4.6 EXAMPLE AND REMARKS

An example where the structure based representation can be seen is the system with phased mission of sect. 2.1 - 2.3. Here we note the following:

- It is possible to define γ_S^{-1} in relation to the kernel of a Boolean mapping. Then the Cartesian product $M_1 \times M_2 \times M_3$ clearly relates to the success trajectories of our system.

  For performability an upper limit can be given which is clearly based on the methods used for fault tree analysis (see (2-3)). Rather than using the minimal cut approach, it is recommendable, to make a repeated expansion over all replicated basic events (see for more details /2/).

4.7 USE OF A GENERALIZED STRUCTURE FUNCTION

It is possible to introduce a generalization of the structure function (see (1-2)) which is useful for certain questions related to performance. Assume a system can work under degradation. We define a function which represents k (or less) executed tasks /13/.

$$\phi_S(\underline{x}) = k \quad k = 1, 2, \ldots, M \qquad \text{4-13}$$

with M maximum number of tasks which can be executed, $\underline{x}$ state vector of system S.
Each single task is executed by a subsystem S_f (f=1, 2, ..., M) which can be described by a structure function

$$\phi_f(\underline{x}) = \begin{cases} 1 & \text{if task is executed} \\ 0 & \text{otherwise} \end{cases} \qquad \text{4-14}$$

Then the (generalized) structure function is given as a sum:

$$\phi_S(\underline{x}) = \sum_{f=1}^{M} \phi_f(\underline{x}) \qquad \text{4-15}$$

This type of system allows to introduce a concept which is similar to coherence but also modular decomposition and minimal paths are possible, to name only a few.
If the states are random variables, the probability to execute k tasks can be given:

$$P(\phi_S(\underline{X}) = k) = P\left((\sum_{f=1}^{M} \phi_f(\underline{X})) = k \right)$$

Assume we have a capability function which defines that at least m < M tasks have to be executed. Thus we obtain a performability measure for a system which can be represented by $\phi_S(\underline{x})$.

If we have, in addition, a phased mission where different phases are time associated (i.e. have a special type of stochastic dependence), the evaluation is more complicated, and only bounds can be given (see /7/).

4.8 REMARKS ON GENERAL SYSTEMS

This was only a small area of the performance methodology. For example, the modelling of a degradable buffer/multiprocessor system with performance Y can be done using parameters of this system with considerations from Markow processes and from Queuing theory. Using an approximate decomposition of the model a closed form solution has been obtained /14/. Moreover, a very interesting field with further methods and numerical techniques has been under consideration in the last few years.

REFERENCES

1. G.G. Weber, Methods of Fault Tree Analysis and Their Limits, KfK 3824, Karlsruhe 1984

2. R.E. Barlow, J. Fussell, N. Singpurwalla (Ed.) Reliability and Fault Tree Analysis, SIAM, Philadelphia, PA, 1978

3. J.D. Murchland, Fundamental Concepts and Relations for Multi-State Reliability in /2/, pp. 581-618

4. G.B. Jasmon, Cutset Analysis of Networks Using Basic Minimal Paths and Network Decomposition, IEEE Trans. Rel. Vol. R-34, pp. 403-407, 1985

5. J.F. Meyer, On Evaluating the Performability of Degradable Computing Systems, Proc. of FTCS-8 (June 1978, Toulouse, France), IEEE, Computer Society, pp. 44-49, Piscataway, N.J. 1978

6. J.D. Esary, H. Ziems, Reliability Analysis of Phased Missions in /2/, pp. 213-236

7. A. Pedar, V. Sarma, Phased Mission Analysis for Evaluating the Effectiveness of Aerospace Computer Systems, IEEE Trans. Rel. R-30, pp. 429-437, 1981

8. R. Fagin, Horn Clauses and Database Dependencies, J. ACM 29, pp. 952-985, 1982

9. A.W. Naylor, On Decomposition Theory: Generalized Dependence, IEEE Trans. Sys., Man, Cybern., Vol. SMC-11, pp. 699-713, 1981

10. R.A. Ballance, J.F. Meyer, Functional Dependence and its Application to System Evaluation, Proc. of 1978 J. Hopkins Conf. on Information Sciences and Systems, Baltimore, MD, pp. 280-285, 1978

11. H. Whitney, On the Abstract Properties of Linear Dependence, Amer. J. Math. 57, pp. 509-533, 1935

12. D.J.A. Welsh, Matroid Theory, Academic Press, London, 1976

13. M. Corazza, Techniques Mathematiques de la Fiabilite Previsionelle des Systemes, CEPADUES Editions, Toulouse, 1976

14. J.F. Meyer, Closed Form Solutions of Performability, IEEE Trans. on Computers, Vol. C-31, pp. 648-657, 1982

TWO NON-STANDARD PARADIGMS FOR COMPUTATION:

ANALOG MACHINES AND CELLULAR AUTOMATA

Kenneth Steiglitz
Dept. of Computer Science
Princeton University
Princeton, New Jersey 08544

1. Introduction

Serious roadblocks have been encountered in several areas of computer application, for example in the solution of intractable (NP-complete) combinatorial problems, or in the simulation of fluid flow. In this talk we will explore two alternatives to the usual kinds of computers, and ask if they provide some hope of ultimately by-passing what appear to be essential difficulties.

Analog computation, the first alternative to standard digital computation was all but abandoned in the early 1960's, but has a long and rich history. (See the textbooks [16,17], for example.) We will use a very general notion of what analog computation means — we will not restrict ourselves to the kind of analog computer that uses operational amplifiers, diodes, and so on. Rather, we will consider any physical system at all as a potentially useful computer, provided only that we can communicate with it. We will then attempt to formulate precisely the following question: Can analog computers solve problems using reasonable (non-exponential) resources that digital computers cannot?

The study of this question leads us to formulate a strong version of Church's Thesis: That any finite physical system can be simulated *efficiently* by a digital computer. By "efficiently" we mean in time polynomial in some measure of the size of the physical system. This thesis provides a link between computational complexity theory and the physical world. If we grant that $P \neq NP$, and Strong Church's Thesis, we can then conclude that no physical device solving an NP-complete problem can do so efficiently. We will propose such a device and explore the physical implications of our argument. While we will be able to make certain statements, the important questions in this field are unresolved, especially those dealing with quantum-mechanical systems.

A *cellular automaton* (CA) is in general an n-dimensional array of cells, together with a fixed, local rule for recomputing the value associated with each cell. CA's were originally proposed by von Neumann as a mathematical model to study self-replication. More recently, they have received attention as possible

J. K. Skwirzynski (ed.), Performance Limits in Communication Theory and Practice, 173–192.

models of general nonlinear phenomena, and have been used for nonlinear image processing in the biomedical and pattern recognition fields [3]. We will mention recent work on the use of CA's to model fluid flow.

Cellular automata offer an ideal vehicle for studying highly parallel, pipelined computation structures, such as systolic arrays. We will describe the design and testing of a custom VLSI chip for implementing a one-dimensional, fixed CA, which has a total throughput of more than 10^8 updates per second per chip [4]. We will then discuss the processor/memory bandwidth problems that arise in higher dimensional cases, which are important in fluid dynamics.

We will then describe certain one-dimensional automata that support persistent structures with soliton-like properties — a phenomenon qualitatively similar to those observed in certain nonlinear differential equations. This suggests ways to embed computation in homogeneous computing media, and leads us to speculate about how we can overcome the limits of lithography in large-scale integration.

2. Measuring Complexity

We begin by reviewing the usual way of measuring complexity for digital computer algorithms, and then discuss the extension of these ideas to the analog world.

A very effective and robust model for digital computation is the *Turing Machine*, which is characterized by the following components:

a) A tape with cells that can hold symbols from a *finite* alphabet. Without changing the essential features of our model, we take this to be the *binary* alphabet, the symbols {0, 1}. An unlimited supply of tape is available, but only a finite number of cells is ever used.

b) A head that moves on the tape. The head sees the symbol at its current position.

c) The machine is at any time in one of a *finite* number of states.

d) A *finite* set of rules, which play the role of the computer program. Given the symbol under the head and the current state of the machine, these rules determine at each time step what the new tape symbol shall be, whether the head then moves left, right, or remains stationary, and what the new state shall be. Of course time is discrete, and each application of the rules is counted as one time step.

A critical feature of this model is the fact that the sets of symbols, of states, and of rules, are all *finite*. This is in sharp contrast with the usual models for analog systems, which are usually differential equations, and which have a continuous time parameter, as well as continuous state variables.

A finite number L of cells at the beginning of the tape are initially set, and these correspond to the program input. The number L is then taken to be the *size* of the input, measured in *bits*. We will measure the input size to an analog device in the same way, as a sequence of bits of length L. It is important to notice at this point that an input of length L can represent a number as large as 2^L.

In the case of a Turing Machine, the only resources used by a computation are time and space, the former being the number of discrete time steps, and the latter being the number of tape cells required by a given computation. In the analog case, the situation is not so clear. There is certainly the time taken for the computation, and the volume occupied by the device. But there is also the energy used, the maximum force employed, the largest electric field, and so on; in other words: any physical quantity that might have a real cost associated with it, and which might be bounded by physical constraints. We call these quantities, collectively, the ***resources*** used by the analog computer. We can then speak of a computation by an analog device as taking ***polynomial*** resources if the resources are bounded by a polynomial function of the input description length L, and similarly for computations that take *exponential* resources. (We usually use the term *exponential* to mean *not polynomial.*)

3. An Analog Computer that uses Polynomial Resources

We now give an example of a simple computation that can be performed by an analog computer using polynomial resources.

The Maze Problem: A graph G is defined on an $n \times n$ array of nodes, and each node can be connected only to the nodes adjacent to it in a column or row. The problem is to determine if there is a path from the upper-left corner node s to the lower-right corner node t. (See Fig. 1a.)

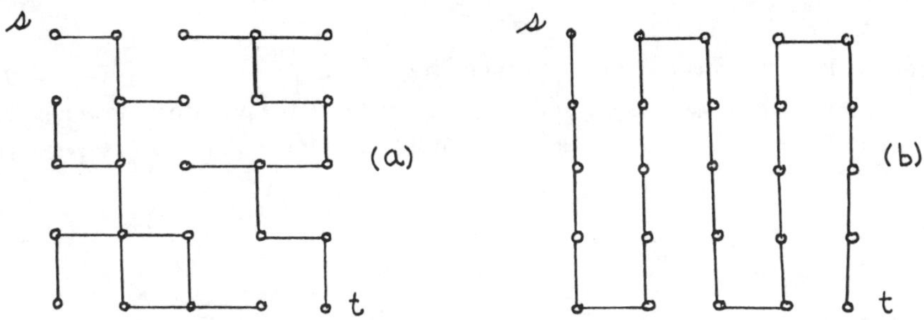

Fig. 1 a) The maze problem; b) A longest s–t path.

This is not a hard problem for a digital computer; in fact, it can be solved in a number of steps proportional to the number of edges. A natural way to approach this problem with an analog computer is to build an electrical network that has a terminal for each graph node, and a wire wherever an edge appears in the graph. We can then apply a voltage source across the terminals s, t, and measure the resulting current flow. The terminals s and t are connected if and only if there is a "positive" current flow.

We need to more precise about what constitutes a positive current, about how long we need to wait to make our decision, and about how expensive the network is. Assume for this purpose that each edge is represented by a wire of rectangular cross-section with width w, length d, and height h, all of which can be a function of n, provided that w doesn't grow faster than d.

Assume first that there is a path from s to t. The longest such path can have $O(n^2)$ edges, as shown in Fig. 1b. The resistance of each wire is proportional to its length, and inversely proportional to its cross-sectional area, so that the total resistance between s and t is

$$R_{cl} \leq k_1 \frac{n^2 \cdot d}{wh} \tag{1}$$

where k_1 is some constant. Notice that if d, w, and h are held constant, the closed-circuit resistance grows as the square of n, or linearly with the number of nodes.

Next, consider what happens when there is no s–t path. The open-circuit (leakage) resistance R_{op} is proportional to d, inversely proportional to the area hd, and in the case leading to the lowest resistance, our worst case, inversely proportional to n^2. This is because there can be in effect no more than n^2 wires in parallel across an s–t cut. We thus have the asymptotic lower bound

$$R_{op} \geq k_2 \frac{1}{n^2 h} \tag{2}$$

where k_2 is another constant, but much larger than k_1.

Combining these two inequalities, we can find an asymptotic lower bound on the ratio between the closed- and open-circuit current:

$$I_{cl}/I_{op} = R_{op}/R_{cl} \geq (k_2/k_1)\frac{w/d}{n^4} \tag{3}$$

If we plan to distinguish successfully between the closed- and open-circuit cases, we need to ensure that this ratio is large enough to make possible reliable measurement. We see from this analysis that the scaling ratio $w/d < 1$ cannot prevent ultimate disaster as n grows indefinitely. What makes this circuit work in a practical situation is the fact that

$$k_2/k_1 >> 1 \tag{4}$$

Thus, the device will work effectively for a large range of n, as we might expect.

However, we also see from this analysis that there is a limit on the size of problem, determined by the technology, beyond which this device simply will not function. It appears that this is an unavoidable limitation in all physical devices that we construct to solve arbitrarily large instances of problems. Suppose, for example, that we try to use water, and model the edges of the maze by pipes. We then need to worry about making the pipes heavy enough to support water pressure that grows indefinitely. If we use microwaves, we need to worry about loss, and so on. But we can accept the fact that there is a large regime in which a device will operate well; the same is true of any real digital computer, even for polynomial-time problems.

Next, consider how long we need to wait (asymptotically) before we can make a reliable decision. This is determined by the RC time constant in the closed-circuit case. The greatest possible capacitance across the terminals of the voltage source can be no worse than proportional to the total surface area between wires, or $O(n^2dh)$, and inversely proportional to the inter-wire distance d, assuming w grows slower than d, for a total capacitance that is $O(n^2h)$. The largest total resistance is $O(n^2d/(wh))$, by (1). The time constant is therefore

$$RC = O(n^4d/w) \tag{5}$$

Thus, letting w grow as d, we find this "analog computer" takes time $O(n^4)$, proportional to the square of the number of nodes. We can also check that the total mass of the machine, and the power consumption, are also polynomial, provided we are in the regime of operation provided by the technology constants k_1 and k_2.

4. Analog Machines that use Exponential Resources

It is all too easy to construct analog computations that require exponential resources, simply because of our previous observation that L bits in the input can encode a number as large as 2^L. An example is provided by A. K. Dewdney's "Spaghetti Analog Gadget" (SAG) [22], for sorting integers, a polynomial-time problem for a digital computer. Instructions for this analog computation follow:

Spaghetti Analog Gadget

1) Given a set of integers a_i to be sorted, cut a piece of (uncooked) spaghetti to the length of each of the numbers.
2) Assemble the pieces in a bundle, and slam them against a flat surface, so that all the pieces are flush at one end.
3) Remove the pieces that extend farthest at the other end, one at a time. The order in which the pieces are removed from the bundle sorts the integers from largest to smallest.

The difficulty here is that the pieces are cut to lengths proportional to the integers in the input data, but these are encoded in binary. The pieces of spaghetti code the numbers in unary, instead of binary. Thus, the machine uses an exponential amount of spaghetti.

We might try to circumvent this problem by using a logarithmic scale to size the spaghetti. This will keep the mass of the machine polynomial, but raises a new problem. We will then need to distinguish between ends of pieces that are exponentially close in size. In the presence of fixed measurement noise, this will require an exponential scaling to keep the measurements reliable, again leading to exponential mass. As proved in [1], there is no way around this difficulty if we insist on representing input numbers by single analog quantities. The digital computer can represent a single number by a (theoretically) unlimited number of registers, and that seems to be an essential difference between digital and analog computation, and the source of the term "analog(ue)".

We now take up the question of whether an analog computer can solve a problem that is intractable for digital computers. To this end we will consider NP-complete problems, problems as difficult as any in the wide class of NP problems, and widely considered to require exponential time on digital computers. (This belief is usually expressed in the conjecture that $P \neq NP$.)

5. Strong Church's Thesis

We now need to emphasize an important distinction between statements about Turing Machines, or digital computers in general, and analog computers. In the former case, our statements can be made and proven with mathematical precision — they are, in fact, mathematical statements. When discussing analog devices, however, we are making statements about what will happen in the physical world, and the correctness of our conclusions depends on the models we choose. For example, we choose to describe the electrical gadget for the Maze Problem by the usual electrical models, including Ohm's law. Our prediction of its speed of operation depends on a differential equation model.

Thus, there is always room to refine our model of analog computers, and it is never guaranteed that predictions of behavior will be correct. For example, we ignored the inductance in the circuit and its effect on its time-domain behavior; we assumed that the noise environment does not interfere with the measurement of current as the size of the device grows to infinity; and so on.

To carry over the ideas of complexity theory from the mathematical realm to the physical, we need something that will relate the two worlds in some way. Such a connection between mathematics and the informal world is provided by Church's Thesis, also called the Church-Turing Thesis [12,13]. The usual way of stating it is that the Turing Machine completely captures the notion of computation; that is, that anything that can be computed at all can be computed by the Turing Machine. In our context, we can view this as stating that the Turing Machine can compute anything that an analog device can.

We now go one step further, and formulate a stronger version of Church's Thesis that relates complexity in the two domains. (A similar idea has been proposed by R. Feynman [14].)

Strong Church's Thesis (SCT) An analog device can be simulated by a Turing Machine in time that is a polynomial function of the resources it uses.

What this says, essentially, is that any piece of the physical world can be simulated by a digital computer without requiring time exponential in any measure of the size of the piece. SCT is not susceptible to mathematical proof; it is in fact a statement about physics that may or may not be true, as is the usual Church's Thesis. Only by postulating a particular model for a piece of the physical world could one hope to prove it. For éxample, it is proved for systems described by a certain class of differential equations in [1].

We are going to consider the possibility of using analog computers to solve NP-complete problems, which are generally considered to be intractable in the sense of requiring exponential time. Furthermore, all the members of the class are equivalent to each other in the sense that if one can be solved in polynomial time, they all can. Thus, either they are all in P (P = NP), or none is (P $\neq$ NP). The reader can find more on this subject in [15].

SCT now allows us to make the following kind of argument. Suppose an analog device solves an NP-complete problem using a polynomial amount of resources. Then the fact that we can simulate it in polynomial time with a Turing Machine means that there is also a Turing Machine that solves the problem in polynomial time, and so P = NP. If we then take as postulates P $\neq$ NP and SCT, we have a metamathematical argument that the given device cannot operate using polynomial resources. We will study such a device in the next section.

6. An Analog Machine for an NP-complete Problem

The NP-complete problem we will solve is called PARTITION [15]:

PARTITION Given a set of positive integers $S = \{\omega_1, \omega_2, \ldots, \omega_n\}$, is there a nonempty subset S' of S with the property that the elements in S' sum to exactly $(1/2)\sum_{i=1}^{n} \omega_i$? That is, does there exist an S' that partitions S into two equal-weight subsets, those in S', and those not?

The analog device we will construct will represent the integer ω_i by a cosine signal with frequency ω_i . We observe first that by using the fundamental operations of addition, subtraction, and squaring of signals, we can form the product of two cosine waves, which also consists of the sum of cosine waves at the sum and difference frequencies. This follows from the following identities:

$$\begin{aligned} 4\cos x \cos y &= 2[\cos(x+y) + \cos(x-y)] \qquad (6) \\ &= (\cos x + \cos y)^2 - (\cos x - \cos y)^2 \end{aligned}$$

From this we see that we can synthesize a signal at the frequency ω with $O(\log\omega)$ such operations, and so the synthesis of these signals requires only a polynomial amount of equipment.

In the same way we can construct the signal

$$\frac{1}{2} \prod_{i=1}^{n} \cos\omega_i = (2^{-n}) \sum \cos(\omega_1 + \sum_{i=2}^{n} \pm\omega_i) \tag{7}$$

where the outer sum is over all 2^{n-1} combinations of + and – in the inner sum. We can then determine if one of these sums is zero by integrating over one period, from $t = 0$ to 2π. In fact, that integral will be zero if and only if one of the inner sums in (7) is zero, and this is equivalent to there being a solution to the partition problem on the numbers $\{\omega_i\}$. Thus we have shown that the following problem is NP-complete, a result due to Plaisted [15, 18]:

COSINE PRODUCT INTEGRATION Given a set of positive integers $S = \{\omega_1, \omega_2, \ldots, \omega_n\}$, does

$$\frac{1}{2\pi} \int_0^{2\pi} \frac{1}{2}\left(\prod_{i=1}^{n} \cos(\omega_i t)\right) dt = 0\ ? \tag{8}$$

Figure 2 shows a sketch of a device that will make this decision, using adders, subtracters, squarers, an integrator, and a threshold detector. While we have tried to make the operation of this PARTITION machine as practical-sounding as possible, the main point of the preceding discussion is that it will not work in practice, in the sense of requiring resources exponential in the length of the input data. Where does the machine founder? Perhaps providing enough bandwidth for the integrator is the problem. Perhaps noise will obscure the distinction between zero and not zero. Perhaps the accuracy requirements on the square-law device are prohibitive. As satisfying as it might be to find the flaw in its operation, it is not necessary to analyze the device; if we accept that P $\neq$ NP and Strong Church's Thesis, it cannot work efficiently in practice.

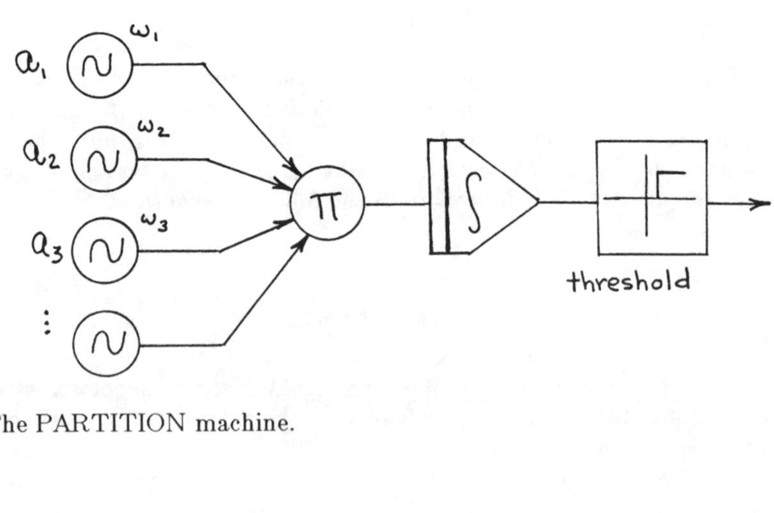

Fig. 2 The PARTITION machine.

In fact, however, it is not hard to locate one flaw that would become evident if we tried to solve large instances of PARTITION with such a machine. Each distinct partition of the ω_i contributes a zero-frequency term to the integrand in (8), and that results in an integral with value 2^{-n}, from (7). Therefore, if there is only one such term we must distinguish between the values 0 and 2^{-n}. If there is a fixed level of noise, we would then need to scale by an exponential factor to accomplish this discrimination. In other words, the machine proposed operates with an exponentially small signal-to-noise ratio. Somehow, it seems that P $\neq$ NP and SCT together imply that there is an irreducible level of noise in the world.

In [1] another analog machine for an NP-complete problem is described. A device constructed with gears, levers, and cam-followers is constructed that ostensibly solves the NP-complete Boolean Satisfiability problem. (Actually, the special case called 3-SAT.) This problem does not have integers in its input description, and the flaw in the operation of the machine is perhaps less obvious than for our PARTITION machine. The link that Strong Church's Thesis provides between the mathematical and physical worlds is strong enough to allow us to make non-trivial statements about certain physical systems.

7. Cellular Automata

We next turn our attention to another way of thinking about computation, *cellular automata.* In one sense this subject does not present the kind of fundamental difficulties we encountered when we studied analog computation. There is nothing that can happen in a cellular automaton that, in principle, cannot be simulated by Turing Machine. Rather, the interest lies in the organization of the computation. A cellular automaton can reflect the way that computations are performed in nature, and so can be a more natural framework than a conventional serial computer for studying certain phenomena — the kinds of phenomena that occur in systems with a large number of identical, locally interacting components. The important features of a cellular automaton are the *uniformity* and *locality* of its computations.

A cellular automaton has three components:

a) A discrete, finite *state space* Σ containing the value 0, usually the additive group mod k on the integers $\{0, 1, 2, \cdots, k\text{-}1\}$. Often $k = 2$, in which case we say the automaton is *binary.*

b) A set of *sites* or *cells* called the *domain* Δ, usually a discrete regular lattice of points in n dimensions, often simply equally spaced points on a line or circle, or a rectangular or hexagonal grid in the plane. Each site in the domain has a value in Σ associated with it, and we refer collectively to the domain and the state values associated with each site as the *state* of the automaton. We also refer to n as the dimension of the automaton.

c) A *rule* Φ which, given the state at discrete time t, yields the state at time t+1. The rule uses as arguments the values in some fixed neighborhood of each site.

Thus, the rule Φ can be thought of as "re-writing" the values at each of the sites, for times $t = 1, 2, 3, \ldots$, given an initial state at $t = 0$. We will always assume that the initial state has only a finite number of sites with non-zero values, and so can be finitely described.

The cellular automaton was invented by von Neumann to study self-replication [19]. Today, there are two main reasons for studying the model: first, as a model for physical phenomena; and second, as a model for computation in regular networks, such as neural networks. We will naturally concentrate here on the latter category, but we should mention some recent, exciting work in the first category, the application of cellular automata to fluid dynamics, as an alternative to the numerical solution of the Navier-Stokes equation [23,24].

The idea behind the fluid dynamic machines is to model the behavior of a fluid by a collection of particles which can move around a lattice, with fixed rules for what happens at collisions. Under certain circumstances, and with the right rules, it can be shown that the average velocity fields obtained do in fact coincide with solutions to the Navier-Stokes equation [25]. So far, the most successful computations use a hexagonal grid in two dimensions, and each site is assumed to have up to 6 particles, possibly one traveling towards each of the 6 neighbors of the site, and possibly a particle at rest at the site. Thus, 7 bits suffice to describe the state at a site. The rules need to be carefully designed so that momentum and mass are conserved. When the automaton is started from an initial random state with a certain population density, numerical velocity fields are obtained by averaging over blocks of sites, typically 48×48 sites square. Results so far illustrate the development of vortex streets, and other qualitative phenomena associated with turbulent flow, for low effective Reynolds Numbers.

Whether this approach is ultimately of practical importance in fluid dynamics depends on whether the calculations required can be made highly parallel. This is one way in which the cellular automaton model shines; its great regularity and simplicity invite deep pipelining and custom VLSI implementations, and the parallelism obtained this way may more than compensate for the primitive nature of the individual computations. This is especially true if one-dimensional pipelining can be used. To illustrate the way in which one-dimensional computation can be pipelined to an arbitrary depth, we will describe briefly a custom VLSI chip that was designed and tested at Princeton.

8. Cellular Automata: Notation and an Example

We first set down some notation to describe cellular automata. Let a_i^t be the state value at time t and position i, $-\infty \leq i \leq +\infty$, and $0 \leq t \leq +\infty$. The next-state rule is of the form

$$a_i^{t+1} = \Phi(a_{i-r}^t, a_{i-r+1}^t, \ldots, a_i^t, \ldots, a_{i+r}^t) \tag{9}$$

We also usually assume that Φ preserves zero,

$$\Phi(0,0,\ldots,0) = 0$$

The next value of site i is a function of the previous values in a neighborhood of size $2r + 1$ that extends from $i - r$ to $i + r$. Given initial states at all the sites, repeated application of the rule Φ determines the time evolution of the automaton.

As an example, we will describe a particular binary, one-dimensional automaton with $r = 2$, denoted by Wolfram [2] the rule-20 totalistic cellular automaton. This is perhaps the simplest automaton that exhibits very complicated behavior, and may in fact be universal in the sense of having the power of a Turing machine. The rule is called *totalistic* by Wolfram [2], because the next-state function Φ depends only on the sum of its arguments. Let that sum at time t and position i be denoted by S_i^t,

$$S_i^t = \sum_{i-r}^{i+r} a_i^t \tag{10}$$

The rule is the following with $r = 2$,

$$a_i^{t+1} = \begin{cases} 1 & S_i^t \text{ even but not } 0 \\ 0 & \text{otherwise} \end{cases} \tag{11}$$

We will call any rule of this form a *parity rule.* It is called *rule-20* because the binary expansion of 20 is 10100, and this string determines which values of S_i^t yield an output value of 1, by the 1's at bits 4 and 2, counting from the right and starting with 0.

Figure 3 shows the evolution of this automaton, and illustrates an interesting feature, namely the presence of persistent structures that we will call *particles.* The figure shows two examples of particles of different speeds colliding destructively. Particles are quite rare in this automaton, and almost all collisions are destructive. Later on we will describe other automata in which particles are much more common, and collide non-destructively.

Based on extensive experimental evidence, Wolfram [2] has classified the behavior of cellular automata into four types:

1) evolution to a homogeneous state (analogous to a *limit point* in nonlinear differential equations),
2) evolution to separated stable or periodic structures (analogous to *limit cycles*),
3) evolution to chaos, and
4) evolution to complex, often long-lived structures.

The last type of behavior, exhibited by the rule-20 cellular automaton, will be most interesting to us, because it will suggest ways in which computation can be embedded into the operation of the automaton.

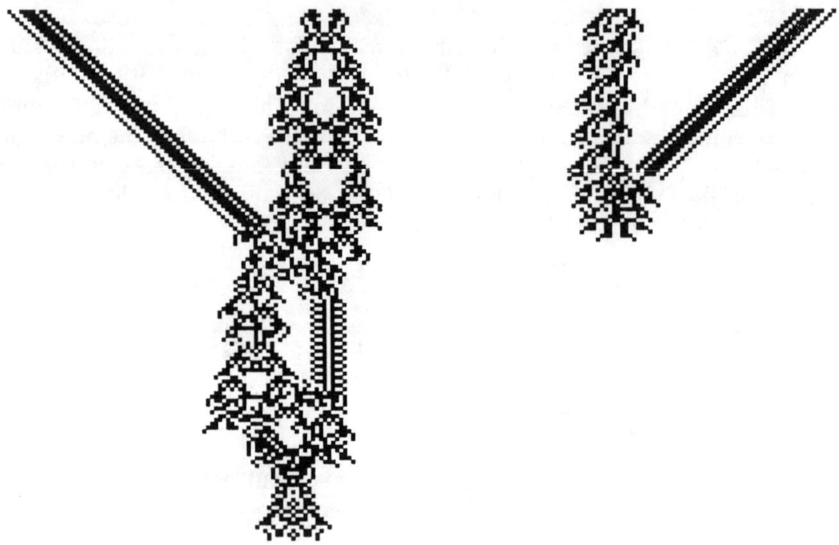

Fig. 3 Destructive collisions in the code-20 cellular automaton.

9. A Custom Systolic Chip for a Cellular Automaton

A chip was designed to implement the rule-20 cellular automata described above. The implementation is interesting from the point of view of the kinds of systolic arrays used to achieve highly parallel computation for signal processing, and we will describe it here briefly.

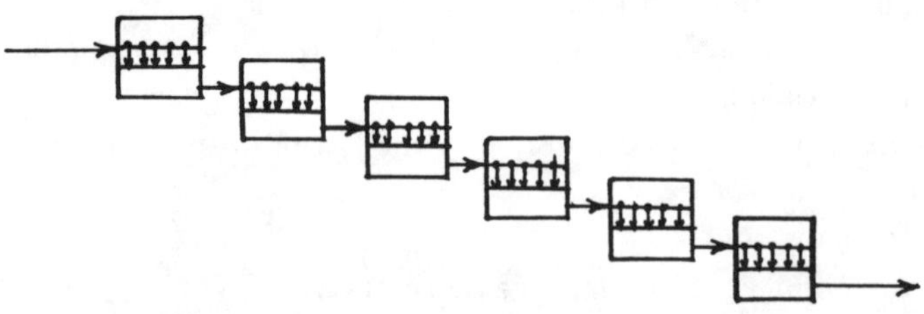

Fig. 4 Bit-serial systolic array for a cellular automaton.

Figure 4 shows how a simple concatenation of bit-serial processors can be used to perform the updating operation. Each processor consists of a $(2r+1)$-bit shift register which holds the argument bits, and fixed combinatorial logic which performs the calculation of the next-state function Φ. The operation is synchronous, and each major clock cycle a new output bit is produced by each processor, and the contents of the shift registers advance one cell.

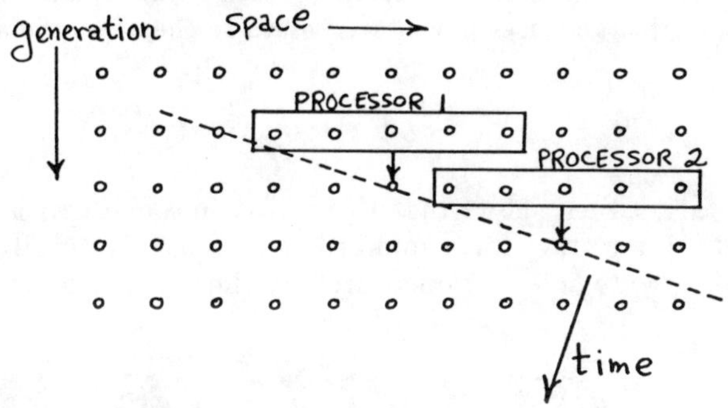

Fig. 5. The computational wavefront.

At any given moment, each processor is fully occupied, and so the array achieves maximum parallelism. Note, however, that it is also true that at any given moment, each processor is working on the generation after the one of the preceding processor, as shown in Fig. 5. Thus, the calculation proceeds along a northwest-to-southeast wavefront. In an actual implementation, the data is circulated through some number N of processors, and each complete circulation results in an update of the one-dimensional array by N generations.

The fixed processor for rule-20 was implemented in 4 μ nMOS, using a 5-bit static shift register and a PLA for the update function Φ. The small chip of 3.1×3.8 μ holds 18 such processors, in 3 columns of 6 processors each, plus the inter-processor wiring and pads. Of the 16 chips that were returned from the MOSIS fabrication facility, 9 were fully functional at speeds of at least 6 Mhz. This implies an effective computational rate of greater than 10^8 site-updates per second per chip, and illustrates the power of the bit-serial systolic approach to one-dimensional computation in such automata.

If this idea is extended to two-dimensional automata, it requires local storage of two full rows of state values at each processor, as opposed to $(2r+1)$ site values in the one-dimensional case. This fact means that many fewer processors can be fit on a single chip, and thus reduces the amount of effective parallelism that can be achieved on a chip. The approach is still very useful in two-dimensional applications such as image processing [26], and is now being studied at Princeton for fluid dynamics applications [27].

10. Parity-Rule Filter Automata

We will now discuss a slightly different kind of automaton, called *Filter Automata* [5]. The original motivation for studying them came from digital signal processing [7]. Noticing that the usual kind of cellular automaton corresponds in its data access pattern to a finite-impulse-response (FIR) digital filter, it is natural to see what happens when the corresponding infinite-impulse-response (IIR) pattern is used. This corresponds to replacing the update rule (9) by one that scans left-to-right, and uses new state values as they become available,

$$a_i^{t+1} = \Phi(a_{i-r}^{t+1}, a_{i-r+1}^{t+1}, \ldots, a_i^t, \ldots, a_{i+r}^t) \tag{12}$$

Some experimentation then shows that these filter automata with the parity rule (11) are especially interesting for a number of reasons. Naturally, we call these the *parity-rule filter automata*, parameterized by the neighborhood width r.

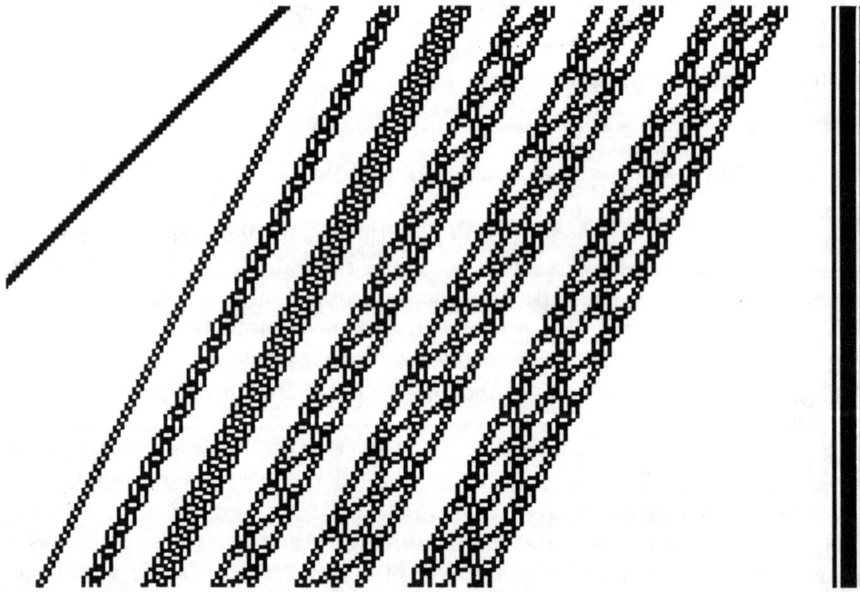

Fig. 6 Some particles in the $r = 2$ parity-rule filter automaton.

The first interesting thing to observe is that these automata support a very wide variety of particles. Figure 6 shows just a few of these for the $r = 2$ parity-rule filter automaton. For each r, there is a unique zero-speed particle (shown rightmost), and all the other particles move to the left. There is also a unique fastest particle, called the *photon*, which consists of $(r+1)$ consecutive 1's, moving to the left at a speed of $(r-1)$.

To classify a particle, we interpret the bit strings in its (periodic) evolution as binary numbers, and define the smallest such number as the ***canonical code*** of the particle. Obviously, this is a unique way of identifying a particle. We also refer to the number of generations before repetition as the particle's ***period***, the number of sites moved left in a period as its ***displacement***, and the ratio of displacement to period as its ***speed***.

The number of distinct particles grows rapidly with r. An exhaustive search program yields the following numbers of particles with canonical code width up to and including 16:

radius	number of particles
2	8
3	198
4	682
5	6534
6	13109

These particles come with a variety of displacement/period pairs.

The next interesting thing about the parity-rule filter automata is that particles pass through each other, sometimes with their identities preserved, and sometimes not. Figure 7a illustrates some identity-preserving collisions for the $r = 3$ automaton, while Fig. 7b shows collisions that don't preserve identity, for the same automaton.

Close inspection of Fig. 7b shows a collision in which two particles come together and interact in such a way as to produce two different particles traveling in parallel (the second and third particles from the left at the bottom). This shows that the parity-rule filter automata are in general not reversible in time.

At this point the similarity of these particles with the solitons supported by differential equations [10,11] becomes apparent. These "solitary waves" occur in nonlinear systems, and are characterized chiefly by the non-destructive nature of their collisions. The close connection with the solitons supported by the nonlinear lattices of Hirota and Suzuki [11] is being studied at Princeton with Nayeem Islam [21].

The particles illustrated in Fig. 6 that appear to be composed of several resonating particles traveling in parallel also have their counterpart in physical solitons, where they are called "puffers".

Many further properties of the parity-rule filter automata have been proved by C. H. Goldberg, in a forthcoming paper [28]. For example, he shows that they are inherently stable, in the sense that an initial state with a finite number of 1's always leads to states that also have a finite number of 1's.

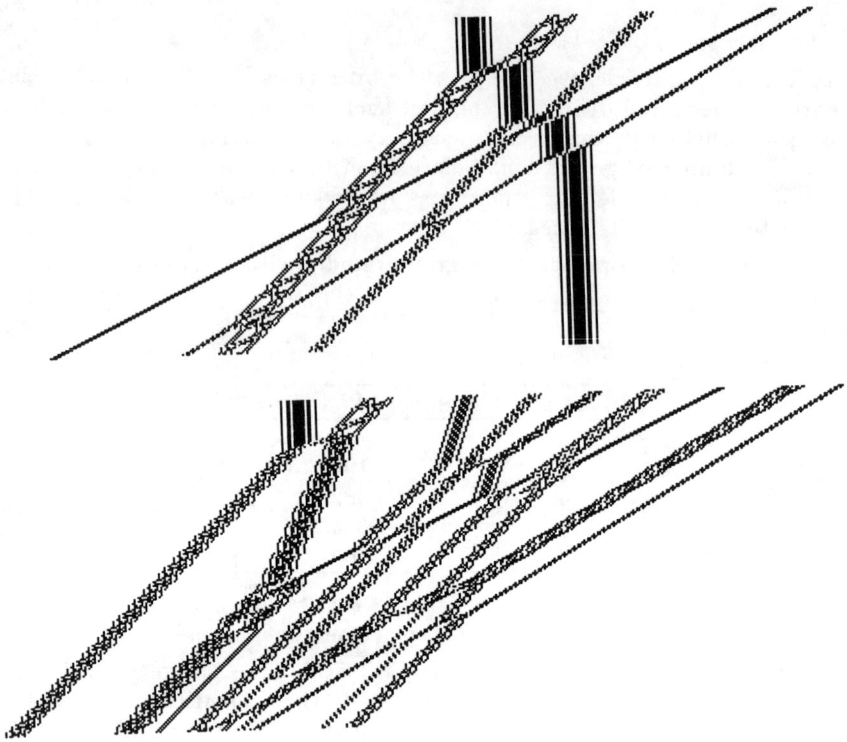

Fig. 7 a) Identity-preserving collisions in the $r = 3$ parity-rule filter automaton; b) Identity-destroying collisions in the same automaton.

11. Phase Shifts and Embedded Computation

We now discuss the possibility of encoding information in the particles supported by these filter automata, and of processing this information by way of the collisions.

As in physical solitons, a collision produces a shift in the trajectory of each particle, which we can think of as a translational phase shift. It always happens that when a fast particle catches up with and passes through a slow particle, the slow particle is thrust backward from where it would otherwise be, and the fast particle propelled forward. This can be verified for the examples shown in Fig. 7. (Incidentally, physical solitons behave in exactly the same way.) We can thus think of the position of a particle relative to some constant-speed reference as representing information that is changed on collision, the *translational* phase of the particle. As a simple example, we might encode a 0/1 bit in the translational phase according to whether its position relative to a reference is even/odd.

There is another kind of phase inherent in the motion of a particle — the relative state of the particle in its periodic path through different configurations. This, too, can be changed by collision, and we call these *orbital* phase changes, as opposed to the translational phase changes. If we liken the motion of the particle to that of a wave packet, we can think of the translational and orbital phases as those of the envelope and carrier, respectively. In what follows, we will refer to the composite translational/orbital phase as simply the *phase* of a particle.

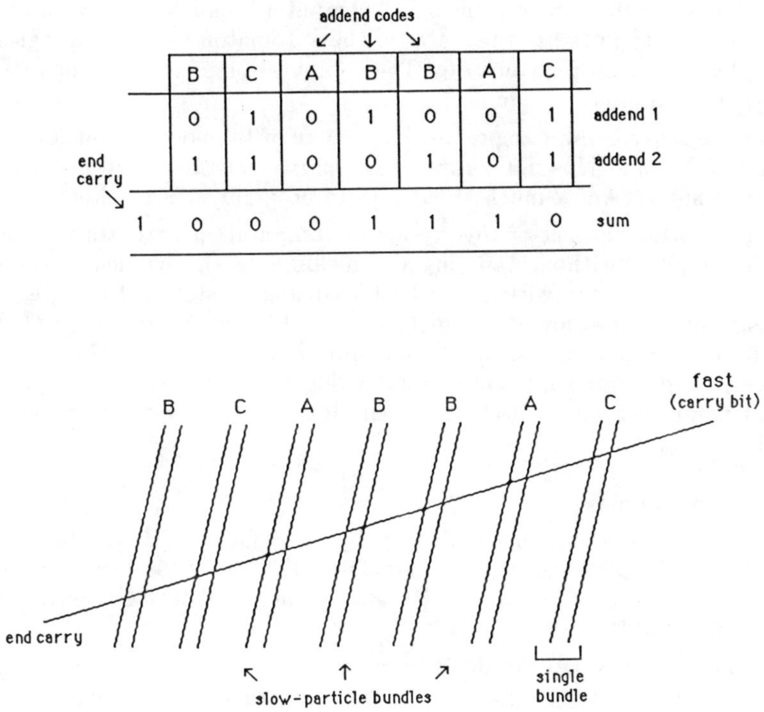

Fig 8. A carry-ripple adder embedded in an $r = 5$ parity-rule filter automaton.

A simple example of a non-trivial computation that can be performed by colliding solitons is described in [20], and we summarize the idea here. The computation is that of a carry-ripple adder, and is illustrated in Fig. 8. The carry bit is represented by the phase of a fast particle entering the scene from the right; the addends are represented by pairs of slow particles. Solutions of this form were found for $r = 5$. Another solution was found for $r = 3$ with 3 particles per addend pair.

Some work is involved in making such a scheme work. We need to make tables of the phase changes that result from collisions, and then we need to search for combinations of particles that behave in the way we want. This can be automated as a search for paths in a constructed graph, and the details are in [20]. We can think of this part of the problem as "programming" the embedded computation.

The idea of using persistent moving structures to do computation is not new; for example, Conway's game of Life [8] is able to mimic the operation of a general purpose computer, and so is capable of universal computation. Similar embeddings are described in [9]. There are two important differences here: first, we are dealing with a one-dimensional "computer", and second, as we have seen, we do not need to initialize the states of the automaton to reflect a logical organization, such as gates, registers, etc. The computation can be encoded very naturally in the serial input stream.

On the other hand, our present knowledge of the power of such computation is very limited, and it is not known if the parity-rule filter automata are universal. We do not yet know much about how to program such machines.

An important aspect of this notion of computation is that a medium can be used to compute, without tailoring the medium to the particular computation, either by designing its "wiring", or by initializing its states. If it were possible to synthesize such a medium at the molecular level [6], we would avoid the problems posed by the limits of lithography, and could have a computer that functioned at that level. The approach also suggests analogous computation with physical solitons [21], and, in a more speculative vein, possible explanations of some kinds of brain function.

Acknowledgements

This work was supported directly by NSF Grant ECS-8414674 and U. S. Army Research Office-Durham Grant DAAG29-85-K-0191; and by computing equipment made possible by DARPA Contract N00014-82-K-0549 and ONR Grant N00014-83-K-0275.

Much of this work was done in collaboration with others, and reported in referenced papers. I am indebted to my co-authors, Brad Dickinson, Nayeem Islam, Irfan Kamal, James Park, Bill Thurston, Tassos Vergis, and Arthur Watson. I also thank the following for helpful discussions: Kee Dewdney, Jack Gelfand, Charles Goldberg, Andrea LaPaugh, Martin Kruskal, Steven Kugelmass, James Milch, Bob Tarjan, Doug West, and Stephen Wolfram.

References

1. A. Vergis, K. Steiglitz, B. Dickinson, "The Complexity of Analog Computation," *Mathematics and Computers in Simulation*, in press.
2. D. Farmer, T. Toffoli, S. Wolfram (eds.), *Cellular Automata*, North-Holland Physics Publishing, Amsterdam, 1984.
3. K. Preston, Jr., M. J. B. Duff, *Modern Cellular Automata, Theory and Applications*, Plenum Press, 1984.
4. K. Steiglitz, R. Morita, "A Multi-Processor Cellular Automaton Chip," Proc. 1985 IEEE International Conference on Acoustics, Speech, and Signal Processing, Tampa, Florida, March 26-29, 1985.

5. J. K. Park, K. Steiglitz, W. P. Thurston, "Soliton-Like Behavior in Automata," *Physica D*, in press.

6. F. L. Carter, "The Molecular Device Computer: Point of Departure for Large Scale Cellular Automata," pp. 175-194 in [2].

7. A. V. Oppenheim, R. W. Schafer, *Digital Signal Processing*, Prentice-Hall, Englewood Cliffs, N. J., 1975.

8. E. R. Berlekamp, J. H. Conway, R. K. Guy, *Winning Ways for your Mathematical Plays, Vol. 2: Games in Particular*, (Chapter 25, "What is Life?"), Academic Press, New York, N. Y., 1982.

9. F. Nourai, R. S. Kashef, "A Universal Four-State Cellular Computer," *IEEE Trans. on Computers*, vol. C-24, no. 8, pp. 766-776, August 1975.

10. A. C. Scott, F. Y. F. Chu, D. W. McLaughlin, "The Soliton: A New Concept in Applied Science," *Proc. IEEE*, vol. 61, no. 10, pp. 1443-1483, October 1973.

11. R. Hirota, K. Suzuki, "Theoretical and Experimental Studies of Lattice Solitons on Nonlinear Lumped Networks," *Proc. IEEE*, vol. 61, no. 10, pp. 1483-1491, October 1973.

12. A. Church, "An Unsolvable Problem of Elementary Number Theory," *Amer. J. Math.*, vol. 58, pp. 345-363, 1936. (Reprinted in [13].)

13. M. Davis, *The Undecidable*, Raven Press, Hewlett, NY, 1965.

14. R.P. Feynman, "Simulating physics with computers," *Internat. J. Theoret. Phys.*, vol. 21, pp. 467-488, 1982.

15. M.R. Garey and D.S. Johnson, *Computers and Intractability: A Guide to the Theory of NP-Completeness*, W. H. Freeman & Co., San Francisco, CA, 1979. See also: D.S. Johnson, "The NP-completeness Column: an Ongoing Guide," *J. Algorithms*, vol.4, pp. 87-100, 1983.

16. A. Jackson, *Analog Computation*, McGraw-Hill, New York, NY, 1960.

17. W. Karplus and W. Soroka, *Analog Methods*, 2nd. ed., McGraw-Hill, New York, NY, 1959.

18. D. Plaisted, "Some Polynomial and Integer Divisibility Problems are NP-Hard," *Proc. 17th Ann. Symp. on Foundations of Computer Science*, pp. 264-267, 1976.

19. A. M. Turing, "On Computable Numbers, with an Application to the Entscheidungsproblem," *Proc. London Math. Soc., Series 2*, vol. 42, pp. 230-265, 1936-7; corrections *ibid*, vol. 43, pp. 544-546, 1937. (Reprinted in [13].)

20. K. Steiglitz, I. Kamal, A. Watson, "Embedding Computation in One-Dimensional Automata by Phase Coding Solitons," Tech. Rept. No. 15, Dept. of Computer Science, Princeton University, Princeton NJ 08544, Nov. 1985.

21. N. Islam, K. Steiglitz, "Phase Shifts in Lattice Solitons, and Applications to Embedded Computation," in preparation.

22. A. K. Dewdney, "On the Spaghetti Computer and other Analog Gadgets for Problem Solving," in the Computer Recreations Column, *Scientific American*, vol. 250, no. 6, pp. 19-26, June 1984. See also the columns for Sept. 1984, June 1985, and May 1985, the last also containing a discussion of one-dimensional computers.

23. J. B. Salem, S. Wolfram, "Thermodynamics and Hydrodynamics with Cellular Automata," unpublished manuscript, November 1985.

24. U. Frisch, B. Hasslacher, Y. Pomeau, "A Lattice Gas Automaton for the Navier Stokes Equation," Preprint LA-UR-85-3503, Los Alamos National Laboratory, Los Alamos, New Mexico 87545, 1985.

25. S. Wolfram, "Cellular Automaton Fluids 1: Basic Theory," preliminary manuscript, Institute for Advanced Study, Princeton, NJ 08540, 1986.

26. S. R. Sternberg, "Computer Architectures Specialized for Mathematical Morphology," pp. 169-176 in *Algorithmically Specialized Parallel Computers*, L. Snyder, L. H. Jamieson, D. B. Gannon, H. J. Siegel (eds.), Academic Press, 1985.

27. S. Kugelmass, K. Steiglitz, in progress.

28. C. H. Goldberg, "Parity Filter Automata," in preparation.

THE CAPACITY REGION OF THE BINARY MULTIPLYING CHANNEL - A CONVERSE

J. Pieter M. Schalkwijk

Department of Electrical Engineering, Eindhoven University of Technology , Den Dolech 2, P.O.Box 513, 5600 MB Eindhoven, the Netherlands.

1. INTRODUCTION

This paper presents a converse establishing the capacity region of the binary multiplying channel (BMC). Blackwell's BMC is a deterministic two-way channel (TWC) defined [1] by $Y_1=Y_2=Y=X_1X_2$, where X_1 and X_2 are the binary input variables, and $Y_1=Y_2=Y$ is the common binary output variable. The BMC is thus an AND-gate.

In [2] the author described a simple coding strategy that allows reliable transmission over the BMC at rate pairs $\underline{R}=(R_1,R_2)$ outside the Shannon inner bound region [1]. Each sender tries to send information that without loss of generality [2] can be taken as a subinterval θ of a [0,1] interval (see Fig. 1). The amount of information sent specifies the length of the subinterval. Hence, the combined information (θ_1,θ_2) of both senders specifies a subrectangle $\theta_1 \times \theta_2$ of the unit square, which is the Cartesian product of the subintervals of the senders. The constructive coding strategy [2] thus successively subdivides the unit square into regions that allow the determination of the second sender's subinterval given that the first

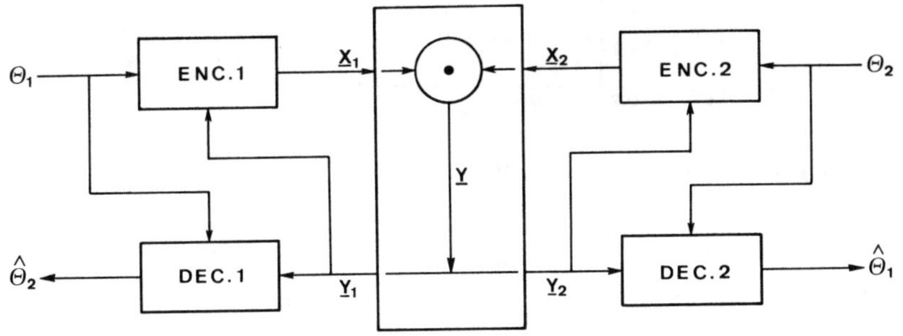

FIGURE 1. Two-way channel configuration.

J. K. Skwirzynski (ed.), Performance Limits in Communication Theory and Practice, 193–205.

sender's subinterval is known. For the case of equal rates in both directions one achieves $R_1=R_2=.61914$ in excess of Shannon's [1] inner bound rate $R_1=R_2=.61695$.

In [3] by bootstrapping the above strategy the achievable rate region was extended to include the point $R_1=R_2=.63056$ rounded to five decimal places. Symmetric $R_1=R_2$ operation also yields the simple equation (8) of [3] for this common rate $R_1=R_2=R$. It is not hard to see [4] that this same equation (8) in vector form also applies to the situation where $R_1 \neq R_2$. Namely, by sending fresh information at rate $\underline{R}^*$ along with the bootstrapping information $\underline{I}_m$, see (7) of [3], so as to make $\underline{I}_m+\underline{R}^*$ colinear with $\underline{R}$, we can substitute for $\eta=(\underline{I}_m+\underline{R}^*)/\underline{R}$ in

$$\underline{R} = \frac{q_i\underline{I}_i+q_m(\underline{I}_m+\underline{R}^*)+q_o\underline{I}_o}{q_i+\eta q_m+q_o} ,$$

which then reduces to (8) of [3] in vector form.

Fig. 2 is a plot of the extended achievable rate region G together with

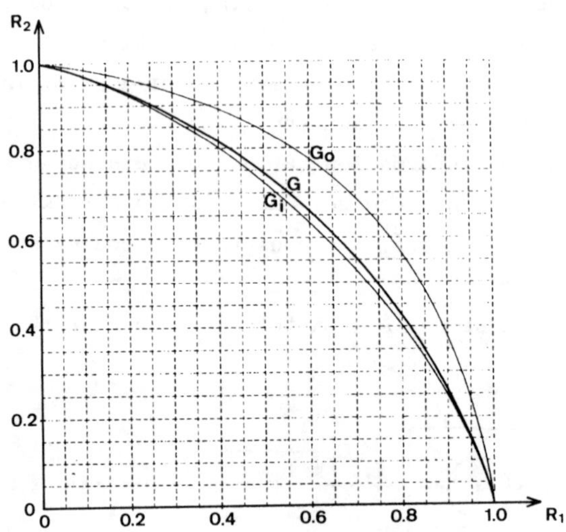

FIGURE 2. The capacity region G of the BMC.

Shannon's inner and outer bound regions G_i and G_o, respectively. The achievable rate region G is given by (8) of [3] in vector form, i.e.

$$\underline{R} = \zeta\underline{I}_i + (1-\zeta)\underline{I}_o, \tag{1}$$

where

$$\underline{I}_i = (\alpha_2 h(\alpha_1),\ \alpha_1 h(\alpha_2)),$$

$$\underline{I}_o = (\beta_2 h(\beta_1),\ \beta_1 h(\beta_2))/(\beta_1+\beta_2-\beta_1\beta_2)$$

with

$$\beta_1 = (1-\alpha_1)/\gamma_1,\ \beta_2 = (1-\alpha_2)/\gamma_2,$$

and

$$\zeta = [1 + (1-\alpha_1)\gamma_2 + (1-\alpha_2)\gamma_1 - (1-\alpha_1)(1-\alpha_2)]^{-1},$$

where $h(\alpha)$ stands for the binary entropy function: $h(\alpha) \triangleq -\alpha\log_2\alpha - (1-\alpha)\log_2(1-\alpha)$. For symmetric $R_1=R_2=.63056$ operation the optimizing parameters [3] are

$$\alpha_1 = \alpha_2 = .69070,$$

$$\gamma_1 = \gamma_2 = .53073.$$

In the remainder of this paper we will show that G as given by (1) is, in fact, the capacity region of the BMC.

Shannon [1] shows that capacity can be attained using fixed length strategies such that the probability of error tends to zero with increasing block length. From Fano's theorem we know that an average error probability of $\varepsilon\to 0$ implies an equivocation no larger than $f(\varepsilon)\to 0$. Now send $L\to\infty$ blocks and eliminate the remaining equivocation using two optimal variable length source codes that are transmitted error free over the BMC in time sharing. It then follows from Shannon's variable length source coding theorem [5] that asymptotically for large block length each fixed length strategy with probability of error tending to zero can be converted into a variable length strategy with probability of error equal to zero that has essentially the same rate! We are thus justified in the sequel to upper bound the rate of variable length strategies with probability of error equal to zero. According to [2] we can represent these coding strategies as strategies for subdividing the unit square.

The crucial idea of the converse can already be found in [3]. There one reads on p.447 that "one can also iterate inward from Shannon's outer bound and arrive at the same achievable rate region G". The converse now proceeds as follows. In subdividing the unit square we consider the average rate of uncertainty reduction for resolutions on the initial t-thresholds. In Section III we will show that this rate must be within the achievable rate region G of (1), i.e. it is upper bounded by .63056 in the symmetric $R_1=R_2$ case. Now the total rate of uncertainty reduction is a weighted average of the above rate of uncertainty reduction for t-resolutions and the rate of uncertainty reduction for non t-resolutions. Initially assuming Shannon's outer bound rate [1] for the average rate of uncertainty reduction of non t-resolutions we can iterate inward, see remark in [3] quoted above, thus establishing G of (1) as an outer bound to the achievable rate region for variable length strategies with probability of error equal to zero. Tolhuizen [6], however, has shown that variable length strategies with probability of error equal to zero can be truncated to obtain fixed length strategies with probability of error tending to zero that have essentially the same rate. This, finally, establishes G of (1) as the capacity region of the BMC for both variable length strategies with probability of error equal to zero, and fixed length strategies with probability of error tending to zero.

2. RESOLUTION STATES

Consider an optimum strategy for the case where each terminal, see Fig. 1, has five equally likeli (same length) message intervals. This optimum strategy, i.e. the optimum resolution of the 5×5 square S, was found by an exhaustive search on the computer. Fig. 3 shows the subdivision of the L-shape, S_0, that pertains upon receiving a 0 on the initial transmission. If the first digit received is a 1 the remaining 3×3 subsquare is subdivided according to Schalkwijk's original strategy with the α,γ parameters of [2] equal to $\alpha=\gamma=\frac{2}{3}$. In Fig. 3 solid and dashed arrows correspond to a 0 and a 1 received, respectively. Likewise, blank and shaded subregions yield 0's and 1's, respectively, on a subsequent transmission. Note that the t-thresholds are used initially in the square S, and later in the subregions S_{000}, S_{011}, and S_{0011}, where S_s stands for the subregion that pertains after receiving the binary $\underline{y}$-string $s\varepsilon\{0,1\}^*$.

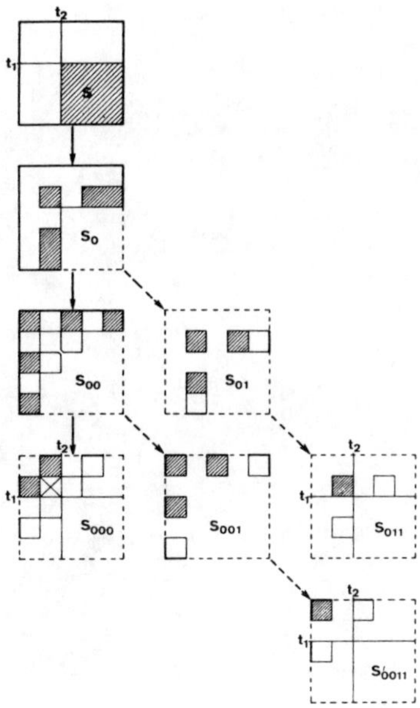

FIGURE 3. Resolution subtree for optimum strategy with five messages at each terminal.

Any coding strategy on an $M_1 \times M_2$ square ($M_1=M_2=M$ in the symmetric $R_1=R_2$ case) can be represented by a Markov chain as in [2], i.e. as soon as a message subsquare has been singled out one returns to the starting state and repeats the resolution process defined by the resolution tree. Note that Fig. 3 represents that part of the resolution tree that pertains after receiving an initial 0. The rate of uncertainty reduction can now be written as

$$\underline{I} = \sum_{s \in \mathbf{S}} q_s \underline{I}_s, \tag{2}$$

where $\mathbf{S}$ is the set of states (nodes of the resolution tree), and q_s, $s \in \mathbf{S}$, are the stationary state probabilities. Without loss of generality we can consider the transmission of information in the 1→2 direction (see Fig.1); the same argument applies to transmission in the 2→1 direction. Split the set $\mathbf{S}$ of states into two subsets, i.e. the subset $\mathbf{S}_{t_1}$ of t_1-resolutions and its complement $\mathbf{S}_{\tilde{t}_1}$ of non t_1-resolutions. For Fig. 3 the set of t_1-resolutions equals $\mathbf{S}_{t_1}=\{S,S_{000},S_{011},S_{0011}\}$. For the 1→2 component I_1 of $\underline{I}=(I_1,I_2)$ in (2) we can now write

$$I_1 = \sum_{s\varepsilon \mathbf{S}_{t_1}} q_s I_{s,1} + \sum_{s\varepsilon \mathbf{S}_{\tilde{t}_1}} q_s I_{s,1}. \tag{3}$$

Define

$$\tau_1 \triangleq \sum_{s\varepsilon \mathbf{S}_{t_1}} q_s, \tag{4a}$$

and

$$I_1^{t_1} \triangleq (\tau_1)^{-1} \sum_{s\varepsilon \mathbf{S}_{t_1}} q_s I_{s,1},\quad I_1^{\tilde{t}_1} \triangleq (1-\tau_1)^{-1} \sum_{s\varepsilon \mathbf{S}_{\tilde{t}_1}} q_s I_{s,1}. \tag{4b}$$

Using (4) one can rewrite (3) as

$$I_1 = \tau_1 I_1^{t_1} + (1-\tau_1) I_1^{\tilde{t}_1}, \tag{5a}$$

and similarly for I_2, i.e.

$$I_2 = \tau_2 I_2^{t_2} + (1-\tau_2) I_2^{\tilde{t}_2}. \tag{5b}$$

We know that $\underline{I}^{\tilde{t}} = (I_1^{\tilde{t}_1}, I_2^{\tilde{t}_2})$ is within the convex Shannon [1] outer bound region G_o. Hence, if we can show in Section III that $\underline{I}^t = (U_1^{t_1}, I_2^{t_2})$ is within the convex region G of (1), then $\underline{I} = (I_1, I_2)$ of (2) has to be within

$$G_1 = \tau G + (1-\tau) G_o, \tag{6}$$

where $\tau = \min(\tau_1, \tau_2)$. Equation (6) is valid because both τ_1 and τ_2 are inversely proportional to the average number $\bar{n}$ of resolutions, i.e. $\tau_i < \frac{1}{2}$, i=1,2. Note that the convex region G_1 is strictly interior to G_o as τ is bounded away from zero by the probability, q_S, of the initial resolution for which we can assume w.l.g. that $S \varepsilon \mathbf{S}_{t_1} \cap \mathbf{S}_{t_2}$. But G_1 being an outer bound it follows that $\underline{I}^{\tilde{t}}$ has to be within G_1^1 for $M_i \to \infty$, i=1,2. Now repeat, i.e.

$$G_2 = \tau G + (1-\tau) G_1 = \tau[1+(1-\tau)]G + (1-\tau)^2 G_o, \text{ etc.}$$

Thus, finally,

$$G = \lim_{k\to\infty} G_k ,$$

is also an outer bound region.

In closing this section one final comment is in order. The crucial step in the previous derivation is the at first sight innocuous statement that $\underline{I}^{\tilde{t}}$ has to be within G_n, n=1,2,... , for $M_1 \to \infty$, i=1,2,. Why is this so? Consider the $\mathbf{S}_{\tilde{t}_1}$ states in the resolution subtree of Fig. 3, i.e. S_0, S_{00}, S_{01}, and S_{001}. For the moment assume t_2 to be resolved, see Fig. 4, where the received $\underline{y}$-digits are indicated in the message subsquares. The associated uncertainty reduction equals $\underline{i}=(i_1,i_2)=(1.32621,\ 1.21936)$. The

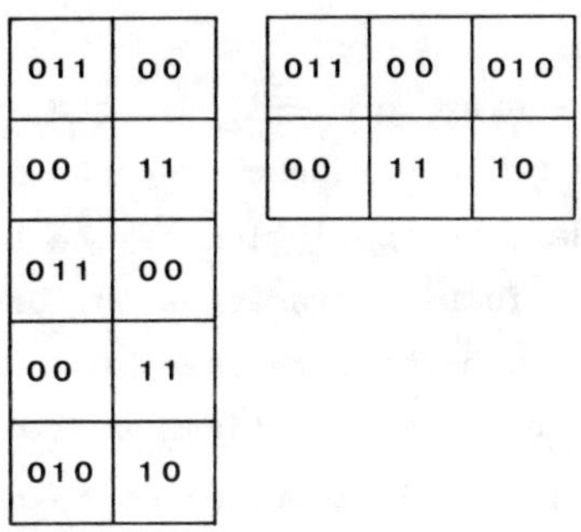

FIGURE 4. L-shape with t_2-threshold resolved.

average $\underline{y}$-sequence length in Fig. 4 equals $\bar{n}=2.31250$. Hence, the rate of uncertainty reduction $\underline{r}=(r_1,r_2)=(.57349,\ .52729)$. The rate point $\underline{r}$ must be within G_1 of (6) because of the absence of dependence between the messages θ_1 and θ_2 in Fig. 4. However, it would appear that the rate point $\underline{r}'=(r_1,r_1)=(.57349,\ .57349)$ in the absence of knowledge about the t_2-threshold could be outside G_1. This can indeed happen for small values of M_i, i=1,2. However, as $M_i \to \infty$, i=1,2, then $\underline{r}'$ will be within G_1 in the limit as the difference i_1-i_2 is bounded by the t_1-uncertainty, i.e. by $\frac{5}{8}h(\frac{3}{5})=.60684$ in the special case of Fig. 4 and, hence, the difference i_1-i_2 divided by $\bar{n}$ vanishes as $\bar{n}$ increases with M_i, i=1,2,. Summarizing, it appears that whereas the rate, $\underline{I}(S_1)$, of uncertainty reduction within subsquare S_1 is confined to G_1 of (6), $\underline{I}^{\tilde{t}}$ is confined to the larger region $G_1+\underline{\delta}$, where the length of the vector $\underline{\delta}$ is inversely proportional to $\bar{n}$. Introducing $\underline{\delta}$ into our iteration (6), we find

$$G_2=\tau G+(1-\tau)(G_1+\underline{\delta})=\tau[1+(1-\tau)]G+(1-\tau)^2G_o+(1-\tau)\underline{\delta},$$

$$G_3=\tau G+(1-\tau)(G_2+\underline{\delta})=\tau[1+(1-\tau)+(1-\tau)^2]G+(1-\tau)^3G_o+(1-\tau)[1+(1-\tau)\underline{\delta}], \text{ etc.}$$

Thus,

$$\lim_{k\to\infty} G_k = G + \frac{1-\tau}{\tau}\underline{\delta}.$$

Now τ is bounded from below by the stationary probability, $q_S=0[(\bar{n})^{-1}]$, that is inversely proportional to the average number, $\bar{n}$, of resolutions. Hence,

$$\frac{1-\tau}{\tau} = 0(\bar{n}).$$

If $|\underline{\delta}|$ is indeed $0[(\bar{n})^{-1}]$ then the offending term $\frac{1-\tau}{\tau}\underline{\delta}$ in the limit expression above does not vanish, and we can no longer conclude that G is indeed an outer bound region. However, consider the resolution of the L-shape S_0 that pertains upon receiving y=0 on the initial transmission. Define for k=1,2,... a sequence of strategies S_k such that the resolution thresholds (including the initial t-thresholds) for S_k are confined to a grid of size 2^{-k}, S_1 is not resolved (as in Fig. 3), S_0 is resolved to a point that yields a maximum value for the rate $I_k(S_0)$, of uncertainty reduction in S_0, and such that we terminate just prior to a final t-resolution. Hence, it follows that the sequence $I_k(S_0)$ is monotonically nondecreasing as I can repeat an earlier coarse grid strategy on the finer grid. Thus $I_k(S_0)$ has a limit $I(S_0)$. Note that for L-shapes S_0 that yield an uncertainty reduction strictly outside the capacity region G upon a single resolution this uncertainty reduction cannot be improved upon by further resolution and, hence, equals $I(S_0)$. Such L-shapes lead together with the initial t-resolution to a Hagelbarger [1] type (two state) strategy that is not optimum. Now think of an L-shape resolution, like the S_0 resolution in Fig. 3, as a Markov chain. A finer grid allows us a better approximation to the optimum threshold parameters (particularly in the later layers of states) of the given Markov chain. However, we cannot exceed the maximum rate of the particular Markov chain strategy. Further improvement in rate can only accrue from the addition of new states, which implies an increase of $\bar{n}_k(S_0)$, the average number of resolutions of the restriction of strategy S_k to the L-shape S_0. Note that a finite S_k cannot be optimum as it contains middle type [2] states, like S_{01} and S_{001} in Fig. 3, the resolution of which can be improved upon [3] by bootstrapping. Define $\bar{n}_k$ to be the average number of resolutions of strategy S_k, then

$$I_k(S) = \frac{1}{\bar{n}_k} t_k h(t_k) + \frac{\bar{n}_k - 1}{\bar{n}_k} I_k(S_0) .$$

Letting $k\to\infty$ it follows (as also $\bar{n}_k\to\infty$) that $I(S_0)=\lim I_k(S)$, and thus $I(S_0)$ is in G_1. As $I_k(S_0)$ is a monotically nondecreasing sequence $I_k(S_0)$ is constrained to G_1 for all $k=1,2,\ldots$. The argument can easily be extended to asymmetrical rate pairs and, thus, we conclude that $\underline{I}(S_0)\varepsilon G_1$ for those L-shapes S_0 to which we have restricted ourselves. But this implies that $\underline{I}^t$ is in G_1, and we are done!

3. INITIAL THRESHOLDS

Let us first take the example of Fig. 3. Without loss of generality we again consider the transmission of information in the 1→2 direction. For a given strategy the uncertainty reduction $I(\theta_1;Y|\theta_2,S_s)$ in subregion S_s, $s\varepsilon\mathbf{S}$, is given by

$$\begin{aligned} I(\theta_1;Y|\theta_2,S_s) &= H(Y|\theta_2,S_s) - H(Y|\theta_1,\theta_2,S_s) \\ &= H(Y|\theta_2,S_s), \end{aligned} \tag{7}$$

because given a strategy the message subsquare (θ_1,θ_2) determines the value of Y and, hence, $H(Y|\theta_1,\theta_2,S_s)=0$. Consider $I_1^{t_1}$ of (4b), where for the example of Fig. 3 we have $S_{t_1}=\{S,S_{000},S_{011},S_{0011}\}$. Hence,

$$\begin{aligned} I_1^{t_1} =& [25H(Y|\theta_2,S)+6H(Y|\theta_2,S_{000}) \\ &+ 3H(Y|\theta_2,S_{011})+ 3H(Y|\theta_2,S_{0011})]/(25+6+3+3). \end{aligned} \tag{8}$$

From Fig. 3 we find by inspection

$$H(Y|\theta_2,S)=\tfrac{3}{5}h(\tfrac{3}{5}), \text{ and}$$

$$H(Y|\theta_2,S_{000})=H(Y|\theta_2,S_{011})=H(Y|\theta_2,S_{0011})=\tfrac{2}{3}h(\tfrac{1}{2}).$$

Upon substitution of these terms into (8) we obtain $I_1^{t_1}=.60984$ as the average rate of uncertainty reduction for t_1-resolutions, whereas $I_1=.59233$ is the actual rate of the symmetric $R_1=R_2$ strategy on the 5×5 square as a whole. Now note the interesting fact that the average rate $I_1^{t_1}=.60984$ of the t_1-resolutions in Fig. 3 is equal to the average rate of the t_1-resolutions of Schalkwijk strategy with parameters [2] equal to $(\alpha,\gamma)=(\frac{3}{4},\frac{4}{5})$, see Fig. 5.

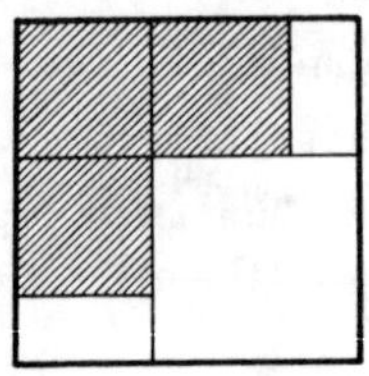

FIGURE 5. Threshold reduction applied to the optimum strategy with five messages at each terminal.

In the next paragraph we will show that if we take any strategy and distribute the probability weight (areas) of L-shape t-resolutions uniformly within each of the three t-quadrants as indicated in the shaded portion of Fig. 5, then the average rate for t-resolutions of the corresponding Schalkwijk [2] strategy is not less than the average rate for t-resolutions of the original strategy. But the average rate of the t-resolutions in a Schalkwijk strategy is given [3] by (1) and, hence, this rate $\underline{I}^t$ has to be within G, see Fig. 2.

Consider an L-shape t_1-resolution in subregion $S_s, s\varepsilon \mathbf{S}_{t_1}\backslash\{S\}$. Let $M_{1,1}$ ($M_{2,1}$) be the number of message intervals above (to the left) of the initial $t_1(t_2)$ threshold, and $M_{1,2}(M_{2,2})$ the number of message intervals below (to the right) of the $t_1(t_2)$ threshold, see Fig. 3. Then the probability weight (area) distribution for an arbitrary subregion S_s, $s\varepsilon \mathbf{S}_{t_1}\backslash\{S\}$ can be written in matrix form, i.e.

$$W_s = \begin{pmatrix} A_s & B_s \\ C_s & Z \end{pmatrix} , \tag{9}$$

where the submatrices A_s, B_s, and C_s have size $M_{1,1}\times M_{2,1}$, $M_{1,1}\times M_{2,2}$, and $M_{1,2}\times M_{2,1}$, respectively, and elements that are 1 or 0 depending on whether or not a particular message subsquare $(\theta_{1i},\theta_{2j})$ is an element of S_s. The matrix Z is an $M_{1,2}\times M_{2,2}$ matrix of all zeros. The uncertainty reduction of a t_1-resolution in S_s, $s\varepsilon \mathbf{S}_t\backslash\{S\}$, now equals

$$I_s^{t_1} = \sum_{j=1}^{M_{2,1}} (a_{.j}+c_{.j})h(\frac{a_{.j}}{a_{.j}+c_{.j}}) , \tag{10}$$

where $m_{\cdot j}$ stands for the sum of the elements of the jth column of a matrix M. If we further let m denote the sum of all elements of a matrix M, it then follows from the convexity of the binary entropy function that $I_s^{t_1}$ of (10) can be upper bounded by

$$I_s^{t_1} \leqq (a_s+c_s)h(\frac{a_s}{a_s+c_s}) \;. \tag{11}$$

Let

$$V_s = \begin{pmatrix} a_s & b_s \\ c_s & 0 \end{pmatrix} , \tag{12}$$

i.e. the probability (area) distribution V_s, $s\varepsilon \mathbf{S}_{t_1} \backslash\{S\}$, is the reduced version of W_s where the fine structure of the distribution within each of the three t-quadrants has been discarded, as in Fig. 5. Using $I^{t_1}(W_s)$ and $I^{t_1}(V_s)$ as shorthand notations for the right hand sides of equations (10) and (11), respectively, we have shown that

$$I^{t_1}(W_s) \leqq I^{t_1}(V_s),\; s\varepsilon \mathbf{S}_{t_1} \backslash\{S\} \;. \tag{13}$$

Now consider the average uncertainty reduction $I_1^{t_1}$ of all L-shape t_1-resolutions in $S_{t_1} \backslash\{S\}$, i.e.

$$\begin{aligned} I_1^{t_1} &\triangleq (\tau_1-q_S)^{-1} \sum_{s\varepsilon \mathbf{S}_{t_1} \{S\}} q_s I^{t_1}(W_s) \\ &\leqq (\tau_1-q_S)^{-1} \sum_{s\varepsilon \mathbf{S}_{t_1} \backslash\{S\}} q_s I^{t_1}(V_s) \\ &\leqq I^{t_1}(V^1), \end{aligned} \tag{14}$$

where

$$V^1 \triangleq (\tau_1-q_S)^{-1} \sum_{s\varepsilon \mathbf{S}_{t_1} \backslash\{S\}} q_s V_s$$

is the average reduced probability distribution of all L-shape t_1-resolutions, and $I^{t_1}(V^1)$ is the corresponding uncertainty reduction of a single t_1-resolution in an L-shape with reduced probability weight distribution V^1. A similar derivation for the information flow in the 2→1 direction yields

$$I_L^{t_2} \leq I^{t_2}(V^2), \; s \in \mathbf{S}_{t_2} \backslash \{S\} \; . \tag{15}$$

It follows from (14) and (15) that the average rate of all L-shape t-resolutions S_s, $s \in (\mathbf{S}_{t_1} \cup \mathbf{S}_{t_2}) \backslash \{S\}$, is dominated by the rate of a single L-shape t-resolution with associated reduced probability distribution V, see Fig.5, where

$$V = \begin{pmatrix} v_{11}^1 = v_{11}^2 & v_{12}^2 \\ v_{21}^1 & 0 \end{pmatrix},$$

and $v_{11}^1 = v_{11}^2$ corresponds to the whole top left hand t-quadrant, see Fig.3, as we require zero probability of error (so no remaining ambiquity about the t-thresholds is allowed). This completes our converse.

4. CONCLUSIONS

The convex region G of (1) is shown to be the capacity region of the BMC for both fixed length strategies with probability of error tending to zero, and for variable length strategies with probability of error equal to zero. The crucial ideas of the converse are 1) that of representing Shannon's coding strategies as strategies for subdividing the unit square, and 2) that of upper bounding the average rate of uncertainty reduction for resolutions on the initial t-thresholds.

One is tempted to conjecture that the capacity region G of an arbitrary TWC has the general form of (1). This conjecture is true in those cases where the resolution subregions of the initial resolution, see Fig. 3, allow subsequent indirect resolutions via bootstrapping [3] that yield new subregions that can themselves be resolved at Shannon's [1] outer bound rate, as is the case with the BMC.

It is interesting to compare the capacity region G of (1), see Fig. 2, with the improved outer bound regions of [7]. This comparison suggests that these new outer bounds can be even further improved. The TWC presents many interesting problems on which only a small group of people has been working recently, see the references.

ACKNOWLEDGEMENT

The author wishes to thank his coworkers J.E.Rooyacker, Tj.Tjalkens, A.J.Vinck, and F.M.J.Willems, and also his students I.v.Overveld, A.P.Hekstra, K.J.v.Driel, E.W.Gaal, and B.J.M.Smeets for informative discussions on the TWC. Thanks are also due to C.M.Bijl-Wind, and H.M.Creemers for their help in prepairing this manuscript.

REFERENCES

1. C.E.Shannon: Two-way communication channels, Proc.4th Berkely Symp.Math. Statist. and Prob., vol.1, pp.611-644, 1961. Reprinted in Key Papers in the Development of Information Theory, D.Slepian, Ed. New York: IEEE; 1974, pp.339-372.
2. J.P.M.Schalkwijk: The binary multiplying channel - A coding scheme that operates beyond Shannon's inner bound region, IEEE Trans.Inform. Theory, vol.IT-28, pp.107-110, Jan. 1982.
3. J.P.M.Schalkwijk: On an extension of an achievable rate region for the binary multiplying channel, IEEE Trans.Inform.Theory, vol.IT-29, pp.445-448, May 1983.
4. J.P.M.Schalkwijk, J.E.Rooyackers, and B.J.M.Smeets: Generalized Shannon strategies for the binary multiplying channel, Proc. 4th Benelux Symp. on Information Theory, Haasrode, Belgium, pp.171-178, May 26-27, 1983.
5. R.G.Gallager: Information Theory and Reliable Communication, Wiley N.Y., 1968.
6. L.Tolhuizen: Discrete coding for the BMC, based on Schalkwijk's strategy, Proc. 6th Benelux Symp.on Information Theory, Mierlo, The Netherlands, pp.207-208, May 23-24, 1985.
7. Z.Zhang, T.Berger and J.P.M.Schalkwijk: New outer bounds to capacity regions of two-way channels, to be published IEEE Trans.Inform.Theory vol.IT-32, pp.383-386, May 1986.
8. G.Dueck: The capacity of the two-way channel can exceed the inner bound, Inform. and Control, vol.40-3, pp.258-266, 1979.
9. G.Dueck: Partial feedback for two-way and broadcast channels, Inform. and Control, vol.46-1, pp.1-15, 1980.
10. T.S.Han: A general coding theorem for the two-way channel, IEEE Trans. Inform.Theory, vol.IT-30, pp.35-44, Jan. 1984.

RECENT DEVELOPMENTS IN CRYPTOGRAPHY

FRED PIPER

ROYAL HOLLOWAY AND BEDFORD NEW COLLEGE

1. INTRODUCTION

This talk is in two halves. In the first we give an introductory overview of the 'classical' encryption techniques and look at their relative merits. This is an abridged form of|Mit2]. Then, in the second half, we look at some recent developments; notably the arrival of public key cryptography. The aim is to illustrate various applications of public key systems and to motivate the lectures of I.F. Blake and Y. Desmedt.

The development over the last 100 years of automatic means for data transmission, and more recently the dramatic evolution of electronic data processing devices, has required a parallel rapid growth of work in **cryptology**, the study of encryption. More and more information of a sensitive nature is both communicated and stored electronically, and so the application for cryptographic techniques are ever increasing. We will attempt to classify and describe the principle techniques for data encryption that have been proposed and used, and to indicate the chief areas of application of these different techniques.

The basic idea behind any data encryption algorithm is to define functions f_k, which transform **messages** into **cryptograms**, disguised forms of the original messages, under the control of secret **keys**, $k \in K$. Thus if we let M denote the set of all possible messages, and C denote the set of all possible cryptograms, we are defining a family of functions: $\{f_k : k \in K\}$, where K is the set of all possible keys, and $f_k(m) \in C$ for every $m \in M$. In order that decryption is always possible, every f_k must be one-to-one (i.e. $f_k(m)=f_k(m')$ implies $m=m'$).

This rather abstract notion of data encryption is not necessarily a good guide to classifying the techniques actually used in cryptographic applications. In general, the idea of a special function being applied to the entire message simultaneously in order to obtain the cryptogram is rarely, if ever, used. In practice all the encryption methods in use involve dividing a message into a number of small parts (of fixed size) and encrypting each part separately, if not independently. This greatly simplifies the task of encrypting messages, particularly as messages are usually of varying lengths.

We shall assume throughout that the message parts are encrypted one at a time in the obvious order; we are thus able to use terms like: "**previous** parts of a message".

First we note that some cipher techniques operate on a single bit at a time, whereas others operate simultaneously on sets of bits, usually called **blocks.** Thus one important property relates to **bit/block operation.** The indivisible set of bits on which the system operates is called a **character.**

Secondly we observe that for some encryption techniques the encryption function which is applied to one piece of the plaintext is independent of the content of the remainder of the message. But for certain other methods, the enciphering function applied to one section of the message depends directly on the results of enciphering previous parts of the message. This property is referred to as **character independence/dependence.**

J. K. Skwirzynski (ed.), Performance Limits in Communication Theory and Practice, 207–223.

In some cipher systems, a message part is encrypted using precisely the same function regardless of its position within the message; in this case the cipher is said to possess **positional independence.** Other systems depend on the fact that different message parts are encrypted according to their position, and are thus **positionally dependent** ciphers.

The final characteristic property which we consider relates to the **symmetry** of the encryption function. This property, discussed more fully later, is the essential difference between conventional **symmetric** or **private key** cryptosystems and the **asymmetric** or **public key** cryptosystems. The fundamental difference is that, in an asymmetric system, encryption and decryption require different keys, and knowledge of an enciphering (or deciphering) key is not in practice sufficient to be able to deduce the corresponding deciphering (or enciphering) key.

Table 1 below illustrates how the different types of cipher system that we discuss here can be characterised in terms of these properties.

CHARACTERISING CIPHER SYSTEMS

Characteristic Property / Cipher Type	Type of character	Character Dependence/ Independence	Positional Dependence/ Independence	Symmetric/ Asymmetric
Stream Cipher	Bit	Independent	Dependent	Symmetric
Block Cipher	Block	Independent	Independent	Either
Cipher Feedback System	Either	Dependent	Independent	Symmetric

Table 1 - Characterisation of Cipher Schemes

We will give more formal definitions of the types of cipher systems and explore some of the advantages and disadvantages of each type of system. However, it is interesting to note at this point how much information is contained in the above table.

For example, it is clear that in any cipher system which has the character dependence property, error propagation will occur, i.e. if any ciphertext bits are corrupted during transmission, then a larger number of plaintext bits will be in error after decryption. Similarly, in any system which has the positional dependence property, if any message parts are lost during transmission, then all subsequent message parts will be decrypted erroneously.

We now discuss each of these three types of cipher system in detail.

2. STREAM CIPHERS AND THE ONE-TIME-PAD

The principle behind the stream cipher technique is that each bit of ciphertext is a function of the value and position of only one bit of plaintext. In order to implement a stream cipher it is first necessary to design a pseudo-random binary sequence generator, the output from which is used to encrypt the plaintext data on a bit by bit basis. More precisely, if the message consists of a sequence of n bits, $m_1 m_2 \dots m_n$ say, then it is combined with n consecutive output bits from the sequence generator, $s_1 s_2 \dots s_n$ say, so that the ciphertext $c_1 c_2 \dots c_n$ is defined by: $c_i = m_i + s_i \pmod 2$ for every i $(1 \leq i \leq n)$.

We can represent the principle diagrammatically as in Figure 1.

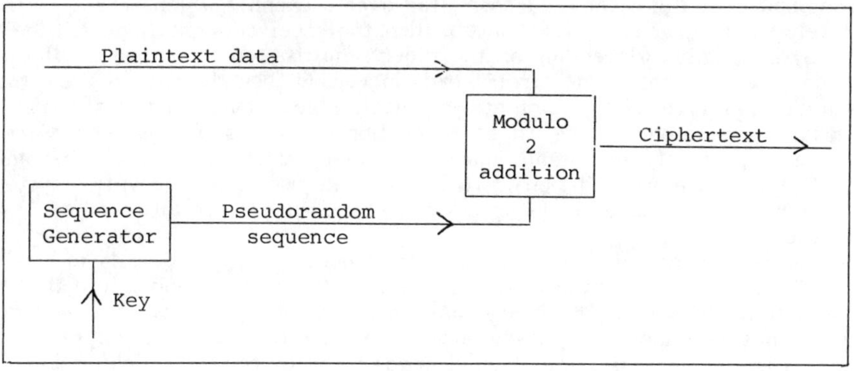

Figure 1 - Stream cipher

The important fact to note is that each bit of ciphertext c_i is a function only of the values of m_i and i. Similarly, when decrypting, each bit of plaintext m_i is a function solely of c_i and i. This has important ramifications as far as error propagation is concerned.

The main inspiration behind the invention of the stream cipher was provided by the one-time-pad. In a one-time-pad system, the sequence generator is not present, and an n-bit key is required to encrypt an n-bit message using modulo 2 bit by bit addition.

The one-time-pad key sequence can only be used to encrypt one message (hence **one-time**-pad) and should be randomly generated. As Shannon showed in his historic paper, [S1], this system offers perfect secrecy, and all unbreakable systems are in some sense analogous to it in the sense that they require as much random key information as data to be encrypted.

Unfortunately, in most situations, the key distribution problem makes the one-time-pad virtually unusable. The stream cipher principle was an attempt to retain some of the good properties of the-one-time pad, but using only manageable amounts of key material.

One additional remark concerns a generalisation of stream ciphers to a situation where the unit of encryption consists of a block of bits rather than a single bit. This would require the use of some mixing function with similar properties to the modulo 2 addition in terms of its invertibility, and the design of a generator of pseudo-random sequences of blocks of bits. Indeed one could regard the Vigenere Cipher as a very trivial example of this (for a discussion of Vigenere Ciphers, see, for example, [BP1]).

A characteristic of stream ciphers, which in certain situations is an advantage, is the fact that they do not propagate errors. Other advantages include ease of implementation and speed of operation. The main disadvantage of stream ciphers is that they require the transmission of synchronisation information at the head of the message which must be received in order that the message can be decrypted. Other disadvantages include the fact that, because they do **not** propagate errors, they provide no protection against message modification by an active interceptor.

The fact that the stream cipher's lack of error propagation can be regarded as both an advantage and a disadvantage, may seem at first rather paradoxical. However it can be explained by considering some typical applications. First consider the situation where the plaintext data consists of a digitised version of a speech signal. If this signal is transmitted over a channel which causes a certain relatively small error rate, say 1 in 50, then the received signal may still be perfectly intelligible (depending on the speech digitisation technique used), but with the property that if the error rate is increased then the speech goes from an acceptable quality to being completely unintelligible. Hence in this situation, which is a genuine one, if the use of encryption causes an increase in the error rate, then a perfectly acceptable channel for clear transmission would become unusable for secure transmissions. This makes the use of an encryption system which does not propagate errors an essential requirement for this type of application.

On the other hand consider an application where encrypted data is to be sent over a channel with a very low error probability (say 10^{-10}), and where it is essential that the data is received completely correctly. This is typical of many computer network applications, where a single bit error may be absolutely disastrous and, as a result, the channel needs to be extremely reliable. In this sort of situation, one error is clearly virtually as bad as 100 or 1000 errors, and so in this application error propagation is not a disadvantage. In fact error propagation may actually be an advantage since 100 or 1000 errors may be much easier to detect than a single error. This idea can be extended to the idea of **Message Authentication Codes** (MACs) where error propagating cipher schemes are used to produce special data sent with a message which can be used to detect both accidental and deliberate alterations to a message in transit.

A further possible disadvantage with stream ciphers is the need for **synchronisation** information. This relates to the fact that, if the same key is used for encrypting two different messages, then the same enciphering sequence will be used to encrypt these messages. This has serious consequences for the security of the scheme, and so it is always necessary to provide an additional randomly selected, **message key**, which is transmitted at the start of the message and is used to modify the encryption key so that different sequences are used for different messages. This information is often referred to as synchronisation information.

Stream ciphers are widely used in military and paramilitary applications for encrypting both data and digitised speech. In this area of application, which was until comparatively recently virtually the only major use for encryption techniques, they probably form the dominant type of cipher techniques.

One reason for their dominance is the relative ease with which good sequence generators may be designed and implemented. However, the chief reason for their dominance in this area is the fact that they do not propagate errors. The sort of channels used for tactical military and paramilitary data and digitised speech traffic have a strong tendency to be of poor quality. So any cipher system which would increase an already relatively high error rate, would almost certainly render channels, which are usable for clear data, unusable for enciphered data. This is normally quite unacceptable.

The main requirement for key stream sequences are that they should all have the following properties: long period, randomness properties and non-linearity (or, to be more precise, large linear equivalence).

The open literature contains a number of suggestions for pseudo-random binary sequence generators for use in stream ciphers. Until relatively recently the use of linear feedback shift registers was suggested as being suitable for this purpose (see, for example, [D1] or [M1]). This idea, although superficially attractive, is fundamentally flawed because of the linearity of the sequence produced, as has been pointed out many times, e.g. [G1] and [M2].

The basic idea of using linear feedback shift registers has not been discarded however, and these sequence generators still form the basis of many of the stream cipher generators in practical use today. Many recent suggestions for sequence generators are based on the idea of combining the outputs from a number of linear feedback shift registers using non-linear logic, sometimes exclusively feed forward, as in the systems of Bruer, [Br1], Geffe, [Ge1] and Jennings, [J1], [J2], and sometimes incorporating elements of non-linear feedback logic, as in the systems of Chambers and Jennings, [C1] and Smeets, [S2]. A particularly significant recent contribution to the theory of stream ciphers is [Reup].

The only national or international standard techniques for stream cipher sequence generation involves use of the **Data Encryption Standard** (DES) block cipher algorithm in what is known as **Output Feedback Mode** (see [ANSI83] and [FIPS81]).

Another suggested method worthy of note is the British Telecom B152 algorithm, which is currently being marketed as the **B-CRYPT** device. Interestingly, this algorithm has been kept secret and looks likely to remain so, although it is not clear whether a secret algorithm will be acceptable as the basis of equipment manufactured by companies other than BT themselves.

3. BLOCK CIPHER TECHNIQUES

For encryption using a **block cipher** the plaintext data is first divided into fixed length blocks, of length m, say. A key dependent function f_k is then employed to transform each m-bit block of plaintext into an m-bit ciphertext block. In order that the decryption be made possible, for each key k the function f_k is in fact a permutation on the set of all m-bit blocks.

Diagrammatically we can represent a block cipher as in Figure 2 below.

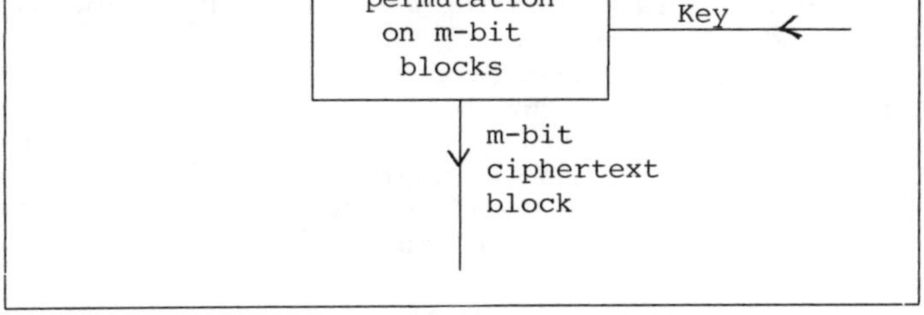

Figure 2 - Block cipher

In order to achieve a reasonable level of security using a block cipher, the plaintext/ciphertext block size, m, must not be small, otherwise it might allow a cryptanalyst to build a useful "dictionary" of plaintext/ciphertext block pairs.

It is important to realise that in a "true" block cipher, every block of message is encrypted in an identical fashion regardless of its position. This is one of their most important distinguishing features from stream ciphers and is the main reason for choosing a large block size. This method of encryption has also been called **Electronic Codebook** encryption, with the idea that this is an automatic version of conventional code book encryption, where each word or phrase has a fixed coded equivalent.

One property that is desirable for any block cipher algorithm is for every ciphertext bit to be a function of all (or virtually all) the plaintext bits in the appropriate block. This requirement is closely related to the Shannon concept of **Diffusion,** see [S1] or Chapters 3 and 4 of [BP1].

The main advantage of using a block cipher is that, unlike the stream cipher situation, an active interceptor of a message cannot alter the ciphertext with known effects on the plaintext. Thus, at least with a well designed block cipher system, small changes to the ciphertext will result in large and unpredictable changes to the corresponding plaintext, and vice versa.

The disadvantages of using block ciphers in this way are quite severe. First and foremost, because of the fixed nature of the encryption, even if the block size m is quite large, say of the order of 50 - 100 bits, a limited form of the "dictionary" attack may still be possible. It is conceivable that a block of bits of considerable size may be repeated in a message. This could result in two identical m-bit plaintext blocks in a message, which in turn would mean two identical ciphertext blocks giving the interceptor some information about the message contents without any work at all.

A further possible disadvantage (depending on the application) of block ciphers is their error propagation properties, a possible problem which really applies to all cipher systems except for stream ciphers. A single bit error in a received ciphertext block will result in an entire block being incorrectly deciphered; in turn this is likely to cause roughly m/2 bit errors in the recovered plaintext.

Because of the major disadvantages of using block ciphers in the way described above, they are not often used to encypt messages in this simple way. However they are widely used as a component in cipher feedback systems. In addition a good block cipher system can be used as part of message authentication schemes.

The most well known and widely used of the block cipher algorithms is the **Data Encryption Standard** (DES) algorithm, originally specified as a US national standard in [FIPS46] in 1977, and subsequently in [ANSI81] in 1981; it now exists as a draft international standard, [DIS8227]. Many commercially available hardware implementations of the DES exist, see, for example, Chapter 3 of [Davies1], [Fai1] and [Mul1]. This algorithm, although widely criticised, particularly for the relatively small size of the key used, e.g in [Dif2], [Hel1] and [Mor1], is still widely believed to be a more than adequate technique for commercial applications. However, according to recent rumours, DES looks to have a very limited future, at least as a US standard method of encryption.

As well as producing the DES algorithm itself, the US standards authorities have also been instrumental in standardising the ways in which encryption algorithms are to be used. The original US standard describing ways in which DES can be used appeared in 1980, [FIPS81], and was followed by [ANSI83] in 1983; again a draft international standard has been produced along the same lines, namely [ISO8372]. These standard **Modes of Operation** apply equally well to any block cipher, and not just to DES.

The modes of operation described in these standards are called: **Electronic codebook** (ECB) **Mode, Cipher block chaining** (CBC) **Mode, Cipher Feedback** (CFB) **Mode** and **Output Feedback** (OFB) **Mode.** The ECB mode relates simply to using

DES as a straight block cipher in the way described above. The CBC and CFB modes are both ways in which a block cipher can be made into a cipher feedback system,and are discussed later. Finally, the OFB mode describes how DES or any other block cipher, can be used to produce a pseudo-random binary sequence for stream cipher applications.

Moving on from DES and its uses, virtually all of the other published techniques for block cipher encryption rely (as DES itself does) on performing variants of a basic block transformation techniques over and over again to the data block in what are usually called "rounds". The idea is that after a number of these rounds, the simple transformations to go make up a much more complex transformation. As in DES, the simple transformations are typically made up of a combination of a permutation and a number of small sub-block substitutions.

Of particular importance is the work which was carried out by IBM in the early 1970s, which included the development both of DES itself and a number of DES like algorithms such as **Lucifer** and the **New Data Seal;** some of this work is described in [Fei2], [Fei1], [Gir1], [Gir2] and [Gro1]. One important and ingenious technique developed during this period, and incorporated into Lucifer, New Data Seal and DES, is the **Feistel Cipher** which is described in [Fei1], (where it is described as an "iterative product cipher using unreversed transformations on alternating half blocks"), and in Capter 7 of [BP1].

Other work making use of the idea of repeated encryption rounds includes that of Even and Goldreich, [Eve0], [Eve1], Kam and Divida, [Kam1], Ayoub, [Ayo2], [Ayo3], [Ayo4], and Gordon and Retkin, [Gor1].

Finally we note that a type of hybrid stream/block cipher system, with some of the best properties of both, may be obtained by combining a stream cipher technique and the use of pseudo-randomly generated permutations. The plaintext to be encrypted is first stream ciphered in the conventional way, and the resulting ciphertext is then divided up into a number of fixed size blocks. Each block is then permuted under a key controlled pseudo-random permutation (preferably a different permutation for each block). The order of these two operations could be reversed without affecting the fundamental properties of the system. Techniques for key controlled permutation selection have been studied in a number of papers, such as [Akl1], [Ayo1] and [Slo1].

The result is a cipher which does not propagate errors, but with the additional property, not possessed by a stream cipher, that an interceptor does not know which ciphertext bit corresponds to which plaintext bit. This latter property makes deliberate changes to the enciphered message, with known effect, much more difficult, if not impossible. Note that this is not a true block cipher however, in that each ciphertext bit is a function of exactly one plaintext bit, not of all plaintext bits.

4. CIPHER FEEDBACK TECHNIQUES

Cipher feedback encryption systems vary widely and thus, in order to incorporate as many of the practical variations as possible, our definition is necessarily rather broad in scope. As for a block cipher, the message is broken into a series of m-bit blocks. A key and ciphertext dependent function $f(\underline{k},\underline{c})$ is then employed to transform each m-bit block of plaintext into an m-bit ciphertext block, where $\underline{k}$ is the key vector and $\underline{c}$ is a ciphertext vector (whose role we describe below). Just as for a block cipher, this function needs to be one-to-one so that decryption is possible, and thus, for fixed $\underline{k}$ and $\underline{c}$ it can be regarded as a permutation on the set of all m-bit blocks. The t-bit ciphertext vector $\underline{c}$ is made up from the s previous m-bit ciphertext blocks, where s is the smallest integer equal to or exceeding t/m.

We illustrate our definition of a cipher feedback encryption system in Figure 3 below.

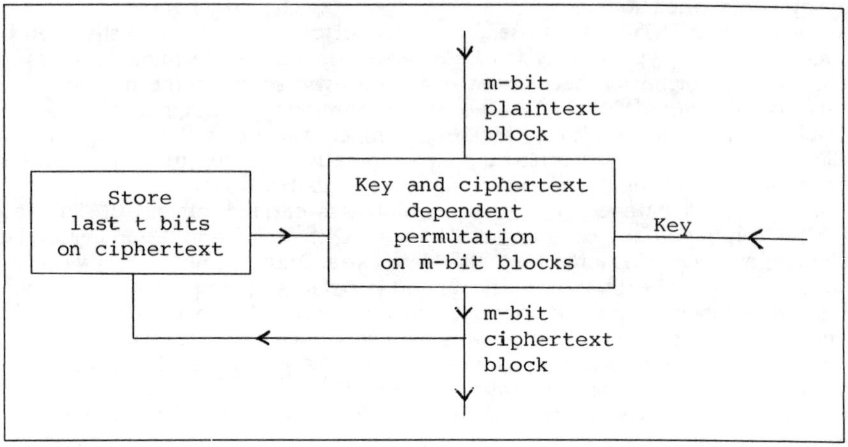

Figure 3 - Cipher feedback system

This general definition of cipher feedback encryption includes, as special cases, a number of different types of commonly used encryption system.

For the general class of **Ciphertext Auto Key** (CTAK) ciphers, we have $m = 1$, and the encryption function takes a single plaintext bit, the key vector and the immediately previous t bits of ciphertext and uses them to produce a single ciphertext bit. In order for decryption to be possible, this function must have the property that for fixed key and ciphertext vectors, the two possible plaintext values (0 and 1) have different corresponding ciphertext values.

A number of other special cases of particular importance exist, including the **Cipher Block Chaining** technique. In this case we have $m = t$, and the m previous ciphertext bits are combined with the next m plaintext bits using bit by bit modulo 2 addition, before applying a block cipher algorithm under key control to obtain the next block of m ciphertext bits. This technique forms one of the four standard modes of use of the DES algorithm described in 3 above.

A further variation of the general technique forms another of the DES standard modes of operation, namely the **Cipher Feedback** (CFB) mode. Suppose that we have a block cipher which operates on r-bit plaintext blocks. Then m (the plaintext/ciphertext block size) is chosen to be any value between 1 and 4, and we describe what is known as m-bit CFB. We let $t = r$ and use the block cipher algorithm, under key control, to encrypt the t most recent ciphertext bits. Of the resulting r-bit block, r-m bits are discarded, and the remaining m bits are bit by bit modulo 2 added to the m-bit plaintext block to form the m-bit ciphertext block. Note that if m is chosen to equal one then this is just a special type of CTAK cipher.

For all these special cases, t must be sufficiently large to prevent "dictionary" type attacks, just as was the case for block ciphers. These attacks are based on the fact that, for any given key, the previous t bits of ciphertext and the m

bits of plaintext completely determine the m-bit ciphertext block, regardless of its position in the message.

There are a number of strong advantages to the use of cipher feedback systems. First and foremost, they can be used to detect message manipulation by active interceptors of messages, both through error propagation and because they can be very easily used as part of schemes to generate Message Authentication Codes.

Second, CTAK systems are often used in place of stream ciphers, and have the significant advantage of not requiring initial synchronisation. This means that, if the start of a message is missed, then successful decryption can still be achieved of most of the remainder of a message (after the correct reception of t consecutive ciphertext bits); this is definitely not the case with stream ciphers. A related advantage is that cipher feedback systems can cope with bits being completely omitted from a message during transmission. In such a circumstance the decryption process will recover and commence decrypting correctly within t bits of the loss of a ciphertext bit.

There are, however, corresponding disadvantages with the use of cipher feedback systems. Most significantly, all cipher feedback systems propagate errors, and a single bit error in transmission will cause in the order of $(1+sm/2)(1+t/2)$ errors after decryption. Thus the requirement for t to be large for cryptographic purposes, on occasion conflicts with the system requirements from the error propagation point of view.

A further disadvantage is that cipher feedback systems are often more difficult both to design and implement than the relatively straightfoward stream ciphers. Significant effort needs to go in to the system design to ensure that there are not sets of t ciphertext bits for which the system is significantly weaker than average.

Cipher feedback systems are very widely used, both in the guise of CTAK schemes, and as CBC or CFB modes of use for block ciphers such as DES. Moreover, these schemes are not only used for message encryption, but also for message authentication of various sorts.

As mentioned above CTAK systems are sometimes used in place of stream ciphers, and they do have the significant advantage of allowing "late entry" to a ciphered transmission, whereas for a stream cipher the synchronisation signal containing the message key must be received in order for any of the message to be decrypted. This is superficially attractive for many applications; for example if the plaintext data is digitised speech, late entry to a secure transmission could be very useful. However, as discussed in [Mit1], in many if not most circumstances the error propagation property is a major drawback, and has meant the continued dominance of stream ciphers in applications where the communications channel is at all noisy.

5. RECENT DEVELOPMENTS

When the security level of a cipher system is assessed it is customary to judge it under the assumption that

(a) the algorithm is known to the attacker
(b) the attacker can intercept the entire transmission
(c) the attacker knows some corresponding plaintext and ciphertext.

The acceptance of these so-called 'worst case conditions' means that the security is totally dependent on the secrecy of the key. This means, for instance, that the number of keys must be large enough that the attacker does not have time to try them all, i.e. to launch an **exhaustive search attack**. Quantifying the expression 'large enough' is often difficult and, clearly, the precise value varies according to the application and the level of security required. However recent advances in computer technology mean that, for almost all systems, there has been a dramatic increase in the minimal acceptable size of a key space. This in turn has necessitated the use of significantly more sophisticated mathematical functions to perform the encryption.

Of course the acceptance that an attacker will probably be able to determine the algorithm does not imply that the user should not attempt to keep it secret. It merely means that the security should not depend upon the algorithm's secrecy. However in two of the most significant recent advances in cryptography, the publication of DES and the introduction of Public Key Systems, all attackers are told the algorithm being used.

We have already mentioned DES. It is a block cipher with 64-bit blocks and a 56-bit key. The algorithm is public and can be found in all the standard textbooks on cryptography. There have been a number of published criticisms and a certain amount of suspicion about the construction of the S-boxes. Since it was first introduced there has been considerable public debate about the feasibility of building a device to perform DES key searches. Not surprisingly the debate has tended to focus on the likely cost of such a device and the time it would need to perform its exhaustive search. One current estimate is that for an outlay of about four million pounds one could build a device to complete the search in about twenty hours. Since the time taken is roughly inversely proportional to the cost, this means that for an outlay of roughly £250000 the time needed would be about one month. If this assessment of the security of DES is accepted then, for any given application, the users have to consider the likely threats to their system and see whether DES is strong enough.

In any cipher system, whether it employs DES or some other algorithm, it is advisable to change the keys often enough that the threat posed by the possibility of an exhaustive key search is nullified. In fact all aspects of key handling are vitally important and good key management is possibly the most important aspect of a secure system. There is absolutely no point in having a strong algorithm with a large number of keys if it is easy for the attacker to obtain the current key. Thus the keys must be protected at all times and it is necessary to have secure procedures for key generation, key distribution, key storage and key destruction.

When we discussed the one-time-pad we mentioned that the key distribution problem made it unusable in most situations. The reason is clear. The key material is as long as the message it is protecting and thus, if the same channel is used for transmitting data and keys, the problem of securing the key is precisely the same as that of protecting the message. If, however, there is an alternative secure link between the two parties then the one-time-pad may be usable. There are a number of situations where such links exist. For instance the senior politicians in certain countries can use diplomatic couriers with armed escorts. This type of secure link is, of course, both slow and expensive but these factors are not really important considerations when keys are being distributed. For systems where the luxury of this alternative secure channel does not exist the task of distributing keys securely can pose many problems.

The Diffie Hellman paper of 1976 [Dif1] is undoubtedly one of the most important contributions to modern cryptography. It is the first publication to contain the idea of a public key system.

The idea behind a public key system is that anyone who wishes to RECEIVE a secret message will publish a key which anyone may then use to encrypt messages for him. This means that anyone who intercepts the cryptogram will not only know the algorithm but will also know the enciphering key. Clearly this puts an extra demand on the enciphering algorithm and most of the standard algorithms, DES for instance, are completely unsuitable for this type of application. Note that for the conventional system, or symmetric key systems, the enciphering key has to be kept secret because knowledge of it is virtually equivalent to knowing the deciphering key. In fact for some systems they are identical. For a public key system it must be computationally infeasible for anyone to deduce the deciphering key from knowledge of the algorithm and the enciphering key.

As with all crypto systems the encipherment process must be easy to implement. Thus the algorithm for a public key system has to be a ONE-WAY FUNCTION, i.e. a function which is easy to perform but difficult to reverse. However this is not sufficient since there is the extra demand that the recipient of the cryptogram must be able to decipher the cryptogram fairly easily. This means that the one-way function must have a TRAPDOOR, i.e. a trick, or piece of knowledge, which enables the receiver to reverse the function.

There have been many proposals for suitable algorithms but only two have been seriously considered for implementation. They are the Merkle-Hellman system, which replies on the difficulty of solving the Knapsack Problem, and the RSA system. The former is the topic of Desmedt's talk and, so, we will use the RSA as our example.

The RSA system

For the RSA anyone wishing to receive secret messages publishes integers n and h where $n = pq$ (p and q large primes) and h is chosen so that $(h,(p-1)(q-1)) = 1$. If message is an integer m then the cryptogram $c \equiv m^h (\text{mod } n)$. The primes p and q are 'secret', (i.e. known only to the receiver), and the system's security depends on the fact that knowledge of n will not enable the inteceptor to work out p and q. Since $(h,(p-1)(q-1)) = 1$ there is an integer d such that $hd \equiv 1 \ (\text{mod } (p-1)(q-1))$. [Note without knowing p and q it is "impossible" to determine d]. To decipher raise c to the power d. Then $m \equiv c^d \ (\equiv m^{hd})(\text{mod } n)$.

The RSA system works because, for any n, $a^{\emptyset(n)+1} \equiv a \ (\text{mod } n)$ for all a and, since $n = pq$ where p and q are primes, $\emptyset(n) = (p-1)(q-1)$.

Small example

$n = 55$ $(p=5, q=11)$ $(p-1)(q-1) = 4.10 = 40$
Choose $h = 3$, so $d = 27$.
Suppose message is 15
then $c = (15)^3 \text{ mod } 55$ i.e. $c = 20$
It is easy to check that $(20)^{27} \equiv 15 \ (\text{mod } 55)$
i.e. that it works.

Although there are various fast algorithms for modular exponentiation, all current implementations of the RSA are too slow for the encryption of data in most applications. One of the fastest claims for a practical RSA implementation is 60 - 70 kbits/sec for a 660 bit modulus. But even this is in the development stage.

For the RSA system the one way function $f(x)$ is given by $f(x) = x^h (\text{mod } n)$, where h and n are large, fixed integers. In general it is hard to determine x from knowledge of $x^h (\text{mod } n)$, h and n. However the trapdoor for the RSA system is the knowledge of $\emptyset(n)$, which is equivalent to knowing the factors of n.

Applications of public key systems

At present there are two main applications for public key systems; (i) as key encrypting keys for distributing keys in systems where conventional algorithms, e.g. DES, are used and (ii) for digital signatures.

Digital signatures

A digital signature is a method by which an individual 'signs' a message in such a way that he cannot later deny that he did so. Public key systems provide this facility and we will use the RSA system to show how.

Suppose I have a public key (n,h) and that my secret key is d. If a third party wants me to sign a message, ℓ say, he sends it to me and I send back

$\ell' = \ell^d \pmod{n}$. He, and anyone else (including a judge!), can look up my public key h and check that $(\ell')^h \pmod{n} = \ell$. (Note that, modulo n, $(\ell^d)^h = \ell$). Since I am the only person with knowledge of how to reverse exponentiation to the power h modulo n, I must have signed the message ℓ. Of course this shows great faith in the difficulty of determining d from knowledge of n and h.

Key encryption keys

We have already mentioned the problem of key management. There are now many systems which use RSA for protecting the session keys in systems which use symmetric key systems for the data encryption. Speed is not usually as crucial for key distribution as it is for transmitting messages and so the fact that RSA is slow does not matter. There are various protocols which use RSA for key management, some use key distribution centres. For a good reference see [Davies1].

The Diffie-Hellman key exchange sceheme

A related cryptographic technique, due to Diffie and Hellman, is based on the difficulty of computing discrete logarithms over GF(q), [Dif1]. The original scheme cannot be used for message encryption, but provides a method for secretly exchanging cryptographic keys for use with other encryption algorithms. Interestingly, Elgamal has recently shown that the scheme can be generalised to produce a full public key cryptosystem for message security [Elg1]. However, for a number of values of q, (e.g. $q = p$ or p^2, p prime, or $q = 2^m$) the method has been successfully attacked, [Blake1], [Blake2], [Cop1], [Elg2] and [Poh1], and although the system has not been broken for every q, the cryptanalytic successes that have been achieved cast doubt over the usefulness of the idea. As with many of the other schemes, generalisations have been suggested, including the recent work of Odoni, Varadharajan, Sanders and Odini, [Odo1], [Var1]. Discrete logarithms will be discussed by Blake.

We end this talk by noting that the use of public key systems often presents special authentication problems, relating to checking whether or not a supplied public key has really been generated by its claimed source. This can be solved by means of certification schemes, such as discussed in [Gor3] and [Mer2].

REFERENCES

[Adi1] B.S. Adiga and P. Shankar, Modified Lu-Lee cryptosystem, Electronics Letters **21** (1985) 794-795.

[Adl1] L.M. Adleman, C. Pomerance and R.S. Rumely, On distinguishing prime numbers from composite numbers, Annals of Mathematics **117** (1983), 173-206.

[Akl1] S.G. Akl and H. Meijer, A fast pseudo random permutation generator with applications to cryptology, preprint.

[ANSI81] ANSI X3.92-1981, Data Encryption Algorithm, American National Standards Institute (New York), 1981.

[ANSI83] ANSI X3.106-1983, Modes of operation of the DEA, American National Standards Institute (New York), 1983.

[ANSI82] ANSI X9.9-1982, Financial institution message authentication, American Bankers Association (Washington, DC), 1982.

[ANSIx] ANSI X9.19-198x (draft standard), Financial institution retail message authentication, American Bankers Association (Washington, DC).

[Ayo1] F. Ayoub, Encryption with keyed random permutations, Electronics Letters **17** (1981) 583-585.

[Ayo2] F. Ayoub, Trapdoors, random structures and encryption functions design, given at the IEE colloquium on techniques and implications of digital privacy and authentication systems, London, 15th October 1981.

[Ayo3] F. Ayoub, Probabilistic completeness of substitution-permutation encryption networks, IEE Proceedings (Part E) **129** (1982) 195-199.

[Ayo4] F. Ayoub, The design of complete encryption networks using cryptographically equivalent permutations, Computers and Security **2** (1983).

[BP1] H.J. Beker and F.C. Piper, Cipher Systems, van Nostrand (UK), 1982.

[Ber1] S. Berkovits, Factoring via superencryption, Cryptologia **6** (1982) 229-237.

[Blake1] I. Blake, M. Jimbo, R.C. Mullin and S.A. Vanstone, Computational algorithms for certain shift register sequence problems, Interim Report, Project no. 307-16, University of Waterloo, Ontario, Canada, 29th November 1983.

[Blake2] I.F. Blake, R. Fuji-Hara, R.C. Mullin and S.A. Vanstone, Computing logarithms in finite fields of characteristic two, SIAM Journal of Algebraic and Discrete Methods **5** (1984) 276-285.

[Bla1] G.R. Blakley and I. Borosh, Rivest-Shamir-Adleman public key cryptosystems do not always conceal messages, Comp. and Maths. with Applications **5** (1979) 169-176.

[Bla2] B. Blakley and G.R. Blakley, Security of number theoretic public key cryptosystems against random attack I, Cryptologia **2** (1978) 305-321.

[Bla3] B. Blakley and G.R. Blakley, Security of number theoretic public key cryptosystems against random attack II, Cryptologia **3** (1979) 29-42.

[Bla4] B. Blakley and G.R. Blakley, Security of number theoretic public key cryptosystems against random attack III, Cryptologia **3** (1979) 105-118.

[Bra1] H. Brandstrom, A public-key cryptosystem based upon equations over a finite field. Cryptologia **7** (1983) 347-358.

[Bri0] E.F. Brickell, A fast modular multiplication algorithm with applications to cryptography, Advances in Cryptology: Proceedings of Crypto 82, Plenum Press (New York) (1983).

[Bri1] E.F. Brickell, Solving low density knapsacks, Advances in Cryptology: Proceedings of Crypto 83, Plenum Press (New York) (1984) 25-37.

[Bri2] E.F. Brickell, J.C. Lagarias and A.M. Odlyzko, Evaluation of the Adleman attack on multiply iterated knapsack cryptosystems, Advances in Cryptology: Proceedings of Crypto 83, Plenum Press (New York) (1984) 39-42.

[Br1] J-O. Bruer, On pseudo random sequences as crypto generators, Proceedings of 1984 IEE International Zurich Seminar on Digital Communications (1984) 157-161.

[Chi1] H.R. Chivers, A practical fast exponentiation algorithm for public key, Proceedings of the International Conference on Secure Communications, IEE, London, February 1984, IEE (1984) 54-58.

[Cop1] D. Coppersmith, Fast evaluation of logarithms in fields of characteristic two, IEEE Transactions on Information Theory **IT-30** (1984) 587-594.

[Cou1] C. Couvreur and J.J. Quisquater, An introduction to the fast generation of large prime numbers, Philips Journal of Research **37** (1982) 231-264.

[Davies1] D.W. Davies and W.L. Price, Security for computer networks, John Wiley and Sons (Chichester, UK) 1984.

[Dav1] J.A. Davis and D.B. Holdridge, Factorization using the quadratic sieve algorithm, Advances in Cryptology: Proceedings of Crypto 83, Plenum Press (1984) 103-113.

[Den1] D.E.R. Denning, Cryptography and Data Security, Addison-Wesley (Reading, Mass.) 1982.

[Dif1] W. Diffie and M.E. Hellman, New directions in cryptography, IEEE Transactions on Information Theory **IT-22** (1976) 644-654.

[Dif2] W. Diffie and M.E. Hellman, Exhaustive cryptanalysis of the NBS Data Encryption Standard, Computer **10 no 6** (June 1977) 74-84.

[D1] E.S. Donn, Secure your digital data, The Electronic Engineer (May 1972) DC5-DC7.

[Elg1] T. Elgamal, A public key cryptosystem and a signature scheme based on discrete logarithms, IEEE Transactions on Information Theory **IT-31** (1985) 469-472.

[Elg2] T. Elgamal, A subexponential-time algorithm for computing discrete logarithms over GF(p^2), IEEE Transactions on Information Theory **IT-31** (1985) 473-481.

[Eve0] S. Even and O. Goldreich, On the power of cascade ciphers (extended abstract), Advances in Cryptology: Proceedings of Crypto 83, Plenum Press (New York) (1984) 43-50.

[Eve1] S. Even and O. Goldreich, On the power of cascade ciphers, ACM Transactions on Computer Systems **3** (1985) 108-116.

[Eve2] S. Even, O. Goldreich and A. Lempel, A randomized protocol for signing contracts, Communications of the ACM **28** (1985) 637-647.

[Fai1] R.C. Fairfield, A. Matusevich and J. Plany, An LSI digital encryption processor (DEP), IEEE Communications Magazine **23 no 7** (July 1985) 30-41.

[Fei2] H. Feistel, Cryptography and Computer Privacy, Scientific American **228 no 5** (May 1973) 15-23.

[Fei1] H. Feistel, W.A. Notz and J.L. Smith, Some cryptographic techniques for machine-to-machine data communications, Proc. IEEE **63** (1975) 1545-1554.

[FIPS46] FIPS PUB 46, Data encryption standard, Federal Information Processing Standards Publication **46**, National Bureau of Standards, US Department of Commerce (Washington, DC), January 1977.

[FIPS81] FIPS PUB 81, DES modes of operation, Federal Information Processing Standard Publication **81**, National Bureau of Standards, US Department of Commerce (Washington, DC), December 1980.

[G1] P.R. Geffe, Open letter to communications engineers, Proc. IEEE **55** (1967) 2173.

[Ge1] P.R. Geffe, How to protect data with ciphers that are really hard to break, Electronics **46 no 1** (4th January 1973) 99-101.

[Gir1] M.B. Girsdansky, Data privacy: Cryptology and the computer at IBM Research, IBM Research Reports **7 no 4** (1971).

[Gir2] M.B. Girsdansky, Cryptology, the computer, and data privacy, Computers and Automation **21** (April 1972) 12-19.

[Goe1] J.-M. Goethals and C. Couvreur, A cryptanalytic attack on the Lu-Lee public-key cryptosystem, Philips Journal of Research **35** (1980) 301-306.

[Goo1] R.M.F. Goodman and A.J. McAuley, New trapdoor-knapsack public-key cryptosystem, IEE Proceedings (Part E) **132** (1985) 289-292.

[Gor1] J.A. Gordon and H. Retkin, Are big S-boxes best?, Cryptography: Proceedings of the workshop on cryptography, Burg Feuerstein, 1982 (1983) 257-262.

[Gor2] J.A. Gordon, Strong RSA keys, Electronics Letters **20** (1984) 514-516.

[Gor3] J.A. Gordon, How to forge RSA key certificates, Electronics Letters **21** (1985) 377-379.

[Gro1] E.K. Grossman and B. Tuckerman, Analysis of a Feistel-like cipher weakened by having no rotating key, IBM Research Report **RC 6375 (#27489)**, IBM Thomas J. Watson Research Centre (Yorktown Heights, NY), January 1977.

[Hel1] M. Hellman, R. Merkle, R. Schroeppel, L. Washington, W. Diffie, S. Pohlig and P. Schweitzer, Results of an initial attempt to cryptanalyze the NBS data encryption standard, Stanford University Report **SEL76-042**, 1976.

[Her1] T. Herlestam, Critical remarks on some public-key cryptosystems, BIT **18** (1978) 493-496.

[Hun1] D.G.N. Hunter, RSA key calculation in ADA, The Computer Journal **28** (1985) 343-348.

[ISO8227] ISO / DIS 8227 (draft international standard), Information processing - Data encipherment - Specification of algorithm DEA 1, International Standards Organisation.

[ISO8372] ISO / DIS 8372 (draft international standard), Information processing - modes of operation for a 64-bit block cipher algorithm, International Standards Organisation.

[ISO8730] ISO / DIS 8730 (draft international standard), Banking - Requirements for standard message authentication, International Standards Organisation, January 1985.

[ISO8731/1] ISO / DIS8731/1 (draft international standard), Banking - Approved algorithms for message authentication - Part 1: DEA-1 algorithm, International Standards Organisation, September 1985.

[ISO8731/2] ISO / DIS8731/2 (draft international standard), Banking - Approved algorithm for message authentication - Part 2: Message authenticator algorithm, International Standards Organisation, February 1986.

[J1] S.M. Jennings, Multiplexed sequences: some properties of the minimum polynomial, Cryptography: Proceedings of the workshop on cryptography, Burg Feuerstein, 1982 (1983) 189-206.

[J2] S.M. Jennings, Autocorrelation function of the multiplexed sequence, IEE Proceedings Part F **131 no 2** (April 1984) 169-172.

[K1] D. Kahn, The Codebreakers: The Story of Secret Writing, MacMillan (New York), 1967.

[Kam1] J.B. Kam and G.I. Davida, Structured design of substitution-permutation networks, IEEE Transactions on Computers **C-28** (1979) 747-753.

[Koc0] M.J. Kochanski, Remarks on Lu and Lee's proposals for a public-key cryptosystem, Cryptologia **4** (1980) 204-207.

[Koc1] M. Kochanski, Developing an RSA chip, paper given at Crypto 85.

[Koc2] M. Kochanski, The FAP4 fast arithmetic processor, preprint, 1985.

[Lag1] J.C. Lagarias, Knapsack public key cryptosystems and diophantine approximations, Advances in Cryptology: Proceedings of Crypto 83, Plenum Press (New York) (1984) 3-23.

[Lu1] S.C. Lu and L.N. Lee, A simple and effective public-key cryptosystem, COMSAT Technical Review **9** (1979) 15-24.

[McE1] R.J. McEliece, A public-key cryptosystem based on algebraic coding theory, DSN Progress Report, JPL **42-44** (Jan/Feb 1978) 114-116.

[M1] R.A. Marolf, 200 Mbit/s pseudo random sequence generators for a very wide band secure communications systems, Proc. NEC **14** (1963) 183-187.

[Mas1] J.L. Massey, A. Gubser, A. Fischer, P. Hochstrasser, B. Huber and R. Sutter, A self-synchronising digital scrambler for cryptographic protection of data, Proceedings of the 1984 International Zurich Seminar on Digital Communications, IEEE (1984) 163-169.

[Mer1] R.C. Merkle and M.E. Hellman, Hiding information and signatures in trapdoor knapsacks, IEEE Transactions on Information Theory **IT-24** (1978) 525-530.

[Mer2] R.C. Merkle, Protocols for public key cryptosystems, Proceedings of the 1980 IEEE Symposium on Security and Privacy, Oakland, April 1980, IEEE (1980) 122-134.

[M2] C.H. Meyer and W.L. Tuchman, Pseudorandom codes can be cracked, Electronic Design **23** (9th November 1972) 74-76.

[Mit1] C.J. Mitchell, A comparison of the cryptographic requirements for digital secure speech systems operating at different bit rates, Proceedings of the International Conference on Secure Communications Systems, London, February 1984, IEE (1984) 32-37.

[Mit2] C.J. Mitchell and F.C. Piper, Data encryption techniques and applications, to appear.

[Moh1] S.B. Mohan and B.S. Adiga, Fast algorithms for implementing RSA public key cryptosystem, Electronics Letters **21** (1985) 761.

[Mok1] N. Mokhoff, Second generation chips displace Data Encryption Standard, Computer Design **25 no 2** (15th January 1986) 35-37.

[Mor1] R. Morris, N.J.A. Sloane and A.D. Wyner, Assessment of the National Bureau of Standards proposed Data Encryption Standard, Cryptologia **1** (1977) 281-291.

[Mül1] C. Müller-Schloer, A microprocessor-based cryptoprocessor, IEEE Micro **3 no 5** (October 1983) 5-15.

[Nam1) K.-H. Nam and T.R.N. Rao, Private-key algebraic-coded cryptosystems, Technical Report NSA-3-85, 5th August 1985.

[Nee1] R.M. Needham and M.D. Schroeder, Using encryption for authentication in large networks of computers, Communications of the ACM **21** (1978) 993-999.

[Nie1] H. Niederreiter, A public-key cryptosystem based on shift register sequences, preprint.

[Odo1] R.W.K. Odoni, V. Varadharajan and P.W. Sanders, Public key distribution in matrix rings, Electronics Letters **20** (1984) 386-387.

[Oka1] T. Okamoto and A. Shiraishi, A fast signature scheme based on quadratic inequalities, Proceedings of the 1985 IEEE Symposium on Security and Privacy, Oakland, April 1985, IEEE (1985) 123-132.

[Ple1] V.S. Pless, Encryption schemes for computer confidentiality, IEEE Transactions on Computers **C-26** (1977) 1133-1136.

[Poh1] S.C. Pohlig and M.E. Hellman, An improved algorithm for computing logarithms over GF(p) and its cryptographic significance, IEEE Transactions on Information Theory **IT-24** (1978) 106-110.

[Pom1] C. Pomerance, J.W. Smith and S.S. Wagstaff Jr., New ideas for factoring large numbers, Advances in Cryptology: Proceedings of Crypto 83, Plenum Press (New York) (1984) 81-85.

[PK1] G.J. Popek and C.S. Kline, Encryption and secure computer networks, ACM Computing Surveys **11** (1979) 331-356.

[Qui1] J.J. Quisquater and C. Couvreur, Fast algorithms for RSA public key cryptosystems, Electronics Letters **18** (1982) 905-907.

[Rao1] T.R.N. Rao, Cryptosystems using algebraic codes, Proceedings of the 11th International Symposium on Computer Architecture, Ann Arbor, Michigan, June 1984.

[Riv1] R.L. Rivest, A. Shamir and L. Adleman, A method for obtaining digital signatures and public key cryptosystems, Communications of the ACM **21** (1978) 120-126.

[Riv2] R.L. Rivest, Remarks on a proposed cryptanalytic attack on the M.I.T. public-key cryptosystem, Cryptologia **2** (1978) 62-65.

[Riv3] R.L. Rivest, Critical remarks on "Critical remarks on some public-key cryptosystems" by T. Herlestam, BIT **19** (1979) 274-275.

[Rub1] F. Rubin, Decrypting a stream cipher based on J-K flip-flops, IEEE Transactions on Computers **C-28** (1979) 483-487.

[Ruep] R.A. Rueppel, New Approaches to Stream Ciphers, Thesis, Swiss Federal Institute of Technology, Zurich (1984).

[Sav1] J.E. Savage, Some simple self-synchronising digital data scramblers, Bell System Technical Journal **46** (1967) 449-487.

[Sha1] A. Shamir, A polynomial time algorithm for breaking the basic Merkle-Hellman cryptosystem, Proceedings of the 23rd Annual Symposium on Foundations of Computer Science (1982) 145-152.

[S1] C.E. Shannon, Communication Theory of Secrecy Systems, Bell System Technical Journal **28** (1949) 656-715.

[Slo1] N.J.A. Sloane, Encrypting by random rotations, Cryptography: Proceedings of the workshop on cryptography, Burg Feuerstein, 1982 (1983) 71-128.

[S2] B. Smeets, A note on sequences generated by clock controlled shift registers, presented at Eurocrypt 85.

[Smi1] D.R. Smith and J.T. Palmer, Universal fixed messages and the Rivest-Shamir-Adleman cryptosystem, Mathematika **26** (1979) 44-52.

[Var1] V. Varadharajan and R. Odini, Security of public key distribution in matrix rings, Electronics Letters **22** (1986) 46-47.

[V1] G.S. Vernam, Cipher printing telegraph systems for secret wire and radio telegraph communications, Journal of the American Institute of Electrical Engineers **45** (1926) 109-115.

[Voy1] V.L. Voydock and S.T. Kent, Security mechanisms in high-level network protocols, Computing Surveys **15** (1983) 135-171.

[Voy2] V.L. Voydock and S.T. Kent, Security in high-level network protocols, IEEE Communications Magazine **23 no 7** (July 1985) 12-24.

[Wil0] H.C. Williams and B. Schmid, Some remarks concerning the M.I.T. public key cryptosystems, BIT **19** (1979) 525-538.

[Wil1] H.C. Williams, An overview of factoring, Advances in Cryptology: Proceedings of Crypto 83, Plenum Press (New York) (1984) 71-80.

[Wun1] M.C. Wunderlich, Recent advances in the design and implementation of large integer factorisation algorithms, IEEE Symposium of Security and Privacy, Oakland, April 1983, IEEE (1983) 67-71.

[Wun2] M.C. Wunderlich, Factoring numbers on the massively parallel computer, Advances in Cryptology, Proceedings of Crypto 83, Plenum Press (New York) (1984) 87-102.

The Role of Feedback in Communication

Thomas M. Cover*
Stanford University

1 Introduction.

Most of us have a reasonably good idea of the role of feedback in control systems. One can drive from Boston to New York with one's eyes open, but not with them closed. Feedback not only makes driving simpler, it makes it possible. Recursive corrections are easier than open loop corrections.

We wish to ask the same questions about the role of feedback in communication. Some possible roles that feedback might play in communication include the following: correct receiver misunderstanding; predict and correct the noise; cooperate with other senders; determine properties of the channel; reduce communication delay; reduce computational complexity at the transmitter or the receiver and improve communication rate.

Information theory will be the primary tool in the analysis because information theory establishes the boundaries of reliable communication. Shannon proved the first shocking result about feedback, which we shall treat in the next section.

2 Feedback for memoryless channels.

Consider the following setup. One has a transmitter X, a receiver Y, and a conditional probability mass function $p(y|x)$. If one wishes to use this channel n times, we shall define a $(2^{nR}, n)$ feedback code as a collection of code words $\mathbf{X}(W, \mathbf{Y})$ of block length n in which the i-th transmission $X_i(W, Y_1, \ldots, Y_{i-1})$ depends on the message $W \in \{1, 2, \ldots, 2^{nR}\}$ as well as the previous received Y's available through feedback. The

*This work was partially supported by NSF Grant ECS82-11568, DARPA Contract N00039-84-C-0211, and Bell Communications Research.

J. K. Skwirzynski (ed.), Performance Limits in Communication Theory and Practice, 225–235.

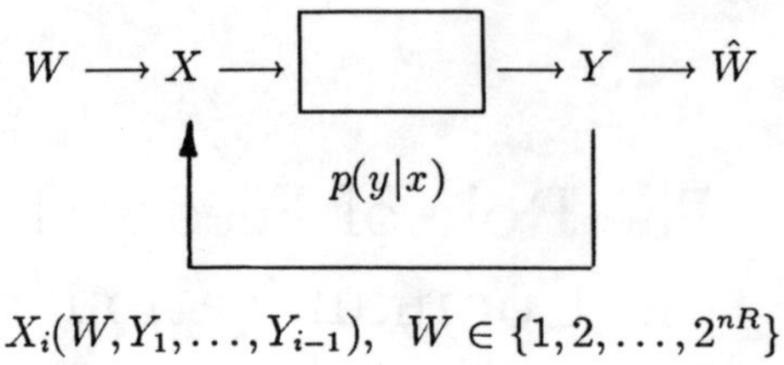

Figure 1: Channel with Feedback.

capacity C of such a channel is the maximum rate R at which one can find $(2^{nR}, n)$ codes such that the probability of decoding error $P_e = \Pr\{\hat{W}(Y) \neq W\}$ tends to zero asymptotically as $n \to \infty$. The first major surprise in the role of feedback in communication is that feedback does not improve the capacity of memoryless channels.

We first recall that the channel capacity for a discrete memoryless channel without feedback is given by

$$C = \max_{p(x)} I(X;Y) \tag{1}$$

(The codewords for such a channel depend on W only and not on the previous Y's). Thus communication rates up to C bits per channel use can be achieved. We now have the following result due to Shannon.[8]

Theorem 1 *Feedback does not improve the capacity of memoryless channels.*

Proof: We take a feedback code, as described above, put a uniform probability distribution on the indices $W \in \{1, 2, \ldots, 2^{nR}\}$ and then observe the following chain of inequalities:

$$\begin{aligned}
nR &\overset{(a)}{=} H(W) \\
&\overset{(b)}{\leq} H(W) - H(W|Y^n) + n\epsilon_n \\
&\overset{(c)}{=} I(W;Y^n) + n\epsilon_n \\
&\overset{(d)}{=} H(Y^n) - H(Y^n|W) + n\epsilon_n \\
&\overset{(e)}{=} H(Y^n) - \sum H(Y_i|W, Y^{i-1}) + n\epsilon_n \\
&\overset{(f)}{=} H(Y^n) - \sum H(Y_i|W, Y^{i-1}, X_i) + n\epsilon_n \\
&\overset{(g)}{=} H(Y^n) - \sum H(Y_i|X_i) + n\epsilon_n
\end{aligned}$$

$$\stackrel{(h)}{\leq} \sum H(Y_i) - \sum H(Y_i|X_i) + n\epsilon_n$$

$$\stackrel{(i)}{=} \sum I(X_i, Y_i) + n\epsilon_n$$

$$\stackrel{(j)}{\leq} nC + n\epsilon_n$$

where $\epsilon_n \to 0$, as $n \to \infty$. Here (a) is the entropy of a uniformly distributed random variable, (b) is Fano's inequality, (c) is by definition of I, (d) is an expansion of I, (e) is the chain rule, (f) uses the fact that X_i is a function of W, Y^{i-1} via feedback, (g) follows from the fact that we have a discrete memoryless channel, (h) is a well known upper bound, (i) is an identity, and (j) follows from the definition of C in (1). Thus we have shown that feedback does not increase capacity.

On the other hand, there are numerous results that indicate that feedback may reduce the coding complexity. The simplest example of this might be the binary erasure channel in which one sends $X \in \{0,1\}$ and receives $Y \in \{0,1,E\}$, where E denotes erasure. Suppose $Y = X$ with probability $1-\alpha$, and $Y = E$, with probability α. The capacity of the channel is $C = 1-\alpha$, but is difficult to achieve. On the other hand, with feedback, there is no problem. Merely retransmit the intended symbol whenever an erasure is received. The expected number of transmissions required to reveal a given symbol X is $1/(1-\alpha)$. Thus $(1-\alpha)n$ transmitted bits require n transmissions and the resulting capacity is $C = 1-\alpha$. In fact, from this example, we see that not only is encoding and decoding simple with feedback, but that the delay in the received transmission is finite and small.

3 Gaussian channels with feedback.

We will now take a look at the special case of Gaussian channels. At this time, we will remove the memoryless condition from the channel and allow the noise to be time-dependent and correlated. For the Gaussian additive noise channel, that simply means that the additive Z process is normal with mean $\mathbf{0}$ and covariance matrix K. Thus the channel is of the form $Y_i = X_i + Z_i$. The X_i's have the previous defined codebook structure, where X_i depends on the index W and the previous Y's. In addition, for the Gaussian channels, there is a power constraint

$$\frac{1}{n}\sum X_i^2(W, Y_1, \ldots, Y_{i-1}) \leq P \tag{2}$$

We immediately have a problem with this because if X_i depends only on W, we can guarantee that its power is limited to P, but if X_i also depends on the received Y's in a nontrivial way, some Y sequences may cause X to burst the power constraint. So in general, we can either talk about an expected power constraint on X averaged over Y or we shall talk about making an error whenever Y causes the power constraint to be violated. Both approaches will yield the same answers below. Let C_{FB} be the feedback channel capacity for the Gaussian channel with a given covariance structure and C_{NFB} be the associated capacity when feedback is not allowed. It is shown in Cover and Pombra [5] that

$$
\begin{aligned}
C_{FB} &= \lim_{n\to\infty} \max_{\frac{1}{n}tr(K_X)\leq P} \frac{1}{2n}\log\frac{|K_{X+Z}^{(n)}|}{|K_Z^{(n)}|} \\
C_{NFB} &= \lim_{n\to\infty} \max_{\frac{1}{n}tr(K_X)\leq P} \frac{1}{2n}\log\frac{|K_X^{(n)}+K_Z^{(n)}|}{|K_Z^{(n)}|}
\end{aligned}
$$

The maximization in the definition of C_{FB} is over K_{X+Z} in which X_i and Z_i are conditionally independent given the past X^{i-1}, Z^{i-1}. Here it is true that $C_{FB} > C_{NFB}$, i.e. feedback *strictly* increases capacity. The reason is due solely to the fact that if the transmitter knows Y_i, he can determine $Z_i = Y_i - X_i$ and therefore all the previous terms in the noise process, and, because the noise is not independent, he can predict where the noise is going and combat it.

Incidentally, the notion of combating the noise is somewhat misleading. In fact, many considerations show that one should not fight the noise, but join it. There is more space between the quills of a porcupine if they point out than if they point in. Thus if one is constrained to add some signal power to noise one should add it in the same direction rather than inwards.

The following two theorems limit the effect of feedback for Gaussian channels.

Theorem 2 *(Pinsker and Ebert).*

$$C_{FB} \leq 2C_{NFB} \tag{3}$$

Proof: Pinsker stated the result and didn't publish it. Ebert [11] published the result in B.S.T.J. A new proof of the result can be found in Cover and Pombra [5].

Theorem 3

$$C_{FB} \leq C_{NFB} + \frac{1}{2} \quad \text{bits per transmission}$$

Proof: (See Cover and Pombra [5].)

Theorem 2 says feedback at most doubles the capacity, and Theorem 3 says that one can at most add one half bit per transmission to the communication. Feedback helps, but not much.

4 Unknown channels.

In the previous section, we showed how feedback is used to predict the noise. Here we will discuss a more extreme case, one in which feedback helps by an arbitrary amount.

Consider the following channel discussed by Dobrushin [10]. The input alphabet is $\mathcal{X} = \{1, 2, \ldots, m\}$ and the output alphabet is $\mathcal{Y} = \{0, 1\}$. All but the i-th input symbol is crushed into $y = 0$, and the i-th symbol is the received as $y = 1$. The transmitter does not know which symbol i survives. This channel stays constant over many uses. (The channel is stationary but not ergodic.)

First we find the capacity with feedback. This is quite simple. The transmitter first cycles through the integers $1, 2, \ldots, m$, sending a test sequence which is received by the receiver and fed back. After m transmissions, both sender and receiver know which symbol i results in $y = 1$. Thereafter, the transmitter simply uses the channel as a binary typewriter sending i or not i accordingly as he wishes to send 1 or 0. Thus the feedback capacity is clearly one bit per transmission.

Now, what about the capacity without feedback? Here is turns out that the receiver actually does know which symbol maps into $y = 1$. The transmitter sends a test sequence $1, 2, \ldots, m$ and the receiver determines the symbol i that maps to $y = 1$. He is then prepared to do very clever decoding in the future. Unfortunately, the transmitter is still in the dark. The best the transmitter can do is to use a uniform distribution over the m letters, in which case he is only sending $Y = 1$ $\frac{1}{m}$-th of the time. The resulting mutual information is $I = \frac{1}{m} \log m + ((m-1)/m) \log(m/m - 1)$, which is approximately $\frac{1}{m} \log m$. Thus, $C_{NFB} \approx \log m/m$.

So here we have it. For large m, the capacity without feedback C_{NFB} is very near zero, while the capacity with feedback C_{FB} equals 1, and the ratio is arbitrarily large. Feedback helps by an arbitrarily large factor.

By changing the example slightly, one can get the additive difference between the two capacities to be as large as possible. So thanks to this example, we see that all is not well in showing that feedback has a bounded effect on the improvement of capacity. Nonetheless, we feel that the factor of two limit and the half bit additive limit from the previous section are typical of practical channels.

5 Multiple access channels.

Until now, we have talked about one sender and one receiver. The multiple access channel, modeling satellite communication, has many senders $X_1, X_2, \ldots, X_m$ and one receiver Y, where Y has the conditional distribution $p(y|x_1, x_2, \ldots, x_m)$. Here one has interference among the senders as well as channel noise.

Theorem 4 *(Ahlswede [3] and Liao [2]). The capacity region for the 2-user multiple access channel is the convex hull of the union of the set of all* (R_1, R_2) *satisfying*

$$\begin{aligned} R_1 &\leq I(X_1; Y|X_2) \\ R_2 &\leq I(X_2; Y|X_1) \\ R_1 + R_2 &\leq I(X_1, X_2; Y) \end{aligned}$$

for some product distribution $p(x_1)p(x_2)$.

Gaarder and Wolf [7] showed that feedback increases the capacity of multiple access channels. Cover and Leung [1] went on to show that the following region is achievable.

Theorem 5 *An achievable rate region for the multiple access channel with feedback is given by the convex hull of the union of the set of* (R_1, R_2) *satisfying*

$$\begin{aligned} R_1 &\leq I(X_1; Y|X_2, U) \\ R_2 &\leq I(X_2; Y|X_1, U) \\ R_1 + R_2 &\leq I(X_1, X_2; Y) \end{aligned}$$

for some distribution $p(u)p(x_1|u)p(x_2|u)$.

However, Theorem 5 is in general only an inner bound to the capacity region. Willems [9] has shown that if a certain condition is satisfied, then the above region is indeed the capacity region. The region in Theorem 5 is larger than the no-feedback region, and we conclude that feedback improves the capacity region of a multiple access channel, even when it is memoryless. See [12, 13, 14] for further results in this direction.

Why does feedback help? The channel is memoryless, so feedback does not help predict the noise. Here capacity is increased because feedback diminishes the interference through cooperation. Each transmitter has a better idea of what the other transmitter is sending than does the receiver Y. To this extent they can cooperate. Whether they help each other or not, they can at least get out of each other's way in some statistical sense. This is the real reason why feedback helps multiple access communications. It allows the senders to cooperate. However, this improvement in capacity is limited.

J.A. Thomas has shown that an arbitrary multiple access channel with two users has its capacity region at most doubled with feedback. He has gone on to show that for $m-$user memoryless Gaussian multiple access channels that feedback at most doubles capacity. He conjectures that capacity is at most doubled with feedback for arbitrary non-Gaussian m-user multiple access channels.

For multiple access channels, feedback helps, but not by very much.

6 Computational complexity for multiple access channels with feedback

Consider the binary erasure multiple access channel in which $X_1, X_2 \in \{0,1\}$, $Y \in \{0,1,2\}$, and $Y = X_1 + X_2$. The capacity region for this channel is shown in Figure 2.

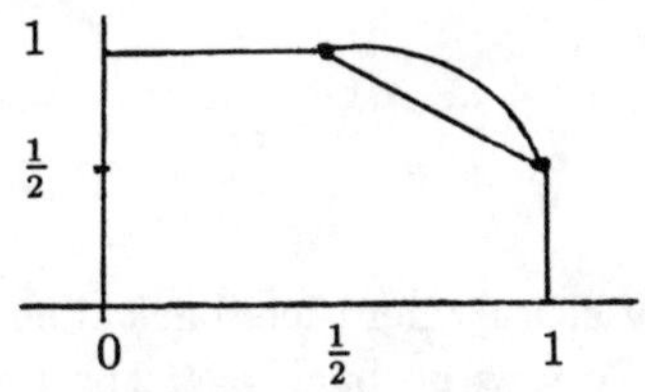

Capacity Region

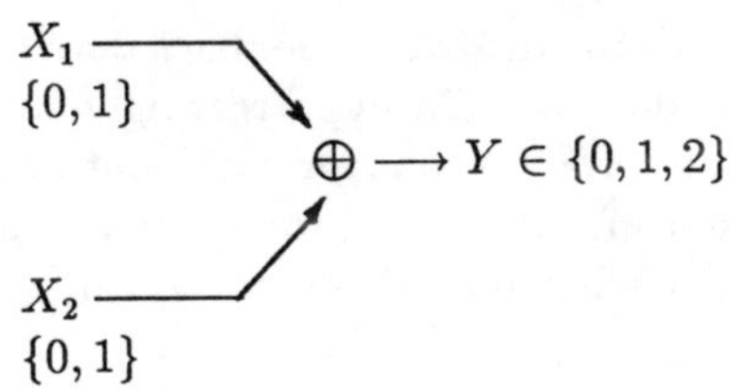

Figure 2: Binary Erasure Multiple Access Channel.

Here it is known that the region in Theorem 5 is indeed the capacity region. This is strictly greater than the no-feedback capacity region. So feedback helps capacity. The question is, how does one send information over this channel using feedback? Suppose X_1 is sending at rate $R_1 = 1$. Thus X_1 sends an arbitrary sequence of zeros and ones, each occurring about half the time. How then can X_2 achieve the rate $\frac{1}{2}$ corresponding to the point $(R_1, R_2) = (1, \frac{1}{2})$ on the capacity boundary? Here is the strategy, relayed to me by J. Massey. Let X_1 and X_2 arbitrarily send their zeros and ones. For example:

$$\begin{array}{rcl} X_1 & = & 0\ 0\ 1\ \ 0\ 0\ 1\ 1 \\ X_2 & = & 0\ 0\ 1\ \boxed{1\ 1\ 1} \\ \hline Y & = & 0\ 0\ 2\ \boxed{1\ 1\ 2} \end{array}$$

Notice that when Y equals 0 or 2, the receiver knows instantly the values of X_1 and X_2. However, $Y = 1$ acts as an erasure. We don't know whether $(X_1, X_2) = (0,1)$ or $(1,0)$. At this point, X_1 continues sending whatever he wishes to send. After all, he is sending at rate $R_1 = 1$. But X_2 now continues to retransmit whatever it was that he sent when the ambiguous $Y = 1$ was received. He continues to do this until Y equals either 0 or 2, thus correcting the erasure. It is then a simple matter for Y to go back and correctly determine previous values of X_1 and X_2. So X_2 has to send on the average two symbols for every one that gets through, achieving rate $R_2 = \frac{1}{2}$.

This point on the boundary of the capacity region could not have been achieved as simply without feedback.

7 Conclusion.

We have now examined many cases where feedback does and does not help the communication of information. We now go back over the previous questions and answer them with respect to these examples.

Possible Roles of feedback:

Correct receiver's misunderstanding? Feedback does not increase capacity for memoryless channels, so it does not aid in correcting Y's misunderstanding. On the other hand, feedback improves the error exponent and it helps reduce the complexity of the communication. Indeed, for additive Gaussian noise channels, the Kailath-Schalkwijck [6] scheme sends correction information to Y and achieves capacity.

Predict and correct the noise? Here feedback helps the capacity if the noise is dependent. On the other hand, the improvement in capacity is less than or equal to a factor of two for Gaussian additive noise channels regardless of the dependence. Also, one does not really correct the noise, but joins it in some sense.

Improve error exponent? Feedback helps.

Cooperation of the senders? Feedback allows cooperation, increases the capacity and lowers the computational complexity.

Determination of the channel? A simple test sequence can be used to determine the entire channel distribution $p(y|x)$. One can then use the code appropriate for that channel and achieve channel capacity as well as if one had known ahead of time what the channel was. Feedback definitely helps – sometimes by arbitrary factors.

Reduction of delay? Feedback can greatly reduce delay. The examples show small delays for many of the channels in which feedback is used. Feedback allows multiple users of satellite and computer networks to share a common channel with minimal delay.

Reduction in computational complexity? Feedback helps.

In summary, feedback helps communication, but not as much as one might think. It simplifies communication without greatly increasing the rate of communication.

Acknowledgment: I would like to thank S. Pombra and J.A. Thomas for contributing their ideas to this discussion. J.A. Thomas also brought Dobrushin's example to my attention.

References

[1] T. Cover and S.K. Leung, "An Achievable Rate Region for the Multiple-Access Channel with Feedback," *IEEE Trans. on Information Theory,* Vol. IT-27, May 1981, pp. 292-298.

[2] H. Liao, "Multiple Access Channels," Ph.D. thesis, Dept. of Electrical Engineering, University of Hawaii, Honuolulu, 1972.

[3] R. Ahlswede, "Multi-way Communication Channels," *Proc. 2nd Int'l Symp. Inform. Theory* (Tsahkadsor, Armenian S.S.R.), pp. 23-52, 1971. (Publishing House of the Hungarian Academy of Sciences, 1973).

[4] J.A. Thomas, "Feedback can at Most Double Gaussian Multiple Access Channel Capacity," to appear *IEEE Trans. on Information Theory,* 1987.

[5] T. Cover and S. Pombra, "Gaussian Feedback Capacity," submitted to *IEEE Trans. on Information Theory.*

[6] T. Kailath and J.P. Schalkwijk, "A Coding Scheme for Additive Noise Channels with Feedback I: No Bandwidth Constraint," *IEEE Trans. on Information Theory,* Vol. IT-12, April 1966, pp. 172-182.

[7] N.T. Gaarder and J. Wolf, "The Capacity Region of a Multiple-access Discrete Memoryless Channel can Increase with Feedback," *IEEE Trans. on Information Theory,* Vol. IT-21, January 1975, pp. 100-102.

[8] C.E. Shannon "The Zero Error Capacity of a Noisy Channel," *IRE Trans. on Information Theory,* Vol. IT-2, Sept. 1956, pp. 8-19.

[9] F.M.J. Willems, "The Feedback Capacity of a Class of Discrete Memoryless Multiple Access Channels," *IEEE Trans. on Information Theory,* Vol. IT-28, January 1982, pp. 93-95.

[10] R.L. Dobrushin, "Information Transmission in a Channel with Feedback," *Theory of Prob. and Applications,* Vol. 34, December 1958, pp. 367-383.

[11] P.M. Ebert, "The Capacity of the Gaussian Channel with Feedback," *Bell System Technical Journal,* Vol. 49, October 1972, pp. 1705-1712.

[12] L.H. Ozarow, "The Capacity of the White Gaussian Multiple Access Channel with Feedback," *IEEE Trans. on Information Theory,* Vol. IT-30, July 1984, pp. 623-629.

[13] G. Dueck, "Partial Feedback in Two-way and Broadcast Channels," *Information and Control,* Vol. 46, July 1980, pp. 1-15.

[14] L.H. Ozarow and S.K. Leung-Yan-Cheong, "An Achievable Region and Outer Bound for the Gaussian Broadcast Channel with Feedback," *IEEE Trans. on Information Theory,* Vol. IT-30, July 1984, pp. 667-671.

THE COMPLEXITIES OF INFORMATION TRANSFER WITH REFERENCE TO A GENETIC CODE MODEL

G A KARPEL
SOFTWARE SCIENCES LIMITED, UK

1 INTRODUCTION

One of the most complex information transfers in the world occurs every day. It is in the replication of genetic material in living cells. Such material contains coded information sufficient to specify the enire host organism. By the use of a series of intermediate steps, the genetic code is passed from parent to daughter cells with little error. By comparison, an Open Systems Interconnection network of computers will transfer information between them via a series of intermediate steps, again with little error. An examination of what such steps are, and their associated processes has been performed with reference to a genetic code model. The details and conclusions are presented below.

2 SYSTEMS MODELLING

The establishment of a formal model by which a complex system can be analysed is of prime importance. It enables the partitioning of elements comprising the system into areas of commonality, referred to as sub-systems. Further partitioning then proceeds until the resulting decomposition provides 'a place for everything and everything in its place'. Specifically there will be two fundamental elements that will be used in this paper to facilitate the correlation of processes between the Biological Model and the Technological Model. These are identified as Entities and Relations.

2.1 Entities

These are the subjects and objects of the system. They have unique identities and associated attributes. By so defining them, a static ('snapshot') understanding of the system's structure is obtained.

2.2 Relations

To provide a understanding of the dynamics of a system Relations are defined. They are the links of commonality between Entities, and identifying the various interactions that occur between a given Entity and its environment (other Entities).

2.3 Models

The two models mentioned earlier will now be described using Entities and Relations. Each model will provide a framework for the understanding of the specific processes involved, and as such requires that all relevant facts are stated, even at the risk of covering familar groundwork.

J. K. Skwirzynski (ed.), Performance Limits in Communication Theory and Practice, 237–265.

3 THE OPEN SYSTEMS INTERCONNECTION (OSI) MODEL

The OSI reference model provide the basis for interconnecting "open" systems for distributed applications processing. That is, it facilitates information transfer from one idiosyncratic system to another across a physical medium. The applications processes may be one or a group of activities that execute a specific set of procedures, according to the demands of their hosts. The OSI reference model[1] was based on the Network Oriented Protocols, which specified the basic functionality that such a network should have[2]. This was in turn based on the ISO seven layer protocol which defined seven network services[3], that would provide interoperability and reduced error rate between communicating systems. Table 1 sets out the OSI seven layer grouping.

Interactions within each layer use a peer protocol that provides the necessary procedures for information transfer. Instructions between layers are facilitated by primitives (requests and responses). Such primitives have associated parameters which convey additional information between layers. Each layer has a management function to monitor and control activation, deactiviation of its service and to monitor and control error propagation. There is a system-wide management function which controls each layer activiation and deactivation, resources and data transfer. This is accomplished by the co-ordination of interaction primitives between the layers.

3.1 The Application Layer

This layer establishes the form of the information transfer, and selects the services necessary to support this form. It initiates and terminates the interconnection between two systems (Applications Processes), and provides the resourcing, synchronisation and tasking and scheduling of those resources.

3.2 The Presentation Layer

The presentation layer provides the service of entry, exchange, display and control of structured data in a form appropriate to each of the communicating applications processes; this involves the representation of the data in a common transfer syntax, for example in data transformation where the character set and code representation are modified or translated. Further modification may be necessary to the layout of the data and reformatting of the information may occur. However, whilst these transformations are taking place, the application processes are not in any way involved. This form of service 'transparency' is the essence of the OSI reference model.

3.2.1 Peer Protocols Within the presentation layer there are three peer protocols which facilitate the services described above. Virtual Terminal protocol allows for different character sets, control characters, and code representations to be used, whilst Virtual File protocol provides for the formatting of file access commands and their code representations. Lastly, Job Transfer and Manipulation protocol (JTMP) provides for the reformatting and control of data storage structure, access commands and data representation.

3.3 The Session Layer

This layer provides the coordination of the interaction between the communicating presentation layers. It establishes a session between them, and its operating parameters. For example, there may be half-duplex or full duplex form of communication. It also provides a buffer to the presentation layer for incoming data, in this way it can act as an error trap. The sending application process realises that the last information transfer was incorrect. It then sends a delete command, which the receiving session layer carries out on the buffered incorrect information.

3.4 The Transport Layer

This provides a path independent transport service to the session layer above it, which is transparent to the underlying network services. It is able to optimise the use of the network since it monitors the state of it, and controls the flow of data through it accordingly. It also provides the connection establishment between one or more sessions set up concurrently, or consecutively between each application process. This layer further provides the multiplexing of sessions through a given route, and is also responsible for detecting errors in the received data. For example: loss, corruption, duplication or misdelivery of data.

3.5 The Network Layer

The network layer provides the means to establish and control switched network connections. That is, there may be a number of different routes in the network which connect each of the application processes together. This layer provides the service of data exchange for the transport layer, between nodes in the network. It also provides notification of uncorrected errors to the transport layer for it to initiate possible error recovery action. It may further provide data delivery confirmation and sequence control of data to the transport layer. A particular feature of this layer is that of multiplexing discrete 'packets' of data along routes between nodes. This provides for high utilisation of the network.

3.6 The Data Link Layer

This layer provides the functional and procedural means to activate, deactivate and maintain data links between nodes in the network. It provides data grouping into 'frames' and synchronises their transfer, by identifying the beginning and end of a frame of data, and setting up certain counters for the frames. It also established an association between nodes of the physical link, and provides flow control of the transferred frames of data.

3.7 The Physical Layer

This provides the physical characteristics to activate, deactivate and maintain a physical connection between application processes. It also provides the physical medium for transferal of information, where it is represented by its fundamental form: the bit. There is also a baisc fault notification (i.e. physical disconnection), but no internal mechanism for recovery.

3.8 Primitives

Primitives are used for co-ordination between layers. There are four such primitives.

3.8.1 Request This is originated by a layer to activate a particular service provided by the next lowest layer.

3.8.2 Indication This is sent from the receiving layer to advise of activation of the requested service.

3.8.3 Response This is sent from the requesting layer in response to the lower layer's indication primitive.

3.8.4 Confirmation The lower layer sends a confirmation primitive to the requesting layer to advise of the completed activation of the requested service.

Each layer provides a set of services to the next higher layer. Request primitives are passed to the adjacent lower layer, and indication primitives are passed up to the adjacent higher layer. Figure 1 illustrates this process occurring between two application processes across a network.

3.9 Information Transfer within the Data Link Layer

This layer has been chosen as an example of the complexities involved in providing a high quality link for information transfer between application processes. It is particularly suitable as it is well defined, and has a number of interesting features in it, which are compared to the Genetic Code model later on.

To provide some control over the physical link which bears a contiguous stream of bits, the data link layer establishes an aggregation of bits known as a frame. A frame is delimited by a special sequence of bits. This sequence comprises a contiguous stream of six '1' bits with a '0' bit at the beginning and end, to make up a field of eight bits (an octet). These delimiters are called flags (F).

The whole frame will have an arbitrary bit pattern. There is an interesting mechanism known as "bit stuffing" which ensures that within the two flag delimiters, there will not be a 'pseudo flag' sequence generated, (i.e. a sequence of six contiguous '1' bits). The mechanism starts monitoring for streams of '1' bits after recognising a flag. After it encounters five contiguous '1' bits, it inserts a '0' bit in the sixth position, thus preventing the 'pseudo-flag' sequence occurring in the frame. The inverse mechanism occurs on receipt of a frame, thus re-establishing the original bit pattern within the frame. If a transmission error had occurred and seven contiguous bits had been received, then the frame is discarded. Furthermore, when fifteen contiguous '1' bits are received, the link is set to an 'idle' state with no information transfer occurring until further link re-establishment. There are three other fields with special purposes, and all are located within the frame as illustrated in Figure 2. The first field after the initial flag is the address field. It is an octet and as such can provide up to 127 (2^8-1) possibilities for the address; an all zero octet is not allowed. The next field is the control field which is also an octet, and provides information for the management of the communications link through which the frames travel. It identifies three types of frame, as illustrated in Figure 3. Unnumbered:- This frame is responsible for the establishment of the mode of communication between the frame source and its destination. In particular, it specifies whether the source, or the destination, or both can regulate the information transfer.

Supervisory is the second type. This is responsible for the acknowledgement of correctly received information, and providing a flow control in the form of requesting retransmission, and indicating impending receiver overload (i.e. wait until ready). There are three supervisory frame types, Receive Ready (RR) which acknowledges receipt of information frames, and/or prompts for the destination node's status. Receive Not Ready (RNR), which indicates that the receiving process is in a busy condition. It also acknowledges information frames and/or prompts for status. Reject (REJ), which implicitly rejects a faulty frame by requesting retransmission. It also acknowledges information frames and/or prompts for status. Lastly, Information frames contain the data that is for the destination's information. It also conveys some supervisory details such as the acknowledgement of received information. Figure 4 illustrates the structure of an I frame. The last field in the frame is the frame Check Sequence (FCS) which provides for the detection of bit transmission errors within the frame. It is a double octet used in a 16-bit polynomial algorithm. This algorithm is performed over all the contents of the frame excluding the FCS; firstly at the source node which places a certain value in the FCS field, then at the destination node which compares its calculated value against the transmitted one. If there is any discrepancy, the whole frame is discarded.

3.9.1 Data Link Level Procedures The link level procedures consist of four phases as depicted in Figure 5. The first phase is the link set-up phase, where the two communicating processes intialise their state variables and prepare to enter the next phase: Information transfer phase. The communicating processes transfer information in such a manner that errors can be detected and action taken to recover from any detected errors. The next phase is Disconnection phase. The communicating processes proceed in an orderly disconnection finishing the information transfer phase. Lastly (and possibly synonomously), the processes move into the Idle phase, monitoring the link for any change of state.

To support the Information transfer phase each communicating process maintains a set of state variables which are used in the control and management of the communications link. There are six such variables: V(S) the send state variable. This is initialised to zero when transmitting or receiving a link set up command. The value is then incremented by one using module 8 addition, every time transmission of an Information (I) frame occurs. The V(S) value is then copied into the next frame and becomes the N(S) value (the send sequence number).

V(R) the receive state variable. This is also initialised to zero on link set up. Upon reception of an I frame, the N(S) value in the frame is compared to the receiving process's V(R) value. If V(R) = N(S) then the I frame is in sequence, the frame is accepted and the V(R) is incremented by one using module 8 addition. If the V(R) does not equal the N(S) then something has gone wrong, and the out of sequence I frame is rejected.

K the number of outstanding (unacknowledged) I frames. This variable is incremented by one each time an I frame is transmitted, and decreased by one each time a previously sent I frame is acknowledged. The K state variable typically has a maximum allowable value of 7.

N1. This variable represents the maximum length for the information field within the I frame. This variable is a session variable and is agreed at the initiation of the data link. It is typically 128 octets of user data plus a network header of 4 octets; a total of 132 octets, as illustrated in Figure 6.

N2 represents the number of retries allowed for any operation before recovery procedures occur. This is also a session variable and is typically a maximum value of 3.

T. This is the timeout value between last receipt of a frame on an active link, to the initiation of recovery procedures. This is typically 100 milliseconds and is dependent on the network propogation delay.

3.9.2 Information Transfer Pathway in the OSI Model The information transfer within the data link layer has been discussed in detail. There are a number of features which support the integrity and timeliness of frames of information. The next layer above, the Network Layer also has similar state variables and special diagnostic packets which provide control and management of information transfer at that level. The Transport layer uses certain variables to monitor and control the routes in the network. The OSI model provides a very flexible and robust way of information transfer, and does this with a certain overhead penalty. Extra information is required to be sent along with the user data as can be seen in Figure 7.

4 THE GENETIC CODE MODEL

A genetic code model is used to illustrate the mechanisms of information transfer as applied to biological systems. In particular, the transfer of genetic information from one generation of organisms to the next will be considered. Figure 8 illustrates this cycle. There has evolved, over the last 3×10^9 years, a very robust and flexible way of genetic information transfer.

Each time a cell replicates it transfers information. This information is sufficient to instruct the daughter cell to build all the necessary proteins it will ever need. There are a series of processes which facilitate this genetic information transfer, and they can be considered to have seven functional groups by analogy to the OSI model, as Table 2 illustrates. Each functional group from both models is now regarded as an entity, and their interaction as relations, to assist in the comparison between the model layers.

4.1 The Gamete

This entity is the highest level of cellular interaction. It is a specialised cell used in the reproduction of organisms. By analogy, the gamete provides the unit of genetic information transfer (user data) as stimulated by the organisms application process. It initiates and controls the information transfer between organisms. In particular organisms of different species cannot successfully transfer gametes, (viruses and bacteria do however, but they bypass the higher level processes and infiltate at the lowest levels of information transfer). The gamete contains the chromosomes of the individual organism, these being the next functional group down in the model. Depending on the type of organism, the gamete may contain as little as one chromosome (e.g. E. Coli bateria), or as many as 23 chromosomes (e.g. humans).

4.2 The Chromosome

This is the 'bearer' system composed of proteins which transport the genes (next level down) to a common meeting place. The chomosomes 'pair up', like with like, at the meeting node. There, occasionally crossovers (chiasmata) occur between the pair of chromosomes and genes are exchanged. It is this process that provides the individual variations in a given population of organisms. However, these variations are not 'durable' enough to be the cause of diversity of organisms in the world. The sequences of genetic material are as readily modified during one cell replication as they are in another. A permanent modification of the genetic material, known as a mutation, is the mechanism by which diversity of organsims occurs, furthermore, it is this diversity that natural selection acts upon, giving rise to the sporadic evolution of the organisms concerned.

4.3 The Gene

Genes are tightly packed bundles of DNA (the next lower level). They are the 'packets' of genetic information, since they contain just enough information to code for the production of a cell's fundamental constituents-proteins (and enzymes), e.g. one gene may code for a specific enzyme. It is interesting to note that genes not only provide the 'data' packets, but also the 'control' packets. Some specific header regions act as switches, enabling or disabling the information transfer. (Figure 10 illustrates.) For example, the gene coding for the production of a particular digestive enzyme is activated when food has been ingested, then de-activated when sufficient enzyme has been produced. A more interesting use of gene control is observed during cell development when differentiaton occurs at the right time and place under normal circumstances, for instance during embryonic development.

4.4 DNA

DNA (Deoxyribonucleic acid) provides the 'database' for the genetic code. It is this material that stores within it the sequence of codons that is the program for the construction of a given organism. However, there are some viruses that use RNA (the next layer down in the model) instead. DNA has a double helix structure first established by Watson and Crick in 1953, and illustrated in Figure 11. Its precise structure imposes certain constraints on the sequence of codons within each of its two strands. Specifically one strand is the complement of the other, and so the double helix comprises one strand of codons and one strand of 'anit-condons'. It appears that only the strand containing the 'anit-codons' (the coding strand) acts as the template for the transcription process carried out by the next layer, mRNA. The non-coding strand is not involved. However, DNA will replicate itself at the initiation of the genetic information transfer process using both strands. Figure 12 shows the codon and anit-codon pairing whilst Figure 13 illustrates the information transfer process.

4.5 RNA

RNA (ribonucleic acid) provides the services of genetic code transcription, transportation and translation. There are two types of RNA available for these services, namely messenger RNA (mRNA) used for code transcription and transportation, and transfer RNA (tRNA) used for code translation and ultimately in protein synthesis. Figure 14 shows the two forms of RNA.

4.5.1 mRNA mRNA exists as single strands of the genetic code, present as codons. When the process or transcription occurs, the DNA double helix unwinds and mRNA pairs with the anti-codon strand, forming the codon strand. This then separates and transports this codon sequence to a special site in the cell (a ribosome) where it is positioned in a convenient way for the next step in the information transfer process to occur.

4.5.2 tRNA and the code translation process The structure of tRNA is somewhat different from mRNA in as much as it is much shorter in length and only has one codon site on it; to be correct it has one anti-codon site which it tries to match to the first codon in the mRNA strand (fundamental unit of proteins) attached to it, which is always associated with a particular anit-codon type tRNA. The process of genetic code translation occurs here. Figure 15 shows this process in a simplified manner. A second tRNA complex locates itself adjacent to the first, and due to their close proximity the amino acids combine to form a dipeptide. The first tRNA moves off, whilst the second tRNA complex remains on the mRNA, with the dipeptide. This sequence of events continues until all the codons are translated, and a polypeptide chain is formed. This chain then folds and convolutes according to chemical bonding principles to produce a specifically structured protein or enzyme.

4.6 Codons

Codons are the unit of the genetic code. A codon comprises a triplet of bases (the lowest layer in the model) in a given sequence. There is no delimiter for codons, so recognition of the triplet depends on establishing the start point of the mRNA sequence of bases; the first three bases are taken to be the first codon and so on until the last three bases are reached. Figure 12 illustrates this.

4.7 Bases

These comprise the lowest layer in the model and are the fundamental entities of genetic information transfer. They are analogans to the 'bits' in the physical layer of the OSI model. However, whereas there can be two different bits (0,1) there are four different bases: Adenine (A), Guanine (G), Cytosine (C) and Thymine (T)/Uracil (U). The last pair T, U are equivalents in the DNA and RNA layers respectively. The codon anti-codon pairing rule introduced earlier is a concomitant of the symmetric base pairing rule. A links to T, and G links to C. Figure 12 illutrates this rule.

4.8 The Genetic Code

A code is required for genetic information transfer as data compaction and integrity is necessary. There are twenty different kinds of amino acid that make up proteins, and only four bases. The way in which a given set of amino acids is sequenced directly affects the type of protein produced, so a coded arrangement of bases influences the structure of a protein, and hence its function. The most economical way of coding for the twenty amino acids is by having a triplet code, that is 4 x 4 x 4 = 64 possible combinations. Figure 16 shows the codon groupings against amino acids in a histogram fashion[4]. It is interesting to note that there is a certain codon redundancy. It is reasonsable to assume that certain amino acids are more safety critical/vital than others and that this is reflected in the differences of redundancy. However, the 'start' codon should be very important. In fact there is not a distinct group of start codons, nor indeed any start codon. It is 'shared' by the amino acid Methionine (Met)-only one codon (AUG). Worse still, under this regime, all proteins and enzymes should start with Met, they do not. After protein synthesis has occurred a post translational process occurs which nibbles off the superfluous Met amino acid. The average redundancy per amino acid is around 3, with the mode at 2. What is interesting is that there are three amino acids namely ARG, LEU and SER which exhibit 6-fold redundancy. These have yet to be shown to be of particular significance. Another possible explantion is that of Thermodynamic consideration[5] in producing certain stable groups of bases which were preferentially taken up by the primordial life in the world. Furthermore, it has been shown that a triplet periodicity for bases is thermodynamically the most stable.

Further work suggests the use of cellular automata to model this genetic code 'evolution' to observe trends in 'island stability' (i.e. codon bias and redundancy). There are DNA sequence computer database libraries available which store around 5 x 10^6 base pairs of sequence (translation) information. To put this translation capability in perspective, the human gamete (layer 7 in the model) contains twenty three chromosomes (layer 6). The DNA (layer 4) within these chromosomes stores around 3 x 10^9 base pairs, and these translate to around 2 x 10^4 proteins.[6] From another viewpoint, every individual in the world has around 10^7 base positions unique to that individual. In the analysis of the genetic code, there inevitably occurs common sequences of bases from organism to organism. Assessing their statistical significance is one important task, particularly when, for example, one base in 5000 is different. Another point about codon bias; different organisms (and different genes) favour certain codons for a given amino acid in particular ways.

4.8.1 Genetic code compactness The data storage requirements for the genetic code are quite impressive. The total mass of DNA representing sufficient genetic information to code for the production of the entire population of the world is just 30 milligrams. For the case of a simple bacterium (E. Coli)[7] which has just one chromosome of around 4×10^6 base pairs, the total unwound length of the DNA would be around 1.4 mm, a thousand times longer than the bacterium itself. The human DNA unwound length would be around 2 m.

4.9 Genetic code and error propagation

As discussed earlier, the evolution of organisms depends on certain 'errors' being propagated. These are mutational errors, and there are four types:

4.9.1 Point mutation of information codons A base is altered of a specific location by its 'noisy' environment (e.g. radiation or chemical), causing a different amino acid to be inserted in the overall protein. This effect is known as a point mutation. There is a certain degree of resilience to this effect since codon redudancy is present, and further, some amino acids have very similar properties so their interchange would not necessarily be noticeable. However, these mutations do occur, for example by a single point mutation in the codon specifying Glutamic Acid, changing it to one specifying Valine.

4.9.2 Base sequence deletion Deletion of some or all of a short base sequence within a given gene may occur, thus no protein synthesis can take place , (e.g. alpha-Thalassaemia, a lethal condition).

4.9.3 Point mutation of control codons The control of code translation is facilitated by start and stop codons. When either of these is corrupted, then protein synthesis either never initiates or spuriously stops, (e.g. beta-Thalassaemia).

4.9.4 Sequence mutations In a sequence of DNA, there are some non-coding portions known as introns. These are possibly examples of pre-historic remnants which are now no longer required. However, introns do become transcribed by the mRNA, and are later 'edited' out. Now when these introns are mutated, the resultant mRNA production is decreased or halted due to the editing process being interfered with (e.g. beta-Thalassaemia).

Some investigations[8] have been carried out on the stereochemical aspects of the amino acids, and it has been observed that there is a certain relationship in the 'handedness' of amino acids and the codon anti-codon symmetry. By establishing a dimer table and comparing the codons it has been observed that there are two genetic code evolutionary errors propagated from pre-history. Specifically, there is a certain organelle which exists within plant cells called a mitochondrion. This object is of great interest, as it generates the 'power' for the cell, (there is an analagous organelle in plant cells, the chloroplast). Furthermore, it contains its own DNA and carries out its own replication process. It is believed that this type of organelle was originally a free ranging bacterium which entered into a symbiotic relationship with a host cell, in prehistory and has stayed that way ever since.

If this is the case then because it has been 'sheltered' from evolutionary forces one would expect to find a close approximation to the original genetic code. It turns out that this is the case, and indeed by considering this codon usage the dimer anomalies disappear. Returning to other error propagation experiments[9] have been carried out in bacteria and their 'parasites' bacteriophages. Errors were induced which resulted in a first generation error rate of around 1 in 10^{-4} errors/translated codon. In this test case the induced errors were made in an RNA polymerase (transcription enzyme) which subsequenly caused a lack of error recognition in the transcription process. Hence error propogation occurred which is considered to imply ageing and ultimately death of the organism. However, further investigation found that when a different error (to a particular codon sequence) was induced, although a faulty RNA polymerase was still produced, this faulty enzyme did not mistranscribe subsequent DNA sequences. This is clear evidence of further resilience in the process of genetic information transfer.

4.9.5 Genetic code error rate To put the genetic code model for information transfer in perspective, and to round off the comparison to the OSI model, there are two interesting parameters to discuss. Firstly the 'data' rate for the OSI model is not fixed, but is typically 4.8 Kbits per second. The 'data' rate in the genetic code model is again not fixed, but consider the simple case of the E.Coli bacterium which has a parent-daughter generation on time of around 40 minutes.

This implies that to achieve this time, about 2000 base parts have to replicated per second, i.e. around 4 Kbases per seconed. To do this, the DNA has to 'Unwind' at a rate of around 10^4 revolutions per minute. Secondly, the overall 'bit' error rate for this bacteria is under normal conditions around 1 in 10^9 (i.e. BER = 10^{-9}). This compares reasonably well with the OSI model; the Eurocom National Networks Standard requires a BER = 10^{-7}.

5 SUMMARY AND CONCLUSION

This paper presented some interim findings in an area of ongoing research. It firstly established a modelling framework in which both the OSI model and the genetic code model were considered. Both have complex subsystems, which attempt to ensure that information transfer is performed with high data integrity, authentication, error control and at reasonably fast data rates. There are some similarities in structure and functionality, but at this stage of research any deeper comparison would be somewhat tenuous.

It is certainly evident, however that this area should be more widely investigated, not only by geneticists and microbiologists, but also by information scientists and communications and computer science specialists. The future progress in more refined communications methods, and secure and more resilient computer systems depends on such diverse investigation into non-standard paradigms.

6 REFERENCES

1 CCIT, Data Communication Networks Open System Interconnection (OSI) System Description Techniques. Recommendations X.200-X.250 Red Book, Volume V111 Fascilcle V111.5 (1984).

2 CCITT, Data Communication Networks Services and Facilities, Terminal equipment and Interfaces. Recommendations X.1-X.29 yellow Book, Volume V111 Fascicle V111.2 (1980).

3 ISO, Information processing Systems - Open Systems Interconnection-Basic Reference Model, IS7498(1982).

4 P.Friedland and L. Kedes, Discovering the Secrets of DNA.ACM/IEEE (November 1985) 49-69.

5 G W Rowe and L E H Trainer, A Thermodynamic Theory of Codon bias in Viral Genes. J. Theor. Biol., 101 (1983) 171-203.
6 R. Williamson, Towards a Total Human Gene Map. Proc. Royal Institution 57 (1986) 45-55.
7 G. H. Fairtlough, Make Me a Molecule: The Technology of Recombinant DNA. Proc. Royal Institution 57 (1986) 57-70.
8 D Grafstein, Stereochemical origins of the Genetic code. J Theor. Biol., 105 (1983) 157-174.
9 R T Libby, Mistranslation in bacteriophage-infected anucleate minicells of E Coli: A test for error propogation.
Mech., Ageing Dev., 26(1984) 23-35.

TABLE 1 The seven functional groupings of the OSI reference model

OSI layer	OSI group	Function
7	APPLICATION	Selects services stimulated by the application process (eg. message authentication, data syntax)
6	PRESENTATION	Provides code and character set conversations (eg. data reformatting)
5	SESSION	Co-ordinates interactions between end application processes (eg. synchronization and orderly release of session dialogues)
4	TRANSPORT	Provides for end to end data integrity and quality of service (eg. error recovery procedures across the network)
3	NETWORK	Provides a transparent transfer of data through the network (eg. network addressing, switching and routing)
2	DATA LINK LAYER	Transfers and controls data between nodes in the network (eg. Synchronization and link control, and error recovery between nodes). Provides 'framing' of data
1	PHYSICAL	Transfers the bit stream through the network medium.

TABLE 2 The seven functional units of the genetic code model

Level	Unit	Function
7	GAMETE	Provides the unit of genetic information transfer; as stimulated by the organism's application process!
6	CHROMOSOME	Provide formatted grouping of genetic material (ie. genes)
5	GENE	Provide the 'unit of heridity'. Controls the interaction of, and the storage of the genetic code.
4	DNA	Provides for 'end to end' data integrity: Stores a whole sequence of genetic code material, specific to a given organisism.
3	RNA	Provides a 'transparent' transfer of a genetic code sequence. Also 'reformats' and translates the genetic code.
2	CODON	Provides the unit of the genetic code, and frames a triplet of data elements.
1	BASE	Provide the fundamental unit of genetic information transfer through the Physical Medium.

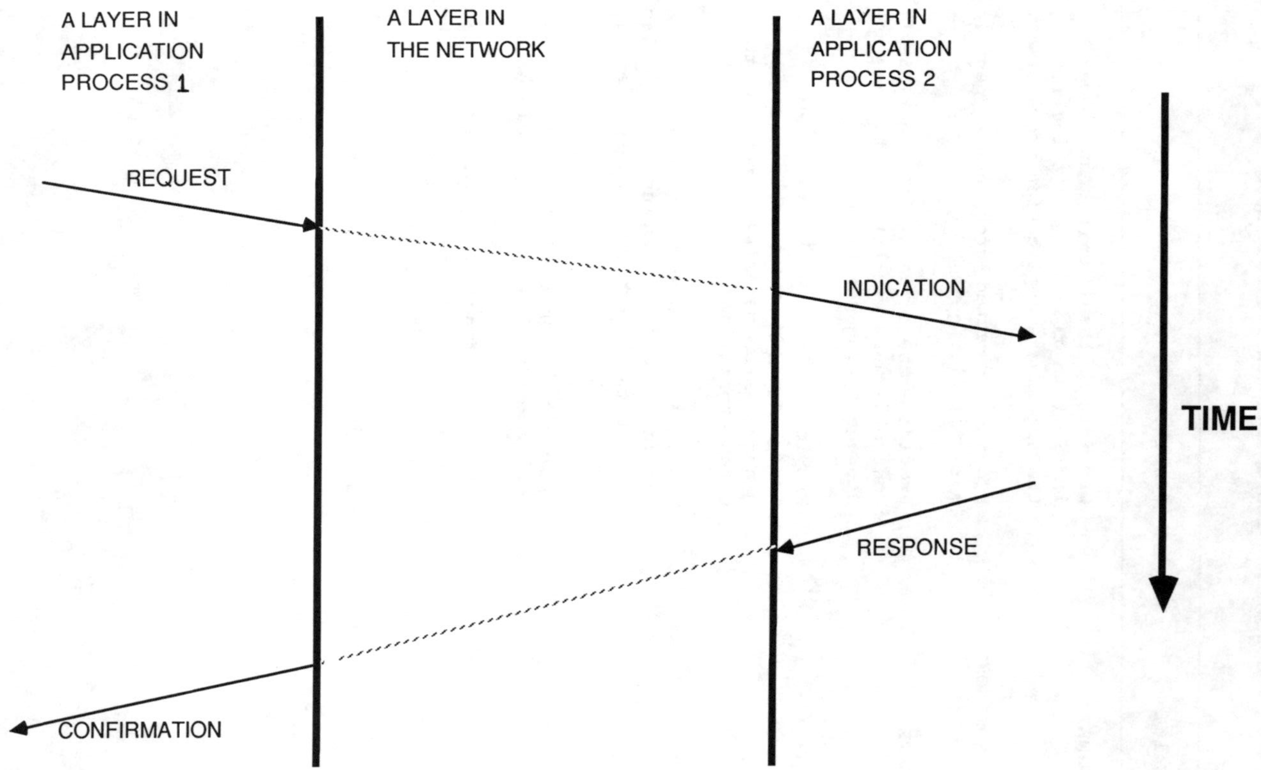

OSI SERVICE PRIMITIVES

Figure 1

Figure 2

DATA LINK LAYER FRAME STRUCTURE

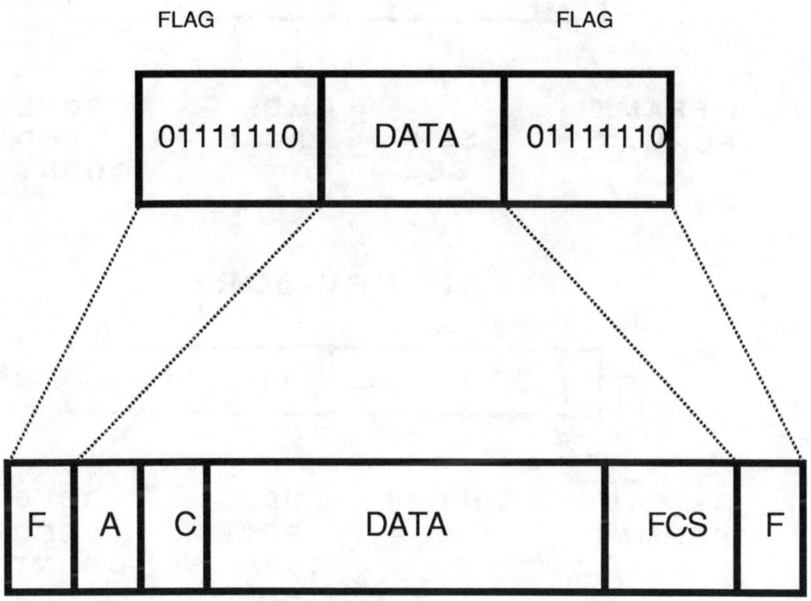

Figure 3

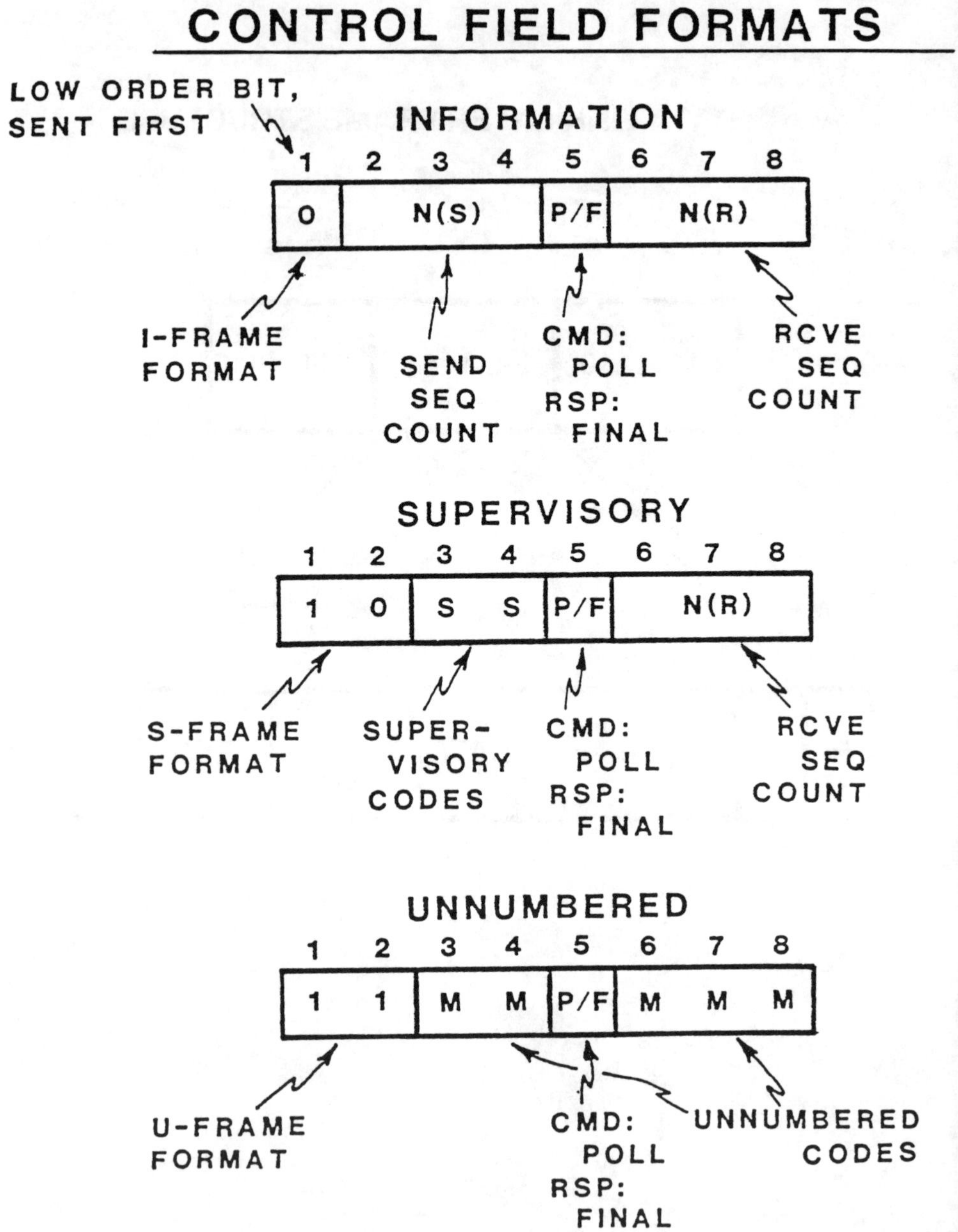
CONTROL FIELD FORMATS
LOW ORDER BIT, SENT FIRST
INFORMATION
1 2 3 4 5 6 7 8
0
N(S)
P/F
N(R)
I-FRAME FORMAT
SEND SEQ COUNT
CMD: POLL RSP: FINAL
RCVE SEQ COUNT
SUPERVISORY
1 2 3 4 5 6 7 8
1 0
S S
P/F
N(R)
S-FRAME FORMAT
SUPER-VISORY CODES
CMD: POLL RSP: FINAL
RCVE SEQ COUNT
UNNUMBERED
1 2 3 4 5 6 7 8
1 1
M M
P/F
M M M
U-FRAME FORMAT
CMD: POLL RSP: FINAL
UNNUMBERED CODES

Figure 4

BIT REPRESENTATION OF AN INFORMATION FRAME

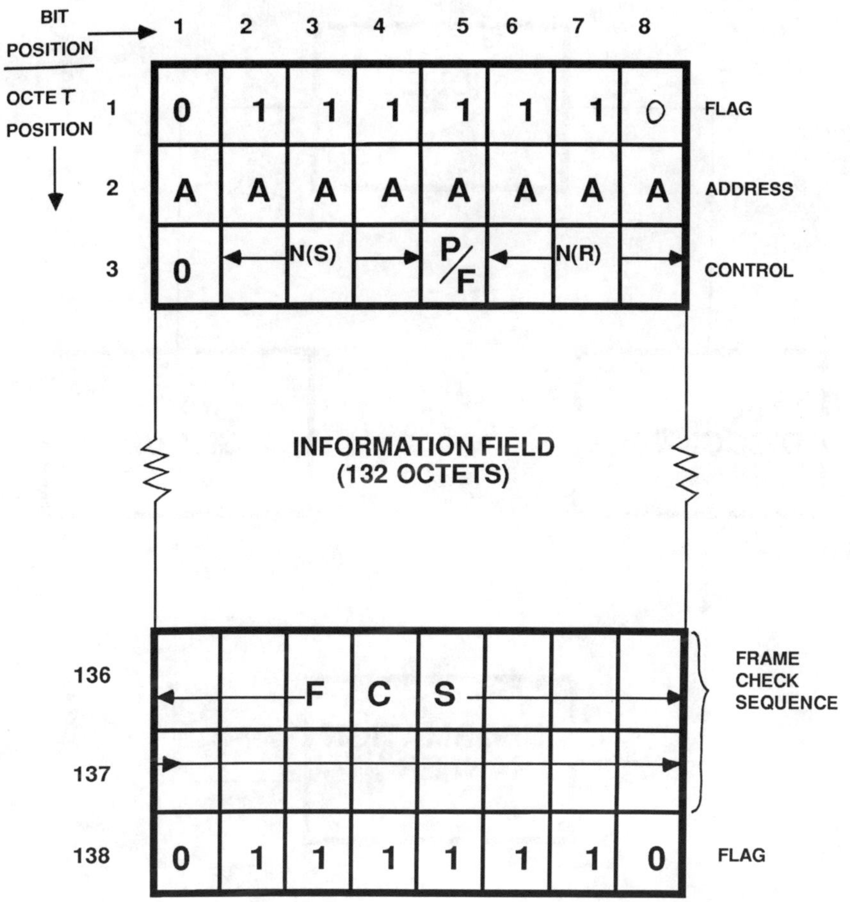

Figure 5

THE INFORMATION TRANSFER CYCLE FOR THE OSI MODEL

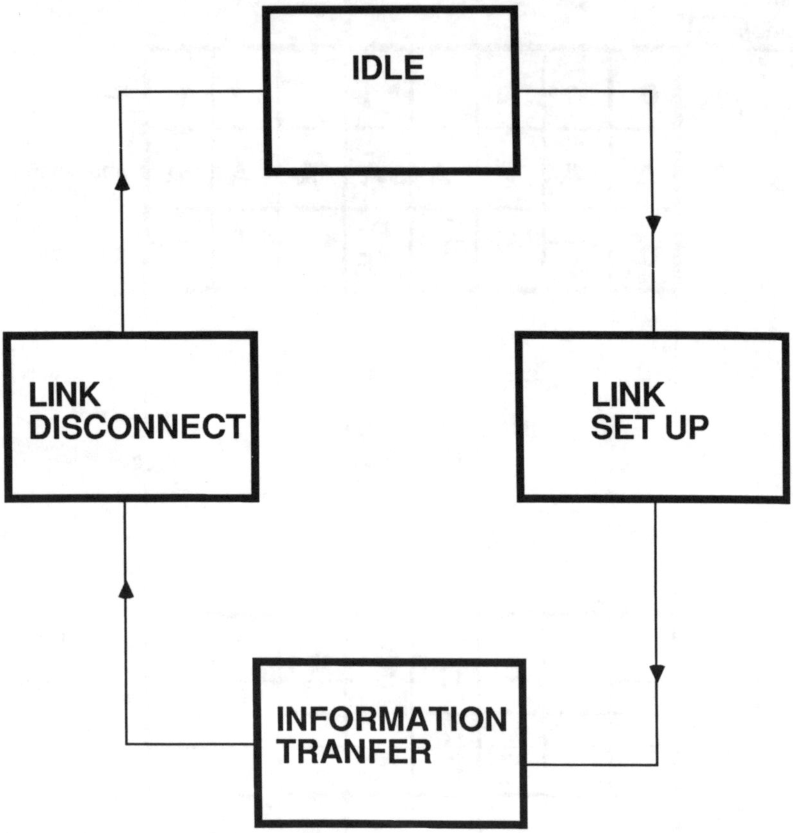

Figure 6

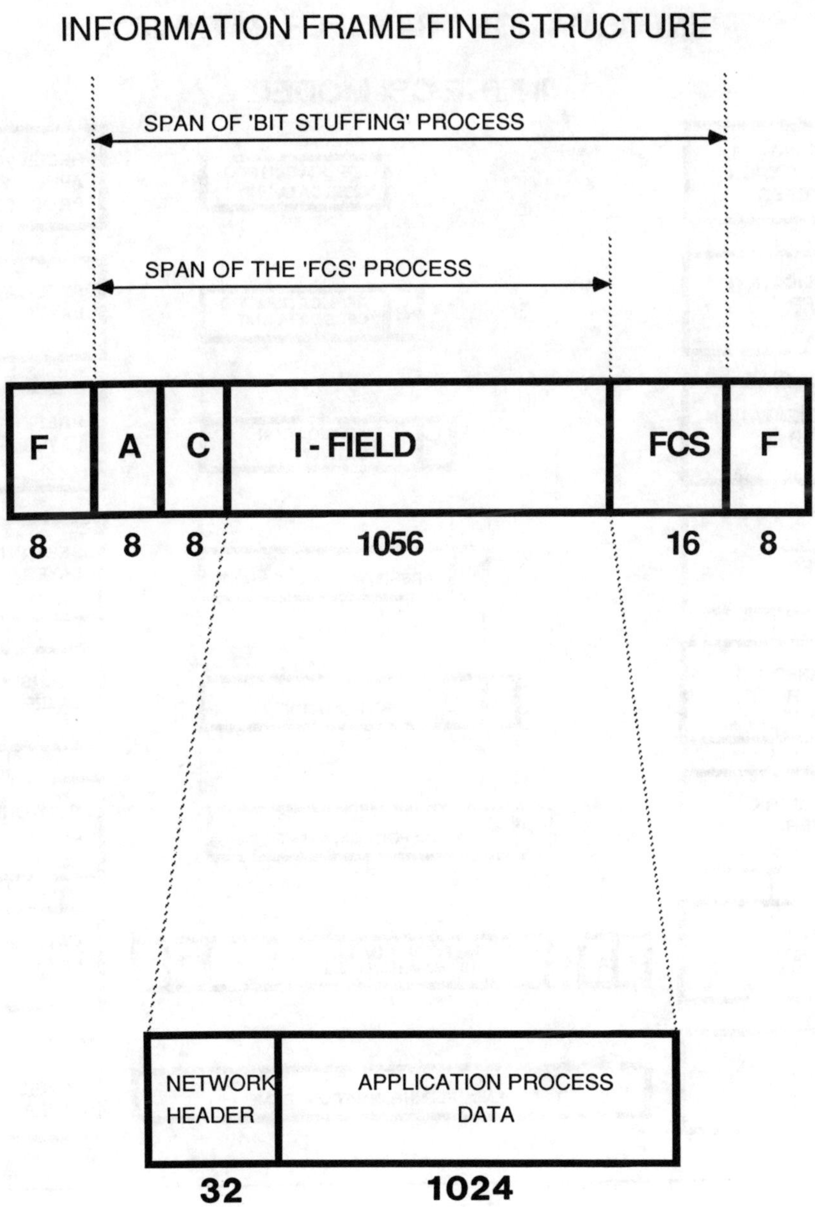

(Numbers refer to maximum field size in bits)

Figure 7

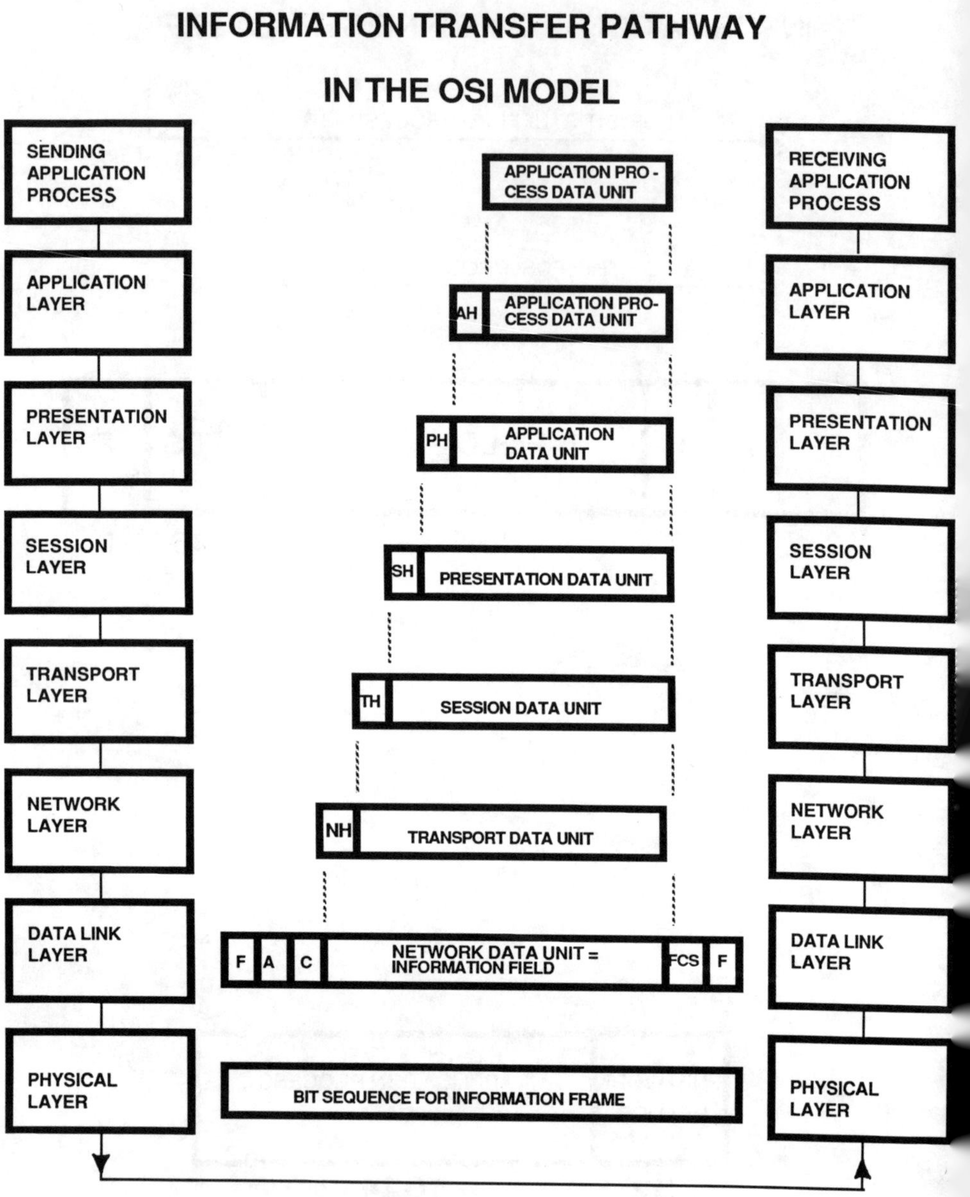
INFORMATION TRANSFER PATHWAY
IN THE OSI MODEL
SENDING APPLICATION PROCESS
APPLICATION LAYER
PRESENTATION LAYER
SESSION LAYER
TRANSPORT LAYER
NETWORK LAYER
DATA LINK LAYER
PHYSICAL LAYER
APPLICATION PRO - CESS DATA UNIT
AH
APPLICATION PRO- CESS DATA UNIT
PH
APPLICATION DATA UNIT
SH
PRESENTATION DATA UNIT
TH
SESSION DATA UNIT
NH
TRANSPORT DATA UNIT
F
A
C
NETWORK DATA UNIT = INFORMATION FIELD
FCS
F
BIT SEQUENCE FOR INFORMATION FRAME
RECEIVING APPLICATION PROCESS
APPLICATION LAYER
PRESENTATION LAYER
SESSION LAYER
TRANSPORT LAYER
NETWORK LAYER
DATA LINK LAYER
PHYSICAL LAYER

Figure 8

THE INFORMATION TRANSFER CYCLE FOR THE GENETIC CODE MODEL

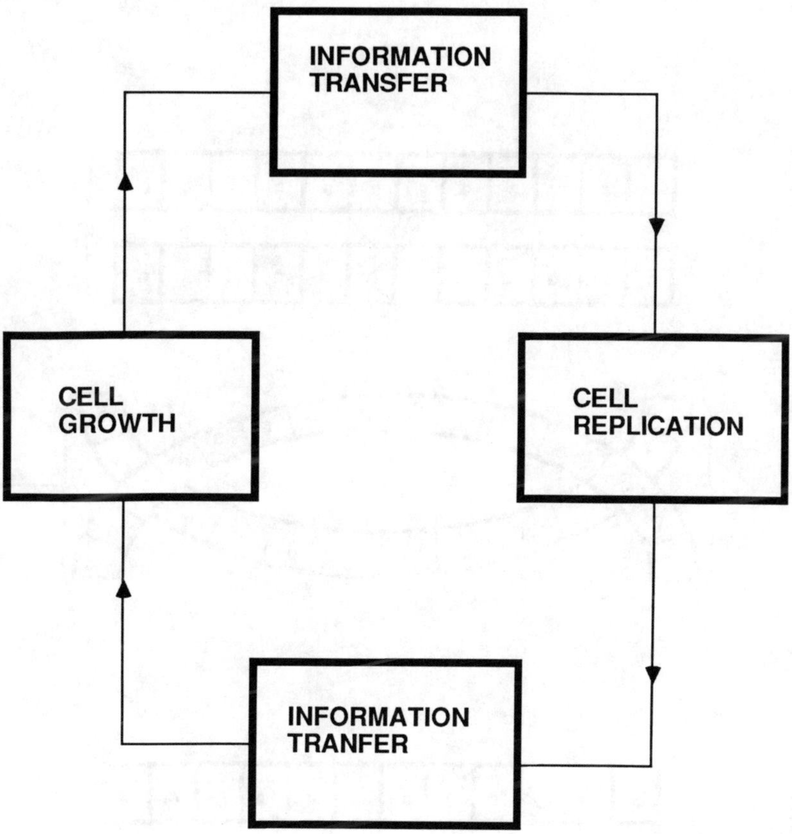

Figure 9

SCHEMATIC OF CHROMOSOME RECOMBINATION

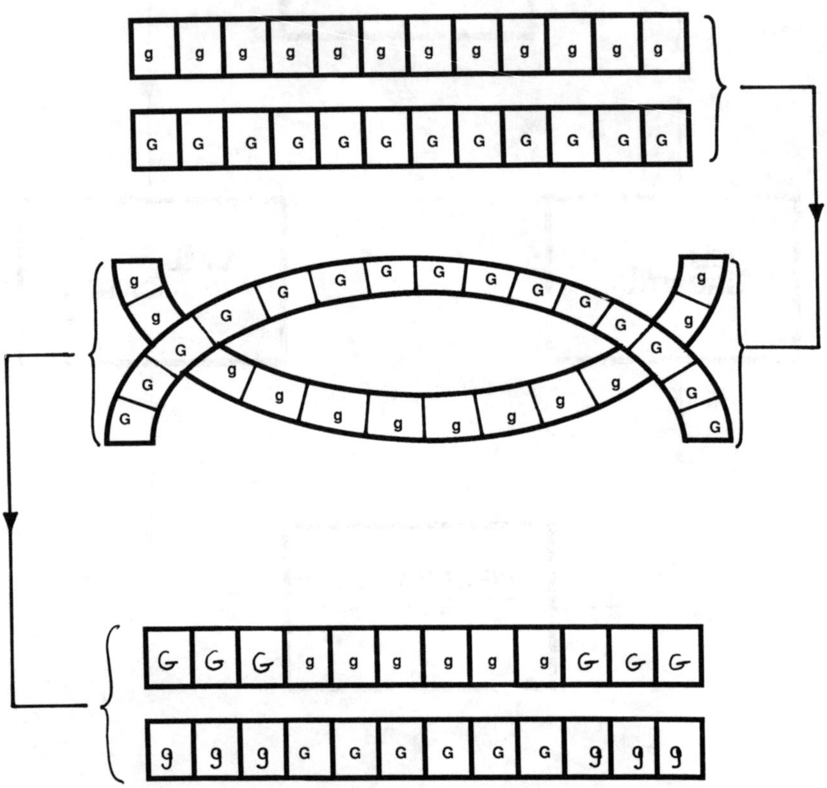

Figure 10

SCHEMATIC OF GENE STRUCTURE

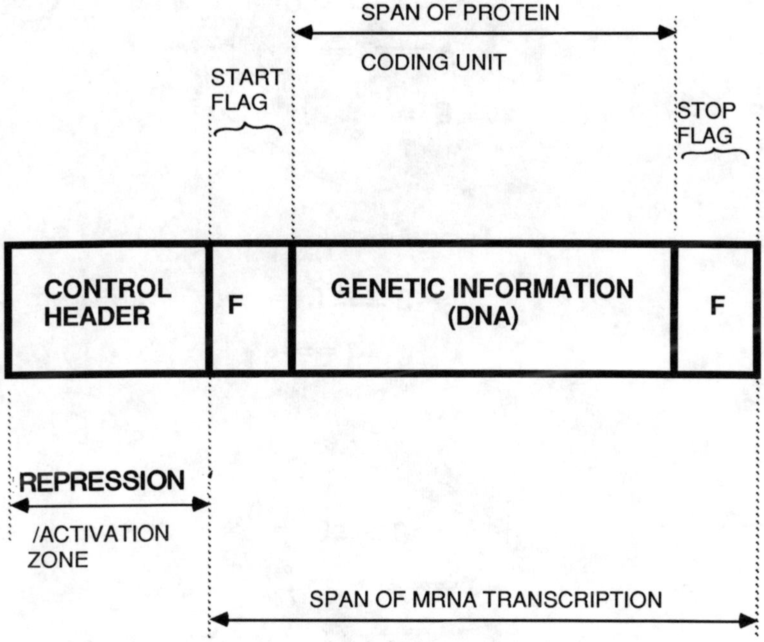

Figure 11

THE DNA DOUBLE HELIX

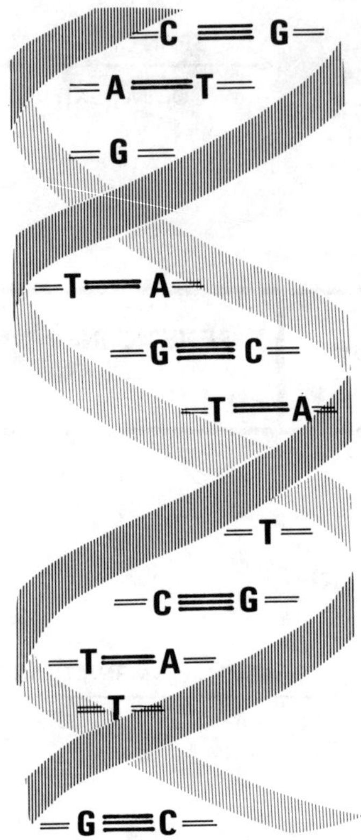

Figure 12

CODON TRIPLETS

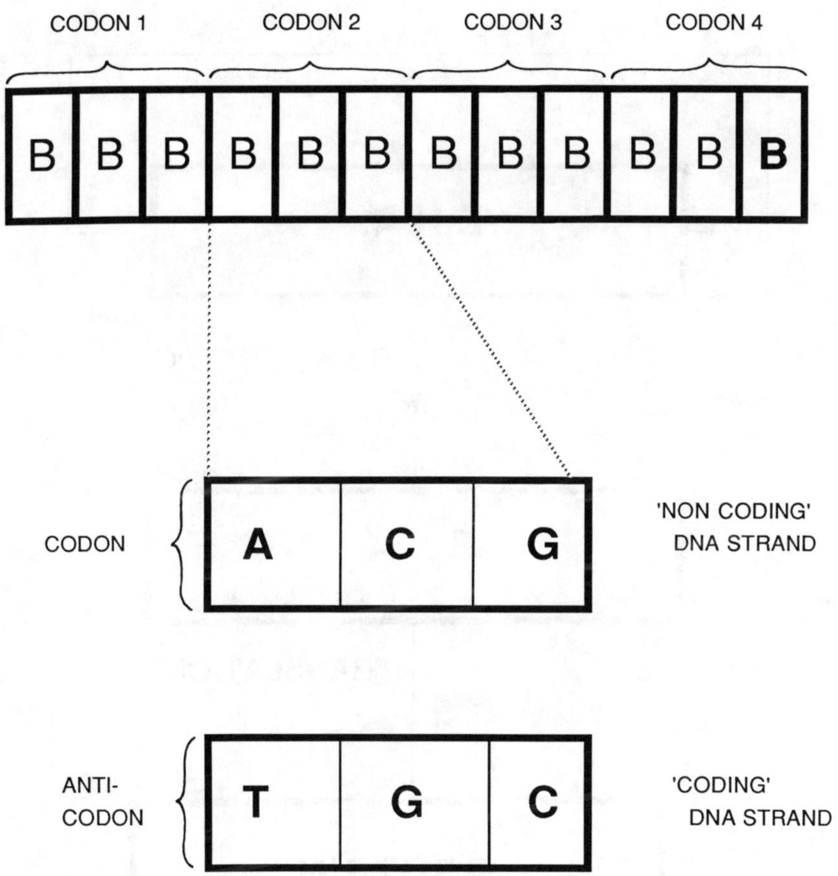

DNA CODON AND ANTICODON BASE PAIRING

Figure 13

GENETIC INFORMATION TRANSFER PROCESSES

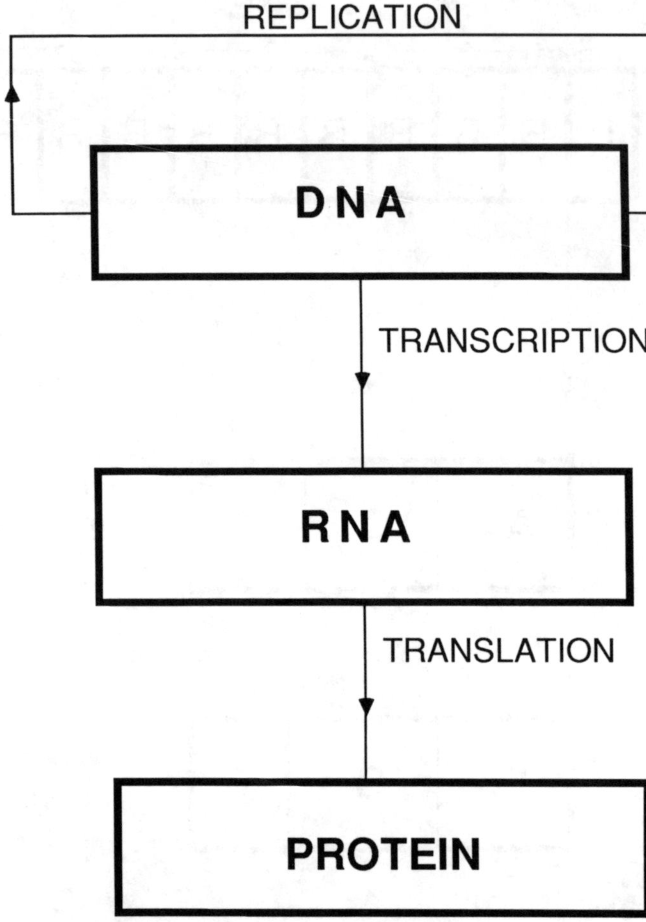

Figure 14

THE TWO TYPES OF RNA

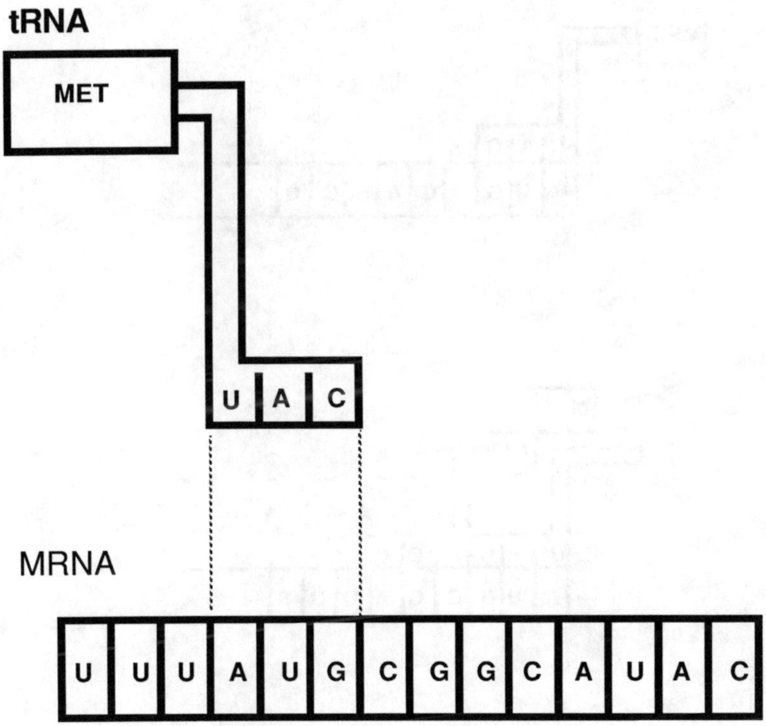

Figure 15

THE GENETIC CODE TRANSLATION PROCESS

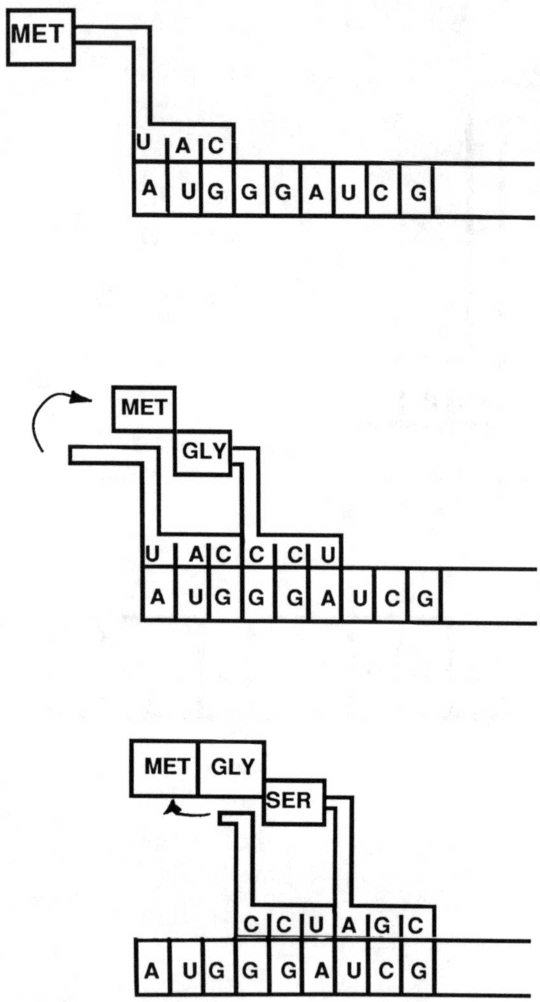

THE GENETIC CODE

Figure 16

STOP	UGA	UAG	UAA			
VAL	GUU	GUG	GUC	GUA		
TYR	UAU	UAC				
TRP	UGG					
THR	ACU	ACG	ACC			
SER	UCU	UCG	UCC	UCA	AGU	AGC
PRO	CCU	CCG	CCC	CCA		
PHE	UUU	UUC				
MET / START	AUG					
LYS	AAG	AAA				
LEU	CUU	CUG	CUC	CUA	UUG	UUA
ILE	AUU	AUC				
HIS	CAU	CAC				
GLY	GGU	CGG				
GLN	CAG	CAA				
GLU	GAG	GAA				
CYS	UGU	UGC				
ASP	GAU	GAC				
ASN	AAU	AAC				
ARG	CGU	CGG	CGC	CGA	AGG	AGA
ALA	GCU	GCG	GCC	GCA		

THE ULTIMATE LIMITS OF INFORMATION DENSITY

KHALED ABDEL-GHAFFAR AND ROBERT J. McELIECE

Department of Electrical Engineering
California Institute of Technology, Pasadena, CA 91125

1. INTRODUCTION

In a conference devoted to studying the ultimate limits of communication systems, we wish to make an information-theoretic contribution. It is surely appropriate to do this, since Shannon's theorem tells us exactly what the ultimate communication limit of a noisy channel is. Nevertheless, it has seemed to us for some time that the usual models of information theory are inadequate for a study of the ultimate limits of many practical communication and information storage systems, because of a key missing parameter. This missing parameter we call the *scaling* parameter. In this paper we hope to remedy this situation a bit by introducing a class of models for *channels with noise scaling*. Rather than give a formal definition immediately, we begin with a "thought experiment" to illustrate what we mean.

Imagine a situation in which it is desired to store information on a long narrow paper tape. The information is binary, i.e., a random string of bits (0's and 1's). We will not specify the read/write process, except to say that once a bit has been written on the tape, when it is read, it may be read in error; a stored 0 might be read as a 1, and vice versa. We assume, in fact, that the write/read process can be modelled as a binary symmetric channel with crossover probability p. If p is too large, coding will be necessary to ensure that the information is stored reliably. This much of the problem is well within the scope of traditional information theory.

However, besides insisting that the information be stored *reliably*, we want it to be stored *compactly*. This means we want to store as many bits per inch (bpi) as possible. For example, suppose we find that we can store 100 bpi reliably without coding, but that when we try to store 200 bpi the (uncoded) error probability is intolerable. If we used a code of rate 3/4, say, then the resulting information density would be 150 bpi. But would this be a *reliable* 150 bpi? That of course depends on whether the capacity of the "200 bpi" channel is greater or less that 3/4. And there is no way to say whether this is the case, unless the model says what the capacity is as a function of the storage density. Thus if x is a scale parameter that measures the physical size (in inches) of each stored bit, we need to know $C(x)$, the capacity of the storage channel at that feature size. Then the maximum number of information bits per inch at "feature size" x will be, by Shannon's theorem, $C(x)/x$, and the "ultimate limit" of storage density will be given by

J. K. Skwirzynski (ed.), Performance Limits in Communication Theory and Practice, 267–279.

$$\text{Ultimate Limit} = \sup_{x>0} \frac{C(x)}{x}.$$

Of course we cannot say more until the physicists determine the function $C(x)$. Or can we? In the next section we will introduce our formal model for a binary symmetric channel with noise scaling, and give several plausible scaling rules for which we can draw some strong conclusions about ultimate limits for storage densities. In Section 3 we will see that orthogonal codes will achieve the ultimate storage limits for a broad class of BSC's with noise scaling.

2. BINARY SYMMETRIC CHANNELS WITH NOISE SCALING

In Figure 1 we show a binary symmetric channel (BSC) whose crossover probability p is a function of a parameter x, which we think of as the amount of an abstract resource available per transmitted bit. In the "paper tape" example of Section 1, the resource would be the length of tape available per bit. More generally, the resource could be energy, time, area, etc. It turns out to be convenient to work not with $p(x)$ directly, but rather with the associated parameter $\delta(x)$, which we define as

$$\delta(x) = 1 - 2p(x). \tag{2.1}$$

We assume that the more resource available per transmitted bit, the better the channel is, and that if no resource is available, the channel is useless. This means that $\delta(0) = 0$, and that $\delta(x)$ is a continuous increasing function of x.

The capacity of a BSC is well-known to be [4]

$$\text{Capacity} = 1 - (p \log_2 \frac{1}{p} + (1-p) \log_2 \frac{1}{1-p})\text{bits}.$$

In terms of the parameter δ introduced in (2.1) this becomes

$$\begin{aligned} C &= \frac{1}{2}((1-\delta)\log_2(1-\delta) + (1+\delta)\log_2(1+\delta)) \\ &= \frac{1}{\ln 2}\left(\frac{\delta^2}{1\cdot 2} + \frac{\delta^4}{3\cdot 4} + \frac{\delta^6}{5\cdot 6} + \cdots\right) \qquad (2.2) \\ &\equiv \frac{1}{2\ln 2}\delta^2 \pmod{\delta^4}. \qquad (2.3) \end{aligned}$$

The infinite series in (2.2) converges uniformly for all $0 \leq \delta \leq 1$.

If we use coding on this BSC to improve its performance, there will be a difference between the resource available per *transmitted* bit and the resource available per *information* bit. Indeed, if we denote the code rate by R, and the resource per information bit by λ , we have

$$\lambda = \frac{x}{R}. \tag{2.4}$$

We will be interested in finding the minimum possible value of λ, i.e., the minimum possible resource needed per information bit, when information must be stored (transmitted) reliably.

According to Shannon's Theorem, reliable storage is possible if and only if

$$R < C(x), \tag{2.5}$$

where $C(x)$ denotes the capacity as a function of the parameter x. Since by (2.4) we have $x = \lambda R$, this means that $R < C(\lambda R)$. Since $C(x)$ is assumed to be an increasing function of x, we can invert the relationship in (2.5) to obtain

$$\lambda > \frac{C^{-1}(R)}{R} \qquad \text{for all } 0 < R \leq 1. \tag{2.6a}$$

Alternatively, since again by (2.4) $R = x/\lambda$, then (2.5) becomes $x/\lambda < C(x)$ and so

$$\lambda > \frac{x}{C(x)} \qquad \text{for all } x > 0. \tag{2.6b}$$

Therefore the minimum needed resource per information bit is given by the following expressions.

$$\lambda_{\min} = \inf_{0<R\leq 1} \frac{C^{-1}(R)}{R} \tag{2.7}$$

$$= \inf_{x>0} \frac{x}{C(x)}. \tag{2.8}$$

In many cases, as we shall see, the "inf" in (2.8) occurs as $x \to 0$, when $\delta(x) \to 0$ too. According to (2.3), in the vicinity of $x = 0$, we have $C(x) \sim \delta^2/2\ln 2$, and so

$$\frac{x}{C(x)} \sim 2\ln 2 \cdot \frac{x}{\delta^2(x)}.$$

This leads us to characterize the noise scaling by defining the parameter k by

$$\lim_{x\to 0} \frac{\delta^2(x)}{x} = k. \tag{2.9}$$

Then combining (2.9) with (2.6) we find that

$$\lambda_{\min} \leq \frac{2\ln 2}{k}. \tag{2.10}$$

In Section 3 we will show that orthogonal codes can always be used to achieve the upper limit on λ given in (2.10).

We conclude this section with six examples of noise scaling. The first three are "toy" examples which illustrate some of the mathematical possiblities. The last three are more practical.

Example 1. $\delta^2(x) = x^{2/3}$. Here we find (Figure 2) that $\lambda_{\min} = 0$ which is achieved as $R \to 0$. Therefore in this case a vanishingly small amount of the resource is needed per information bit for reliable storage. Such a scaling law is plainly not likely to occur in practice.

Example 2. $\delta^2(x) = x$. Here we find (Figure 3) that the limit of λ as $R \to 0$ is $2\ln 2$. However, $\lambda_{\min} = 1$ which is achieved by having $R = 1$, i.e., with no coding. This is an example in which coding requires more resources for reliable storage than what is required without coding.

Example 3. $\delta^2(x) = x^{4/3}$. Here we find (Figure 4) that $\lambda \to \infty$ as $R \to 0$. In this example, $\lambda_{\min} = 1$, which is again achieved by having $R = 1$, i.e., with no coding.

Example 4. Binary FSK: In this example, $\delta(x) = 1 - e^{-x/2}$, where x denotes the symbol signal-to-noise ratio [5, Eq. (7.68)]. Here we find (Figure 5) that $\lambda_{\min} \approx 6$, which is achieved by a code of rate $R \approx 0.5$.

Example 5. Binary PSK with binary output quantization: In this example, $\delta(x) = 1 - 2Q(\sqrt{2x})$, where $Q(x) = \frac{1}{\sqrt{2\pi}} \int_x^\infty e^{-t^2/2} dt$, and again x denotes the symbol signal-to-noise ratio [5, Eq. (4.78)]. Here we find (Figure 6) that $\lambda_{\min} = \frac{\pi}{2} \ln 2$ which is achieved as $R \to 0$. This result should be compared to the better-known result (see [4, Prob. 5.3]) that with no output quantization, the minimum achievable bit signal-to-noise ratio is $\ln 2 = -1.59$ dB.

Example 6. Thermal noise in VLSI memory chips: Each cell in a memory chip contains either "0" or "1." The energy required to alter the contents of a cell is called the switching energy. This switching energy, E, is a function of x, the area of a single memory cell, which represents the resource per stored bit. A widely accepted rule in designing VLSI memory chips is to scale the switching energy such that the ratio $E(x)/x^{3/2}$ is kept constant [1,3].

Due to the random motion of electrons induced by thermal noise, a memory cell may change its content, causing an error. The probability of error due to thermal noise is given by $p(x) = Q\left(\sqrt{\frac{E(x)}{kT}}\right)$, and hence $\delta(x) = 1 - 2Q\left(\sqrt{\frac{E(x)}{kT}}\right)$, where k is the Boltzmann's constant and T is the temperature. Since $E(x)/x^{3/2}$ is constant, then $\delta(x) = 1 - 2Q\left(cx^{3/2}\right)$ for some constant c. Using typical values for the switching energy for the 64K RAMs, we find that $c = 10^4$, where $x = 1$ is the area of a single cell in a 64K RAM chip. Figure 7 gives $C^{-1}(R)/R$ as a function of R. In this case, we find that $\lambda_{\min} \approx 1.2 \times 10^{-5}$, which is achieved by a code of rate $R \approx 0.5$. From these numerical values, we conclude that it is impossible to produce a reliable chip with the same area as that of the 64K RAM chip, and whose information content exceeds 5.4 Gigabits! Of course, this bound is very optimistic as we have ignored many other sources of errors as alpha particles, cosmic rays, quantum effects, etc.

3. THE UNIVERSALITY OF ORTHOGONAL CODES

In this section we will show that for all BSC's with noise scaling, the family of orthogonal codes can be used to achieve the value for resource per stored information bit given by (2.10). To do this, we will begin with a general discussion of decoder error probability on a BSC, and then specialize to orthogonal codes.

Thus consider a BSC with crossover probability p and δ defined as in (2.1) by $\delta = 1 - 2p$. Let $\mathbf{C}$ be a code of length n with M codewords, viz.

$$\mathbf{C} = \{0, \mathbf{x}_1, \mathbf{x}_2, \ldots, \mathbf{x}_{M-1}\}.$$

We assume that $\mathbf{0}$, the all 0's codeword, is stored, but that what is retrieved is $\mathbf{z} = (z_1, z_2, \ldots, z_n)$, where $\mathbf{z}$ is a noise vector whose components are independent, identically distributed random variables with common distribution

$$\Pr(z = 0) = 1 - p$$
$$\Pr(z = 1) = p.$$

For future reference we note that the mean and variance of the components of the noise vector are given by

$$E(z_i) = p = \frac{1-\delta}{2} \tag{3.1}$$

$$\operatorname{Var}(z_i) = p(1-p) = \frac{1-\delta^2}{4}. \tag{3.2}$$

A maximum-likelihood decoder works as follows. Given the retrieved vector $\mathbf{z}$, for each codeword $\mathbf{x}_i$, it calculates the Hamming weight of the vector $\mathbf{z} \oplus \mathbf{x}_i$, where "$\oplus$" denotes mod 2 addition. It then selects that codeword for which this Hamming weight $|\mathbf{z} \oplus \mathbf{x}_i|$ is smallest. Therefore if $\mathbf{0}$ is the codeword that was actually stored, the decoder will make a correct decision if $|\mathbf{z}|$ is less than $|\mathbf{z} \oplus \mathbf{x}_i|$, for all $i = 1, 2, \ldots, M-1$. If P_c denotes the probability of correct decoding, then we have

$$P_c = \Pr\{|\mathbf{z}| < |\mathbf{z} \oplus \mathbf{x}_i| \ , \ i = 1, 2, \ldots, M-1\}. \tag{3.3}$$

(In fact, P_c is slightly larger than this, since in case of a tie for the smallest Hamming weight there is a chance that the decoder will still make the correct decision. If necessary, then, one can think of (3.3) as describing the performance of a decoder which is slightly worse than optimal.) The following lemma will allow us to put (3) into a more convenient form.

Lemma 1: If $\mathbf{x}$ and $\mathbf{z}$ are two binary (0's and 1's) vectors of length n, then $|\mathbf{z}| < |\mathbf{z} \oplus \mathbf{x}|$ if and only if $\langle \mathbf{z}, \mathbf{x} \rangle < \frac{1}{2}|\mathbf{x}|$. (Here $\langle \mathbf{z}, \mathbf{x} \rangle$ denotes the real inner product $\sum z_i x_i$ of the vectors $\mathbf{z}$ and $\mathbf{x}$.)

Proof: We illustrate the proof with the specific vector $\mathbf{x} = (101011)$ of length 6. In this case

$$|\mathbf{z}| = z_1 + z_2 + z_3 + z_4 + z_5 + z_6$$
$$|\mathbf{z} \oplus \mathbf{x}| = \bar{z}_1 + z_2 + \bar{z}_3 + z_4 + \bar{z}_5 + \bar{z}_6,$$

where $\bar{z}_i$ denotes the complement of z_i, i.e., $1 - z_i$. Then $|\mathbf{z}| < |\mathbf{z} \oplus \mathbf{x}|$ if and only if

$$\begin{aligned} z_1 + z_3 + z_5 + z_6 &< \bar{z}_1 + \bar{z}_3 + \bar{z}_5 + \bar{z}_6, \\ z_1 + z_3 + z_5 + z_6 &< (1 - z_1) + (1 - z_3) + (1 - z_5) + (1 - z_6), \\ 2(z_1 + z_3 + z_5 + z_6) &< 4 \\ z_1 + z_3 + z_5 + z_6 &< \frac{1}{2} \cdot 4. \end{aligned}$$

But $z_1 + z_3 + z_5 + z_6 = \langle \mathbf{z}, \mathbf{x} \rangle$, and $|\mathbf{x}| = 4$. This proves the lemma in one special case. The general proof follows exactly similar lines. ∎

It follows from (3.3) and Lemma 1 that if we define the random variables S_i by

$$S_i = \langle \mathbf{z}, \mathbf{x}_i \rangle \quad , \quad i = 1, 2, \ldots, M - 1 \tag{3.4}$$

and the integers w_i by

$$w_i = |\mathbf{x}_i| \quad , \quad i = 1, 2, \ldots, M - 1, \tag{3.5}$$

that

$$P_c = \Pr\{S_i < \frac{1}{2} w_i \quad , \quad i = 1, 2, \ldots, M - 1\}. \tag{3.6}$$

We next consider the first- and second-order statistics of the random variables S_i. It follows immediately from the definition (3.4), together with (3.1) and (3.2), that

$$E(S_i) = w_i \cdot \frac{1 - \delta}{2} \tag{3.7}$$

and

$$\mathrm{Var}(S_i) = w_i \cdot \frac{1 - \delta^2}{4}. \tag{3.8}$$

To compute the covariance $\mathrm{Cov}(S_i, S_j)$ for $i \neq j$, we use the following lemma.

Lemma 2: $E(S_i S_j) = (w_i w_j - w_{ij}) \left(\frac{1-\delta}{2}\right)^2 + w_{ij} \left(\frac{1-\delta}{2}\right)$, where $w_{ij} = \langle \mathbf{x}_i, \mathbf{x}_j \rangle$.

Proof: Let $S_i = \sum_{\alpha \in I} z_\alpha$, $S_j = \sum_{\beta \in J} z_\beta$. Then

$$S_i S_j = \sum_{\alpha \in I} \sum_{\beta \in J} z_\alpha z_\beta$$

$$E(S_iS_j) = \sum_{\alpha\in I}\sum_{\beta\in J} E(z_\alpha z_\beta). \tag{3.9}$$

But since $z_1, z_2, \ldots, z_n$ are i.i.d. (0, 1) random variables satisfying (3.1), we have

$$E(z_\alpha z_\beta) = \begin{cases} \left(\frac{1-\delta}{2}\right)^2 & \text{if } \alpha \neq \beta \\ \frac{1-\delta}{2} & \text{if } \alpha = \beta. \end{cases}$$

But of the $|I \times J| = w_i w_j$ terms in the sum (3.9), exactly $|I \cap J| = w_{ij}$ of them have $\alpha = \beta$. This completes the proof of the lemma. ∎

It follows from Lemma 2 and (3.7) that

$$\begin{aligned} \mathrm{Cov}(S_i, S_j) &= E(S_iS_j) - E(S_i)E(S_j) \\ &= E(S_iS_j) - w_i w_j \left(\frac{1-\delta}{2}\right)^2 \\ &= w_{ij} \cdot \frac{1-\delta^2}{4}. \end{aligned} \tag{3.10}$$

Therefore if we define a new set of random variables $\mathbf{T} = (T_1, T_2, \ldots, T_{M-1})$ by

$$T_i = S_i - w_i \left(\frac{1-\delta}{2}\right), \tag{3.11}$$

then the T_i all have mean zero, and the covariance matrix $\mathrm{Var}(\mathbf{T})$ is given by

$$\mathrm{Var}(\mathbf{T})_{ij} = \frac{1-\delta^2}{4} \cdot w_{ij}, \tag{3.12}$$

and furthermore, from (3.6)

$$P_c = \Pr\{T_i < \frac{\delta}{2} w_i \ , \ i = 1, 2, \ldots, M-1\}. \tag{3.13}$$

Equation (3.13) is as far as we can take a general analysis. We now consider the special case of *orthogonal codes*, which are codes for which n is a multiple of four, $M = n$, and

$$w_i = M/2, \tag{3.14}$$

$$w_{ij} = \begin{cases} M/4 & \text{if } i \neq j \\ M/2 & \text{if } i = j. \end{cases} \tag{3.15}$$

The codewords in such an orthogonal code can be taken as the rows of an $M \times M$ Hadamard matrix, in which the $+1$'s have been changed to 0's and the -1's have been changed to 1's. (Figure 8 illustrates the case $M = 8$.)

If we specialize the covariance matrix (3.12) to the class of orthogonal codes, we find (using (3.15)) that (illustrated for $M = 4$)

$$\mathrm{Var}(\mathbf{T}) = \frac{1-\delta^2}{16}\cdot M \cdot \begin{bmatrix} 2 & 1 & 1 \\ 1 & 2 & 1 \\ 1 & 1 & 2 \end{bmatrix}. \tag{3.16}$$

If M is large, the central limit theorem guarantees that the T_i's will be approximately normal. The next lemma will allow us to produce a tractable model for the T_i's.

Lemma 3: If $X_0, X_1, \ldots, X_{M-1}$ are M i.i.d. normal random variables, each with mean 0 and variance 1, then if we define

$$Y_i = X_0 + X_i \quad , \quad i = 1, 2, \ldots, M-1,$$

the Y_i's are mean zero normal random variables with covariance matrix (illustrated for $M = 4$)

$$\mathrm{Var}(\mathbf{Y}) = \begin{bmatrix} 2 & 1 & 1 \\ 1 & 2 & 1 \\ 1 & 1 & 2 \end{bmatrix}.$$

Proof: A routine verification which we omit. ∎

It follows from (3.16) and Lemma 3 that we can take as a model for the random variables $T_1, T_2, \ldots, T_{M-1}$, the normal random variables $\tilde{T}_i$ defined by

$$\tilde{T}_i = \sqrt{\frac{1-\delta^2}{16}M}\ (X_0 + X_i), \tag{3.17}$$

where $X_0, X_1, \ldots, X_{M-1}$ are i.i.d. normal random variables with mean zero and variance 1. Thus from (3.13) the probability of correct decoding is approximated by

$$\begin{aligned} P_c &\cong \Pr\left\{X_0 + X_i < \frac{\delta}{2}\cdot\frac{M}{2} \Big/ \sqrt{\frac{1-\delta^2}{16}M} \ , \ i = 1, 2, \ldots, M-1\right\} \\ &= \Pr\left\{X_0 + X_i < \frac{\delta\sqrt{M}}{\sqrt{1-\delta^2}} \ , \ i = 1, 2, \ldots, M-1\right\} \\ &= \Pr\left\{X_i < \frac{\delta\sqrt{M}}{\sqrt{1-\delta^2}} - X_0 \ , \ i = 1, 2, \ldots, M-1\right\}. \end{aligned} \tag{3.18}$$

Therefore if we define

$$Z(x) = \frac{1}{\sqrt{2\pi}}e^{-x^2/2}, \tag{3.19}$$

and

$$P(x) = \int_{-\infty}^{x} Z(t)dt, \tag{3.20}$$

we can write (3.18) as

$$P_c \cong \int_{-\infty}^{\infty} P\left(\frac{\delta\sqrt{M}}{\sqrt{1-\delta^2}} - x\right)^{M-1} Z(x)dx. \tag{3.21}$$

Equation (3.21) gives us an explicit expression (modulo the central limit theorem) for the probability of correct decoding of an orthogonal code of length M on a binary symmetric channel with crossover probability $p = (1-\delta)/2$, assuming the all-zero codeword is sent. However, using the same techniques, and taking into consideration the structure of orthogonal codes, it can be shown that (3.21) represents the probability of correct decoding no matter which codeword is sent. We now wish to investigate the limit, as $M \to \infty$, of the expression (3.21), it being understood that δ is a decreasing function of M. The goal is to discover necessary and sufficient conditions for $P_c \to 1$ in terms of the relationship between M and δ.

Since $P(x)$ is an increasing function of x and approaches 1 as $x \to \infty$, plainly a necessary condition for $P_c \to 1$ is that

$$\lim_{M\to\infty} \delta\sqrt{M} = \infty. \tag{3.22}$$

If (3.22) holds, then for each finite x,

$$\lim_{M\to\infty} P\left(\frac{\delta\sqrt{M}}{\sqrt{1-\delta^2}} - x\right)^{M-1} = \lim_{M\to\infty} P\left(\delta\sqrt{M}\right)^{M}. \tag{3.23}$$

It follows then from (3.23) and (3.21) that

$$P_c = \lim_{M\to\infty} P\left(\delta\sqrt{M}\right)^{M}. \tag{3.24}$$

It is known [2, Eq. (26.2.12)] that $P(x) \sim 1 - Z(x)/x$ as $x \to \infty$, so that

$$P\left(\delta\sqrt{M}\right)^{M} \sim \left(1 - \frac{Z(\delta\sqrt{M})}{\delta\sqrt{M}}\right)^{M}. \tag{3.25}$$

Combining (3.23) and (3.25), we find that

$$\begin{aligned} \ln P_c &\sim M \ln\left(1 - \frac{Z(\delta\sqrt{M})}{\delta\sqrt{M}}\right) \sim -\frac{\sqrt{M}}{\delta} Z(\delta\sqrt{M}) \\ &\sim -\frac{1}{\sqrt{2\pi}} \cdot \frac{\sqrt{M}}{\delta} \cdot e^{-\delta^2 M/2}. \end{aligned} \tag{3.26}$$

Thus $P_c \to 1$ if and only if

$$\lim_{M\to\infty} \frac{\sqrt{M}}{\delta} \cdot e^{-\delta^2 M/2} = 0. \tag{3.27}$$

In Section 2 we assumed that $\delta^2 \sim kx$, where $x = \lambda R$, R being the code rate. For an orthogonal code, we have $R = \dfrac{\log_2 M}{M}$, and so as $M \to \infty$

$$\delta^2 \sim \lambda k \cdot \frac{\log_2 M}{M}, \tag{3.28}$$

and so

$$\delta^2 M \sim \lambda k \cdot \log_2 M. \tag{3.29}$$

Thus when we evaluate the limit in (3.27), we find that

$$\begin{aligned}\frac{\sqrt{M}}{\delta} e^{-\delta^2 M/2} &\sim \frac{\sqrt{M}}{\sqrt{k\lambda \cdot \frac{\log_2 M}{M}}} \quad e^{-\frac{k\lambda}{2}\cdot \log_2 M} \\ &= \frac{M}{\sqrt{k\lambda \log_2 M}} \quad M^{-\frac{1}{2}\frac{k\lambda}{\ln 2}} \\ &= \frac{M^{1-\frac{1}{2}\frac{k\lambda}{\ln 2}}}{\sqrt{k\lambda \log_2 M}}\end{aligned}$$

For fixed k, λ, this expression will approach 0 as $M \to \infty$ if and only if $k\lambda \geq 2\ln 2$. It follows then that for any λ satisfying

$$\lambda > \frac{2\ln 2}{k},$$

it is possible to achieve reliable storage using at most λ units of resource per information bit. But $\frac{2\ln 2}{k}$ was shown in (2.10) to be the ultimate minimum resource needed as $x \to 0$. We have therefore proved the following theorem.

Theorem. *For a BSC whose noise scales as described in Eq. (2.9), the family of orthogonal codes achieves a minimum resource per information bit of*

$$\lambda_{\min} = \frac{2\ln 2}{k}.$$

If the infimum in (2.7) occurs at $R = 0$, then this value is the absolute minimum for the given BSC.

References

1. Abdel-Ghaffar, K., and McEliece, R., "Soft-error correction for increased densities in VLSI memories," Proc. 11th Annual International Symposium on Computer Architecture (June 5-7, 1984), Ann Arbor, MI, pp. 248-250.
2. Abramowitz, M., and Stegun, I., eds., *Handbook of Mathematical Functions*. New York: Dover, 1965.
3. Dennard, R., Gaensslen, F., Yu, H., Rideout, V., Bassous, E., and LeBlanc, A., "Design of ion-implanted MOSFETs with very small physical dimension," IEEE J. Solid State Circuits, v. SC-9, Oct. 1974, pp. 256-268.
4. McEliece, R., *Theory of Information and Coding*, Reading, MA: Addison, Wesley, 1977.
5. Wozencroft, J., and Jacobs, I., *Principles of Communication Engineering*. New York: John Wiley, 1965.

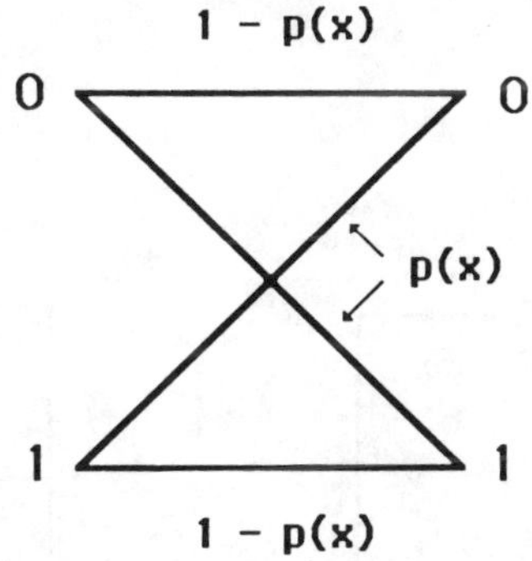

Figure 1. A BSC whose crossover probability depends on x, the "resource" per channel bit.

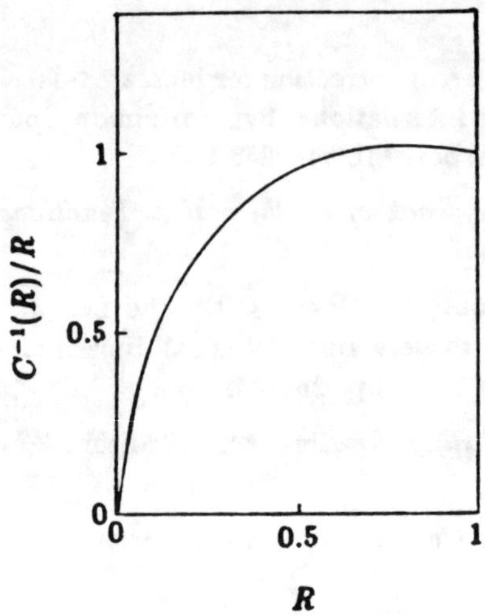

Figure 2. $\delta(x) = x^{1/3}$

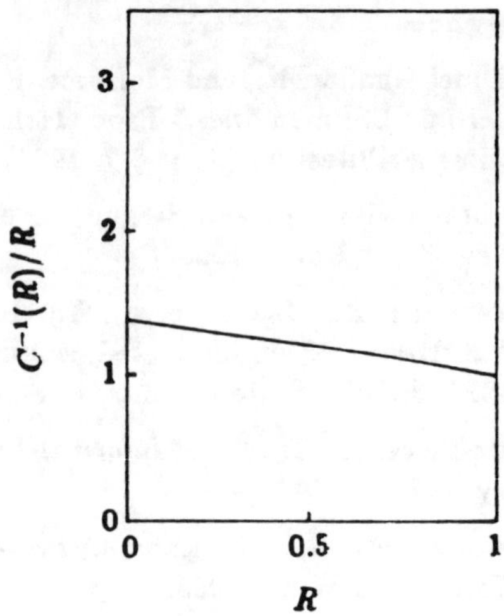

Figure 3. $\delta(x) = x^{1/2}$

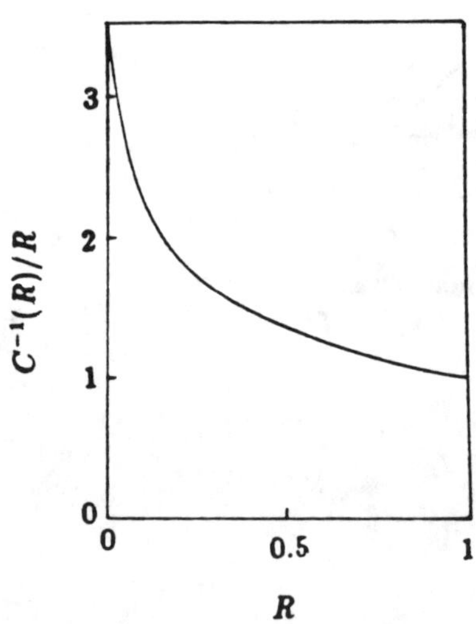

Figure 4. $\delta(x) = x^{2/3}$

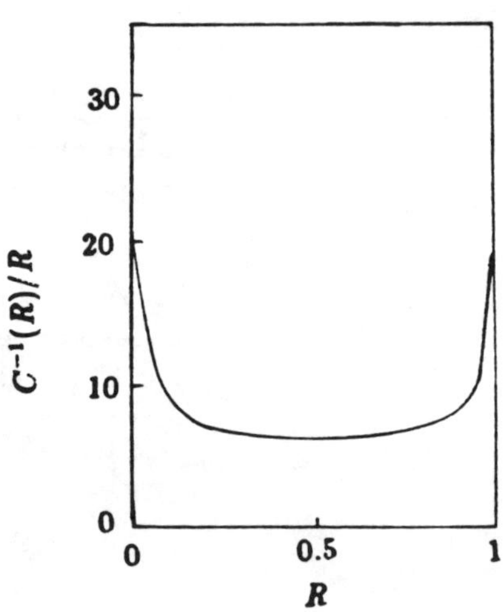

Figure 5. $\delta(x) = 1 - e^{x/2}$

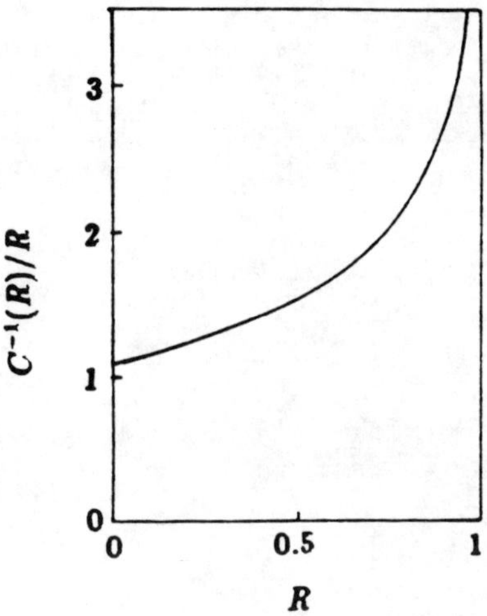

Figure 6. $\delta(x) = 1 - 2Q(\sqrt{2x})$

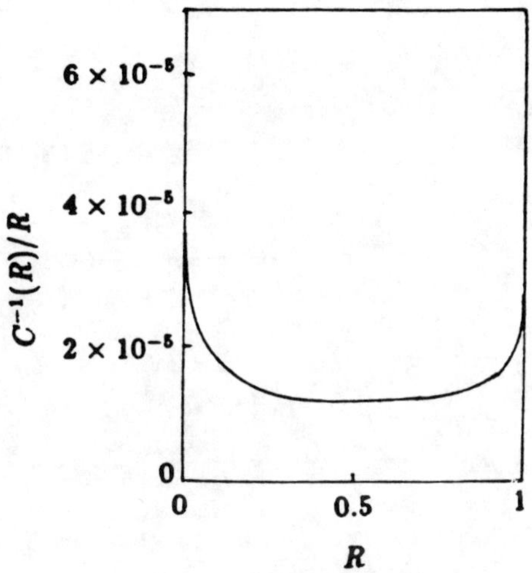

Figure 7. $\delta(x) = 1 - 2Q(10^4 x^{3/2})$

$$\begin{array}{cccccccc} + & + & + & + & + & + & + & + \\ + & - & + & - & + & - & + & - \\ + & + & - & - & + & + & - & - \\ + & - & - & + & + & - & - & + \\ + & + & + & + & - & - & - & - \\ + & - & + & - & - & + & - & + \\ + & + & - & - & - & - & + & + \\ + & - & - & + & - & + & + & - \end{array} \qquad \begin{array}{cccccccc} 0 & 0 & 0 & 0 & 0 & 0 & 0 & 0 \\ 0 & 1 & 0 & 1 & 0 & 1 & 0 & 1 \\ 0 & 0 & 1 & 1 & 0 & 0 & 1 & 1 \\ 0 & 1 & 1 & 0 & 0 & 1 & 1 & 0 \\ 0 & 0 & 0 & 0 & 1 & 1 & 1 & 1 \\ 0 & 1 & 0 & 1 & 1 & 0 & 1 & 0 \\ 0 & 0 & 1 & 1 & 1 & 1 & 0 & 0 \\ 0 & 1 & 1 & 0 & 1 & 0 & 0 & 1 \end{array}$$

Hadamard Matrix — Orthogonal Code

Figure 8. Hadamard Matrices and Orthogonal Codes, Illustrated for $M = 8$.

LIMITS OF RADIO COMMUNICATION - COLLABORATIVE TRANSMISSION OVER CELLULAR RADIO CHANNELS

P. G. FARRELL,* A. BRINE,* A. P. CLARK[o] and D. J. TAIT*
*** Department of Electrical Engineering, University of Manchester, Manchester, M13 9PL, U.K.**
o Department of Electronic and Electrical Engineering, University of Technology, Loughborough, LE11 3TH, U.K.

1. INTRODUCTION

Communication by the radiation of electromagnetic waves - radiocommunication - is potentially the simplest and most flexible means of transmitting information over distances of more than a few metres. A transmitter, a receiver, and two antennas is all that is needed. Its flexibility lies in that it is wireless - pun intended! - especially if either the transmitter, or the receiver (or both) are mobile, when radio is the only feasible means of communication. But this very convenience and flexibility also makes radio the interesting and challenging field of endeavour that it is in practice. The difficulty is the all-pervasive transmission medium. Radio waves can propagate - as air-waves, space-waves, and so on - to many places where they are unwanted, whereas "wire" communication is guided (normally) only to where it is required. More subtly, the purpose of communication is to put people in contact, by means of suitable connecting arrangements and protocols. This element of organisation is built in to wire communications, if only because of the need to provide a transmission medium between all those who may wish to communicate; often it has been overlooked or neglected in the case of radiocommunication.

Cellular mobile radio overcomes both the above difficulties. By suitable choice of frequency band, transmitter power, and antenna structure, radio waves are constrained to propagate over a well-defined area, or cell, with minimum interference in appropriately distant cells which re-use the same frequency channels. Proper control arrangements permit dynamic allocation of the frequency channels within each cell, and also enable communication to continue without interruption when a mobile crosses a boundary between cells (hand-off). Additional traffic can be accommodated by splitting cells into smaller ones; in this way, a very large number of users can efficiently share a relatively small number of frequency channels. The cellular concept, coupled with recent advances in electronic technology and the release of fairly substantial tranches of spectrum, has led to the active implementation and installation of commercial cellular mobile radio systems, which have developed rapidly in response to a very buoyant demand.

There are problems, of course. Uniform propagation over a cell is often difficult to arrange, particularly if the terrain is hilly or very built-up. Equally, co-channel interference can be difficult to control, especially under anomalous propagation conditions. Cells cannot be made as small as may be required for dense traffic areas, if only because of the frequent incidence of hand-off across cell boundaries that this would imply, together with other associated control problems. It is not clear whether adequate service for portable (personal) radios can co-exist with an efficient vehicular mobile service. In particular, it seems that the capacity of a fully developed cellular system, as presently conceived, will not be able to

J. K. Skwirzynski (ed.), Performance Limits in Communication Theory and Practice, 281–307.

meet the predicted demand for service. A number of additional techniques will have to be explored, perhaps including more direct forms of frequency sharing. The ultimate aim must be to achieve a mobile communication service which is as ubiquitous and interconnected as the worldwide telephone network, wherever the user may be, and at whatever speed he or she may be moving!

This contribution begins by introducing some of the important features of cellular mobile radio. There now exists an extensive literature on cellular systems; a few useful papers describing various aspects are listed in references (1-8) and (13-15). Two efficient forms of frequency sharing for digital cellular systems are then reported. These sharing schemes permit an increase in the capacity of a basic cellular system without the need to increase the number of frequency channels allocated to it (9-12). Finally, the performance of the second of the two schemes is assessed. This involves characterising the bursty shared digital cellular channel, and devising an efficient error control scheme for it.

2. CELLULAR MOBILE RADIO

2.1 Efficiencies of Mobile Radio Schemes

2.1.1 Fixed area coverage. An area of A km^2 is covered by a single transmitter providing F frequency channels, each of bandwidth B. For a given grade of service (i.e. blocking probability) on average M mobiles will be able to use each channel. Each mobile in the area is assigned to a fixed channel. Then, the coverage efficiency is

$$\eta_{af} = \frac{M.F}{A} \quad \text{mobiles per unit area,}$$

and the spectral efficiency is

$$\eta_{sf} = \frac{M.F}{F.B} = \frac{M}{B} \quad \text{mobiles per unit bandwidth}$$

2.1.2 Trunked area coverage. This system is the same as the previous one, but now each mobile is dynamically assigned a frequency channel. In this case many more mobiles, on average, will be able to use each channel, for a given grade of service; i.e.,

$$M_t = KM$$

where K is a factor found to lie in the range 5 to 10 (the trunking improvement factor). Thus the coverage efficiency becomes

$$\eta_{at} = \frac{F.M_t}{A} = K\eta_{af}$$

and the spectral efficiency becomes

$$\eta_{st} = K\eta_{sf}$$

in both cases 5 to 10 times better than those of a fixed system.

2.1.3 Cellular coverage. The same area A is now divided into C clusters of cells, each cell being covered by its own transmitter, with N cells per cluster. F frequency channels with bandwidth B are allocated to each cluster. These F channels will be assumed to be allocated equally between the cells (i.e., F/N channels per cell), though in practice the allocation will reflect the traffic in each cell. Because the channels in each cell are re-used C times within the area, then

$$\eta_{ac} = \frac{M_c \left(\frac{F}{N}\right) C}{A} = K . \frac{C}{N} . \eta_{af} = \frac{C}{N} . \eta_{at}$$

where $M_c = M_t = KM$ and $\eta_{sc} = \frac{C}{N} . \eta_{st}$

In theory, any number of users (mobiles) can be accommodated by making C/N as large as is needed, thus overcoming the limitations of the fixed and trunked systems. In practice, however, maximum C is limited by the minimum feasible cell radius (1 km?) and minimum N by the need to maintain a workable signal-to-interference ratio (SIR). Efficiencies many times greater than those of fixed or trunked systems are quite possible, though.

2.2 Cell Geometry and Signal-to-Interference Ratio

Assuming hexagonal cells (unobtainable in practice!), then the distance between cell centres, X, is given by $X = R\sqrt{3}$, see figure 1, where R is the cell "radius". If the cell is N times larger in area, then the distance between centres is $\sqrt{N}$ times larger. Hence, see figure 2, the relationship between D, the frequency re-use distance, and R is given by $D/R = \sqrt{3N}$. Now the signal strength at a distance R from the transmitter in a given cell is proportional to $1/R^n$, where n = 3-4, as determined by the propagation conditions (e.g. the terrain). The strength of interference from the six other cells nearest to the given cell that use the same set of frequencies is proportional to, in the worst case, $6/(D-R)^n$. Thus the SIR is approximately

$$\frac{1}{6}\left(\frac{D}{R} - 1\right)^n = \frac{1}{6}\left(\sqrt{3N} - 1\right)^n$$

Useful practical values of SIR are in the range 15-20dB, so N = 7, 9 or 12 is required.

2.3 Problems and Enhancements

2.3.1 Uniform illumination. The effectiveness of a cellular scheme depends upon controlled illumination within each cell. The terrain may help to constrain propagation as required, but it often does not. Under ideal conditions (flat earth), for example, if a cell is illuminated so as to adequately cover 90% of cell locations, then 5% of locations in a cell at a distance D away in an N = 4 scheme is also illuminated. Power control helps to determine the mobile-to-base path, but is trickier for the base-to-mobile path. Sectoring, where directional (e.g., 120°) instead of omni-directional antennas are used, also helps to alleviate the difficulties. In this case it is effective to trunk the sectored channels.

2.3.2 Co- and adjacent-channel interference. Co-channel interference has already been discussed, but additional difficulties may arise from anomalous propagation (ducting, refraction, reflection). Adjacent channel interference must also be taken into account. Now the use of relatively wide channels can lead to a reduction in the required SIR (e.g., as in the case of higher deviation FM improving capture properties). But wider channels (e.g., as much as 32kHz for 25kHz channel spacing is optimum in some cases) leads to adjacent channel interference difficulties. This may mean not being able to allocate adjacent frequency channels within the same cell. There is clearly a balance to be found between co- and adjacent-channel interference performance (4). Power control (reduction of transmited power as a mobile nears the base station antenna) is also used to mitigate co- and adjacent-channel interference.

2.3.3 Minimum cell size. The above illumination and interference difficulties, together with the hand-off problems noted below, all limit the minimum feasible cell size. It is difficult to control illumination and interference in very small cells, and infrastructure complexity and cost become excessive (7). Cell radii of less than 1-2Km are felt to be impracticable for the 900MHz and lower frequency bands. At much higher frequencies (60GHz?) much smaller dimensions may be possible.

2.3.4 Portables. During the early stages of development of a cellular system, before the installation of overlay or expansion cells, portables, (particularly personal radios) suffer from a degraded service because of the relatively large areas of low signal strength. On the other hand, a dense concentration of portables in one cell could easily overload the system. It is questionable, in fact, whether a scheme optimised for vehicle mobiles is ever efficient for portable mobiles, and vice-versa.

2.3.5 Hand-off. When a mobile crosses the boundary between two cells, its presence, and any call in progress, must be transferred to the new cell. There are two related problems here; firstly, cell boundaries are often ill-defined; and secondly, hand-off must not occur too often. The problems are related because at the edges of a cell the signal strength is very variable, and can cross the hand-off threshold several times during a traverse of the boundary region. If cells are small, and the mobile is travelling fast, then hand-off will also occur unacceptably often. Hence in practical cellular schemes several criteria are used to judge the correct moment for hand-off; not just signal strength, but also jitter on a supervisory audio tone (SAT) or digital "quality" sounding signal, identity of mobile (is it far from its usual position?), and location of mobile. This last, in its most sophisticated form, would enable precise cell boundaries to be maintained, taking into account the vagaries of terrain and propagation.

2.3.6 Wide area coverage. Cellular schemes rely on effective sharing of frequency channels over fairly small areas: very large cells are not very efficient. This makes it difficult to cover areas of low population density. Good service in such areas may only be possible if the cellular scheme is integrated with a land mobile area coverage system, or a satellite mobile system (8).

2.3.7 Cell splitting. As the demand for service in a cellular system rises, it can be accommodated by splitting or sub-dividing cells. It is important to avoid having to move transmitters, and to minimise the number of new transmitters required. Depending on the terrain, overlay or expansion cells can be used, the transmitted power at the "old" cell sites being or re-directed as appropriate (6). Sectoring of transmitter antennas is also useful as a means of splitting cells (7). The process of system expansion should as far as possible be planned ahead, based on traffic forecasts and demographic data, so as to maximise the cost-effectiveness of the system, and the service it provides. A practical cellular system has a range of cell sizes; larger where the traffic is not so dense, and smaller where the traffic is denser. The number of channels per cell (F/N) should vary to reflect the concentration of traffic in each cell and the cluster to which it belongs (13).

2.4 Limitations on Capacity

Even the most sophisticated first generation cellular system is unlikely to be able to meet the projected demand for service, which could easily rise to 5000 subscribers per km^2. Existing systems can handle up to about 1000 users per km^2, so an improvement in capacity by a factor of 5 to 10 is required, particularly in congested cells and clusters. Plans for second generation cellular systems incorporate various enhancements designed to increase the system capacity. These include: cell queuing (14); use of more spectrally efficient modulation methods (e.g. SSB instead of FM), which allows the use of smaller channel spacings (such as 12.5 or 6.25 kHz instead of 25 or 30 kHz (13); trunking between cells, and possibly clusters, which permits adaption to time-varying traffic concentrations (e.g. rush hour congestion); off-air call set-up (OACSU), and other control arrangement improvements; reduction in cell size and re-use distance by making use of higher frequency bands (e.g. 60 GHz); diversity on mobiles as well as at base stations; and converting to digital modulation. This last enhancement is potentially capable of having a major impact on cellular system performance and capacity, and offers considerable other advantages, including (15): the ability to provide a truly integrated service (e.g. voice and data); the ability to take full advantage of digital signal processing, coding, signalling, switching and control techniques; better area coverage and range; higher link capacity and system throughput because of faster speed of transaction; and the ability to offer private and secure communication by means of cryptographic techniques.

Even with the above enhancements, however, two factors remains to limit the capacity of cellular system. These are the minimum workable cell diameter, and the number of frequency channels that can be allocated to the system. There is thus considerable interest in any technique which would in effect increase the number of channels available in each cell. This is particularly important for mobile-to-base transmission, because it is relatively easy, using broadcast channel modulation and coding techniques (16), to increase the capacity of base-to-mobile transmission.

3. COLLABORATIVE TRANSMISSION OVER DIGITAL CELLULAR CHANNELS

3.1 A Multi-User Collaborative Coding Scheme

Collaborative coding is a technique which permits potentially efficient simultaneous transmission by several users in a shared frequency channel. In the context of digital mobile radio, the use of collaborative coding multiple access (CCMA) could more than double the capacity of a cellular system, for example. The advantage of CCMA over other forms of multiple access, such as TDMA, FDMA and CDMA (spread spectrum), is that it allows a substantial increase in the number of users, leading to a higher combined information rate and hence a potentially more efficient system.

Collaborative coding technqiues can be applied to both mobile-to-base and base-to-mobile transmission. Information and coding theory aspects of these two cases (denoted the multiple access channel (MAC) and the broadcast channel (BC), respectively) have been studied for several years, and many papers have appeared in the literature (16). In the MAC case, each user is provided with a code which enables the receiver to unscramble the individual information streams, by detecting the resulting combined signal. In the BC case, a combined coded signal is transmitted, and each receiver is able to detect and decode the information destined for it. The codes used, in addition to providing the 'unscrambling' function, can also be extended to incorporate a degree of error detection and correction, to protect the messages from the imperfections (noise, fading, interference) of the mobile radio channel. Two particular forms of signal combination have been extensively studied, mainly for the MAC. In the first form, each mobile transmits (or receives) a binary signal, and the base station detects (or transmits) the sum of all the signals (e.g. in the 2-user case, transmitted signals representing 0 or 1 are detected as 0, 1 or 2; this is the adder MAC (or BC)). In the second form, each user is allocated a sub-set of a set of tone frequencies and the base station is capable of detecting (or transmitting) all of them; this is the T-user M-frequency MAC (or BC), and is an example of an OR-channel (17,18). In both forms, quite simple codes (16) permit the 'unscrambling' to be done (even in the absence of bit and block synchronisation); thus the complexity of a collaborative coding scheme is concentrated at the base station, and the mobile transmitters and receivers need to be no more complex than those required in the absence of collaboration.

Adder MAC collaborative codes clearly apply quite straightforwardly to baseband channels. When applied to carrier-modulation channels, then in addition to the existing channel imperfections noted above, there will also be a detection problem related to the 'cancellation fading' that will occur if signals are received with carriers out of phase. Since it will not be practical to lock the carriers, this condition could occur frequently. The remainder of this sub-section presents a possible solution to this problem by proposing a simple carrier modulation scheme wich provides the required adder channel operation even when the carriers are arbitrarily phased.

3.1.1 The baseband adder channel. Figure 3 shows a communications system in which T independent data sources are simultaneously transmitting to their respective destinations via a common discrete channel. Data from the i-th source (i = 0,1,...,T-1) is encoded by the i-th encoder according to a uniquely assigned block code, C_i The resulting binary codeword, X_i, is then transmitted over the channel where it combines with the other T-1 codewords (codeword synchronisation is assumed) to produce a composite codeword Y. It is the task of the single decoder to decode the received codeword into the T users' original data. If all the user codes are binary block codes and the channel provides arithmetic addition of the users' transmitted codeword symbol values (S_0, $S_1,\ldots,S_{T-1}$), then the resulting multilevel adder channel output symbol during any symbol interval is given by

$$S_r = \sum_{i=0}^{T-1} S_i$$

An illustrative example of such a coding scheme is the 3-user code proposed by Chang and Weldon (19) shown in table 1. The code is formed from the three half-rate constituent codes C_0 = {(00),(11)}, C_1 = {(10,(01)} and C_2 = {(00),(01)}. It can be seen that each of the eight possible composite codewords - resulting from symbol-wise addition on the channel - is unique, and can therefore be unambiguously decoded into its constituent codewords, and thence into the corresponding users' original data bits (shown in square brackets).

	X_0	00[0]	00[0]	11[1]	11[1]	(C_0)
X_2	X_1	10[0]	01[1]	10[0]	01[1]	(C_1)
00[0]		10	01	21	12	
(C_2) 10[1]		20	11	31	22	

Table 1 A 3-user code

Such a code is attractive because it offers a rate-sum of 1.5 bits per received symbol which is greater than the unity rate-sum achieved by simple time division.

Implementing the necessary addition is reasonably straightforward at baseband, if each user transmits an on-off keyed voltage according to his binary codeword symbols. However, extending this to its bandpass equivalent - each user transmitting on-off keyed carrier - is normally only viable if the user carriers are phase-coherent at the receiver. Without carrier coherence, cancellation fading occurs and the required symbol addition is not achieved. By deliberately inducing fading in a controlled manner, the bandpass adder channel shown in figure 4, and described next, provides the required symbol addition with arbitrarily phased carriers.

3.1.2 Bandpass adder channel implementation. To implement the T-user discrete channel shown in figure 4, assign to the i-th user (i = 0,1,...,T-1) the carrier

$$f_i(t) = \frac{2}{T_s} \cdot \cos\,[\omega_0+i(2\pi/T_s))t+\phi_i]$$

(where T_s is the symbol interval, ω_0 is the 0-th user's carrier frequency, chosen such that $(\omega_0 T_s)/\pi$ is an integer, and ϕ_i is an arbitrary carrier phase angle), such that during the symbol interval, $0 < t \leqslant T_s$, the i-th user transmits the on-off keyed carrier signal

$$s_i(t) = S_i \cdot \frac{2}{T_s} \cdot \cos\,[(\omega_0+i(2\pi/T_s))t+\phi_i]$$

(where $S_i \in [0,1\}$ is the i-th user's constituent codeword symbol). The resulting demodulator input, assuming no losses on the channel, is the received composite signal

$$s_r(t) = \frac{2}{T_s} \cdot \sum_{i=0}^{T-1} S_i \cos\,[(\omega_0+i(2\pi/T_s))t+\phi_i]$$

from which we wish to recover the collaborative codeword symbol, S_r. This is achieved by squaring $s_r(t)$ and then integrating over the symbol interval. By virtue of the mutual orthogonality of the users' transmissions afforded by the specified carrier separation, $2\pi/T_s$, all cross-product terms resulting from the squaring process integrate to zero, and it can be shown that

$$\int_0^{T_s} [s_r(t)]^2 \, dt = \sum_{i=0}^{T-1} S_i = S_r$$

Example demodulator waveforms for a 3-user system based on the Chang and Weldon code are shown in figure 5. The waveforms are a result of the simultaneous transmission of the constituent codewords corresponding to the users' data sequences: U_0 = {101}, U_1 = {100} and U_2 = {011}. Using table 1, it can be seen that the integrate-and-dump output samples (figure 5(e)) form the composite codewords that unambiguously decode into the users' original data sequences, U_0, U_1 and U_2. The demodulator shown in figure 4 would be preceded by a bandpass filter to remove out of band noise components, and the integrator output samples (no longer discrete values) would require (T+1)-level detection. A receiver filter bandwidth of at least $(T+1)/T_s$ Hz would be required to pass the T overlapping main lobes of the transmitted spectra, assuming each to have a (null-to-null) bandwidth of $2/T_s$ Hz.

This simple carrier modulation scheme provides the symbol addition required by Collaborative Code Multiple Access schemes operating over mobile radio channels. The key features of the scheme are its ability to cope with arbitrarily phased carriers and its simple noncoherent detection of the received composite signal. Perhaps inevitably, these advantages are gained at the expense of overall bandwidth requirements which, whilst competitive for 2 or 3 users, become excessive thereafter. It is also clear that the decision thresholds in the receiver would require dynamic adjustment in order to achieve adequate performance under fading conditions. Possibly, the use of power control combined with the diversity reception would significantly assist this adaptive demodulation process.

3.2 Simultaneous Collaborative Transmission of Two Spectrally Efficient Signals

A simultaneous transmission scheme (20,21) is now reported which is more efficient, in terms of channel rate and system capacity, than the previous scheme. The multi-user collaborative coding scheme of section 3.1 has a per channel (per user) rate of 0.5, corresponding to a redundancy of 50%. The scheme being considered in this section has a redundancy of only about 10% per channel (user).

In the assumed model of the radio system, with carrier frequencies close to 900M MHz, adjacent carrier frequencies are spaced at 25 kHz and, for every channel, full raised-cosine spectral shaping is used for the demodulated baseband signal at the receiver. A four-level (quarternary) quadrature amplitude modulated (QAM) signal is transmitted over each channel at 12,000 bauds, to give a transmission rate of 24,000 bit/s per channel and a nominal bandwidth of 24 kHz. Thus adjacent channel interference is avoided. The bandwidth efficiency of the system, for signals transmitted from the mobiles to the base station, can now be doubled by permitting each channel to be used by two different mobiles. The independent random fading of the two signals occupying one channel enables these to be correctly detected at the receiver, for the large majority of the time (22). In a cellular implementation of such a system, and with sufficiently small size of cell, it is possible to achieve both element-timing and frame-timing synchronisation of the signals transmitted by all mobiles in a cell. This enables the phase of the sampling instants in the modem receiver at the base station to be optimised simultaneously for all received signals, thus avoiding intersymbol interference in any individual sampled baseband signal

(22). The channel estimation process required before detection is now greatly simplified.

The selected QAM signal can also be considered as a four-level PSK (QPSK) signal having a considerable envelope ripple caused by bandlimiting. It is generated by separate in-phase and quadrature suppressed carrier amplitude modulation, to provide a truly linear modulation method. The signal can be bandlimited without degrading its tolerance to noise or impairing the element-timing and carrier-phase synchronisation processes at the receiver. Thus a better tolerance to noise for a given efficiency in use of bandwidth, together with more effective element-timing and carrier-phase syncrhonisation, can be obtained than is possible with a truly constant envelope signal (23). Techniques are now available which enable a high power amplifier in a mobile to handle the large envelope ripple in the QAM signal,

3.2.1 Model of system. The model of the data-transmission system is shown in figure 6, where all signals are baseband and complex valued. The signals $\{s_{0,i}\delta(t-iT)\}$ and $\{s_{1,i}\delta(t-iT)\}$ originate from different mobiles, and the data-symbols $\{s_{0,i}\}$ and $\{s_{1,i}\}$ are statistically independent and equally likely to have any of the four values $\pm 1 \pm j$. Each transmission path is a linear baseband channel that introduces the baseband equivalent of flat Rayleigh fading. This is a useful but perhaps rather idealised model of the fading likely to be introduced into a 900 MHz carrier in an urban environment (23). The two channels fade independently and the fading rate here is typically up to around 100 fades a second. A transmission path may also introduce a Doppler shift. Stationary white Gaussian noise is added to the sum of the two fading signals, and the resultant signal is filtered to give the bandlimited noisy baseband waveform r(t). The resultant transfer function of the transmitter and receiver filters is raised-cosine in shape and such that, with the appropriate phase Δ in the sampling instants $\{iT + \Delta\}$, there is no intersymbol interference in the samples $\{r_i\}$, where $r_i = r(iT + \Delta)$ (24). Thus the received sample, at time $t = iT + \Delta$, is

$$r_i = s_{0,i}\, y_{0,i} + s_{1,i}\, y_{1,i} + w_i$$

where r_i, $y_{0,i}$, $y_{1,i}$ and w_i are complex valued. The lowpass filter C is such that the real and imaginary parts of the noise components $\{w_i\}$ are statistically independent Gaussian random variables with zero mean and fixed variance (24). The quantities $y_{0,i}$ and $y_{1,i}$ may vary quite rapidly with i and each represents the attenuation and phase-change introduced into the corresponding signal by the transmission path.

The optimum detector, that has prior knowledge of $y_{0,i}$ and $y_{1,i}$, takes as the detected values of $s_{0,i}$ and $s_{1,i}$, the possible values $s'_{0,i}$ and $s'_{1,i}$ for which $s'_{0,i}\, y_{0,i} + s'_{1,i}\, y_{1,i}$ is at the minimum distance d_i from r_i, where

$$d^2_i = |\, r_i - s'_{0,i}\, y_{0,i} - s'_{1,i}\, y_{1,i}\, |^2$$

and $|x|$ is the absolute value (modulus) of x. Unfortunately, one of the major problems in the design of the digital modem is that of obtaining sufficiently accurate estimates of $y_{0,i}$ and $y_{1,i}$. For example, with a transmission rate of 12,000 bauds (elements/second) and a fading rate of 100 fades/second, there are typically 60 received samples $\{r_i\}$, between a peak and an adjacent trough in the fading of either of the two received signals. Thus, over a sequence of only some ten $\{r_i\}$, considerable changes may take place in either or both $y_{0,i}$ and $y_{1,i}$. Furthermore, the changes are of too random a nature to be predicted reliably or accurately over more than about one quarter of a cycle of a fade.

3.2.2 Channel estimation for two signals. A simple technique for estimating $y_{0,i}$ and $y_{1,i}$ is the gradient or steepest descent algorithm (25,26), which operates, together with the predictor and detector, as follows. On the receipt of r_i, at time $t = iT + \Delta$, the detector uses the predictions $y'_{0,i,i-1}$ and $y'_{1,i,i-1}$, that were formed from $y_{0,i}$ and $y_{1,i}$, respectively, at time $t = (i-1)T + \Delta$, to determine the detected values of the data-symbols $s_{0,i}$ and $s_{1,i}$. The detected data-symbols $s'_{0,i}$ and $s'_{1,i}$ are determined by the detector (figure 6) as their possible values ($\pm1\pm j$) that minimise d^2_i, where $y_{0,i}$ and $y_{1,i}$ are now replaced by $y'_{0,i,i-1}$ and $y'_{1,i,i-1}$, respectively.

The gradient estimator uses the detected data-symbols $s'_{0,i}$ and $s'_{1,i}$ to form the estimate

$$r'_i = s'_{0,i}\, y'_{0,i,i-1} + s'_{1,i}\, y'_{1,i,i-1}$$

of the received sample r_i, and then the error signal

$$e_i = r_i - r'_i$$

The updated estimates of $y_{0,i}$ and $y_{1,i}$ are now given by

$$y'_{0,i} = y'_{0,i,i-1} + be_i(s'_{0,i})^*$$

and

$$y'_{1,i} = y'_{1,i,i-1} + be_i(s'_{1,i})^*$$

where b is an appropriate small positive real-valued constant. $(s'_{0,i})^*$ and $(s'_{1,i})^*$ are the complex conjugates of $s'_{0,i}$ and $s'_{1,i}$, respectively. The errors in the predictions $y'_{0,i,i-1}$ and $y'_{1,i,i-1}$ are then taken to be

$$\epsilon_{0,i} = y'_{0,i} - y'_{0,i,i-1} = be_i(s'_{0,i})^*$$

and

$$\epsilon_{1,i} = y'_{1,i} - y'_{1,i,i-1} = be_i(s'_{1,i})^*$$

respectivley. Finally, the prediction $y'_{0,i+1,i}$ of $y_{0,i+1}$, given by the appropriate least-squares fading-memory polynomial filter, is as shown in table 2(25,27). The terms $y''_{0,i+1,i}$ and $y'''_{0,i+1,i}$ here are functions of the first and second derivatives of $y'_{0,i+1,i}$ with respect to time, and are considered in further detail elsewhere (27). Relationships, exactly corresponding to those in table 2, hold also for $y'_{1,i+1,i}$, and $y''_{1,i+1,i}$ and $y'''_{1,i+1,i}$. Having determined the predictions $y'_{0,i+1,i}$ and $y'_{1,i+1,i}$, the detected data-symbols $s'_{0,i+1}$ and $s'_{1,i+1}$ are determined from r_{i+1}, at time $t = (i+1)T + \Delta$, ready for the next estimation process, and so on.

3.2.3 Combined estimator and detector. The weakness of the estimation processes considered so far is that they rely very heavily on the correct detection of the data symbols. This suggests an estimation process that considers several different possible values of each detected data-symbol in an appropriate combination of estimation and detection, as follows.

Degree of polynomial	One-step prediction at time $t = iT + \Delta$
0	$y'_{0,i+1,i} = y'_{0,i,i-1} + (1-\theta)\epsilon_{0,i}$
1	$y''_{0,i+1,i} = y''_{0,i,i-1} + (1-\theta)^2\epsilon_{0,i}$
	$y'_{0,i+1,i} = y'_{0,i,i-1} + y''_{0,i+1,i} + (1-\theta)^2\epsilon_{0,i}$
2	$y'''_{0,i+1,i} = y'''_{0,i,i-1} + 0.5(1-\theta)^3\epsilon_{0,i}$
	$y''_{0,i+1,i} = y''_{0,i,i-1} + 2y'''_{0,i+1,i} + 1.5(1-\theta)^2(1+\theta)\epsilon_{0,i}$
	$y'_{0,i+1,i} = y'_{0,i,i-1} + y''_{0,i+1,i} - y'''_{0,i+1,i} + (1-\theta^3)\epsilon_{0,i}$

Table 2 Least-squares fading-memory prediction using a polynomial filter

Consider a composite data-symbol given by a two-component vector

$$q_i = [q_{0,i} \quad q_{1,i}]$$

where $q_{0,i}$ and $q_{1,i}$ take on the possible values of $s_{0,i}$ and $s_{1,i}$, respectively. Thus q_i has 16 different possible combinations of $s_{0,i}$ and $s_{1,i}$. Just prior to the receipt of r_i, at time $t = iT + \Delta$, the detector holds in store k different n-component vectors $\{Q_{i-1}\}$, where

$$Q_{i-1} = [q_{i-n} \quad q_{i-n+1} \quad \cdots \quad q_{i-1}]$$

Each vector Q_{i-1} represents a different possible pair of the sequences

$$[s_{0,i-n} \quad s_{0,i-n+1} \quad \cdots \quad s_{0,i-1}]$$

and

$$[s_{1,i-n} \quad s_{1,i-n+1} \quad \cdots \quad s_{1,i-1}]$$

Associated with each vector Q_{i-1} is stored its cost c_{i-1} (to be defined presently), which is a measure of the likelihood that the vector is correct, the lower the cost the higher being the likelihood.

On receipt of the signal r_i, each vector Q_{i-1} is expanded into m vectors $\{P_i\}$, where

$$P_i = \{q_{i-n} \quad q_{i-n+1} \quad \cdots \quad q_i]$$

and m either has the same value, say 4, for each vector Q_{i-1}, or else m decreases as the cost of Q_{i-1} increases. In each group of m vectors $\{P_i\}$ derived from any one vector Q_{i-1}, the first n components $\{q_{i-h}\}$ are as in the original Q_{i-1} and the last component q_i takes on m different values.

Each of the resulting m vectors $\{P_i\}$ has the cost

$$c_i = \phi c_{i-1} + |r_i - q_{0,i}\, y'_{0,i,i-1} - q_{1,i}\, y'_{1,i,i-1}|^2$$

where ϕ is a real-valued constant in the range 0 to 1. The quantity c_{i-1} is the cost of the vector Q_{i-1} from which P_i was derived, such that

$$c_{i-1} = \Sigma\ \phi^{h-1}\, |r_{i-h} - q_{0,i-h}\, y'_{0,i-h,i-h-1} - q_{1,i-h}\, y'_{1,i-h,i-h-1}|^2$$

It is assumed that transmission began at time t = 0, so that

$$q_{0,i} = q_{1,i} = s_{0,i} = s_{1,i} = 0$$

for i < 0. the nearer ϕ approaches to zero, the smaller is the effect of earlier costs on c_i, thus reducing the effective memory in c_i.

The detected values $s'_{0,i-n}$ and $s'_{1,i-n}$ of the data-symbols $s_{0,i-n}$ and $s_{1,i-n}$ are now given by the value of q_{i-n} in the vector P_i with the smallest cost. Any vector P_i whose first component q_{i-n} differs in value from that of the above q_{i-n} is then discarded, and from the remaining vectors $\{P_i\}$ (including that from which $s_{0,i-n}$ and $s_{1,i-n}$ were detected) are selected the k vectors having the smallest costs $\{c_i\}$. The first component q_{i-n} of each of the k selected vectors $\{P_i\}$ is now omitted (without changing its cost) to give the corresponding vectors $\{Q_i\}$, which are then stored, together with the associated costs $\{c_i\}$, ready for the next detection process. The discarding of the vectors $\{P_i\}$, just mentioned, is a convenient method of ensuring that the k stored vectors $\{Q_i\}$ are always different, provided only that they were different at the first detection process, which can easily be arranged (28). When ϕ = 1 the algorithm becomes a direct development of a conventional near-maximum-likelihood detector (28). To prevent now a possible overflow in the value of c_i over a long transmission, the value of the smallest c_i must be subtracted from each c_i, after the selection of the k vectors $\{Q_i\}$, so that the smallest cost is always zero.

In the version of this technique that has been tested by computer simulation, k = 4 and m has the values 4, 2 and 1, respectively, for the four $[Q_{i-1}\}$, when arranged in the order of increasing costs and starting with the lowest cost vector. Thus, on the receipt of r_i, the first, second, third and fourth vectors $\{Q_{i-1}\}$ are expanded into four, two and one vectors $\{P_i\}$, respectively. There are now ten vectors $\{P_i\}$, from which are selected four vectors $\{Q_i\}$, as previously described. In most tests, ϕ has been set to unity, which generally seems to give the best performance.

With no intersymbol interference, as is the case here, and a single estimation and prediction process, no advantage would be gained by the arrangement just described over a simple detector (section 3.2.1). However, in the system tested, each of the four stored vectors $\{Q_{i-1}\}$ is associated with its own separate estimator and predictor, which may operate as previously described and which take the received sequences of data-symbol values $\{s_{0,i-h}\}$ and $\{s_{1,i-h}\}$ to be those given by the corresponding vector Q_{i-1}. Thus there are four separate estimation and prediction processes operating in parallel. When a vector Q_{-1} is expanded into m vectors $\{P_i\}$, the same predictions of $y_{0,1}$ and $y_{1,i}$ are used for each of the m vectors $\{P_i\}$, but these predictions normally differ from those associated with any of the other three vectors $\{Q_{i-1}\}$. After the selection of the four vectors $\{Q_i\}$ from the ten vectors $\{P_i\}$, the prediction errors $\epsilon_{0,i}$ and $\epsilon_{1,i}$ are evaluated separately for each Q_i. Then, for each of these vectors, $\epsilon_{0,i}$ is applied to the appropriate prediction algorithm of table 2, to give the one-step prediction $y'_{0,i+1,i}$ of $y_{0,i+1}$, and $\epsilon_{1,i}$ is handled similarly, to give the one-step prediction $y'_{1,i+1,i}$ of $y_{1,i+1}$. Thus, since the four $\{Q_i\}$ are different, so also, in general, are the predictions of $y_{0,i+1}$ and $y_{1,i+1}$ associated with the four $\{Q_i\}$. By considering more than one possible value of each detected data symbol, the estimator is more tolerant to errors in detection, and there is a reduced probability of a complete failure in the estimation and detection processes, when these are operating together.

3.2.4 Assessment of the scheme. Figure 7 shows the error-rate performance of the combined estimation and detection algorithm, when the data on each of the two channels is transmitted in packets of 60 symbols (120 bits), consisting of 54 symbols (108 bits) of information followed by a 6 symbol (12 bits) retraining sequence. The retraining sequence enables the estimator to continue tracking the signals even in the presence of fading, carrier phase shifts and self-interference. It is clear that even when there are four expansion vectors ($m = 4$), the residual error rate is unacceptably high when only one receiving antenna is used (no diversity). When dual diversity is used, however, the performance is much improved, even for $m = 2$. It is in practice quite feasible to achieve a correlation of less than 0.7, thus giving a diversity improvement, with a vertical separation between antennas at the base-station of about 15λ (approximately 15-20m at 900 MHz) (29,30). In the dual diversity case joint detection is used, see figure 8.

4. CHARACTERISATION AND CHANNEL CODING FOR THE DIGITAL CELLULAR CHANNEL

The error statistics of digital mobile radio channels, both with and without collaborative transmission, are predominantly bursty. It is therefore inappropriate to characterise the channel in terms of bit error rate only, and to design and evaluate error control (channel) codes on the basis of the Hamming distance metric. The latter may be appropriate if interleaving is being used to essentially randomise the error bursts, but a knowledge of the burst characteristics is required in order to determine the interleaving depth. Until recently the characterisation of bursty channels was not well understood, because the classical definition of a burst (a section of the channel error sequence, starting and ending with an error, where the density of errors is higher than in the sections preceding and following the burst) does not conveniently lead to a meaningful and useful statistical analysis. This is particularly so when the bursts are of relatively low density and the gaps between them have a sprinkling of apparently random errors (diffuse error bursts), which is most often the case. Even if it is possible to identify bursts, and to calculate burst rates and average burst and gap lengths, this information is not of much use for designing a suitable error control code and estimating its performance. More recently, the concepts of the multi-gap distribution (31-33) and the burst-b distance metric (34-36) have been developed to overcome these difficulties of modelling bursty channels. The multi-gap distribution, derived from the error sequence record, contains all the information needed to obtain relevant and useful statistics, such as the burst-b rate, the bit and burst-b correlations, and the probability of obtaining m bursts of length b in a block of length n. The burst-b distance replaces Hamming distance (burst-b distance when $b = 1$) as a metric for determining the burst-control power of a code, for decoding the code, and for estimating its performance, in a way almost exactly analogous to that for random error control codes. The burst-b definition of a burst is that it is an error pattern which starts with an error, is of length b digits, but not necessarily ends with an error, so that the bursts of a given length contain all shorter bursts as well.

4.1 Correlation Functions for Collaborative Transmission

Figures 9-11 show various correlation functions of an error sequence record obtained during simulation tests of the second collaborative transmission scheme (section 3.2). The symbol rate is 12 Kbaud, corresponding to 24 Kbit/s for each user. The fading rate is about 100 fades/sec, corresponding to a vehicle speed of 60 mph. Figure 9 is the bit error correlations (b = 1) for the two users, given by (33):

$$R(k) = \frac{1}{1-p}\left[\sum_{m=0}^{k} M(m,k) - p\right]$$

where p is the bit error probability, and M(m,k) is the mutli-gap distribution function; i.e. the probability that a multi-gap consisting of m gaps will have total composite length k. R(k) can also be computed directly from the error sequence record. For user A, p = 0.0239, and for user B, p = 0.0292. Figure 9 shows that the errors are correlated over quite long separations (> 60 bits), corresponding to the presence of long bursts, as expected. Truly random errors would be a correlation function that was substantially zero and flat for k > 1. The figure therefore indicates that an interleaving degree of at least 60 would be required in order to properly randomise the error bursts.

Figures 10 and 11 are the burst-b correlations for users A and B, respectively, with the same error sequence and bit error rates as in figure 9, for b = 2, 4, 8, 16 and 32; given by (36):

$$R_b(k) = \frac{P(k/o) - p_b}{1-p_b}$$

where pb is the burst-b rate, and P(k/o) is the probability that the next burst starts k bits after the end of a burst. $R_b(k)$ can also be obtained from the multi-gap distribution. These figures (which are almost the same) show that as the burst length b is increased, the incidence of bursts becomes more and more random (less correlated). Thus bursts of length 32 are virtually random, and could be corrected by a burst-error-code with b = 32. Alternatively, a shorter burst length code, combined with burst length interleaving, could be used to control the errors. for example, for b = 16, interleaving of 16-bit bytes to depth 8 would effectively randomise the 16-bit bursts.

4.2 Channel Coding for Collaborative Transmission

Many different types of code can be used to correct errors on digital cellular channels, including block codes, binary convolutional codes and multi-level convolutional codes (trellis codes), both with and without interleaving. In order to indicate the effectiveness of the random burst-b correlation model, however, examples will be given of the use of burst-error-correcting ½-rate block codes, which might be shortened and modified forms of standard binary burst-correcting, Reed-Solomon, or array (37) codes.

The bit error rate after decoding is approximately given by

$$P_o \triangleq \frac{(2t+1)b}{2n} \cdot \frac{n(n-b) \ldots (n-tb)}{(t+1)!} \cdot p_b^{t+1}$$

where t is the number of bursts of length b that the code can correct.

Table 3 gives calculated values of P_o for the values of p_b corresponding to the error record of user A (figure 10), for various combinations of t and burst length interleaving depth i.

p	b	p_b	t	n	i	ni	P_o
0.024	32	0.0041	1	128	1	128	0.0387
0.024	32	0.0041	2	256	1	256	0.0056
0.024	16	0.0057	1	64	8	512	0.0187
0.024	16	0.0057	2	128	8	1024	0.0019
0.024	8	0.0082	1	32	16	512	0.0097
0.024	8	0.0082	2	64	16	1024	0.0007

Table 3 Bit error rates after decoding

ni is the effective block length with interleaving; note that any advantage gained by shortening the burst length is lost because of the increase in effective block length (e.g. in terms of decoding delay). Note also that it is assumed that interleaving does randomise the bursts; this may not be true in practice. This strengthens the advisability of using a double instead of single-burst-correcting code.

5. CONCLUSIONS

Cellular mobile radio systems are now well established, and enhanced second generation systems are being planned and developed. The cellular concept has, for the first time in mobile radio communications, begun to satisfy the potential demand for convenient and reliable service, linking mobile subscribers to the public switched telephone network. The future of cellular probably lies in being part of an integrated personal communications service. The research reported here is a contribution towards the development of such a second or third generation system.

The biggest problem facing current cellular systems (and also some second generation systems) is that of congestion due to lack of capacity. This will remain a problem, in spite of the various enhancements that are being incorporated, even if digital transmission is used. The use of digital techniques, however, makes it possible to consider the use of simultaneous collaborative transmission by more than one user in the same frequency channel. This will at least double the capacity of a cellular system.

Two such collaborative transmission schemes are reported in this contribution. The second scheme, in particular, with its high bandwidth efficiency, is very interesting. Until now, if spectrally efficient transmission was required (i.e. ruling out spread spectrum transmission), only one user could be allocated to a frequency channel at any given time. The reported scheme, however, suggests that two users can operate on one channel, with satisfactory performance and reasonable complexity of implementation. One of the limits on wireless communication appears to have been substantially extended.

Another limit on digital comunication over bursty channels has been the difficulty of characterisng modelling and assessing such channels. The result reported in section 4 indicate the value and relevance of the random burst-b model, and indicate a direction for further extension and refinement.

REFERENCES

1. Young, W. R. et al: Issue on Advanced Mobile Phone Service, BSTJ, Vol. 58, No. 1, Jan. 1979.
2. Kemp, L. J.: A Technical Description of the UK TACS Cellular Radio System, Mobile Phone Tech. Memo. 85/1, British Telecom, Jan. 1985.
3. Oetting, J.: Cellular Mobile Radio - An Emerging Technology, IEEE Comms. Mag., Nov. 1983, pp. 10-23.
4. Halpern, S. W. et al: Adjacent-and Co-Channel Interference in Large-Cell Cellular Systems, Telecommunications, March 1984, pp. 112-116.
5. Cooper, G. R. and Nettleton, R. W.: Cellular Mobile Technology: the Great Multiplier, IEEE Spectrum, June 1983, pp. 30-37.
6. Wells, J. D.: Cellular System Design Using the Expansion Cell Layout Method, IEEE Trans. on Veh. Tech., Vol. IT-33, No. 2, May 1984, pp. 58-66.
7. Stocker, A. C.: Small-Cell Mobile Phone Systems, IEEE Trans. on Veh. Tech. Vol. VT-33, No. 4, Nov. 1984, pp. 269-275.
8. McDonnell, M. and Georganas, N. D.: Combined Mobile Telephone and Dispatch Services in a Cellular Land-Mobile Radio System, Proc. IEE, Vol. 131, Pt. F, No. 4, July 1984, pp. 357-363.
9. Brine, A. and Farrell, P. G.: Bandpass Adder Channel for Multiuser (Collaborative) Coding Schemes, Elec. Letters, Vol. 21, No. 22, pp. 1030-31, Oct. 24th, 1985.
10. Brine, A.: A Bandpass Adder Channel for Collaborative Code Multiple Access Schemes, Comms. Res. Group Report, Department of Elec. Eng. University of Manchester, Jan. 1985.
11. Clark, A. P.: Channel Estimation for Digital Land-Mobile-Radio Systems, IEE Colloq. on Modems for Radio Communication, Savoy Place, London, May 1986.
12. Clark, A. P.: Modems for Cellular Mobile Radio, IEE Colloq. on Digital Mobile Radio, Savoy Place, London, pp. 8.1-6, Oct. 1985.
13. Hata, M. and Sakamoto, M.: Capacity Estimation of Cellular Mobile Radio Systems, Elec. Letters, Vol. 22, No. 9, pp. 449-450, April 24th 1986.
14. Guerin, R.: Queuing and Traffic in Cellular Radio, Ph.D. Thesis, Calif. Inst. of Technology, 1986.
15. Farrell, P. G.: Mobile Radio - The Future, Chapter 14 In Land Mobile Radio Systems, ed. R. J. Holbecke, Peter Peregrinus/IEE, 1985.
16. Farrell, P. G.: Survey of Channel Coding for Multi-User Systems, in New Concepts in Multi-User Communications, ed. J. K. Skwirzynski, Sijthoff and Noordhoff, 1981, pp. 133-159.
17. Chang, S. C. and Wolf, J. K.: On the T-uyser M-frequency Noiseless Multiple-access Channel With and Without Intensity Information, IEEE Trans. Vol. IT-27, No. 1, pp. 41-48, Jan. 1981.
18. Healy, T. J.: Coding and Decoding for Code Division Multiple Users Communication Systems, IEEE Trans. Vol. COM-33, No. 4, pp. 310-316, April 1985.

19. Chang, S. C. and Weldon, E. J.: Coding for T-user Multiple-access Channels, IEEE Trans. on Information Theory, Vol. IT-25, No. 6, pp. 684-691, Nov. 1979.
20. Clark, A. P.: Digital Modulation for Cellular Radio Systems, Telecommunications (USA), to be published.
21. Clark, A. P.: Channel Estimation for Digital Land-Mobile-Radio Systems, IEE Colloq. on Modems for Radio Communications, Savoy Place, London, pp. 6/1-7, May 1986.
22. Clark, A. P.: Modems for Cellular Mobile Radio, IEE Colloq. on Digital Mobile Radio, London pp. 8/1-6, Oct. 1985.
23. Clark, A. P.: Digital Modems for Land Mobile Radio, IEE Proc. Pt. F, Vol. 132, pp. 348-362, August 1985.
24. Clark, A. P.: Principles of Digital Data Transmission, Second Edition (Pentech Press, 1983).
25. Clark, A. P. and McVerry, F.: Channel Estimation for an HF Radio Link, IEE Proc. Pt. F, Vol. 128, pp. 33-42, Feb. 1981.
26. Magee, F. R. and Proakis, J. G.: Adaptive Maximum Likelihood Sequence Estimation for Digital Signalling in the Presence of Intersymbol Interference, IEEE Trans. Vol. IT-19, pp. 120-124, Jan. 1973.
27. Morrison, N.: Introduction to Sequential Smoothing and Prediction, (McGraw-Hill, 1969).
28. Clark, A. P., Harvey, J. D. and Driscoll, J. P.: Near-Maximum-Likelihood Detection Processes for Distorted Digital Signals, Radio and Electron. Eng., Vol. 48, pp. 301-309, June 1978.
29. Parsons, J. D. et al: Diversity Techniques for Mobile Radio Reception, Radio and Elec. Engr., Vol. 45, No. 7, pp. 357-367, July 1975.
30. Clark, A. P., Farrell, P. G., Parsons, J. D. and Xydeas, C.: Improved Methods for the Transmission of Data and Digital Speech for Use in Cellular Systems, Progress Report on SERC Collab. Res. Project, Univ. of Liverpool, Loughborough and Manchester, UK, July 1985.
31. Adoul, J-P. A.: Error Intervals and Cluster Density in Channel Modelling, IEEE Trans. Inform. Theory, Vol. IT-20, pp. 125-129, Jan. 1974.
32. Kanal, L. N. and Sastry, A. R. K.: Models for Channels with Memory and Their Applications to Error Control, Proc. of the IEEE Vol. 66, No. 7, pp. 724-744, July 1978.
33. Draft Final Report on a Study and Development of Error Control Techniques for Mobile Communications, ESTEC Contract No. 6176/85/NL/GM, Racal-Decca Adv. Dev. Ltd. and Manchester University, April 1986.
34. Wainberg, S. and Wolf, J. K.: Burst Decoding of Binary Block Codes on Q-ary Output Channels: IEEE Trans. Vol. IT-18, No. 5, Sept. 1972, pp. 684-686.
35. Farrell, P. G. and Daniel, J. S.: Metrics for Burst-Error Characteristics and Correction; IEE Colloq. on Interference and Cross-Talk in Cable Systems", London, Dec. 1984.
36. Brysh, H. and Farrell, P. G.: Burst-b Characterisation of a Mobile Radio Channel: presented at IEEE Int. Symp on Information Theory, Brighton, UK, June 23-28th, 1985.
37. Daniel, J. S.: Double Burst Correcting Array Codes - Generation and Decoding, presented at IEEE Int. Symp. on Information Theory, Brighton, UK, June 1985.

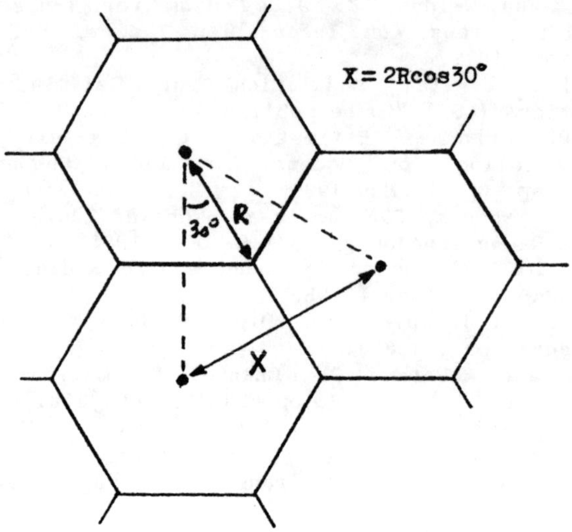

Fig. 1: Cell Geometry

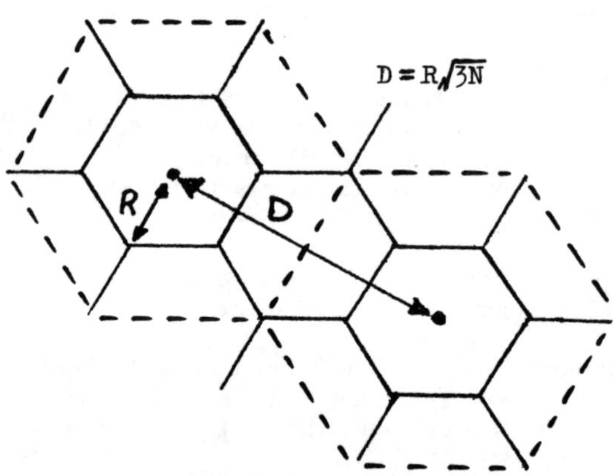

Fig. 2: Re-use Distance

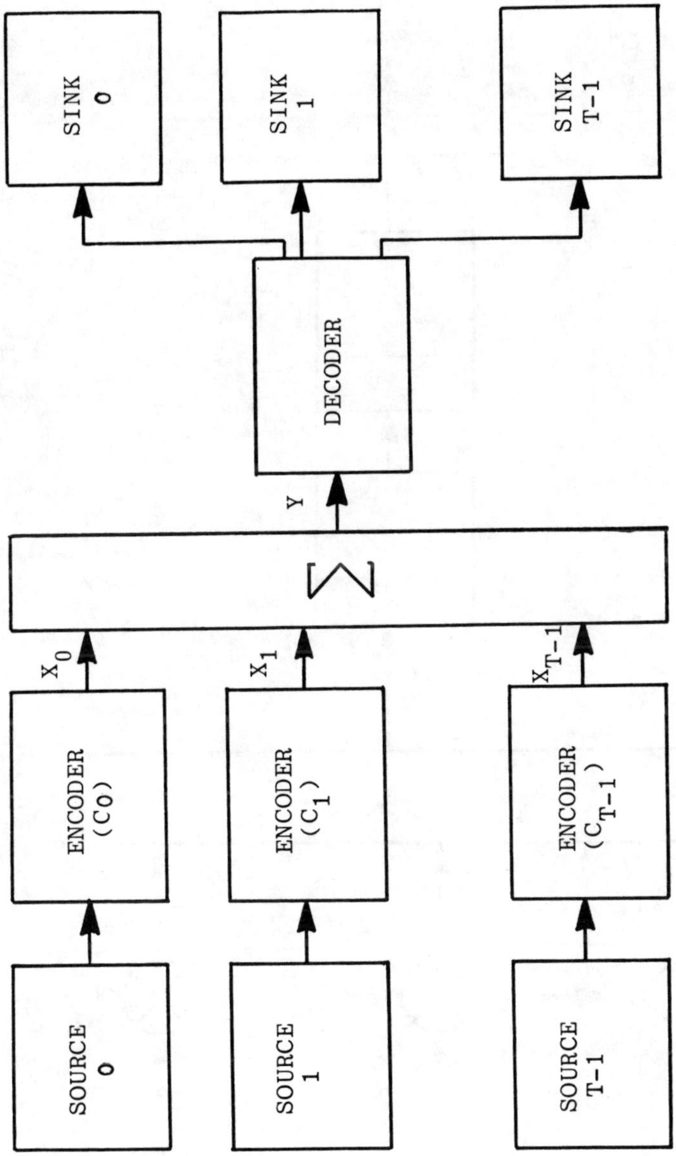

Fig. 3: A T-User Collaborative Coding Scheme

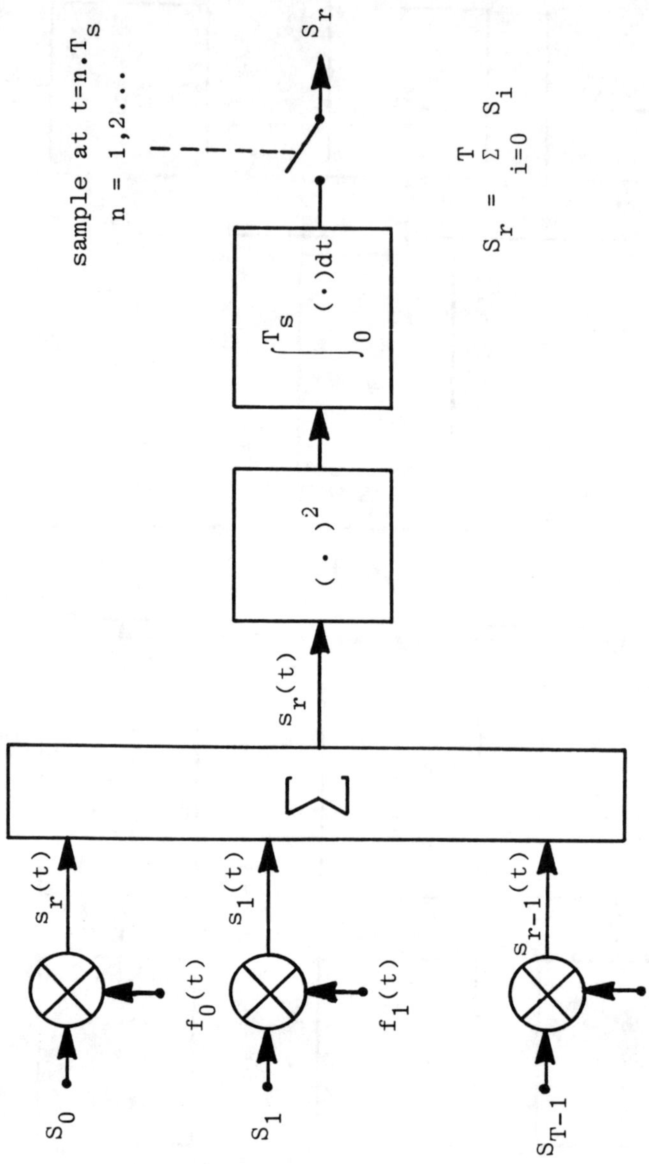

Fig. 4: Bandpass Adder Channel Implementation

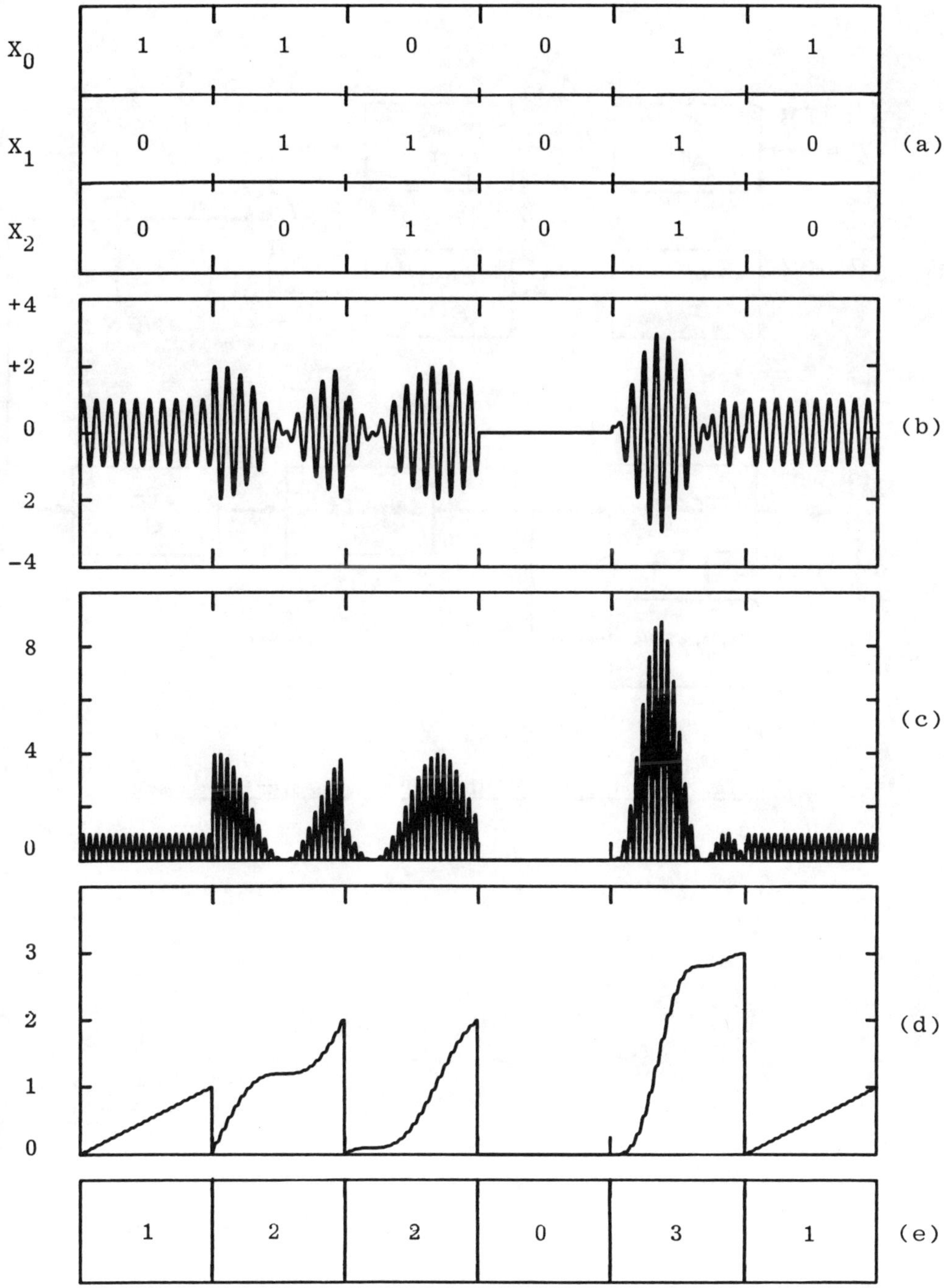

Fig. 5: Demodulator Waveforms for the T= 3 Scheme

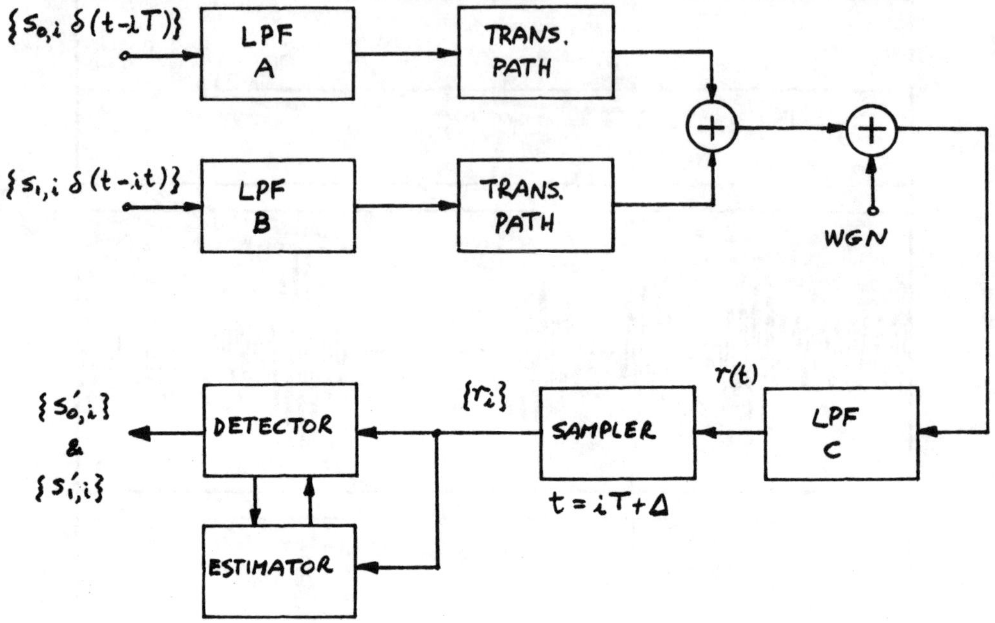

Fig. 6: Model of Collaborative Transmission Scheme

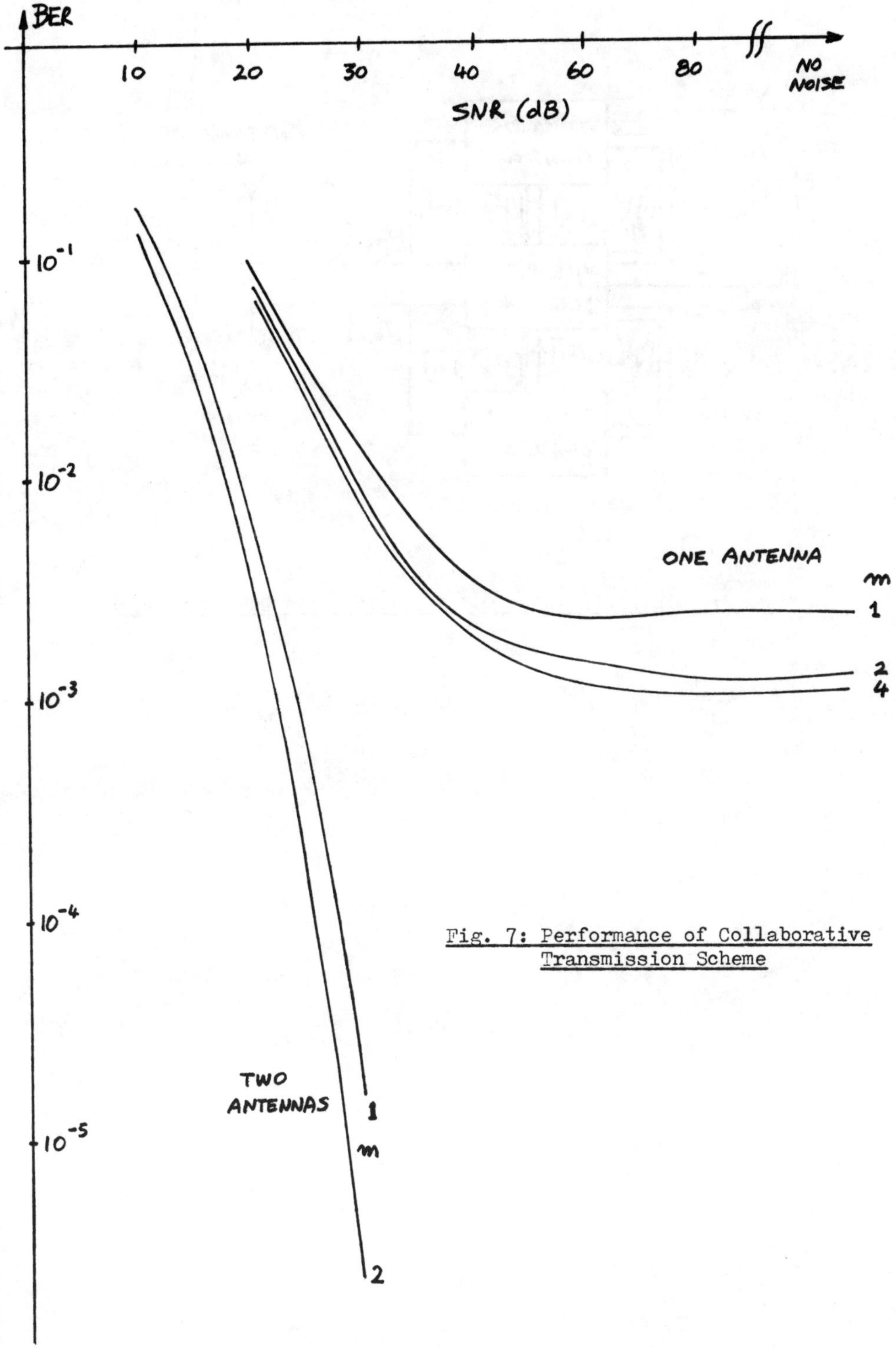

Fig. 7: Performance of Collaborative Transmission Scheme

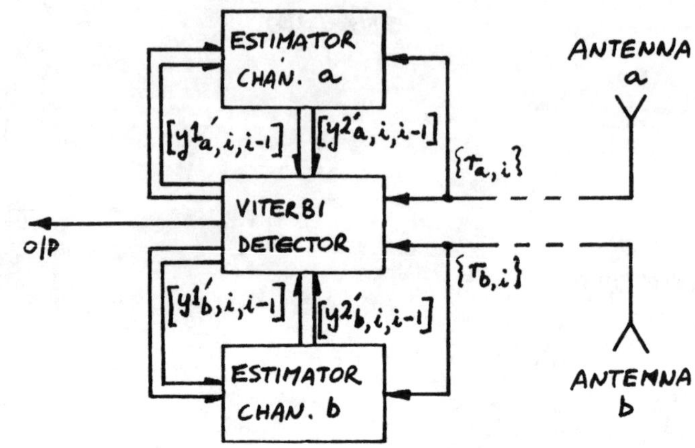

Fig. 8: Joint Estimation and Detection with Dual Diversity

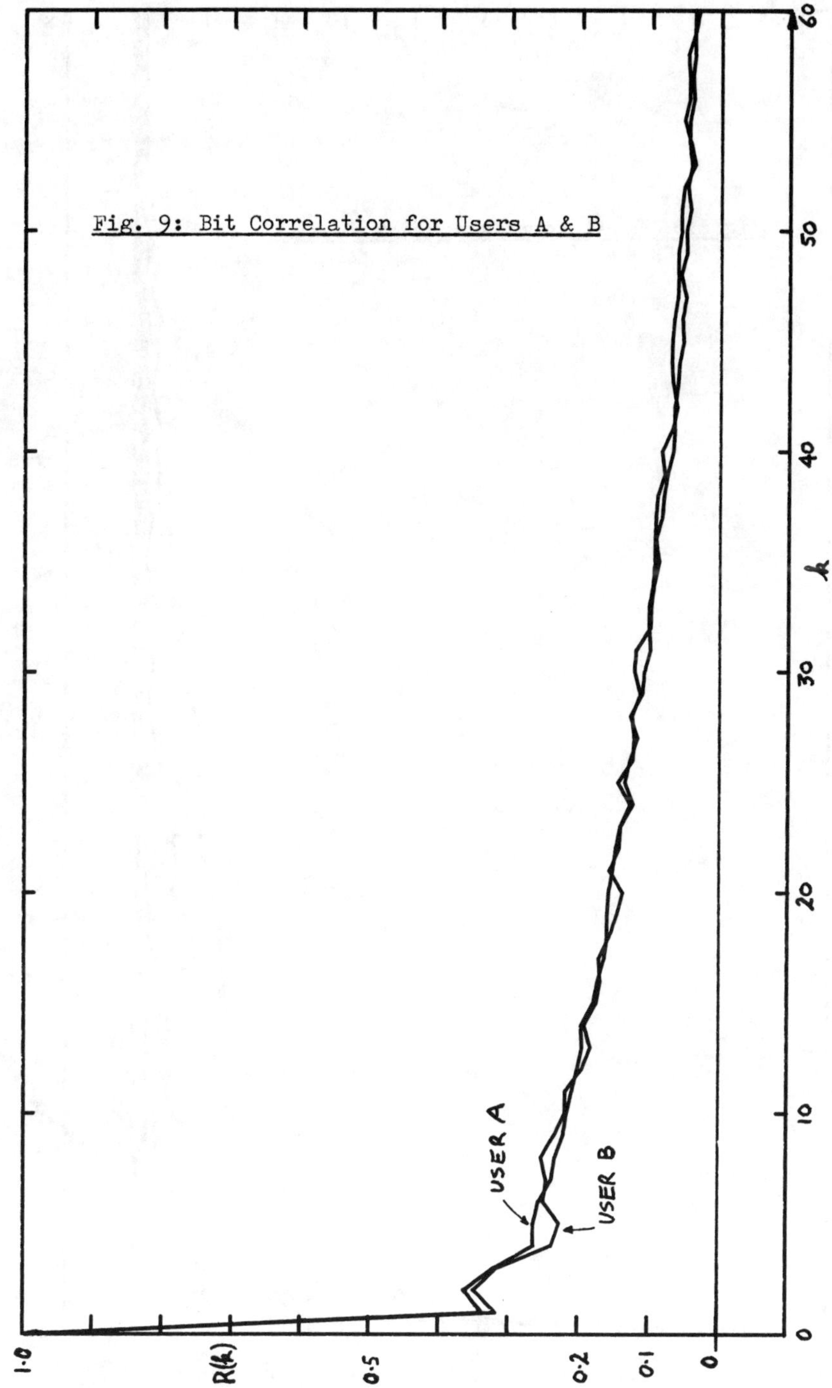

Fig. 9: Bit Correlation for Users A & B

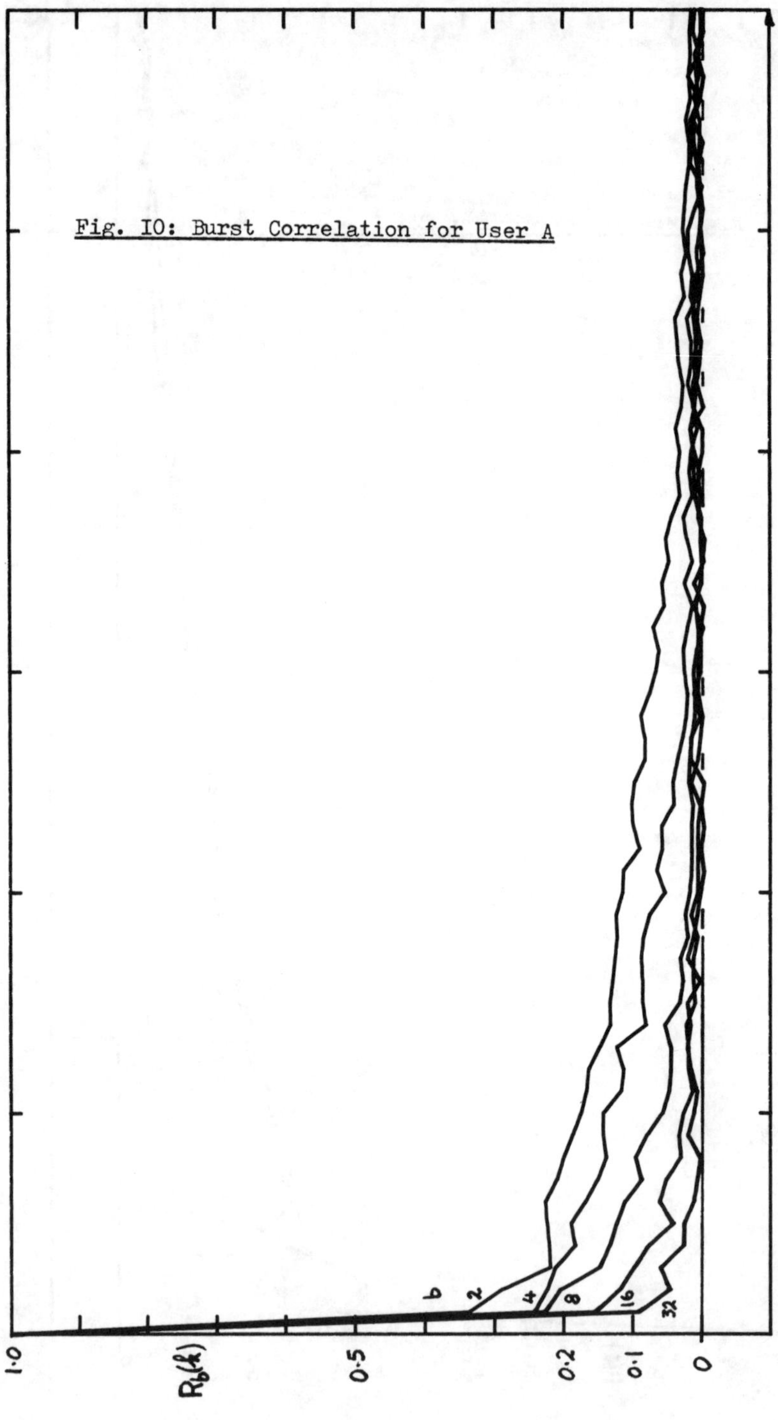

Fig. IO: Burst Correlation for User A

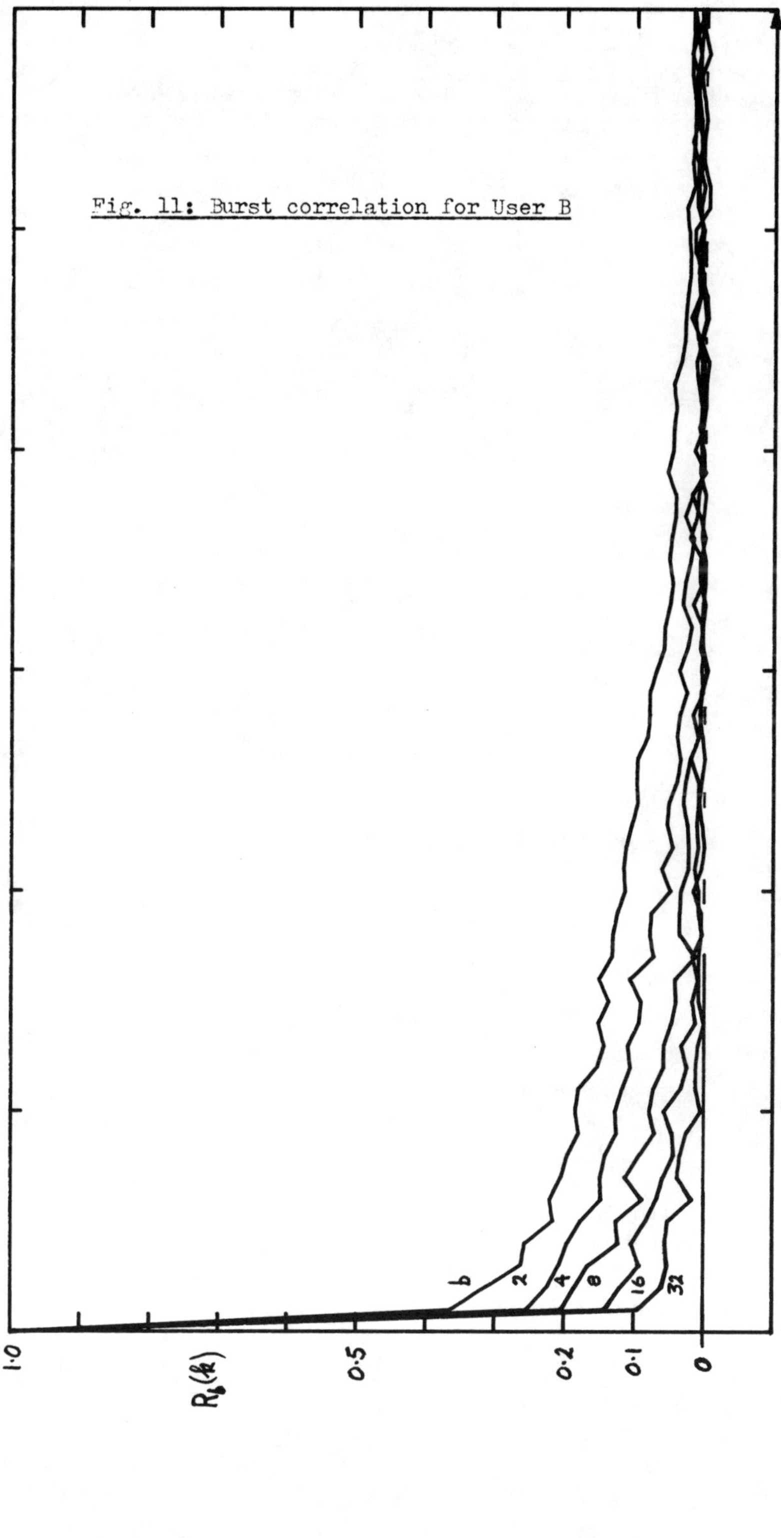

Fig. 11: Burst correlation for User B

PERFORMANCE BOUNDARIES FOR OPTICAL FIBRE SYSTEMS

J E MIDWINTER

University College London
Torrington Place
London
WC1E 7JE

I. HISTORICAL INTRODUCTION

Optical fibre transmission systems have developed rapidly from a "gleam in the eye" in 1966 to the 1st Generation Production Systems in 1980 and on to the 2nd or 3rd Generation Systems of today which offer huge performance already and promise far more. We will briefly trace this historical development and then concentrate on the single-mode fibre technology, both in terms of its present production form and of its future potential. Following from that, we will then suggest some networking implications of this radical new technology.

The first proposal to seriously use optical fibres for telecommunication transmission stem from 1966 (Ref.1) but it was 1970 before the target attenuation of 20dB/km was achieved (Ref.2). This stimulated major interest world wide and by 1975, fibre attenuations of a few dB/km had been reported and dispersion figures looked acceptable for system use. During this period, graded-index fibres were used (Ref.3) and immense effort was devoted to identifying and producing the "optimum index profile" to achieve a low level of multi-path dispersion. Pulse spreadings of less that 0.1ns/km have been reported but in practice, most production graded-index fibre has been nearer to 1ns/km because of profile imperfections.

Such values allowed the first systems to enter service at the turn of the decade in about 1980. Typically, they operated at a wavelength of about 850nm where fibre attenuation due to Rayleigh Scattering alone would be of order 2dB/km so that repeater section lengths in the range 5 to 10 km were typical at bit rates in the range 8 to 140 Mbit/s (the CEPT European digital heirarchy is used throughout the paper!).

The design of graded index fibre links is either rather approximate or exceedingly complex. We note that to predict with accuracy the pulse spreading, the following data is needed:-

- attenuation vs mode number for each fibre
- group delay vs mode number for each fibre
- mode coupling vs every mode pair for each fibre
- launched modal power distribution
- power redistribution at each splice, both for guided

J. K. Skwirzynski (ed.), Performance Limits in Communication Theory and Practice, 309–321.

and unguided modes.
- launch pulse shape
- material dispersion characteristics of fibre
- temporal spectral variation of source

Obviously, in the practical case of a fibre supporting 250-500 guided modes, very little of this will be known so that "rules of thumb" have been developed that give rise to approximate results that can be used provided suitably large margins are left. However, for high performance systems, this rapidly becomes untenable and hence we find now an increasing use of single mode fibre. In retrospect it is interesting to note that the first proposals (Ref.1) were for single mode systems and that in the late 1970s, interest revived in this technology as the shear complexity of graded-index propagation became apparent.

Two other factors hastened the shift to single mode fibre. During the 1970s, the studies of attenuation mechanisms in silica based fibres made clear that over much of the wavelength range of interest (800-1600 nm), the attenuation was dominated by Rayleigh Scattering. Hence, from a best value of about 1.2 dB/km at 850nm, it falls away to as low as a mere 0.2 dB/km at 1300nm although in most fibre design, values about 1.5 times greater are observed. Nevertheless, there was an obvious advantage in shifting to the longer wavelength. Secondly, was the fact the material dispersion in silica falls to zero close to 1300nm at the point of inflection in the refractive-index/ wavelength curve.

These developments heavily favoured the extension of repeater separations from the 1st generation values of 5-10km. Moreover, the very low dispersion associated with the zero of material (and total waveguide) dispersion in the region of 1300nm has allowed this to be coupled with higher data rates so that today we find the widespread deployment of systems operating at bit rates in the range 140 to 1200 Mbit/s and with repeater spacings of 25 to 30km. In addition, the economic advantage given by this technology means that single mode optical fibre is now the favoured long haul cable medium and other cable types have largely ceased manufacture.

We are now in a position to examine some of the limiting factors that control the use of this exciting transmission medium. In discussing this, we will first examine the fibre alone in terms of attenuation, dispersion and non-linear properties and then discuss some of the device properties to see how they interact with the fibre properties. Following this, we will examine the physical limitations so presented in a systems context and look to see what new opportunities this leads to from a networking point of view.

II FIBRE ATTENUATION

The attenuation limits for silica based optical fibres are fairly well understood today. Two effects are universally fairly well understood today. Two effects are universally known, Rayleigh Scattering from the silica glass matrix and Infra-Red absoption from the Si-O, Ge-O and other bonds associated fundamentally with the glass.

Rayleigh scattering depends on composition, with pure silica being superior to doped silica, so that some detailed optimisation in the fibre design process allows the effect to be minimised. For silica, a value of about 0.6dB/km at 1000nm wavelength is appropriate, scaling as wavelength to the minus fourth power.

The Infra-Red absorption appears as a sharply rising edge towards longer wavelength from about 1600nm onwards although again, the exact value depends on the composition of the glass. Values for Germania doped Silica are very similar while Boron doping is much worse.

Taking these two together leads to a (predicted) minimum attenuation of about 0.12 to 0.14 dB/km in the region of 1600nm wavelength (Ref.4). The best experimental value reported to date is 0.15/0.16 dB/km but more typically single mode fibres have attenuations of 0.2 to 0.25 dB/km at 1550-1600nm. Detailed optimisation continues on the fibre design and manufacturing process to combine the most favourable attenuation figures with the other necessary design features such as appropriate dispersion, mode field (spot) size and ease of manufacture and cabling.

Many factors can degrade the attenuation of a fibre. Some of the most common are water contamination giving rise to the O-H bond absorption spectrum with a strong line at 1370nm and many weaker lines (Ref.3), gaseous Hydrogen inclusion giving rise to a similar spectrum, loss arising from micro or macro bending of the fibre during cabling, effects of nuclear radiation etc. However, in properly designed and fabricated cables in normal telecommunications environments, these can all be held to a low level.

The most complex attenuation to optimise so far described has been one that is due to an interplay between many different fabrication parameters including core dopant concentration, core dopant profile, preform pulling speed and pulling temperature. The effect is most serious in fibres designed to have a large index difference between core and cladding and has led to a number of novel design approaches (Ref.5).

III FIBRE DISPERSION

The move to single mode fibre led to an immediate and dramatic decrease in the pulse spreading observed compared to graded-index fibres where it was dominated by multi-mode or multi-path dispersion, the difference in transit time between different guided modes. In the single mode fibre, this effect is absent and one is left with effects which depend upon the spectral (modulated) linewidth of the source convolved with spectral (modulated) linewidth of the source convolved with the variation in fibre transit time with optical wavelength or frequency. Detailed analysis (Ref.6) demonstrates the presence of three effects:-

- Material Dispersion
- Waveguide Dispersion
- Profile Dispersion

Material Dispersion arises from the variation of group velocity in the (bulk) fibre glass with wavelength. It is linearly proportional to the second differential of refractive index against wavelength. This is often the dominant effect.

The Waveguide Dispersion arises from the guiding structure and has no analogue in a bulk material. As the wavelength changes, so too does the spot size and hence the fraction of guided mode energy in the core. Since the core and cladding materials have slightly different optical properties, this leads to a change in "average" group velocity with wavelength. If the fibre is constructed with two significantly different composition glasses (high NA) and is operated at a wavelength where the fraction of energy in the core is changing rapidly with wavelength (close to $V = 1.2$), then this effect is large and can over-ride material dispersion. Otherwise, it can be regarded as a secondary correction factor. Fibres with large waveguide dispersion are constructed to deliberately change the dispersion characteristics away from those of the material, for example to shift the zero of dispersion from 1300 to 1500nm wavelength.

Finally, Profile Dispersion arises from the change in Index Difference between core and cladding materials with wavelength and is normally a very small correction factor. The total fibre dispersion is the simple sum of these three effects and is normally expressed in units of picoseconds pulse spreading per nanometre source linewidth per kilometre distance. Typical values range from 0 to +/- 20 in the 1300-1600 minimum attenuation region.

To give these values some context, consider a worst case of 20ps/(nm.km), operated at 1500nm wavelength in a 100km link with a poor quality laser source having a linewidth of 5nm. Such values are typical of todays production components. Then the pulse spreading or jitter will be in the region of (20x100x5ps =) 10ns, evidently a problem for a high bit rate system. Operating the same fibre near to the zero if dispersion, say less than 2ps/(nm.km) would lead to 2ns jitter under the same conditions; still very significant.

However, one should note that the source linewidth need not be greater than that produced by applying the signal modulation to a monochromatic oscillator. Suppose we apply modulation with frequencies extending up to 10GHz to a laser, giving a source linewidth of 20GHz. Then, on a single frequency optical carrier centred at 1500nm wavelength this corresponds to a carrier centred at 1500nm wavelength this corresponds to a linewidth of 0.015nm. (In the case of the normal laser, the linewidth is set by the frequency instability of the laser and the many different oscillating modes). Over the same link and taking the worst case 20ps/(nm.km) dispersion, this would now lead to a temporal separation of the different spectral components of 30 picoseconds corresponding to a maximum of 0.3 of a cycle at the highest modulation frequency. At other regions in the transmission spectrum 1300-1600nm, much lower dispersion can be expected. Evidently the medium has great bandwidth potential when used with narrow spectral linewidth sources.

Detailed fibre design optimisation involves multiple variables. From the intended system designer come specifications on the source and modulation characteristics to be used together with the desired transmission distance. The fibre structure can then be optimised in terms of core profile, index difference and core diameter to yield acceptable attenuation and dispersion. We have already alluded to the fact that simply varying core diameter and index difference changes the waveguide dispersion and this allows the zero of dispersion to be shifted from 1300 to 1550nm.

This ability meets virtually all todays requirements. However, it has also been shown both theoretically and experimentally that by resorting to more complex designs in which the central guiding core is surrounded by rings of differing index material, the dispersion curve can be flattened and held very close to zero throughout a broad wavelength region of interest. Presumably, the use of still more complex designs would allow the dispersion curve to be further flattened and held even closer to zero over a broader range. However, this has become an increasingly academic pursuit since the values already achievable so far exceed what is needed. Moreover, increasingly ther complexity of the fibre design tends to increase its production cost and reduce its yield.

A useful analysis of dispersion in single mode fibre having the normal dispersion characteristics has been carried out by Lin & Marcuse (Ref.7) which includes the unmodulated source linewidth and the broadening due to the modulation process. These are then convolved with the dispersive properties of the fibre to generate an output pulse width as a function of the input pulse width. This shows a clear minimum for any particular set of conditions of centre wavelength (and hence dispersion) and fibre length. What is striking are the numbers deduced which suggest a dispersion limited transmission capability at rates approaching 100Gbit/s over an unrepeatered section of 50-100km. Such rates are not practicable for many reasons, not least the limitations of the electronics in the regenerators or terminals but do suggest the scope remaining for further engineering development.

We should also note in passing that a modern optical fibre offers a spectral window of enormous spread for frequency multiplexed transmissions. The transmission window centred on about 1500nm wavelength is 100 to 200 nm in extent. Converting this to frequency terms leads to a spectral window of more than 20 000 GHz! In principle, this spectral width could be accessed in 20 000 units each of 1GHz in parallel and simultaneously.

IV. TRANSMITTER & RECEIVER

For reasons of engineering practicality, virtually all fibre systems use semiconductor Laser or LED transmitters and either PIN or Avalanche Photo-diode detectors. Sources fabricated from the GaAlAs/GaAs materials systems operate in the waveband 850-900nm while those fabricated from the quaternary system GaInAsP grown on InP are used to span the 1200-1600 nm wavelength region by suitable choice of composition. Typical output powers for Lasers are in the range 1 to 10mW although higher powers have been reported. Since this power can be emitted from good laser structures into a single electromagnetic mode, a high proportion can be coupled to a single mode fibre (perhaps 3dB down) while better can be achieved with multimode fibres.

LED sources are characterised by the brightness, in watts per (steradian-sq.cm) usually measured for both polarisations together. This can be converted to power per mode by multiplying by the wavelength squared. The maximum power launched into a fibre is then obtained by multiplying this figure by the number of guided modes (making due allowance for polarisation), a result which leads to launched powers for single mode fibres that are typically 20 to 30 dB down on multi-mode fibres (typically -20dBm).

LED sources find favour as cheap sources where launch power is not at a premium. They also launch temporally and spatially incoherent light, thus minimising some unwanted noise effects that are related to interference and occur particularly in multimode systems. However, Laser sources are obligatory for high performance systems. The trends in development are clearly towards higher launch powers and narrower spectral linewidths. However, this combination can be confidently expected to lead to problems of non-linear response in the fibre, as discussed below.

At the receiver, the semiconductor photodiode receivers typically require between 1000 and 2000 photons per ONE bit for 1E-9 error rate detection in a digital system. Poisson statistics would lead one to expect a sensitivity of about 21 photons per ONE bit (and zero per ZERO bit) for a perfect detector, when direct intensity OOK detection is used. The large difference arises because of noise in the photodiode and following amplifier. There is thus some 20dB improvement in receiver sensitivity available through better receiver design.

One route that has already been followed successfully in the laboratory is to exploit either heterodyne or homodyne reception, using a second laser as a local oscillator in the receiver and beating it with the incoming signal on the photodiode. This also opens up new modulation techniques in addition to the commonly used ASK/OOK, such as PSK and FSK. Of these PSK is predicted to be the most sensitive and PSK homodyne receivers have been demonstrated at 1500nm operating very close to the quantum limit (Ref.8).

The obvious problem with such systems is that they require a narrow spectral linewidth and very stable frequency optical laser source. These have only been fabricated using semiconductor lasers very recently (Ref.9) and it is still a subject of intensive research to establish precisely what constitutes the optimum engineering solution to a given system need. Simple FSK Heterodyne systems have operated using simple single-chip lasers of the DFB (distributed feedband) variety with electronic control. At the other extreme, PSK Homodyne systems generally require very sophisticated external cavity lasers with electronic servo controls operating on several variables. Such lasers only exist in one or two laboratories.

From an engineering viewpoint, homodyne reception would be desirable for a very high bit rate system since it results in a baseband signal rather than one on a high carrier frequency with the latter requiring receivers of exceptionally high frequency response. A very recent development suggests a possible way forward. The "local oscillator" laser was used to generate gain in the incoming signal fibre through Stimulated Brillouin scattering at the carrier frequency. This amplified the residual carrier on the (weak) incoming homodyne signal to generate a strong local oscillator self phase-locked to the remote transmitter, thus overcoming very elegantly the most difficult single problem control problem in engineeing PSK Homodyne systems (Ref.10). To achieve substantially better "receiver sensitivities", resort would have to be made to more efficient coding techniques, codes such as PPM, that associate more than one bit of data with each optical pulse.

In summary, it seems likely that we will see optical receivers in engineered systems within a few years that operate very close to the quantum limits of detection and typically 15-20 dB more sensitive than todays production direct-detection receivers. Transmitter powers can also rise by 10 to 20dB provided that non-linear effects in the fibre remain under control which is open to considerable doubt. However, fibre systems with power budgets greatly extended from todays values of 25-35dB can be confidently predicted.

V. NON LINEAR EFFECTS IN THE FIBRE

In pushing system performance, it is natural to seek greater transmitter power together with narrower spectral linedwidth to minimise dispersion effects, implying much more sophisticated laser sources. Such a route leads quickly to other problems with the fibre transmission medium. In the desirable 1500nm transmision window, a transmitter power of

2-3mW from a narrow linewidth source leads to stimulated Brillouin Scattering (Ref.11). This manifests itself as the backscattering into the transmitter of any additional power above the threshold power, with the result that the effective launch power is hard limited to the threshold level. However, it has been shown that there are modulation techniques for preventing this and one in particular of systems interest is coherent PSK. However the same effect can be used to advantage as noted above to generate a self phase-locked local oscillator for an homodyne detection system.

Another non-linear effect of systems interest is Raman scattering. A single high power laser beam in a fibre generates gain over a broad band of wavelengths extending some 200-400nm to longer wavelengths. Other signals in the waveband will be amplified at the expense of the first. If just two signals are present, one is attenuated and the other amplified, leading directly to crosstalk. If many wavelengths are present, as might be the case in a system designed to access the 20000 GHz spectral band discussed above, then serious cross talk could easily occur unless proper attention is given to the system design, particularly in terms of the source powers involved since all the effects are non-linear in power (Ref.12).

Another class of effects arises from the dependence of the refractive-index (or dielectric constant) of the glass on light intensity. The effect is small but can be significant for high power pulses travelling long distances and can lead to non-linear dispersive effects such as soliton propagation and intensity dependent pulse broadening. These can either decrease or increase the effective fibre dispersion depending on the sign of the normal dispersion. Perhaps more seriously, they also lead to situations in which multiple pulses travelling serially along the fibre can either exchange positions or coalesce (Ref.13). In general, such effects are expected to occur at peak pulse powers in the +10 to +30 dBm range and with pulse durations of 1 to 30 ps.. Ultimately, however, these effects will set an upper limit to the data rate that can be transmitted along the fibre. To this authors knowledge, there has not yet been any attempt to provide a unified analysis of the overall effect of all the non-linear mechanisms in establishing a fundamental upper limit to information flow.

VI. LONG HAUL SYSTEMS

The limiting characteristics of long haul fibre systems are readily deduced from the above data by applying the notions of attenuation and bandwidth limited distances (assuming linear medium response). In the interests of simplicity, we give below several "rule of thumb" guidelines that can be easily used to suggest limiting values.

Taking the typical value of 2000 photons/bit mean power for todays direct detection receivers, it leads to a receiver power in dBm of (-66+log(M)) where M is the bit rate in MBit/s and the log is to base 10. The power ratio for the system is then obtained by subtracting this from the transmitter power and by subtracting suitable amounts for launch efficiency, splice loss, margin etc. If we assume a mean transmit power of 0dBm and no other degradation, then dividing minus one times the receiver sensitivity figure by the attenuation in dB/km leads to the maximum loss limited range. For example, if we take fibre attenuation of 0.2 dB/km, then the maximum repeater spacing in kms is given by:-

$$R = (66 - \log(M))/0.2 \text{ kms.}$$

For 100Mbit/s and 1Gbits this leads respectively to 230 and 180 kms, values that are typical of many experimental reports at the present. To stretch further, we can postulate higher transmitter, perhaps +20dB, assume that we control non-linear effects and perhaps use coherent detection (homodyne PSK) at 20 photon/bit sensitivity. We would then have, for the same fibre attenuation,:-

$$R = (106 - \log(M))/0.2 \text{ kms}$$

which evaluates for the same bit rates to 430 and 380 kms. For an ultimate physical limit using silica based fibre, we might further postulate and attenuation of 0.15 dB/km which leads to distances of 573 and 506 kms respectively. Since the figures include no allowance for margins and are placing all components close to or beyond their physical limits, we may reasonably regard these targets as unlikely to be achieved. Nevertheless, more hard headed calculations have led to serious engineering endeavour to achieve repeater sections in the region of 300km for unrepeatered undersea systems use.

In the above calculations we have taken no account of dispersion effects. We ustify this on the basis that by suitable choice of fibre and source, dispersion effects can be reduced to negligible levels for all bit rates of practical interest. In practice, this may be an unduly sweeping assumption since it may well be desirable to use a less sophisticated fibre with higher dispersion or a laser having less than ideal spectral characteristics. However, to discuss these factors in any detail would involve a complex engineering design problem, weighing many factors against one another, to arrive at an estimate of the cost/benefit ratio of any particular component combination.

In the design of ultra long repeater spacing systems, such as for unrepeatered undersea systems, limitations are rapidly met in terms of the available source output power. High power lasers at the desired wavelength are generally not available although steady advance is occuring in the performance of the semiconductor sources used and no hard physical limits have been approached at present. Given superior sources, the non-linear fibre effects will become increasingly important but detailed system optimisation in such a regime has yet to be carried through.

In summary, the advent of long wavelength (1300/1500nm) single mode fibre technology has produced a revolution in cable transmission, dramtically reducing costs and opening up new opportunities for systems development. The technology already available within research labortories will provide the basis for many more years of engineering development before the fundamental physical component limits become a major factor. As a result, research interest has shifted to two other broad areas, one detailed below under "Wideband Networks" and one discussed in the following lecture under the title "Digital Optics".

VII. WIDEBAND NETWORKS

The rapid dominance by optical technology of long haul cable transmission together with dramatic advances in optical component technology have led to a surge of interest in the possibility of providing wideband switched services to private customers. Wideband could imply anything greater than the standard ISDN channel of 64kb/s but is generally assumed to mean 2Mbit/s or more in digital form and usually includes either a digital or analogue video capability. Thus one is concerned with a variety different network forms spanning a vast range of plausible new services. The network types studied include the conventional tree distributive networks for CATV applications, switched star-format equivalents for CATV with dedicated wideband services, digital rings and highways as well as more novel formats. The services spanned can vary from packet format information on time-shared rings with highway rates from 10Mbit/s to as high as several Gbit/s, to Conference TV at rates of 2Mbit/s, full colour broadcast 625 line (PAL) TV at 6.5MHz analogue or 65-140Mbit/s according to coding on through to high definition TV displaying at 1100-1200 line resolution and carried in analogue or digital rates of perhaps 20MHz or 565Mbit/s respectively.

In facing this bewildering array of present and future services, the network designer has an awesome design optimisation problem. There is no doubt that optical transmission technology can provide such services. The problems are related to finding the most cost effective technical solution offering a sufficiently attractive mix of services so that the customer will buy it. More sophisticated networks and components offering more capability will generally costing more as will the lower penetration more specialised services. Added to these uncertainties are also difficult planning questions concerned with finding ways for such new networks to evolve gracefully from the present networks which are predominantly optimised for either telephony or CATV only.

Looking at the wideband network by component, we start with the fibre cable which has low attenuation, huge bandwidth but is not particularly cheap except in cost/performance terms. In long haul applications, the opto-electronic interfaces are spaced far apart so that, although they are expensive, they do not dominate the system cost. In a wideband local network, the cable length may be a few 100 metres up to a few kilometres long, so that these costs are much more significant. Moreover, since the equipment with which they interface is normally designed for an electronic input (computer, TV, Hi-Fi etc) opto-electronic conversion is essential whereas if the cable were metallic, it might be possible to make a direct connection (as with coaxial cable into a TV set).

The emphasis on cost effectiveness generates two lines of attack, one aimed at reducing component cost and the other at seeking new opportunities inherent in optical technology that might give a competitive edge. A good example of such an opportunity lies in the "Versatile Network" concept. This can take many forms but basically rests upon the recognition that an optical network can embrace not just transmission but also switching or routing. Given that the signal remains in optical form throughout, then in many cases a transparent highway would be established between sender and receiver both in the restricted communications meaning of the word and in the wider general meaning. Moreover, in such cases, the network will be perfectly capable of carrying simultaneously many different communications in different formats and at different data rates.

To construct such networks, a variety of new components is required. From Integrated-Optics, planar guided wave technology (Ref.14), we have a variety of electro-optically controlled filters and switches which allow optical signals to be directed, under electrical control, through different guided wave pathways. Hence it is becoming possible to construct optical switching cross-point arrays or nodes that will either route all or some wavelengths within a given fibre to another. Semiconductor lasers can be made to operate at many different wavelengths and can be tuned, so that a single source can become tunable. Likewise, tunable receivers can be constructed so that it starts to be possible to exploit the optical spectrum in similar manner to the radio spectrum.

At the system level, we note that a fibre technology that allows 100-200km repeater spacing implies a "lossless infinite bandwidth" medium over short distances. As a result, it is possible to envisage a large fraction of the available system power ratio being dissipated in components such as filters, switches etc rather than more conventionally in the transmission medium.

As an example of such a future possibility, envisage a star format network with two fibres per terminal, one inwards and one outward. Each terminal is assigned a unique transmission wavelength and each receiver is tunable to allow reception of each wavelength. At the network's central node, a power splitter is arranged so that the incoming power from each transmitter is split equally between all output fibres. Network control and routing is exercised by tuning the individual receiver to the required channel. Such a network now offers some novel features.

Since the network is totally transparent from Transmitter to Receiver, only the communicating terminals need to "speak the same language". Other terminals operating at the same time can use other data rates, formats, etc. The network can support simultaneously many different forms of communication. For example, if we number the terminals and use T to indicate Transmitter and R Reciever, then setting up the following connections:- Tl to R2; T2 to Rl; T3 to R4; T4 to R3; gives two conventional bidirectional wideband connections between 1 and 2 and between 3 and 4.

Now set up another set of connections as follows:- T5 to R6; T6 to R7; T7 to R8; T8 to R5. This has established a "ring" network between 5,6,7 & 8 while 1/2 and 3/4 are still maintaining bidirectional links. Finally, we can add another form of communication to the network (using Hex numbering for our terminals to 15):- T9 to RA; T9 to RB; T9 to RC; T9 to RD. Terminal 9 is operating in a broadcast mode to terminals A, B, C & D.

Such networks would offer a new flexibility to a campus like area, giving a freedom to graft on new services to selected terminals as or when necessary simply by adding the new terminal equipment, informing the central control of its existence but making no reconfiguration of the network itself. With some semiconductor laser transmitters already tunable to many hundreds of discrete wavelengths and with 20-25dB available for power splitting, such a network could be both powerful and realistic, supporting perhaps 100-300 terminals. However, it is likely to be 5 to 10 years before such ideas have been evaluated and become established. In the meantime, a wealth of new opportunities await exploration.

VIII CONCLUSIONS

The low attenuation and wide bandwidth of single mode fibre has allowed dramatic advance to occur very rapidly in long haul high bit-rate cable system development. The technology that is already either in production or well down the development path will allow further dramatic advances in transmission data rate and repeater spacing so that most engineering activity is now devoted to establishing the optimum engineering solution to each of the many different long-haul transmission system requirements.

The use of fibres in a local network context raises a quite different set of problems and is currently the subject of intensive study world wide. Here the problems are very much concerend with the evolution of totally new network designs to carry a hitherto unavailable range of wide and narrow band services. Potential solutions range from straight developments using optical components of existing network formats to radically new network formats that fully exploit the opportunities offered by optical technology.

In most cases, the true "physical limits" of optical transmission systems are not yet closely approached, so that the design of the systems offered commercially is dictated more by considerations of cost effectiveness than by physical component limitation. It will take many years before this situation has changed radically.

REFERENCES

1. K C Kao & G Hockham, Proc.IEE Vol.113, 1151, (1966)
2. F P Kapron, D B Keck & R D Maurer, Conference on "Trunk Telecommunications by Guided Waves", IEE, London, 29 Sept.1970
3. J E Midwinter, "Optical Fibers for Transmission", Pub.Wiley Interscience, New York, 1979.
4. J E Midwinter, Proc.Roy.Soc. Vol.A392, 247-277, (1984)
5. B J Ainslie et-al, Electronics Letters Vol.18, 842, (1982)
6. M J Adams, "An Introduction to Optical Waveguides", Pub.Wiley-Interscience, (1981)
7. C Lin & D Marcuse, IEEE.Jour.Quant.Electron, Vol.QE-17, 869-878, (1981)
8. D J Malyon et-al, Electronics Letters Vol.19, 144-146, (1983)
9. R Wyatt et-al, Electronics Letters Vol.18, 292-293, (1982)
10. C G Atkins et-al, Electronics Letters, to be published.
11. D Cotter, Electronics Letters, Vol.18, 495-496 & 638-640, (1982)
12. J Hegarty et-al, Electronics Letters Vol.21, 395-7, (1985)
13. K J Blow & N Doran, Electronics Letter Vol.19, 429-430, (1983)
14. See for example "Integrated Optics", Ed. R C Hunsperger, Springer Series in Optical Sciences

DIGITAL OPTICS & OPTICAL COMPUTING

J E MIDWINTER

University College London
Torrington Place
London WC1E 7JE

I. INTRODUCTION

There has recently been a surge of interest and public speculation on the possibility that photons might start to replace electrons in computers, hence the so-called "Optical Computer". We will look here at some of the reasons that lie behind this speculation and try to assess the position from an engineering viewpoint.

We should note first that "Optical Computing" has a long history. For example, in the late 1950s and 1960s, there were large programmes, particularly in the US defence sector, studying the use of analogue optical data processing in applications such as processing high-resolution, sideways-looking airborne radar data (Ref.1). Impressive results were obtained but the widespread application of the technology was prevented by the very rapid advance of digital electronic techniques and by the difficulties of feeding data from electronic sensors into the optical processor.

Work on this class of applications continues today and finds application in the solution of particular processing problems but few digital electronic system designers feel their jobs are under threat as a result! The most likely new applications seem to lie in the area of pattern recognition or robot vision and use analogue image processing. These optical processors generally exploit the Fourier Transform relationship that exists between the near-field and far-field radiation patterns in a coherently illuminated optical system to carry out spectral analysis or correlation in some form.

More recently, optical "two-state" devices have been described in the "Physics" literature which exhibit properties reminiscent of an electronic two state logic gate. These seem to offer the prospect of digital optical processors in which the logical operations such as AND, OR, NAND & NOR are connected in serial and/or parallel to build up more complex functions as is done in the digital electronic computer. If this is to happen, then the digital optic technology must compete head-on with digital electronics and displace it, if not from all applications, then at least from some. Thus we start by recognising that digital electronics is extremely well established and advancing in speed, complexity and overall capability. Hence we must seek some powerful reasons

J. K. Skwirzynski (ed.), Performance Limits in Communication Theory and Practice, 323–332.

to support digital optics if it is to be taken seriously.

II. WHY DIGITAL OPTICS?

To answer this question, we first examine those applications where (digital!) optic systems have already firmly established a civil market niche. Two stand out clearly.

First, the Compact-Disk Hi-Fi player and Digital-Disk ROM demonstrate elegantly an optimum mix between optical storage and read/write technology on the one hand and digital electronic processing on the other, to offer a new and very superior product. Second we observe that optical fibre transmission systems have become the first choice on both cost and performance criteria for high data-rate long-haul transmission.

The CD player highlights one feature relevant to digital optics, namely the ability to communicate through space using a simple lens system. More generally, we all know from common experience that a good lens can image many million "resolvable points" from one two-dimensional array (object) to another (image). Since each "connection" from object to image can be regarded as a wideband communication channel, we see in this ability something that optics can do for which there is no electronic equivalent. Developing this theme a little further, we see that in principle, a large array of optical logic gates can be imaged to a large array of memory elements in parallel through space. This is in contrast to the electronic situation in which a single processor addresses a large bank of memory through a time multiplexed highway, one element at a time. This has led to the much discussed "Von Neumann bottleneck" and to numerous developments aimed at operating electronic processors in parallel to bypass it.

Optical fibre transmission systems demonstrate the other feature of optical systems, namely exceptionally high bandwidth or low dispersion, and remind us that the shortest pulses generated are optical and substantially sub-picosecond (circa 8 femto second). Electronic media are severely limited by comparison. Very significant pulse dispersion occurs when sub-nanosecond pulses propagate short distances on metal conductors whilst radiation and cross talk problems plague the design of complex very wideband circuits. Optics seems custom designed to provide a solution here, provided that a suitable technology can be established. Thus, although it is beyond doubt that the speed and complexity of digital electronic circuits will continue to advance, it is already clear that the processing power that can be packaged on a single chip or board is increasingly limited by communications problems. As a result, the efficiency with which single logic gates are used is falling and an increasing fraction of the chip area is being taken up by communication highways, pin connections, drivers and interfaces.

Against this background, we see two broad strengths of optics vis-a-vis electronics. The ability of simple optical systems to carry light through space and provide high resolution imaging offers an exceptional wideband parallel-highway opportunity whilst the low dispersion and high (optical) carrier frequency of light makes possible exceedingly fast pulsing, in the sub-nanosecond to sub-picosecond range. It seems certain that one if not both of these attributes must be exploited if optics is to compete with electronics and win.

To provide some points of reference, we note that a Cray computer achieves 0.1 to 1.0 G.FLOP/sec. Associating of order 100 discrete logic gate operations with a single FLOP and noting that such machines already involve much parallel operation leads value of 1E11 logic gate processing operations per second before any allowance in made for I/O. In the same manner, we note that microprocessors typically operate at down to 100ns cycle times and again, associating of order 100 logic gate operations with a single 32 bit operation, we suggest such a device implies an gate logic rate of order 1E9 per second. Systolic array processors achieve much higher rates at the expense of some versatility by using many (100-1000) simple processors in parallel. Thus electronic processors are alady pushing into the region of 1E12 logic gate operations per second and we suggest that, other things being equal, optical processors would need to achieve similar values to compete. At the same time, fibre systems have already been reported carrying 20Gbit/s of data per fibre and it is already clear that the fibre is intrinsically capable of much more.

The very short optical wavelength of light also leads directly to the very low cross talk that we already associate with optical signal channels. However, the fact that optical interactions are normally linear means that it is relatively difficult to make optical signals interact when required. By contrast, electrical signals interact strongly through the wide variety of very non-linear elements that we associate with digital electronics.

The intermediate case of light interacting with electronics or vice versa is much easier. Excellent photo-detectors abound. Electro-optic modulators with a wide range of speed and power characteristics exist, ranging from the slow but very low power liquid crystal devices through to relatively high power consumption but high speed devices based upon the electro-optic effect. Very recently, these have been supplemented by a new range of high speed modulators based upon semiconductor material which rely upon the electro-absorption effect for their action and are particularly interesting because of their potential for monolithic integration in close association with electronic components.

III. DIGITAL OPTICAL MATERIALS AND DEVICES

If we examine the relationship between the optical non-linearity of a given material and the characteristic response time for the same material, we find that the ratio of non-linear coefficient to response time varies by rather small amounts (typically 1-2 orders of magnitude about a mean over the range of interest) while for different materials, the response time may vary by as much as ten orders of magnitude (Ref.2). The result of this is that low optical switching power tends to equate with low speed and vice versa. Hence a challenge is presented to the optical logic designer to obtain devices with low speed-power product so that a powerful logic system can be established within a reasonable power budget. At present there appears to be a large gap between what is desirable and what can be achieved when any real system is contemplated.

The typical all-optical logic gate, performing the logic operations of OR, NOR, AND or NAND, consists of a Fabry-Perot etalon formed by two partially reflecting mirrors with a non-linear medium between them which exhibits intensity-dependent refractive-index (Ref.3). The device relies for its operation upon the interaction of several effects. A passive etalon has the characteristic that when resonant with the applied optical frequency, the optical field between the mirrors is greatly enhanced and the device transmits well, whilst when off resonance, the internal field is small and power is reflected. If the intensity-dependent refractive-index effect operates in such a manner as to pull the device towards resonance. This increases the intensity of the incident light which in turn increases the internal optical field, pulls the device towards resonance and further enhances the internal optical field. Positive feedback can then occur under the correct conditions leading to switching action, from OFF to ON (or vice versa on reducing the incident intensity).

Of the devices that have attracted significant interest, the non-linear mechanisms fall into three broad classes, reactive and absorptive electronic effects and thermal effects (Ref.4). The reactive materials are used at wavelengths for which they exhibit negligible absorption. The non-linear change in index then arises from a multi-photon effect involving virtual states with very short lifetimes, often of a few femto-seconds. The result is that they are capable of extremely fast switching but generally exhibit rather small non-linear coefficients. A particularly interesting class of such materials is found in the organic crystal Poly Di-Acetylene and its derivatives.

The next class of materials involves absorption of light by real electronic levels to produce a change in the carrier distribution and optical absorption within the material. A wide variety of semiconductor materials come into this category, materials such as InSb, InAs, GaAs etc (Ref.5,6,7). The changes in refractive index produced depend upon the number of photons absorbed per carrier recombination lifetime. The effect thus increases as the wavelength moves deeper into the absorption edge of the material and the absorption increases. Relatively large changes in refractive index are obtained but these are characterised by decay times that are linked to carrier recombination times and are typically many nanoseconds or microseconds. Hence these devices cannot be used for high speed (ON/OFF) switching unless some measures are taken to remove the carriers more rapidly. Doing this will reduce the effective non-linear coefficient. The materials of greatest interest in this class seem to be the III-V semiconductors because of their close association with sources and detectors and because of the wealth of growth and fabrication technology already developed in the electronics industry.

The third class of materials relies upon the change of refractive index produced by laser beam heating to provide switching. The most widely studied material in this class is ZnSe in sputtered thin-film form (Ref.8). Once again, a small region of material can be rapidly heated by absorption from a laser beam input to produce switching ON. Switching OFF relies upon the conduction of that heat away from the switched region which fundamentally implies heating adjoining material. Hence the engineering problems involved in fabricating large arrays of such devices look formidable and the useable switching times are probably in the region of 1 to 10 microsecond although a single isolated devices could probably be switched ON and OFF once in a few 10s of nanoseconds.

These devices still have formidable problems associated with their practical use. In general, they are slow and heavily consumptive of optical power by electronic standards, with optical switching energies per gate operation of greater than a pico-joule and often greater than a nano-joule. For processors operating at rates in excess of 1E12 logic gate operations per second, optimistically assuming 10% power efficiency and recognising the need for several layers of optical logic and memory, these would imply minimum optical powers in range 10-100 watts to kilowatts which is rather fearsome. In addition, the all-optical gates are effectively alltwo terminal threshold devices, leading one towards an optical analogue of "tunnel-diode" logic. The true multi-terminal optical device does not appear to have been seriously described at present. Furthermore, study of the properties of the individual gate elements indicates a wide range of rather fearsome control problems, even for single gates let alone for large arrays of devices that must work simultaneously and in close proximity. There is thus much work still to be done.

IV. HYBRID DIGITAL OPTICS

The devices described above are "all optical" in operation in the sense that they involve no overt electronic effects although in reality, many involve free carriers directly but without invoking electronic transport. Whilst for devices that must switch faster than 10ps, there is likely to be little choice but to use such an all-optical logic, for lower speeds, there may be great advantage in a hybrid approach. Here we imply by "hybrid" the deliberate and conscious involvement of purely electronic compconents in support of the digital optic elements. We note that discrete electronic devices can operate at speeds approaching 1ps. and it is primarily interconnection problems that forbid their use in complex logic. This inevitably leads one to study hybrid solutions which exploit electronic components for those operations that they are good at using optics for functions at which it excels.

At its simplest, this approach can be regarded as simply providing optical interconnection between or within otherwise all electronic chips; a line of study that is being followed by many groups (Ref.9). At a more sophisticated level, it can be viewed as the interleaving of optical and electronic technology at "chip" and/or "system" level to give radically new processing architectures. This latter approach seems to be particularly powerful and forms the basis of our own programme at UCL.

One basis for a whole family of devices is the GaAs electro-absorption modulator using Multiple Quantum Well material. This exhibits a strong and sharp exciton resonance absorption at room temperature whose central wavelength is sensitively dependent upon applied electric field (Ref.10). As a result, small electric fields can produce large changes in absorption at the band-edge and hence large changes of refractive index. This leads directly to a variety of different absorption and phase modulators which can be driven electrically either alone or via a simple gain stage and photo-detector to form a hybrid element with a highly non-linear (optical) input-output response. Invoking electronic elements opens up numerous possibilities for enhanced sensitivity and control, yet since the electrical circuit is only very local, it does not imply that the problems of wideband interconnect implicit in fast all-electronic logic will be present, since the optical beam can be used to provide wideband communication between logic gates.

This technology is still in its infancy but already some interesting results have been reported and more can be confidently expected in the near future. A particular attraction of the technology is its expected ability to allow the logical interface between optical and electrical processing to move (at the design stage) according to system need. It also promise a wide range of true multi-terminal logic devices with excellent i/o isolation, the involvement of electronic gain offers improved optical sensitivity whilst the full range of electronic control technology can be brought to bear upon the problems of set-point and sensitivity control.

V. OPTICAL WIRING

Although optics provides wideband, cross-talk free, parallel interconnect rather easily, it less easily provides complex and "designed to order" wiring patterns such as are possible on printed circuit boards or chips. Moreover, in the general optical computer case, it is very unclear how programming and

I/O will be performed. Some general ideas are emerging but it appears that many major advances must occur before anything that could be regarded as a general purpose optical computer can emerge that might be competitive against the electronic opposition.

Assuming that large arrays of optical logic gates could be established, then optics offers the possibility of performing a simple function simultaneously on a very large number of (pairs of) inputs. For example, given an array of a million AND gates, then on a single clock pulse, one could AND the corresponding points in two data arrays each consisting of 1 million (single bit) data points. Repeating this at a modest repetition rate (1 microsecond) leads to a large computation rate. However, this would be a highly inflexible processor and to reconfigure the operation from clock cycle to clock cycle looks to be a difficult task. More difficult still would be to reconfigure individual gates within the logic array. A further consequence of using such processors seems inevitably to be the need for radically new approaches to programming. What form these will take remains to be seen although some interesting pointers perhaps exist, for example the work of Huang on symbolic substitution (Ref.11).

Against this inflexibility, one has the benefit that each logic gate can, in parallel, address its own dedicated RAM, so that for our hypothetical example above, a million bits of data can also be stored in RAM in a single clock pulse. In practice it has been suggested that three such planes of active memory and logic elements should be present to avoid problems of unwanted interference and to assist in the logical timing (Ref.12). This represents a clear advantage over the normal electronic processor and is closer in concept to many electronic systolic array or pipeline processors.

Bit operations per second is not the only measure of computing power. The extent to which data can be associated is another measure and involves the product of the individual gate operations per second with the degree of fan-in or fan-out. Thus a massively parallel electronic processor can achieve a large computing power not just because data can flow through the parallel processors at great speed but also because a high level of interconnection exists between the different processors. In principle, this is possible to an exceptional degree in optics, since optical beams can cross in space without cross talk (Ref.13). However, designing an optical element to produce such a controlled yet extremely complex interconnect is more challenging. It is postulated that computer generated holograms may provide this facility but it remains to be demonstrated that useful levels of beam-formation efficiency and definition coupled with low stray-light scatter-level can be achieved, particularly in a coherent optical system where small stray light levels will cause very large intensity fluctuations.

An alternative approach to optical wiring is to look for

algorithms that are readily transformed into the optical domain rather than to try to force the optical domain to copy the electronic. One example of such an algorithm uses the perfect-shuffle wiring pattern (Ref.2). In terms of a linear array of data points, subscripted 1 to 2n, the perfect shuffle operation reorders them in the sequence:-

1,n+1, 2,n+2, 3,n+3, ...n,2n

The observant reader will note that is the result of splitting the original sequence at mid point and interleaving the two half sequences. This operation is carried out optically on a two dimensional data array by magnifying the array by a factor of two in one dimension, shearing the image in two and overlaying the two half images.

This can be visualised in image terms through the following set of "data array" patterns.

1. Original data input image	abcdefgh ijklmnop qrstuvwx	
2. Expand and split image	a b c d i j k l q r s t	e f g h m n o p u v w x
3. Shear and overlay image to yield shuffled array	aebfcgdh imjnkolp qurvswtx	

The power of this "wiring operation" was first realised many years ago by Stone (Ref.14) who studied its application to Sorting, Fast Fourier Transform and Matrix operations. In the optical domain, apart from being readily implemented using lenses and prisms which are capable of very high resolution and low stray light levels, it interacts with logical processing modules that never require more than nearest neighbour data exchange and which are thus ideal for ultra-fast optical implementation. Moreover, because of the imaging properties of lenses, we note that the shuffle wiring pattern involves zero time skew across the image field, something that would be virtually impossible with a large electronic data array. Note also that this process associates logical data address with spatial image position and in this way exploits 2D imaging optics. There are many other similar operations in 2D matrix processing that are equally easily reduced to "hard wired" imaging operations, such as shifting, rotating and inverting and which con be done on a large data set in a single (optical) clock pulse.

The existence of the shuffle "wiring operation" encourages one to seek applications for Sort or FFT processors. An obvious example for the former would be an ultra-fast packet switch, for use in a very high data rate packet transmission network (Ref.2). Packets might enter the "switch" with their optical address headers in MSB-first order. Imaging the address bits onto the first row of the logic array allows the optical logic to compare the addresses, bit by bit, on each pair of input fibres and to perform the necessary straight through or exchange operation on the following data in the packet so that they are sorted by address. After streaming through the perfect shuffle optics, the packets impinge on the second logic row and again an address comparison is made and logical exchange/bypass operation performed. For 128 input fibres, 64 logical sub-assemblies are required per row with 43 rows interspersed by 42 perfect shuffle cycles before the packets emerge sorted into the 128 output fibre array in ascending order of address. Given that the individual fibres might carry data at 10.Gbit/s per fibre, the power of this simple processor concept is readily apparent! Needless to say, much of the optical infra-structure to assemble such a system does not yet exist although the individual component specifications can be set down to a limited extent and appear to be within sight, say in five years. Such a processor is a direct analogue of many electronic pipe-line processors but has the attraction that it lends itself to optical implementation using a folded optical system.

VI. CONCLUSIONS

We cannot conclude that optics will displace electronics either in digital logic or in computing in the foreseeable future, nor is it easy to see how optics can seriously challenge the general purpose digital electronic processer, largely because of the restrictions imposed by optical interconnects and the problems of programming such machines which have not yet been seriously attacked. We can confidently predict that optical interconnects will be used extensively to relieve communication bottlenecks within or between chips on processor boards and we believe that hybrid opto-electronic processors will carry that process further, moving the centre of gravity of the processing system more into the optical domain by the use of the type of technology suggested in Paragraph IV. We also conclude that specialised hard-wired processors may emerge in response to special needs, the postulated packet switch for example or a specialist matrix processor. However, in all cases that we have identified to date, we conclude that a modest level of optical parallelism (say 100-1000) operated with clock speeds at the frontiers of electronic technology must be achieved before a competitive package has much chance of emerging.

VII. REFERENCES

1. See for example Preston K, "Coherent optical Computing", McGraw-Hill, New York, 1972.
2. Midwinter J E, "Light Electronics, Myth or Reality", Proc.IEE Pt.I Opto-electronics, to be published.
3. Smith S D, "An introduction to optically bistable devices and photonic logic", In the volume "Optical Bistability and Photonic Logic", Ed.S D Smith, B Wherrett & A Miller,

Pub.Royal Society, London 1984.

4. Smith P W, "Applications of Optical Switching and Logic", Bell.Syst.Tech.J. Vol.61, p.1975, 1982
5. A C Walker, et-al, "InSb devices: transphasors with light gain, bistable switches and sequential logic gates". As Ref.3.
6. E Garmire, C D Poole & J A Goldstone, "Bistability experimentally observed at 3mW in InAs and theoretically predicted for a new class of nonlinear dielectrics". As Ref.3.
7. H M Gibbs et-al, "Semiconductor non-linear etalons". As Ref.3.
8. M R Taghizadeh, I Janossy & S D Smith, "Optical bistability in bulk ZnSe due to increasing absorption and self focussing", Appl.Phys.Letters Vol.46, p.331, 1985.
9. J W Goodman et-al, "Optical interconnections in micro-electronics", Proc.IEEE Vol.72, p.850, 1984
10. T H Wood et-al, "High speed optical modulation with GaAs/GaAlAs MQWs in a pin diode structure", Appl.Phys. Letters Vol.44, p.16, 1984
11. C H Brenner & A Huang, "Optical processor based on Symbolic Substitution", Topical Meeting on Optical Computing, Incline Village, Nevada, March 18-20, 1985, Proceedings published by OSA 1985.
12. B Wherret, Heriot-Watt University Private communiction
13. D Psaltis & N Farhat, "Optical computing and the Hopfield model", As Ref.11.
14. H S Stone, "Parallel processing with the Perfect Shuffle", IEEE Trans.Comp. Vol.C-20, p.153, 1971
15. For general review of Optical Logic materials and devices, see the book by H M Gibbs, "Optical bistability, controlling light with light", Pub.Academic Series on Quantum Electronics, principles and applications, New York, (1985)

Part 3.
Limits in Modelling and of Characterisation of Communication Channels

ROBUSTNESS AND SENSITIVITY OF COMMUNICATION MODELS

K.W.CATTERMOLE

Dept. of Electronic Systems Engineering, University of Essex, Colchester, England

1. The deficiencies of models

For the purposes of analysis and design, communication channels and networks are customarily represented by models which are, deliberately, abstractions from and simplifications of the real entities. We may variously consider a model to be an ideal which real systems should copy as closely as possible, or an approximate representation of a pre-existing real system. Whether we take the Platonic or the Aristotelian view, an important property of the model is that it shall correspond with the real system in respect of those features which are of dominant importance in defining qualitative aspects of behaviour and quantitative measures of performance. Another important property is that the model shall be sufficiently definite and sufficiently tractable to admit mathematical analysis, computation or simulation from which properties and performance may be estimated. In pursuit of clarity and tractability, it is likely to depart from reality in either or both of two ways.

Firstly, factors believed to be of secondary importance may be ignored in order to simplify the model. There is a problem of circularity here: only with the aid of some insight into the workings of the system can we identify the important and the unimportant factors. Moreover, the relative importance depends on the aspect of performance which is being estimated. As an obvious example, one might model a transmission line as lossless when designing a matching stub, treat the overall loss as dominant when analysing a long-distance transmission system, and partition the loss into dissipation and radiation when investigating an interference problem.

Secondly, some factors may be variable or uncertain. If they are believed to be of secondary importance, they may be ignored; or perhaps some specific assumption will be made, as a basis for analysis. If they are known or suspected to be significant, then several approaches are possible, none without its problems.

(a) Make a specific assumption. If the model is to be useful, then two conditions must obtain: the assumption must be a reasonable approximation to practice, and it must not be too critical, i.e. the performance measure of interest should not vary radically in response to small departures from the assumed state of affairs. Either or both of these conditions may be difficult to verify.

(b) Assume a set of conditions (for example, a range for a parameter) and require the system to meet any condition in that set. This is a very common practice; a communication channel is often required to convey any signal waveform within a given frequency range and not exceeding a given peak amplitude. The main problem is that design for the notional worst case, which might never occur, can be unduly expensive and perhaps less than optimal for cases of common occurrence.

(c) Assume a probability distribution for the variable quantities and make statistical estimates of performance measures. This is very commonly done in communication engineering, and indeed statistical concepts are firmly embedded in familiar measures such as signal/noise ratio, mean error rate, etc.

J. K. Skwirzynski (ed.), Performance Limits in Communication Theory and Practice, 335–352.

There are two problems with this approach. Firstly, a statistical specification of performance may be insufficient: perhaps we would also like to set some bound on 'worst cases'. Secondly, there may be some higher-order uncertainty. Everything we have said about uncertainty and variability of determinate properties also applies to statistical distributions. The form of distribution may be unknown or uncertain: even with a plausible form, the parameters may be liable to fluctuation. So if any specific assumption be made about the distribution of a variable quantity, we would like to know that it is a reasonable approximation to the practical state of affairs, and that departures from it are not too critical.

(d) The model may be given more variable parameters that we ultimately wish to incorporate; some of these parameters may be estimated from practical data, and thereby eliminated. These parameters may be determinate quantities, or parameters of probability distributions. The difficulty with this approach is that quite a large amount of data may be needed to make a reliable estimate, especially if we are concerned with tails of probability distributions. Given that the parameter estimates may themselves have a substantial variance, we would again like to know whether the parameter values are critical.

The obvious practical question , how well does a given model fit the real situation, is usually difficult to answer. Often it may be replaced by the more tractable question: what happens if some assumption of the model be relaxed or varied? If it turns out that the results of interest are not too sensitive to the most likely kinds of departure from the assumed properties, we may feel some confidence in applying the model to real situations with some fortuitous variation.

2. Sensitivity and extension of models

Our subject is, then, the effect on a model of relaxing some assumed requirement, or changing a parameter or other property. Much of the discussion in the previous and present sections would be equally applicable to many fields of science and technology. One distinctive feature of communication engineering must however be noted. Theoretical communication engineering makes extensive use of statistical analysis. Our channels are subject to random perturbations in the form of noise or fading: the symbolic signals they convey are drawn unpredictably from a known repertoire and are well modelled as random variables: the traffic demand on a system, often drawn from many independent sources, is well modelled as a random process. With this background, it is not surprising that many of our exemplary problems are equivalent to the question: what happens if some probability distribution in the model be changed?

Before citing the many and varied examples which carry the main weight of exposition, we summarise the principal techniques which appear to be useful in extending or modifying simple models.

(a) A model first established in a restricted case may be valid over a much wider range of conditions. There are several examples in the theory of traffic and queueing, two of which are cited in section 3. It is, of course, very common in mathematics that a theorem first established in some special case turns out to be more general.*

* For examples and discussion see Polya(1) and Lakatos(2).

(b) In a model whose exact properties depend on a choice of probability distribution, it may be possible to establish distribution-free bounds. For this to be most useful, we need upper and lower bounds which are not too far apart. Not all problems are susceptible to this approach; but for those which are, the moment-space technique (which usually gives both bounds) may be more fruitful than methods giving a single bound or an asymptotic approximation. The are many examples in the theory of digital signals and channels; one is given in section 4.

(c) It may happen that a problem can be solved not just in one special case but in two or more. It may then be possible to conflate two models or to interpolate between them. One example, discussed in section 5, is the combination of thermal noise and quantum noise in an optical communication channel.

(d) A well-established model may be taken as a central case from which small perturbations are made, and the sensitivity to these perturbations calculated. This is an especially common and valuable approach. It underlies many techniques of approximate estimation by means of power series or series of orthogonal functions, for example in the analysis of non-linear systems in terms of their departures from linearity (section 6). As another and perhaps more complex example, consider the problem of queueing networks. The ideal Markovian network with Poisson arrivals and departures is readily soluble. What if the arrivals and departures are not quite Poisson processes? This is a rather likely practical condition, for the Poisson process is a limiting process, approached by the superposition of many independent point processes of more general form. We can investigate the effect of perturbation in the distribution of service time or interarrival time, with interesting results (section 7). The perturbation approach may be useful if some parameter of the model is being estimated from a data sample (as discussed in section 1 above): for then the estimated model may be taken as the central case, and the effect of perturbation gives directly the sensitivity to inaccurate estimation. An estimation problem in a cellular radio system is discussed in section 8.

3. Generalisation of a model: examples from traffic theory

Many traffic distributions are equilibrium state probabilities of a Markov-process model with states $i = 0, 1, 2, \ldots$, state probabilities p_i, upward transition probability densities b_i (from state i to state $i+1$) and downward transition probability densities d_i (from state i to state $i-1$).* These constant transition probability densities imply negative-exponential distributions of interarrival times and service times. It is easy to show from this model that, with an unrestricted number of states, a constant arrival probability and a departure probability proportional to the state number (which may represent, say, the number of telephone calls in progress), the state number has a Poisson distribution. However, a different model predicts this same distribution if we postulate constant service times: for then the current state number is just the number of arrivals during a service time. In fact, the state distribution is independent of service time distribution, as may be shown**

* Described in most texts on traffic theory, for instance ref.(3) section 5.2 or ref. (6) section 3.8.

** Ref. (4), volume 2 p. 102.

by yet another model, namely random filtering of a Poisson point process. Similar remarks apply to the Markov-process model with the transition probabilities cited above and a finite number of states, which yields the Erlang distribution

$$p_r = \frac{A^r/r!}{\sum_{i=0}^{n} A^i/i!} \tag{1}$$

This is a curtailed Poisson distribution, and approaches Poisson as the number of states n increases. Erlang's original derivation was based on negative-exponential holding times: he himself derived some results for constant holding time by a completely different argument: and the Erlang distribution was eventually recognised as independent of holding time distribution, for a given mean. *

As another example of generalisation of conditions attached to a model, we note that historically many results of queueing theory originally derived for first-in-first-out queues were later found to apply to work-conserving queues with other disciplines: the main requirements being that service time should be independent of service order, and service order independent of service time. **

4. Distribution-free bounds: moment space and digital transmission

Many communication-system models are statistical, and consequently incorporate some assumption about a probability distribution taken to describe a signal source, or a noisy or fading channel. Sometimes, as with the assumption of additive Gaussian noise, the distribution is a standard one and its validity is well attested. In dealing with some classes of signal, however, there is no consensus as to the best distribution, and none of the rival candidates is really well validated. The amplitude distribution of speech signals is a good example. *** Reputable authorities have variously used a Gaussian distribution: a negative-exponential: a gamma distribution of order ½: and a compound distribution based on an ensemble of Gaussians. These are only the most popular distributions: other examples could be found, and moreover it would be agreed that the best-fit distribution depends on the period of observation. There is the further problem that voice-band channels are often used for transmission of non-voice signals with yet other statistics. Clearly, therefore, one would like to be able to analyse and design a speech channel without too critical dependence on the assumed statistics of speech waveforms.

This problem arose at an early stage in the development of pulse code modulation systems for telephony. The analysis of then novel forms of distortion, such as the quantising error in a companded system, appeared to require some assumption about the speech amplitude distribution. With a truly logarithmic compandor,

* Ref. (5) reprints an English translation of Erlang's paper of 1917 (pp.138-155); Brockmeyer's commentary (p.103) elucidates the assumptions about holding time.

** The earliest publication on this topic is by Kleinrock (7): for a good textbook presentation see ref. (6) section 3.7.

*** For a discussion and experimental evidence see Richards(8).

signal/noise ratio would be constant; but of course a truly logarithmic compandor is not a physical possibility. Many investigators spent time and effort on analysing various approximately-logarithmic characteristics with various postulated speech distributions.* Eventually, it became apparent that with suitable near-logarithmic encoders the signal amplitude distribution was not too critical, and distribution-free bounds were discovered which will be presented below.

Firstly we should note, however, that a significant practical problem is unlikely to be so non-critical that truly distribution-free bounds will be close enough to be useful. A more realistic hope is that the effect of the distribution will be reduced to one parameter; in this example the signal/noise ratio is bounded by two functions of signal power. The moment-space technique lends itself very well to this type of bound.** It is perhaps less widely known than it deserves, so we describe it briefly.

Let $m_i = E[g_i(X)]$, $i = 1, 2, \ldots n$ be statistics of a random variable X with some probability distribution F(x). For any given F(x) the coordinates $\underline{\underline{m}} = (m_1, m_2, \ldots m_n)$ define a point in an n-dimensional Euclidean space. The set of points corresponding to all possible F(x) is clustered in a convex set, and it is this property which enables bounds to be defined. Let the coordinates $y_i = g_i(x)$ define a curve in Euclidean space, with x as its parameter. Then all points $\underline{\underline{m}}$ corresponding to probability distributions lie in the convex hull of the curve, i.e. the smallest convex set containing the curve. If we then take any moment m_i as parameter, the possible values of the other m_j are bounded above and below.

For the p.c.m. example, let $g_1(x)$ define a companding characteristic, say

$$g_1(x) = \left[\frac{1 + \log A}{A}\right]^2 \qquad 0 \leqq x \leqq 1/A$$

$$= x^2(1 + \log A)^2 \qquad 1/A \leqq x \leqq 1 \tag{2}$$

in which case $E[g_1(X)]$ is the mean squared quantising noise. Let $g_2(x) = x^2$, so that $E[g_2(X)]$ is the mean squared signal. Figure 1 shows the parametrised curve, which is also the lower bound to noise, and the upper bound A is found by completing the convex hull. From these curves,

$$\frac{\text{Noise power}}{\text{Signal power}} \leqq \frac{(1 + \log A)^2}{3N^2}\left[1 + \frac{1 - S^2}{A^2S^2}\right]$$

$$\geqq \frac{(1 + \log A)^2}{3N^2} \max\left[1, \frac{1}{A^2S^2}\right] \tag{3}$$

where S^2 is mean signal power and N is the number of quantum levels.

* Experto crede : I was among them, before I found the bounds given in ref. (9) section 3.2. This book gives many examples, further references and a full discussion.

** The moment-space technique was originated by Dresher (10) and introduced into communication engineering by Yao (11). For a textbook treatment see ref. (4) volume 2 pp.47-51.

Upper bound B is the distribution-free bound first established by other means.*

For other examples of moment-space bounds see (4), (11) and their further references. It is likely that the method could be widely applied to signals and perturbations of uncertain statistical properties.

5. Conflation of two models: an example from optical transmission

Suppose that the solution to a problem depends on a probability distribution, and that results are known for two specific distributions. How far is this a guide to the solution for other distributions? There are several possible cases.

(a) If two radically different distributions give the same solution for some statistic, this suggests that the solution may be distribution-free. We have quoted in section 3 the case of Erlang's solution for telephonic blocking probabilities, which was demonstrated first for negative-exponential holding times, then for constant holding time, and finally for an arbitrary distribution.

(b) The negative-exponential distribution and the constant time interval have coefficients of variation of 1 and 0 respectively. In traffic and queueing theory, results for these distributions may be considered as bracketing the results likely to be obtained with intermediate values of the coefficient of variation. Sometimes it is possible to derive an exact or approximate interpolatory theory with coefficient of variation as a parameter; we discuss this further in section 7.

(c) In more complex cases, the difference in distribution may arise because one is contemplating a fundamentally different model. Interpolation is then less obvious. However, it may still be possible to conflate the models in a suitable way. We illustrate this by reference to the problem of an optical receiver which is required to achieve optimal performance in the presence of both thermal and quantum noise.

A signal conveyed over a channel with additive noise has a classical solution for its optimal receiver: the matched filter. ** A signal element of waveform $h(t)$ with white additive noise yields the optimum signal/noise ratio if the receive filter has a time-reversed impulse response of the form $h(-t)$. Specifically, a maximum-likelihood decision between two signal elements $h_1(t)$, $h_0(t)$ at a time epoch T requires a filter with response $g(t)$, where

$$g(T-t) = h_1(t) - h_0(t) \qquad (4)$$

A signal conveyed over an optical channel is, fundamentally, a variation of mean rate imposed on a stream of random photons; that is, a non-homogeneous Poisson process. The randomness leads to a signal-dependent

* Ref. (9) pp.135-137. This reference also gives other bounds for quantised systems which could now be tightened somewhat using the moment-space technique: especially the hyperbolic compandor on p.145, which I then recognised as 'probably not the strongest bound'.

** Described in many texts on communication theory, including ref.(4) volume 2 pp.159-165.

quantum noise. A maximum-likelihood detection statistic may be formed by a filter with a time-reversed logarithmic characteristic

$$g(T-t) = \log h_1(t) - \log h_0(t) \tag{5}$$

where again $h_1(t)$, $h_0(t)$ are binary element waveforms. * What is the optimum filter in the presence of both thermal and quantum noise? A solution, due to House,** uses a Chernoff bound to the probability of binary errors induced by noise. This is expressed in terms of the log moment-generating function $\psi(s)$ of the total noise, and is

$$p_e \leq \exp[\tfrac{1}{2}\psi_0(s) + \tfrac{1}{2}\psi_1(-s)] \tag{6}$$

where ψ_0, ψ_1 pertain to the two signal values respectively. These functions depend on the receiver's impulse response g(t). It may be shown that the bracketed expression in (6) is

$$\int \left\{ h_0(t)[e^{g(t)}-1] + h_1(t)[e^{-g(t)}-1] + \tfrac{1}{2}N[g(t)]^2 \right\} dt \tag{7}$$

where the first two terms in the integrand relate to the quantum noise and the third term to additive white noise of spectral density N. The filter can be chosen to minimise error by differentiating (7) with respect to g and equating to zero,

$$h_0(t)\,e^{g(t)} - h_1(t)\,e^{-g(t)} + N g(t) = 0$$

The solution appears very clearly as an interpolation if we plot against g the two quantities

$$y_1 = h_1 e^{-g} - h_0 e^{g} \quad , \quad y_2 = Ng$$

as in Figure 2: the intersection $y_1 = y_2$ defines a value of g. If N is very small, the y_2 line is almost horizontal and thesolution tends towards the Poisson solution (5): if N is very large the y_2 line is almost vertical and the solution tends towards the matched filter (4). Similar methods can be extended to deal with more complex cases, including non-white thermal noise and compound quantum noise due to photo-multipliers.

* Originally derived by Bar-David (12). See also ref. (13)section 3.2.

** Ref.(14): also House's doctoral thesis (University of Essex, 1979) and ref. (13) chapter 3.

6. Sensitivity to perturbations: orthogonal series

A classical method of dealing with small perturbations from a central case assumed to be well known is to expand some characteristic in series. If the effect of each term can be estimated, and the series does in some relevent sense converge rapidly, useful results may be obtained. Much analysis of small non-linearities in approximately linear systems is done in this way, using for example a power-series representation of the non-linearity. Other types of series, notably series of orthogonal functions, may be more powerful, and in particular capable of dealing with more drastic non-linearities.*

The general principle of orthogonality is well known. A set of functions $q_i(x)$, $i = 0, 1, 2, \ldots$ is orthogonal under the weight function $w(x)$ if

$$\int q_i(x)\, q_j(x)\, w(x)\, dx \begin{cases} = 0 & j \neq i \\ = k_i \neq 0 & j = i \end{cases} \tag{8}$$

and is also orthonormal if $k_i = 1$. Orthogonal functions are known for many weight functions including the Gaussian distribution (which leads to Hermite polynomials), the negative-exponential distribution (Laguerre polynomials) and several other probability distributions of common occurrence. A function $f(x)$ may be approximated in mean square by a superposition of orthogonal terms

$$f(x) \approx \sum_i c_i q_i(x)$$

$$c_i = \int q_i(x)\, f(x)\, w(x)\, dx \tag{9}$$

There are two well-known applications of such orthogonal expansions. Suppose firstly that we wish to calculate some statistics of a random variable X with distribution $w(x)$. Any statistic is the expectation of a function of X. Let it be $E[f_1(X)f_2(X)]$ where f_1, f_2 can be represented by series of the form (9) with coefficients b_i, c_i respectively. Then

$$E[f_1(X)f_2(X)] = \sum_i \sum_j b_i c_j \int q_i(x) q_j(x) w(x) dx$$

$$= \sum_i b_i c_i \tag{10}$$

where we have eliminated terms with $i \neq j$ by orthogonality. This is very

* An extensive treatment of this approach, with many examples and references and a full list of orthogonal functions, is given in ref.(3) chapter 9.

useful for estimating the effects of a non-linearity on a random process; for we can take X as the input variable, and f(X) as the output variable from an instantaneous non-linear device. Moreover, using the theory of diagonal processes (which are random processes whose bivariate distributions have simple orthogonal expansions) we may estimate the joint statistics of two samples from such outputs, and hence derive the autocorrelation and power spectrum of the output. *

The second well-known application of orthogonal expansions is to the modelling of probability distributions. Suppose we have a distribution y(x) which differs slightly from a standard distribution w(x). Perhaps it can be expressed as a series

$$y(x) = \sum_j b_j q_j(x)\, w(x) \tag{11}$$

where the $q_j(x)$ are orthogonal with weight w(x). If w(x) is Gaussian and the $q_j(x)$ are Hermite polynomials, this approach leads to the Gram-Charlier and the Edgeworth series, which are similar within a rearrangement and differ somewhat in their convergence properties. They have been used to model empirical distributions, and also to investigate the convergence of sums of random variables towards the Gaussian central limit. ** This latter topic is highly relevent to perturbation analysis. One of the main reasons for the common occurrence of the Gaussian distribution is its property of being a limiting distribution: the sum of many small random variables tends towards the Gaussian. But one of the main practical causes of perturbation is that a random variable arises as the sum of a number of contributions which is large but finite; so that the distribution is near-Gaussian but perceptibly different. One would like to know how far this difference affects the statistics derived from the model. For central statistics, the orthogonal series is probably the best approach: for tail probabilities, the Chernoff bound is useful.***

It is very convenient for perturbation theory that these two applications of orthogonal expansions, to the representation of a statistic E[f(X)] and a distribution y(x), can be conflated. Suppose that the function f(x) has an expansion (9), and the distribution y(x) has an expansion (11). Then

$$\begin{aligned} E[f(X)] \text{ under distribution } y(x) &= \int f(x)y(x)dx \\ &= \sum_i \sum_j \int c_i b_j q_i(x) q_j(x) w(x) dx \\ &= \sum_i c_i b_i \end{aligned} \tag{12}$$

* Diagonal process theory was introduced by Barrett and Lampard (15): see also Lampard's doctoral thesis (Cambridge University, 1954) and ref. (3) chapter 9 which gives a full list of diagonal processes.

** Ref. (16) sections 17.6-7: ref. (17) chapter 16.

*** Ref.(18) is the original publication on the Chernoff bound. There are several textbook presentation including ref.(4) volume 2 pp. 45-47; while recent extensions to the technique appear in refs.(19) and (20), and ref. (3) section 7.6.

where we have used the orthogonality of the $q_i(x)$. Small and smooth departures from some common distribution $w(x)$ may be expected to produce a tractable summation.*

7. Sensitivity to perturbations: queueing analysis

In queueing analysis, the central case with which all others can be compared is the single-server queue with Poisson arrivals (negative-exponential interarrival time) and negative-exponential service time distribution, M/M/1 in Kendall's notation. As Kleinrock remarks ** 'the M/M/1 queue is the simplest interesting queueing system....it is the classic example, and the analytical techniques are rather elementary. Whereas these techniques do not carry over into more complex queueing systems, the behaviour of M/M/1 is in many ways similar to that observed in more complex cases.'

This queue is easily analysed by means of the Markov chain model outlined in section 3, with constant arrival rate $b_i = A$ and constant departure rate $d_i = 1$ (normalised to unit mean holding time. A standard result is that the mean time in the queue for a delayed call is (in units of mean holding time) ***

$$E(W_d) = \frac{1}{1 - A} \tag{13}$$

The most natural, and historically the earliest, generalisation is to the M/G/1 queue, with Poisson arrivals but a general distribution of independent service times. The mean queueing delay then becomes

$$E(W_d) = \frac{1 + c_s^2}{2(1 - A)} \qquad (14)^{****}$$

where c_s is the coefficient of variation of the service time, i.e. the ratio (standard deviation / mean).

Another type of generalisation is to non-Poisson arrivals, i.e. to G/M/1 or G/G/1 queues. There is an extensive literature on this, but before discussing the topic further we consider why the Poisson process is often considered to be a good model for arrivals, and what is the nature of likely perturbations from it.

Let us suppose that the arrival process is generated as the superposition of many independent contributary arrival streams. Let each stream have a distribution of independent interarrival times, not necessarily negative-exponential but with no coincident events, i.e.each tributary is a renewal

* However, we should not overlook the problems of convergence of the Gram-Charlier series, discussed in ref.(16) sections 17.6-7

** Ref. (21) volume 2 p.10. This is an excellent general text on queueing.

*** See any text on traffic or queueing, including (3), (6), (21).

**** This is the Pollaczek-Khintchine formula, discovered independently by these authors circa 1930. See the above references for details.

process. Then as the number of tributaries increases, the pooled process approaches Poisson statistics. * That is, the Poisson process is a limiting point process, in much the same way as a Gaussian distribution is a limiting distribution. It is also a stable process: if the tributaries are Poisson, so is the pooled process.

A practical queueing system or network will usually have a large but finite number of contributors. How rapidly does their pooled arrival stream tend towards Poisson? A sequence of events at uniform intervals ($c = 0$) is about as far as one can get from the randomness of the Poisson process: suppose we take n independent contributors of this type. The distributions of intervals and of waiting time to next event are shown for $n = 1, 2, 3, 4$ in Figure 3, along with the limiting case (the broken line): by subjective impression, convergence is obvious save, of course, for tails beyond n time units. For a quantitative measure, consider the coefficient of variation c_n of the interarrival time for n tributaries: it is given by

$$c_n^2 = \frac{n-1}{n+1} \qquad (15)^{**}$$

As another example with a moderate degree of randomness in the tributaries, suppose the latter have a gamma distribution of order 2 ($c^2 = \frac{1}{2}$) for interarrival times. Then for the pooled process

$$c_n^2 = \sum_{i=1}^{n} \frac{n!}{(n-i)!} \frac{1}{(2n)^i}$$

$$\approx \frac{n+1}{n+3} \qquad (16)^{**}$$

These and some other examples are consistent with the conjectural approximate expression

$$c_n^2 = \frac{(n-1) - (n-3)c^2}{(n+1) - (n-1)c^2} \qquad (17)$$

where c is the coefficient of variation for each of n tributaries.

These examples suggest that convergence occurs but is rather slow. Other published analyses also suggest that convergence is slow. For instance, Whitt (23) gives the formula

$$c_a^2 = w \sum_i [\lambda_i / \sum_k \lambda_k] c_i^2 + (1-w) \qquad (18)$$

* Proved in ref. (4), volume 2 pp.73-75.

** Unpublished analysis by Dr. J.P.Eade and the author

where c_a is the coefficient of variation of the pooled arrivals, λ_i and c_i are the mean and coefficient of variation for stream i, and w is an empirical weighting of the form

$$w = \frac{1}{1 + kn} \tag{19}$$

where k depends on the imbalance of the streams and the overall traffic level. This agrees reasonably well with equations (15) and (16) above if we choose k to be about ½. Sriram and Whitt (25) give an analysis which suggests that, for a wide range of tributary statistics, convergence is ultimately approximated by equation (15). One of their examples has tributaries with very large c, such as might arise from packetised voice signals; c_n^2 diminishes rapidly for n increasing from 1 to about 10, then goes below unity and follows equation (15) from n = 20 onwards. The interval distribution is a good fit to negative-exponential for $n \geq 20$. Other examples of pooled processes are given in the literature, usually in terms of coefficient of variation. * Convergence of short-term statistics towards the Poisson limit seems to be well attested; however, one must beware of long-period correlations in the tributaries, which if fairly consistent may still appear in the pooled process.

Having established that the coefficient of variation of the arrival stream c_a is likely to be near 1 but not precisely 1, we consider what effect this has on a queue which must now be the general case G/G/1. Exact theory is difficult, but there have been several studies leading to approximate results which are reasonably accurate for c_a, c_s in the vicinity of 1. The mean queueing delay (analogous to equations 13 and 14 for the simpler systems) is approximately

$$E(W_d) = \frac{k(c_a^2 + c_s^2)}{2(1 - A)} \tag{20}$$

where k is a factor

$$k = 1 \qquad c_a \geq 1$$

$$= \exp\left[-\frac{2(1-A)}{3A} \cdot \frac{(1-c_a^2)^2}{c_a^2 + c_s^2}\right] \qquad c_a < 1 \tag{21}**$$

The general effect is that mean delay increases with variability, either of arrival or of service times. In general the classic M/M/1 queue remains a remarkably good central model from which moderate perturbations can be

* See especially refs. (22)-(27).

** This is quoted from Whitt(23) . A similar expression is given by Kramer and Lagenbach-Belz (28), but their factor k departs slightly from unityfor $c_a > 1$. Lavenberg (34) quotes an expression like (20) with k = 1 as a bound.

estimated.

We now turn to queueing networks. The so-called Markovian network with M/M/1 queues and independent routing on each link has some beautiful properties which facilitate analysis. Firstly, the departure process of such a queue is Poisson.* Secondly, from the stability of the Poisson process, the superposition of any number of outputs (and also any number of Poisson inputs from external sources) gives a Poisson input to another queue. So the whole network has negative-exponential distributions at every point, and each queue exhibits the statistics implied by its total traffic. Jackson's theorem (30) is that in a Markovian queueing network the joint occupancy distribution of the queues is the product of their marginal distributions: in other words, the statistics behave as if the queues were independent. Many recent studies have extended the range of product form solutions to deal with a variety of queue disciplines, state-dependent service time, multiple classes of customer etc.. ** In general, Markovian queues are analytically tractable even with some complicating factors thrown in.

For investigation of non-Markovian queueing networks, two approaches have been used. (a) Simulation, which is often a useful resort if a problem is otherwise intractable (b) Computational models based on representation of each distribution by two parameters, namely mean and coefficient of variation.*** As we have implied above, this is the minimal supplementation of Markovian theory, allowing c_a and c_s to depart from the classic value of 1, and is most accurate if these perturbations are not too great.

8. Parameter estimation: an example from cellular radio

We have noted in section 2 that the performance of a system may vary with a parameter which can be estimated from empirical data. This may mean that an accurate estimate can effectively eliminate the parameter, at least in evaluation and perhaps in actual performance. However, any practical estimator has some variance. According to the standard theory of sampling statistics****, this variance will diminish monotonically as the amount of data increases. It is likely to be non-negligible if a parameter is being estimated in real time for the purposes of adaptation so that the estimation is in effect part of the model. One may then be faced with two sensitivities: to the real value of the parameter, and to the value of its estimator. Moreover, this joint sensitivity is not necessarily a simple function of the estimator error. We illustrate this by an example drawn from some current work on the performance engineering of cellular mobile radio systems.*****

In cellular mobile radio, some mobiles have adequate radio paths to two or more base stations. The traffic capacity can be enhanced if the choice of path used is made so as to optimise the joint traffic pattern of a set of contiguous cells. The first problem in evaluating the performance of such a system is to estimate the proportion of mobiles with the desired

* Originally proved by Burke(29). See also ref.(3) pp.102-103 for a slight generalisation.

** References (31)-(34) are recommended samples from what is now a very large literature.

*** An excellent account of the latter approach is given by Whitt (23)(24). Lavenberg's book (34) is a valuable compendium relevent to analysis and simulation of queueing networks.

**** See any statistical textbook, or ref. (4) volume 2 pp. 133-138

***** A preliminary report is in ref. (35): detailed papers in preparation.

flexibility. This is a good example of the technique cited in section 5, of bracketing by two models. It is difficult to quantify the path quality experienced by a moving terminal, since both temporal and spatial fluctuations occur and are partially averaged out in defining a criterion: consequently, it is difficult to specify when two paths are acceptable alternatives. However, it is possible to estimate the proportion of mobiles with two paths differing by less than a stated tolerance (a) in mean level, making no allowance for fading (b) in instantaneous level making full allowance for Rayleigh or lognormal fading. Intuitively, it seems clear that these bracket the proportion of flexible traffic. Fortunately the two estimates differ by only about 14% in a typical hexagonal cell array.

For any given adaptive proportion, the overall performance of two contiguous cells (expressed as a mean blocking probability) depends on a parameter b representing the imbalance in the cell traffic levels A_i

$$b = \frac{A_1 - A_2}{A_1 + A_2} \tag{22}$$

For most of the practical range, blocking is a quadratic function of b, say

$$P_b = k_0 + k_2 b^2 \tag{23}$$

where the k_i depend both on the proportion of flexible traffic and the routing strategy. Now the obvious strategy is to direct flexible traffic preferentially to the more lightly loaded cell: but how do we know which this is? The system controller will register the instantaneous state (x_i channels busy in cell i). It can make estimates of A_i based on short-term averages. The most efficient routing algorithms utilise both the x_i and the A_i, and in this case the blocking approximation includes another term. In general, flexible traffic is directed to cell i if $x_i < x_j$, but if $x_i = x_j$ (and sometimes also in other states) it is directed to cell i if $A_i < A_j$: that is, according to the sign of b. Under these conditions, we may have

$$P_b = k_0 - k_1 |b| + k_2 b^2 \tag{24}$$

where $k_1 > 0$ if the sign of b is correctly estimated (minimising the blocking probability within the constraints of the system). If the sign of b is wrongly estimated, then the sign of k_1 is reversed, with detrimental effects on blocking. So performance depends both on the true value of the parameter b, and on the sign of the estimator.

Complex and critical though this may appear, at first sight, on analysis it turns out to be satisfactory. From (22) it is clear that sampling errors in A_1 and A_2 are likely to lead to an error in the sign of b when $|b|$ is small, but progressively less likely as $|b|$ increases. So the 'best' algorithm, including practical imperfections, yields a mean blocking probability not worse than (23) for all b, and a typical blocking probability (most of the time) nearer to (24) especially for large $|b|$ where improvement is most important.

9. Concluding comment

Most models have an implicit dependence on some assumed condition or distribution. There are some general techniques of sensitivity analysis, but in this paper we have taken the view thatpractical problems and models are so diverse that it is helpful to exhibit a variety of examples rather than purely abstract methods. Often we can either

(a) show the sensitivity to be small or even zero

(b) show it to be within usefully close bounds

(c) make a good estimate of a non-negligible effect.

Such investigations have an important role in the invention and evaluation of models.

References

1. Polya,G : Induction and analogy in mathematics. Oxford University Press 1954

2. Lakatos,I : Proofs and refutations: the logic of mathematical discovery. Cambridge University Press 1976

3. Cattermole,KW and O'Reilly,JJ : Problems of randomness in communication engineering. Pentech Press 1984

4. Cattermole, KW : Mathematical foundations for communication engineering. Pentech Press, Volume 1 (1985) and Volume 2 (1986)

5. Brockmeyer,E Halstrom,HL and Jensen,A : The life and work of A.K.Erlang. Acta Polytechnica Scandinavica, 2nd edition 1960

6. Kobayashi, H : Modeling and analysis: an introduction to system performance methodology. Addison-Wesley 1978

7. Kleinrock,L : A conservation law for a wide class of queueing disciplines Naval Research Logistics Quarterly 12, 181-192 , 1965

8. Richards,DL : Telecommunication by speech. Butterworth, 1973

9. Cattermole,KW : Principles of pulse code modulation. Iliffe, 1969

10. Dresher,M : Moment spaces and inequalities. Duke Math.J. 20, 261-271, 1953

11. Yao,K : Moment space error bounds in digital communication systems. In J.K.Skwirzynski(ed): Communication systems and random process theory, 1978

12. Bar-David,I : Communication under the Poisson regime. Trans. IEEE IT-15, 31-37, 1969

13. Cattermole,KW and O'Reilly,JJ : Optimisation methods in electronics and communications. Pentech Press, 1984

14. House,KE : Filters for the detection of binary signalling: optimisation using the Chernoff bound. Trans.IEEE COM-28, 257-259, 1980

15. Barrett,JF and Lampard,DG : An expansion for some second order probability distributions and its application to noise problems. Trans. IRE IT-1, 10-15 , 1955

16. Cramer,H : Mathematical methods of statistics. Princeton, 1946.

17. Feller,W : An introduction to probability theory and its applications. Volume 2, Wiley, 1966

18. Chernoff,H : A measure of asymptotic efficiency for tests of a hypothesis based on a sum of observations. Ann. Math. Stat. 23, 493-507 1952

19. Prabhu,VK : Modified Chernoff bound for PAM systems with noise and interference. IEEE Trans. IT-22, 95-100, 1982

20. Rocha, JRF and O'Reilly,JJ : A modified Chernoff bound for binary optical communication. Electronics Lett. 18, 708-710, 1982

21. Kleinrock,L : Queueing systems, Volumes 1 and 2. Wiley, 1976

22. Whitt,W : Approximating a point process by a renewal process: two basic methods. Oper. Res. 30, 125-147, 1982

23. Whitt,W : The queueing network analyser. BSTJ 62, 2779-2815, 1983

24. Whitt,W : Performance of the queueing network analyser. BSTJ 62, 2817-2843, 1983

25. Sriram,K and Whitt, W : Characterising superposition arrival processes and the performance of multiplexers for voice and data. Globecom 85, paper 25.4, 1985

26. Kuehn, PJ : Analysis of complex queueing networks by decomposition. 8th International Teletraffic Congress, paper 236, 1976

27. Kuehn, PJ : Approximate analysis of general queueing networks by decomposition. IEEE Trans. COM-27, 113-126, 1979

28. Kramer,W and Lagenbach-Belz,M : Approximate formulas for delay in the queueing system GI/G/1. 8th International Teletraffic Congress, paper 235, 1976

29. Burke, PJ : The output of a queueing system. Oper. Res. 4,699-704,1956

30. Jackson,JR : Networks of waiting lines. Oper.Res.5, 518-521, 1957

31. Baskett,F Chandy,KM Muntz,RR and Palacios,FG : Open,closed and mixed networks of queues with different classes of customers. JACM 22, 248-260, 1975

32. Reiser,M and Lavenberg,SS : Mean value analysis of closed multi-chain queueing networks. JACM 27, 313-322, 1980

33. Kelly,FP : Reversibility and stochastic networks. Wiley, 1979

34. Lavenberg,SS : Computer performance modeling handbook. Academic Press 1983

35. Cattermole,KW Yacoub,MD and Munoz Rodriguez,D : Alternative routing in cellular mobile radio. 3rd UK Teletraffic Symposium, 1986

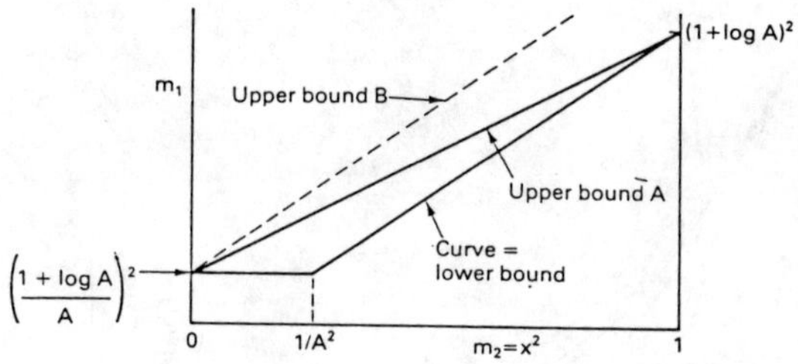

<u>Figure 1</u> Moment-space bounds to quantising noise (after ref. 4)

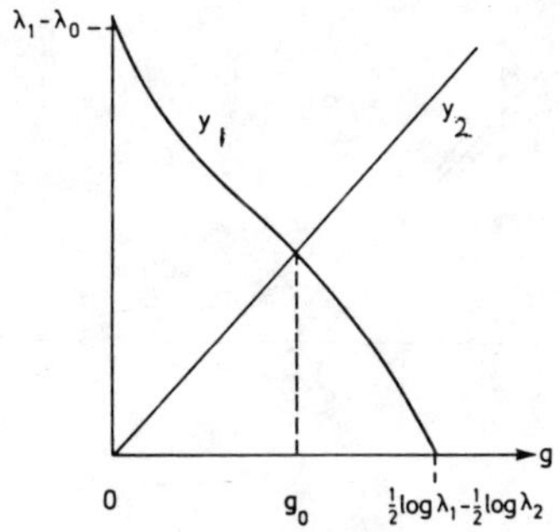

<u>Figure 2</u> Optimal filtering for combined thermal and quantum noise (after ref. 13)

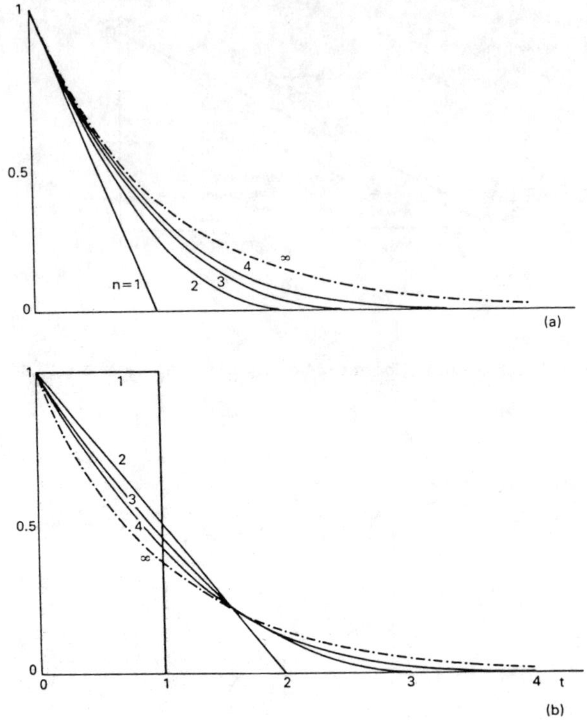

Figure 3 Pooled point process example (a) waiting time tail (b) interval tail (after ref. 4)

MODULATION AND CODING FOR THE MAGNETIC RECORDING CHANNEL

Jack Keil Wolf
Center for Magnetic Recording Research
University of California, San Diego

INTRODUCTION

Modern communications theory has played a major role in increasing the efficiency and reliability of communication systems. For example, whereas a few years ago the highest speed modems for transmitting data over the telephone network operated at 2400 bits per second, now a new generation of modems have been announced which transmit at 19,200 bits per second [1]. These modems employ a clever use of error correcting codes and modulation coupled with maximum likelihood (Viterbi) decoding [2].

Our aim in this research is to achieve analogous increases in efficiency and reliability for the storage and retrieval of digital data in magnetic recording systems. Specifically, here the aim is to increase the packing density of the digital data, that is, the amount of data that can be stored in a unit area of the magnetic media. This can be achieved by increasing the linear density (i.e., the number of bits per inch along a track) and/or by increasing the density of the tracks (i.e., the number of tracks per inch). All of this must be done without sacrificing the reliability of the retrieved data.

The areal packing density of digital recording on magnetic media has been steadily increasing over the last decade. Most of these gains, however, have been achieved by means other than coding and modulation. Improved magnetic materials, improved heads and improved tracking systems have been some of the methods used to achieve these increases. It is our belief, however, that still further substantial increases in packing density can be realized by advanced modulation and coding techniques. This is the goal of our research.

The notion of treating the magnetic recording channel as a communications channel is not a new one [3-4]. Many clever modulation and coding schemes have been proposed and implemented in magnetic recording systems [5-6]. However, many of the techniques which have been successfully implemented in communication channels have not yet found their way

J. K. Skwirzynski (ed.), Performance Limits in Communication Theory and Practice, 353–367.

into magnetic recording systems. In particular, the combined modulation/coding systems, which were the basis for the new high speed modems that were discussed in the first paragraph, have not been employed in any of the present magnetic recording systems.

In this paper, we will discuss several techniques for utilizing modulation and coding for increasing the packing density for magnetic recording systems. This work was performed with several graduate students at the Center for Magnetic Recording Research at the University of California, San Diego.

MODULATION CODES

Modulation codes are a type of coding unique to recording technology which convert the information to a form which matches specific constraints of the channel. One specific type of binary modulation code is the so-called (d,k) code whereby the input stream to the channel must be a sequence of binary digits where the runs of zeros between ones are at least of length "d" and are at most of length "k" [7]. For example, the IBM 3380 magnetic storage system uses a (2,7) code where the runs of zeros are always at least 2 and at most 7 in length [8]. Alternatively, the compact disk optical storage system uses a (2,10) code [9]. This is a special case of a more general type of code called run length limited (RLL) codes [10].

The reason for the use of (d,k) codes in magnetic recording systems will next be explained. In two level (or saturation) recording for magnetic media, binary data is written on a track by magnetizing the media in one of two (opposite) directions. The transition between two adjacent areas magnetized in opposite directions is called a "flux change". A given system is often characterized by specifying the maximum allowable flux changes per inch (i.e. fci) along a track. Current high density magnetic tape systems operate at about 50,000 fci while experimental systems have been built that operate at about twice that density. Magnetic disk systems operate at lower densities. If uncoded binary information were to be directly recorded, no matter how the information digits were mapped into the magnetization, the density of information bits per inch along a track could be no more than the maximum allowable number of flux changes per inch. We will see that if we use (d,k) codes with $d > 0$ then it is possible to have more information bits per inch than flux changes per inch.

Before we discuss how (d,k) codes increases the linear packing density, we must describe how zeros and ones are mapped into the pattern of magnetization along a track. We assume a scheme called NRZI recording whereby a one is written as a flux reversal and a zero is written as no flux reversal. Let F(max) be the maximum number of flux reversals per inch for the media. Since in a (d,k) code there are at least "d" zeros between adjacent ones then we can write these digits on a track at a density of [F(max) * (d+1)] binary digits per inch without violating the condition that there be no more than F(max) flux changes per inch. This argument appears to say that we have increased the density of recording information digits by a factor of (d+1) with (d,k) codes. This is not the case, however, since we cannot map, on a one to one basis, an arbitrary string of information digits into a string of channel digits satisfying the (d,k) constraint. In particular, there is a parameter, called the capacity of the (d,k) code, and denoted C(d,k), which describes the maximum ratio of information digits to (d,k) digits. That is, for any (d,k) code, the average number of information digits per channel digit is upper bounded by C(d,k) where $C(d,k) \leq 1$. The density of recording can thus be improved by a factor which is upper bounded by (d+1) * C(d,k). We shall see later that it is not difficult to find (d,k) codes at rates close to the capacity C(d,k). Thus, ignoring problems due to the difficulty of locating these transitions in the presence of noise, a density increase close to this upper bound can be achieved.

Next we discuss the reason for, and choice of, the parameter "k" in (d,k) codes. In reading signals from a magnetic media using the usual type of read heads, a flux reversal on a track causes a pulse at the output of the read head. A detector circuit must decide on the presence or absence of a pulse for each time interval that could contain such a pulse. Thus an accurate timing circuit is required as part of the detector. This is a major problem because for magnetic media, and especially for magnetic tape systems, the time between pulses can vary in an unpredictable manner. Thus the timing must be derived from the signals read from the magnetic media. In order to guarantee that there is a steady stream of pulses from which timing can be derived we must insure that there is not a long period of time during which there are no output pulses. Since in a (d,k) code there is a maximum of "k" zeros between ones there must be at least one output pulse for every (k+1) recorded symbols.

We have found several new results on binary (d,k) codes. First we have a new derivation on the capacity of these codes. The classical derivation [11] follows from writing a difference equation for the number of sequences of length n satisfying the (d,k) constraint and then solving the difference equation for large n. The alternative derivation starts by considering the code as the concatenation of phrases of binary digits, each string consisting of a set of zeros followed by a single one. The phrases are of minimum length (d+1) and of maximum length (k+1). Then we compute the maximum amount of information per channel symbol. We show that this information density is a maximum when the phrases are statistically independent and chosen in accordance with a specific distribution. The maximum information density is the capacity of the code.

Details of both the difference equation approach and the new approach follow. The details of the difference equation approach are taken (almost verbatim) from what are believed to be unpublished notes of Claude Shannon [12].

We define Q_n as the number of different channel sequences satisfying the (d,k) constraint whose total length is greater than n -(d + 1) but not greater than n and which begins with at least d and at most k zeros and which ends in the symbol 1. This definition leads to a better behaved function of n than if one considers sequences of length exactly n. For example, with the present definition, Q_n is monotone increasing. Then Q_n satisfies the difference equation:

$$Q_n=Q_{n-(d+1)} + Q_{n-(d+2)} + \cdots + Q_{n-(k+1)} . \qquad (1)$$

Furthermore, we take as the boundary conditions, $Q_n=0$ for $n < 0$ and $Q_n=1$ for $0 \leq n < (d+1)$. This difference equation is described by the following characteristic equation:

$$1 = x^{-(d+1)} + x^{-(d+2)} + \cdots + x^{-(k+1)} . \qquad (2)$$

The right hand side of (2) is strictly monotone decreasing. It varies from infinity to zero as x varies from zero to infinity. Thus, this equation has a unique positive root which we will write as X. Note that $X \geq 1$.

One can show that for all $n \geq 0$, $Q_n \leq X^n$ and $Q_n \geq X^{(n-k-1)}$. We give the proof for the upper bound. The proof for the lower bound is similar. The proof will be based upon induction on increasing intervals of n, each interval of length (d+1). The upper bound is true for $0 \leq n \leq (d+1)$ since $X \geq 1$ and $Q_n = 1$. Assume the upper bound holds for all $n \leq n_o$. Then since for $n_o \leq n \leq n_o + (d+1)$ equation (1) holds, we have

$$\begin{aligned} Q_n &\leq X^{(n-d-1)} + X^{(n-d-2)} + \cdots + X^{(n-k-1)} \\ &= X^n . \end{aligned} \qquad (3)$$

This shows that the upper bound holds for the increased interval up to $n_o + (d+1)$ and by induction this shows that the bound holds for all n.

Taking logarithms (base 2) of the upper and lower bounds and dividing by n results in the equation:

$$\log(X) - (k+1)/n \leq 1/n \log(Q_n) \leq \log(X) . \qquad (4)$$

We now define C(d,k) as:

$$C(d,k) = \lim 1/n \log (Q_n)$$

where the limit is as n approaches infinity. Thus we see that C(d,k) is equal to log(X) and that the approach to this limit is at least as fast as 1/n.

Before proceeding we note that equation (2), after multiplication by x^{k+1} and some simplification, can be written as

$$x^{(k+2)} - x^{(k+1)} - x^{(k+1-d)} + 1 = 0 \qquad (5)$$

and that X is the largest root of this equation. We now derive this result in an entirely different manner.

A binary sequence satisfying the (d,k) constraint can be parsed into phrases where each phrase consists of a sequence of zeros which ends with a single one. Let Z_i be a random variable which takes on values equal to the length of the i-th phrase. Note that Z_i is an integer greater than or equal to (d+1) and less than or equal to (k+1). We know that the amount of information that can be carried by the sequence Z_1, $Z_2,\ldots,Z_t$ is given by the maximum value of the joint entropy of the sequence $H(Z_1,Z_2,\ldots,Z_t)$ where the maximum is taken over all joint distributions for the random variables Z_1, $Z_2,\ldots,Z_t$. However, the capacity C(d,k) is the maximum <u>information</u> <u>rate</u> and is given by the formula

$$C(d,k) = \lim\sup \frac{H(Z_1,Z_2,\ldots,Z_t)}{E[Z_1 + Z_2 + \ldots + Z_t]} \qquad (6)$$

where E[ ] stands for statistical expectation, the limit is as t approaches infinity and the supremum is over all joint density functions of the t random variables. But if A_1, A_2, ..., A_t and $B_1,B_2,\ldots,B_t$ are positive real numbers then

$$\frac{A_1+A_2+\ldots+A_t}{B_1+B_2+\ldots+B_t} \leq \max_j \frac{A_j}{B_j} \quad . \qquad (7)$$

Then since $H(Z_1,Z_2,\ldots,Z_t) \leq H(Z_1) + H(Z_2) + \ldots + H(Z_t)$,

$$C(d,k) \leq \max \frac{H(Z)}{E(Z)} = \max \frac{\sum_{i=d+1}^{k+1} p_i \log(1/p_i)}{\sum_{i=d+1}^{k+1} i\, p_i} \quad . \qquad (8)$$

The right hand side of this equation is a maximum when $p_i = 2^{-iK}$, i=d+1,d+2,...,k+1, and a careful examination of this argument reveals that C(d,k) = K. Equation (5) is obtained by summing p_i over all i and setting the results equal to 1.

Both upper and lower bounds can be obtained for C(d,k) by a slight modification of the previous argument. By choosing the joint distributions of the $Z_1, Z_2, \ldots$ to be a product distribution of i.i.d. random variables each of which takes on the values (d+1), (d+2),...,(k+1) with equal probability, we obtain the lower bound

$$C(d,k) \geq \frac{\log(k-d+1)}{((k+1)+(d+1))/2} \quad . \tag{9}$$

An upper bound is obtained by choosing a uniform distribution for the numerator in the right hand side of (8) while choosing a distribution for the denominator that puts all the weight on the smallest length. This results in the upper bound:

$$C(d,k) \leq \frac{\log(k-d+1)}{(d+1)} \quad . \tag{10}$$

A simple method of achieving a (0,k) code is a technique we call "bit-stuffing". In the encoding process, a one is stuffed" (or inserted) after every instance of "k" consecutive zeros irrespective of whether the next binary digit is a zero or a one. In decoding, one removes the next binary digit after the occurrence of every sequence of "k" consecutive zeros. If this technique is used with binary i.i.d. data where the probability of a data zero is p and the probability of a data one is q = 1 -p then the average rate for this technique would be $m\backslash(m+1)$ where $m = (1-p^k)/qp^k$. A comparison of the average code rate for bit stuffing and the capacity of the (0,k) code is given in the following table.

k	Average Rate Bit Stuffing	C(0,k)
1	.6667	.6942
2	.8571	.8792
3	.9333	.9468
4	.9677	.9752
5	.9841	.9881

The above derivation for the capacity of (d,k) sequences can be applied to a different class of constrained sequences where both the lengths of the strings of ones and zeros must satisfy minimum and maximum values. These sequences may have applications in optical recording systems. Specifically, assume that we are concerned with channel sequences for which all of the following conditions are satisfied:

1. The minimum number of zeros between adjacent strings of ones is d.

2. The maximum number of zeros between adjacent strings of ones is k.

3. The minimum number of ones between adjacent strings of zeros is e.

4. The maximum number of ones between adjacent strings of zeros is m.

We will call sequences which satisfy these four constraints (d,k;e,m) sequences.

As before, we treat such channel sequences as a concatenation of phrases where each phrase consists of a sequence of zeros followed by a sequence of ones. The leading sequence of zeros in each phrase has length at least d and at most k and the number of ones in the trailing string is at least e and at most m.

Again let us assign the random variable Z_i to the i-th phrase. Now Z is a two dimensional (vector) random variable whose first component denotes the length of the string of leading zeros and whose second component denotes the length of the trailing ones.

The maximum rate for (d,k;e,m) sequences is then given by a formula analogous to equation (6). By a derivation similar to that described previously, one finds that the maximizing probability is one that leads to i.i.d. phrases such that the probability that a phrase consists of i zeros $(d \leq i \leq k)$ and j ones $(e \leq j \leq m)$, $p_{i,j}$ is given as:

$$p_{i,j} = 2^{-C(i+j)} \quad . \tag{11}$$

Here C is the capacity of the (d,k;e,m) sequences. Summing this probability over i and j and setting the result equal to 1 yields an equation for which the logarithm of the largest root yields the capacity. This equation (after some simplification) is:

$$(x^{k-d+1} - 1)(x^{m-e+1} - 1) - x^{k+m}(x - 1)^2 = 0 \quad . \tag{12}$$

We have also derived the power spectral density of these sequences [13]. For the sake of brevity, however, this result is omitted.

THE CAPACITY OF (d,k) CODES WITH RANDOM ERRORS

In the previous section, we calculated the maximum rate for (d,k) codes when there were no errors in the system. In data storage systems, however, we must take into account the effects of errors. In this section we consider a very simple model for the channel errors, namely the binary symmetric channel. In this model the errors occur randomly and the same probability of error holds for a recorded zero and for a recorded one. The advantage of this model is its simplicity: a single parameter is all that is required to describe the channel. The disadvantage of this model is also its simplicity: i.e., it is far too simple to be a good model for the actual channel behavior.

In this section we are interested in the capacity of the binary symmetric channel (BSC) under the constraint that the input sequences are (d,k) sequences. Although the channel is memoryless, the input sequences are not. Thus each output of the channel depends both on the corresponding input and on past inputs and outputs.

One can produce sequences satisfying the (d,k) constraint by a (k + 1) state machine with states labeled 0, 1, 2, . . ., k. When the machine is in state i, $0 \le i \le (d - 1)$, the machine produces a symbol "zero" and then goes to state (i + 1). When the machine is in state j, $d \le j \le (k - 1)$, the machine either produces a "zero" symbol and goes to state (j + 1) or the machine produces the symbol "one" and returns to the 0 state. When the machine is in state k it always produces the "one" symbol and returns to the 0 state. Indeed, for the case of no errors, if one puts appropriate probabilities on the branches of this state diagram, one will produce (d,k) sequences of maximum rate. Such a state diagram with probabilities describes a first order Markov process.

Unfortunately, we do not know that a first order Markov process will produce the input sequences that will maximize the mutual information for the binary symmetric channel. Thus, in what follows, when we refer to the state diagram that produces (d,k) input sequences, we are not necessarily referring to a first order Markov process. Specifically, when the machine is in state j, $d \le j \le (k - 1)$, the probability that it produces a "zero" or a one may depend on the entire past history of the states and not just on the current state.

Let $\underline{Y}_N = (y_1, y_2, \ldots, y_N)$ denote a sequence of binary digits satisfying the (d,k) constraints. Furthermore, let us denote by $\underline{S}_N = (s_1, s_2, \ldots, s_N)$ the sequence of N states of the finite state machine which produced the sequence $\underline{Y}_N$. The channel is assumed to be a binary symmetric channel (BSC) with transition probabilities 1-p and p. The output of the channel is denoted by the sequence $\underline{R}_N = (r_1, r_2, \ldots, r_N)$. The channel capacity is equal to

$$C = \lim_{N\to\infty} \sup_{P(Y_N)} \frac{I(\underline{Y}_N;\underline{R}_N)}{N} = \lim_{N\to\infty} \sup_{P(S_N)} \frac{I(\underline{S}_N;\underline{R}_N)}{N} , \qquad (13)$$

where the sup is taken over all probabilities $P(\underline{Y}_N)$ for the sequence $\underline{Y}_N$ which satisfy the (d,k) constraint or over all probabilities $P(\underline{S}_N)$ for the sequence of states $\underline{S}_N$ of the finite state machine corresponding to the (d,k) sequence. The last equality in (13) results from the one-to-one mapping from a sequence $\underline{Y}_N$ to the corresponding state transitions sequence $\underline{S}_N$, for $N > k$. In the previous section we showed that for $p=0$, capacity was achieved for a Markovian finite state machine. However, as stated previously, for $p > 0$ we do not know this is the case. If we take the sup over all probabilities $P(\underline{Y}_N)$ which are stationary and Markovian and satisfy the (d,k) constraint, then we will have a lower bound on the capacity. The following lemma enables us to find a lower bound on the capacity of the channel. For convenience we define $s_0 = \phi$.

Lemma 1

For a stationary and Markovian source and memoryless channel we have,

a. $$H(\underline{Y}_N \mid \underline{R}_N) = H(\underline{S}_N \mid \underline{R}_N)$$

$$\sum_{i=1}^{N} H(s_i \mid s_{i-1}, r_i, \ldots, r_N), \text{ for } N > k , \qquad (14)$$

b. $$H(s_i \mid s_{i-1}, r_i, \ldots, r_N) \geq H(s_j \mid s_{j-1}, r_j, \ldots, r_N),$$

$$\text{for } i > j \geq 2 \quad . \qquad (15)$$

Proof

a. By definition

$$H(\underline{S}_N \mid \underline{R}_N) = H(s_1 \mid r_1, \ldots, r_N) + H(s_2 \mid s_1, r_1, \ldots, r_N) +$$

$$H(s_3 \mid s_1, s_2, r_1, \ldots, r_N) + H(s_N \mid s_1, \ldots, s_N, r_1, \ldots, r_N) \quad .$$

By using the assumptions that the source is Markovian and the channel is memoryless we obtain (14).

b. $$H(s_i \mid s_{i-1}, r_i, \ldots, r_N) = H(s_2 \mid s_1, r_2, \ldots, r_{N-i+1}) \geq$$

$$H(s_2 \mid s_1, r_2, \ldots, r_{N-j+1}) = H(s_j \mid s_{j-1}, r_j, \ldots, r_N) \quad .$$

The two equalities follow from the assumption that the process is stationary and Markov, while the inequality is due to the fact that conditioning cannot increase the entropy.

We are now ready to derive a lower bound on the capacity.

Lemma 2

The capacity of the binary symmetric channel under the constraint that the inputs are (d,k) sequences is lower bounded by the equation:

$$C \geq \sup_{P(s_1,s_2)} I(s_2;r_2|s_1) \quad . \tag{16}$$

Proof

In our bound we assume that the source is stationary and Markovian. Then

$$\lim_{N\to\infty} H(\underline{S}_N)/N = \lim_{N\to\infty} H(s_N|s_1,s_2,\ldots,s_{N-1})$$

$$= \lim_{N\to\infty} H(s_N|s_{N-1}) \quad . \tag{17}$$

Therefore it remains to find a bound on the conditional entropy $H(\underline{S}_N|\underline{R}_N)/N$. By combining Equations (14) and (15) we obtain:

$$H(\underline{S}_N|\underline{R}_N)/N = H(s_1|r_1,\ldots,r_N)/N$$

$$+ \sum_{i=2}^{N} H(s_i|s_{i-1},r_i,\ldots,r_N)/N$$

$$\leq H(s_1|r_1,\ldots,r_N)/N$$

$$+ (1-1/N)H(s_2|s_1,r_2) \quad . \tag{18}$$

Since the inequality is for any N, taking the limit of (18) as $N\to\infty$, the following upper bound on the conditional entropy is obtained:

$$\lim_{N\to\infty} H(\underline{S}_N|\underline{R}_N)/N \leq H(s_2|s_1,r_2) \quad . \tag{19}$$

Therefore we obtain

$$C = \lim_{N\to\infty} \sup_{P(S_N)} \frac{I(\underline{S}_N:\underline{R}_N)}{N} = \lim_{N\to\infty} \sup_{P(S_N)} \frac{H(\underline{S}_N) - H(\underline{S}_N|\underline{R}_N)}{N}$$

$$\geq \sup_{P(s_2,s_1)} \{H(s_2|s_1) - H(s_2|s_1,r_2)\} = \sup_{P(s2,s1)} I(s_2:r_2|s_1)$$

where $P(s_2,s_1)$ is the joint stationary probability distribution of states s_2 and s_1. We now put this bound in a form convenient for numerical evaluation.

Lemma 3

The capacity of the BSC channel under (d,k) constraints of the source is lower bounded as

$$C \geq \max \sum_{j=0}^{k} p(s=j)h(p + p_j - 2pp_j) - h(p) \quad , \tag{20}$$

where p_j is the conditional probability of a transition from state j to state 0. The max is taken over all possible distributions which satisfied the (d,k) constraints and $h(x) = -x\log x - (1-x) \log (1-x)$.

Proof

By definition

$$I(s_2;r_2|s_1) = H(s_2|s_1) + H(r_2|s_1) - H(s_2,r_2|s_1)$$

$$= \sum_{j=0}^{k} p(s=j)\{h(p_j) + h(p+p_j-2pp_j) - [h(p_j) + h(p)]\} \quad .$$

Now the capacity of the channel is lower bounded by the maximum of $I(s_2;r_2|s_1)$ over all possible source distributions which satisfy the (d,k) constraint. Plots of this bound for three different choices of the pair (d,k), namely (0,7), (1,3) and (2,7), are given in Figure 1.

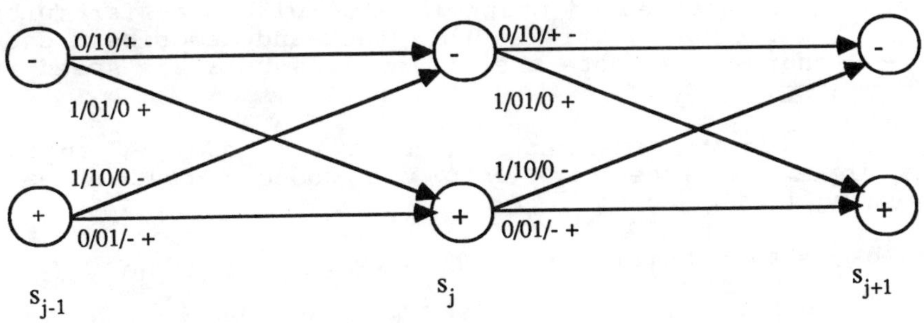

State labels $\quad a_j/ b_{2j}, b_{2j+1}/ z_{2j}, z_{2j+1}$

a_j = information digit

b_{2j}, b_{2j+1} = encoder outputs.

z_{2j}, z_{2j+1} = noiseless channel outputs.

FIGURE 1. Trellis diagram for code.

TRELLIS CODES FOR PARTIAL RESPONSE CHANNELS

Measurements indicate that if two level (saturation) recording is utilized with a magnetic recording channel, the channel can be modeled as a linear system. That is, assume that an isolated transition from one saturation level to the other applied at time t_o, $u(t - t_o)$, results in the output signal $s(t - t_o)$. Then an arbitrary sequence of the transitions of alternating signs, say

$$f(t) = \sum_i a_i\, u(t - t_i),\quad a_i = +1 \text{ or } -1,$$

will result in the output signal

$$e(t) = \sum_i a_i\, s(t - t_i) .$$

Furthermore, these measurements indicate that the transfer function of the system closely approximates a class 4 partial response channel [14]. This similarity was exploited by several investigators [3,5,15]. In particular, Wood [15] has explored the possibility of using a Viterbi detector to take advantage of the inherent redundancy in the output signal. In a previous paper [16], a general procedure was described for synthesizing codes for use with partial response channels. Here we illustrate this technique by describing a very simple code with good distance properties and which can be easily decoded.

We first note that a class 4 partial response channel with discrete time transfer function $1 - D^2$ can be considered as two time interleaved channels each with transfer functions $1 - D$. We choose to utilize this fact and describe a code for a $1 - D$ channel. We note that this gives us a 2:1 speed factor in decoding.

The code we consider is a rate 1/2 code. We encode the data input sequence $\{a_i\}$ into the coded output stream $\{b_j\}$ according to the rule:

$$b_{2j} = a_j + b_{2j-1} + 1$$

$$b_{2j+1} = b_{2j} + 1 .$$

Here the inputs and outputs of the encoder are assumed to be from the set $\{0,1\}$ and all additions are modulo 2.

Note that the encoder has only two states: the content of the single delay element that stores the previous output. Thus, a Viterbi decoder matched only to the code also would have only two states. However, the output of the encoder must eventually pass through a $1 - D$ channel so that one would think that a decoder matched to both the code and the channel would require four states. As we shall see, this is not the case. That is, we will find that a two state decoder is all that is required even for a 1 - D channel. Furthermore, note that the code is not a convolutional code but rather is a coset of a convolutional code.

Before entering the channel, the encoded binary digits are converted to real numbers from the set (-A/2,+A/2}. This results in three level outputs from the 1 - D channel: outputs from the set {-A,0,+A}. Note that these three level outputs cannot occur in any order. Specifically, nonzero outputs must occur in a sequence with reversal of signs.

A trellis diagram which shows the information input, the encoder output and the channel output is shown in Figure 2. The state of the trellis is the sign of the signal stored in the 1 - D channel. Note there is a one-one correspondence between this signal and the symbol stored in the encoder. Thus, the trellis has only two states and we require only a two state decoder.

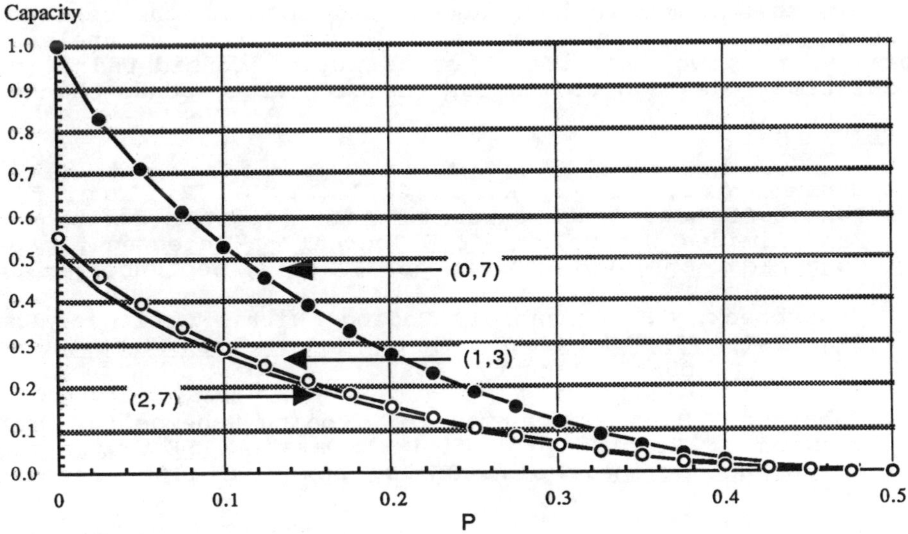

FIGURE 2. Lower bound on the capacity of BSC channel under (d,k) constraint.

Two important properties of code that can be seen from this trellis are:

1. Any two paths in the trellis which start in a common state and end in a common (but perhaps different state) have squared Euclidean distance between channel output sequences at least $24A^2$. This should be compared with the situation

for the uncoded 1 - D channel where the value is $8A^2$. This yields a 5db coding gain (or 2db if one takes into account the 3db loss due to the rate 1/2 code) for an additive Gaussian channel.

2. The maximum number of consecutive zero symbols at the output of the 1 - D channel is one. Thus, timing recovery is not a problem.

The Viterbi decoder for this code is no more complicated than the Viterbi decoder for the uncoded 1 - D channel. Because of its simplicity, it can be implemented at very high speeds. Simulation results show that about a .3db penalty occurs when one uses a 4 bit quantizer.

More complicated codes (with more complicated decoders) exhibit greater coding gains. The interested reader is referred to the references [16] for details.

SUMMARY

In this paper we have summarized some of the results on coding and modulation for magnetic recording channels . In particular, two specific topics were considered: (d,k) modulation codes and coding for the partial response channel.

REFERENCES

1. Forney, Jr., G. D., R. G. Gallager, G. R. Lang, F. M. Longstaff, and S.U. Qureshi. "Efficient Modulation for Band-Limited Channels", IEEE Journal on Selected Areas in Communications, vol. SAC-2, pp. 632-647, September, 1984.

2. Ungerboeck, G. "Channel Coding with Multilevel/Phase Signals", IEEE Transactions on Information Theory, vol. IT-28, pp. 55-67, January, 1982.

3. Kobayashi, H. "A Survey of Coding Schemes for Transmission or Recording of Digital Data", IEEE Transactions on Communications, vol. COM-19, pp. 1087-1100, December, 1971.

4. Mallinson, J. C. "A Unified View of High Density Digital Recording Theory", IEEE Transactions on Magnetics, vol. MAG-11, pp. 1166-1169, September, 1975.

5. Price, R., J. W. Craig, H. E. Melbye, and A. Perahea. "An Experimental, Multilevel, High Density Disk Recording System", IEEE Transactions on Magnetics, vol. MAG-14, pp. 315-317, September, 1978.

6. Lin, S. and D. J. Costello. Error-Control Coding. Prentice Hall Inc., Englewood Cliffs, NJ, 1983, (Chapter 16).

7. Siegel, P. H. "Recording Codes for Digital Magnetic Storage", IBM Research Report, RJ 4720 (SO 198), June, 1983.

8. Adler, R. L., D. Coppersmith, and M. Hassner. "Algorithms for Sliding Block Codes - An Application of Symbolic Dynamics to Information Theory", IEEE Transactions on Information Theory, vol. IT-29, pp. 5-22, January, 1983.

9. Watkinson, J. R. "Channel Code and Disc Format-1", Electronics and Wireless World, pp. 27-28, May, 1985.

10. Norris, K. and D. S. Bloomberg. "Channel Capacity of Charge Constrained Run-Length Limited Codes", IEEE Transactions on Magnetics, vol. MAG-17, pp. 3452-3455, November, 1981.

11. Shannon, C. E. and W. Weaver. The Mathematical Theory of Communications. Urbana, Illinois, 1963.

12. Shannon, C. E. "Some Miscellaneous Results in Coding Theory", unpublished notes, not dated.

13. Zehavi, E. and J. K. Wolf. "On Run-Length Codes", submitted for publication, IEEE Transactions on Information Theory.

14. Kretzmer, E. R. "Generalization of a Technique for Binary Data Communication", IEEE Transactions on Communications, vol. COM-14, pp. 67-68, February, 1966.

15. Wood, R., S. Ahlgrim, K. Hallamasek, and R. Stenerson. "An Experimental Eight-Inch Disk Drive with One Hundred Magabytes per Surface". IEEE Transactions on Magnetics, vol. MAG-20, pp. 698-702, September, 1984.

16. Wolf, J. K. and G. Ungerboeck. "Trellis Coding for Partial Response Channels", IEEE Transactions on Communications, vol. COM-34, August, 1986.

Modelling of and Communication Limits for non-gaussian noise.

F.L.H.M.Stumpers,Philips Research Laboratories,scientific advisor(retired)

Abstract. The information theory approach to the capacity of communication channels considers mainly gaussian(white or colored noise) Neither natural noise of terrestrial origin nor man-made noise fall into this category.The simplest model treats noise pulses occurring randomly or periodically in time.They may be caused by switching,atmospherics,ignition,electric motors,radio and television receivers,industrial,scientific and medical equipment.Techniques to improve reception under such circumstances include clipping,hole punching,blanking and "smear-desmear filtering",adaptive approaches and signal processing.Statistical modelling of ignition noise was performed by Spaulding and extensively by Middleton.This led to the construction of receivers,which approach optimality as the signal becomes small.

Introduction. As radio broadcast reception became more general in the early thirties and wellknown symphony orchestras joined the programme(e.g.the Amsterdam Concertgebouw orchestra in the Netherlands),the presence of noise and interference was considered obnoxious.After a first meeting in 1933,in 1934 the International Special Committee on Radio Interference was formed by National Committees of The International Electrotechnical Committee,the International Broadcasting Union and several other interested Societies.Its first interest was to cancel,as far as possible noise and interference at its source and in order to reach agreement measuring instruments had to be constructed so that limits for radiated and conducted interference could be agreed upon.The Subcommittee on Limits was chaired by mr.Braillard and the Subcommittee on Measurements by dr.Van der Pol.Fairly soon agreement was reached on the quasi-peak

J. K. Skwirzynski (ed.), Performance Limits in Communication Theory and Practice, 369–381.

receiver,defined by the charge and discharge time-constant of a detector in such a way,that the annoyance effect on the human ear was taken care of.The work of the Committee was suspended during the war years.In 1955 several working groups were formed for the study of different sources of noise and interference,and one of them not only was interested in interference caused by radio and television receivers,but also in the immunity of these receivers to interference.This turned out to be one of the more difficult subjects,and agreement on a Report was reached only recently.

We will now treat some of the methods invented to reduce the effect of impulsive noise in radio receivers.

1.Audio clipping and blanking.

In a system described by Chow[1] the impulse noise in the antenna triggers the gating signal generator.This generator produces a gating signal which in turn controls a balance gate.The func tion of the gate is to block the transmission during the gatin period.The delay in the rf part and the first mixer is such, that the gating signal arrives at the balanced gate,just befor the desired signal.In another possible arrangement the gating signal generator not only produces one gating signal,but a chain of gating signals evenly spaced with respect to time.If the repetition rate in Hz is larger and beyond the audio band of the receiver,the noise cannot be heard and the desired info mation is recovered without distortion.(This suggests,that the pulses arrive periodically.)Experimental results give the improvement at the audio output ranging from 14 to 18 dB.In an analogous application Gosling[2] used a 64 usec glass delay line for the signal in order to let the pulse blanker arrive at the right time.He did not use repetitive blanking,for which Chow claims an improvement of 4 dB over the improvement made by one shot gating.In blanking wanted signal and interference are blanked out during the duration of an impulse.In clipping any signal exceeding a predetermined threshold is clipped.Signal insertion is a further development of the blanking concept.In first order blanking the signal magnitude is maintained at a constant value,and in second order blanking the signal slope i maintained at a constant value.Clipping,first- and second order blanking have been tried by Martin[3],but the difference is of the order of 2 dB for double sideband AM,and double sideband AM with suppressed carrier.Clipping is the least effective of the three techniques,but the easiest technique to apply.Hepp[4] used a second order blanking technique to suppress interference in fm radio receivers.During stereo transmission,which includes a pilot tone,a parallel resonant circuit is arranged in series with the storage capacitorin order to obtain substantially interference free reception.

2."Smear-desmear"filter techniques

Wainwright[5] tried to overcome impulse-noise problems in data systems by performing a linear operation on the signal before the impulse noise is added,and then to perform the inverse of this operation(restoring the original signal) before detection in such a way that the noise energy is spread out in time at the detector and hence has a low amplitude and is not necessarily impulsive at any given time.Transmit and receive filters with complimentary delay vs frequency characteristics can be employed for this purpose.A differential time delay of 7.5 ms over the frequency band 500 to 3000 Hz was used.The desmear filter is most effective when operating on single impulses or a small group of impulses that occur far enough apart such that the stored energy in the desmear network from previous impulses has subsided to a significant degree.The smear and desmear filters each employed 27 pass-poles and a pair of first order corrector poles.There were problems with amplitude sensitivity of the time-delay characteristics,and with the independent adjustment of amplitude and delay characteristics,(formidable but not insurmountable).Al_Araji and Abdullah[6] used the smear-desmear technique to reduce the effect of impulsive noise in mobile radio receivers.The reduction was achieved by selecting only three bands(325-575Hz),(950-1200Hz)and (1875-2125 Hz) out of the speech frequency spectrum(300-3000Hz).These bands were delayed by 0,4 and 8 msec using digital delay line.At the receiver,the bands were delayed in the opposite sense hence smearing the impulsive noise.This achieved an improvement of 17dB in peak signal to peak noise ratio(confirmed experimentally).

Beenker, Claasen and Van Gerwen[7] used digital filters and based their method on time domain considerations. They either placed their filters before the modulator and after the demodulator (operating on baseband signals) or between those two (passband signals). Their first merit factor $F_1 = \frac{\text{max. amplitude of impulse noise}}{\text{max. amplitude of smeared impulse noise}}$, under the constraint, that the transmitter power is fixed: $P_u = 1$. The insertion of the two filters in the transmission system causes intersymbol interference. If the impulse response function of the transmission system without smearing filters is $h(n)$, and with smearing filters $h(n)+\beta(n)$, then the second merit factor $F_2 = \frac{\max_n |h(n)|}{\left\{\sum_n \beta^2(n)\right\}^{1/2}}$

The coefficients of the desmearing filter are constrained by the requirement that the maximum amplitude of the equivalent impulse response functions of the transmission system without and with smearing filters, respectively, are both equal to unity

$$\max_n |h(n)| = \max_n |h(n)+\beta(n)| = 1$$

Optimizing one of the two merit factors must be paid for with a rather poor performance for the other one.

First the authors show, that the maximum smearing efficiency $F_1 = (N)^{\frac{1}{2}}$, and this is true for all binary sequences. Now we try to find a large value for $F_2 = \frac{N}{\left\{\sum_{k\neq 0} R^2(k)\right\}^{1/2}} = \frac{N}{\left\{2\sum_{1}^{N-1} R^2(k)\right\}^{1/2}}$, where $R(k) = \sum_{l=0}^{N-1} b(l)\, b(l+k)$ is the k th autocorrelation coefficient of the sequence $b(n)$, where in the summation $b(l) = 0$, if $l < 0$ or $l \geq N$.

Golay has found suboptimum results for all odd values of N, $N \leq 59$. He has shown, that for large values of N, F_2 3.51 (10.9 dB).

F_2 can be further increased by taking a longer smearing filter than the desmearing filter. The Merit Factors of several Smearing-desmearing filters for baseband signals are given, as follows:

N	M	$N^{\frac{1}{2}}$	F_1	F_2	F_1 in dB	F_2 in dB
13	29	3.61	3.53	12.48	10.95	21.92
27	67	5.20	4.92	10.50	13.83	20.43
59	123	7.68	6.97	10.54	16.86	20.46

A disadvantage of the use of baseband smearing filters is that the modem design needs to be adapted. Smearing filters can also be applied with existing modems if the smearing filter is placed after the modulator and the desmearing filter before the demodulator.The discussion is focussed on a 4800 bit/s modem,using 8-phase modulation;for this case the first Merit Factor is about 10 dB and the second Merit Factor about 20 dB.Ternary filters of length 64 give good results.The results given refer to laboratory experiments and the influence of a practical environment still needs to be determined.

For his dissertation Engel[8] studied different forms of modulation in the presence of impulsive noise,double-sideband AM,frequency shift keying,phase shift keying,with coherent or differentially coherent detection.He also used complementary delay filters, with linear or sinussoidal envelope delay characteristics and found improvements in dB of about 5 dB(depending on the delay in Bauds).Limiting is discussed in his paper only as a possible source of further improvement.Baghdady[9] has suggested ,that clipping of impulsive disturbances,followed by simple linear cancellation circuits,will lead to almost complete elimination of the harmful effects of impulse noise.Watt-Carter and Wheeler[10] showed,that the effects of radio noise and interference on the performance of hf radio telephone circuits are reduced by the introduction of true compandor action,which is effected by controlling the expanders through a separate channel.The Lincomplex system proved capable of providing a very considerable improvement in radio telephony on hf channels.

3.Adaptive Noise Cancelling .

In 1959 Widrow and Hoff at Stanford University devised the least-mean square adaptive algorithm and the pattern recognition scheme known as "Adaline"("adaptive linear threshold logic element") With 8 colleagues and students Widrow[11] gave an extensive survey of Principles and Applications of Adaptive Noise Cancelling.His method uses a "primary"input containing the corrupted signal and a "reference" input containing noise correlated in some unknown way with the primary noise.The reference input is adaptively filtered and subtracted from the primary input to obtain the signal estimate.If there is full correlation between the noise in the corrupted signal and the reference noise,the adaptive filter causes the noise to be perfectly cancelled;uncorrelated noise components in the signal remain uncorrected.

Adaptive filters are required to "learn"the statistics of the noise and to track them if they vary slowly.The steady-state performance of adaptive filters closely approximates that of Wiener filters,and Wiener theory provides a convenient method of mathematically analyzing statistical noise cancelling problems.In certain situations a primary input is available consisting of a signal component with an additive undesired sinussoidal interference.One can use a notch filter,but an adaptive canceller tracks the exact frequency of the interference.A major problem in electrocardiography is "the appearance of unwanted 60Hz interference in the output".The result of adaptive filtering shows complete adaptation and absence of interference after half a second.The cardiac transplantation technique of Shumway involves the working at the same time of the old and the new heart.With adaptive filtering the pulses of the two hearts are easily separated.An adaptive transversal filter con taining 48 weights was used.The sampling rate was 500 Hz.In fetal cardiography the mother's heartbeat interferes in the observation of the fetal heartbeat.The mother's heartbeat is recorded at the chest and abdominal leads record the combined heartbeats.In the noise canceller output the maternal heartbeat is almost completely absent and the fetal heartbeat is clearly discernible.A pilot communicating from the cockpit of an aircraft where a high level of engine noise is present has a second microphone placed ar a suitable location in the cockpit.An adaptive filter with 16 hybrid analog weights,digitally controlled by a computer reduces the noise output by 20 to 25 dB.Convergence time is of the order of seconds.Strong unwanted signals incident on the sidelobes of an antenna array can severely interfere with the reception of speaker signals in the main beam.Adaptive sidelobe cancelling is combined here with noise cancelling.The playback of speech or music in the presence of tape hum or turntable rumble has no external input free of signal available.If a fixed delay is inserted of sufficient length to decorrelate the signal components,the interference components,because of their periodic nature,will remain correlated.This allows cancelling of this type of noise.Experimental data demonstrate the ability of adaptive noise cancelling greatly to reduce additive periodic or stationary random interference in both periodic and random signals. Plotkin and Plotkin[12)] used an adaptive transversal Notch Filter designed in the time-domain. If the interference is not precisely known(e.g.periodical) it is necessary to estimate the instants of impulsive noise occurrences. A delay is necessary to obtain the estimated values of the filter parameters before the actual filtration. The proposed linear structure of the NF makes it possible to track and suppress powerful impulsive interference in the presence of both the desired signal and random noise. Plotkin, Roytman and Swamy[13)] considered the suppression of finite length pulse interference using a modified linear predictor, under the assumption that the pulse interference $\eta(i)$ can be represented as being generated by a linear system driven by a a train of randomly weighted noise impulses, where this system has a finite dimensional(N) linear prediction model

$$\eta(i) = -\sum_{k=1}^{N} b_k \, \eta(i-k) + B\,u(i), \quad i \in (0,M) .$$

where b_k and B are the parameters of the generating linear system and u(i) is the signal applied to it. The adaptive system can take care of deviations from the predicted situation. The advantages of this modified linear predictor are demonstrated by two computer simulated examples.

4. Coding.

If a message is well coded it can be made resistant to the effects of noise. Already in 1946 van Duuren showed, that a simple 4mark-3spaces code can find most errors in telex messages on radio. An ARQ(automatic repetition request) then asks for repetition, and usually the correct message will come through with a certain delay. Hagelbarger found codes for the correction of burst errors. Gramophone records can reproduce excellent sound on long-play records. A serious drawback is, that the mechanical tracking of the grooves causes wear and damage due to operating errors cannot always be avoided. For these and other problems the compact disc system offers a solution. Optical scanning means that the disc is not susceptible to damage and wear. Digital signal processing corrects the great majority of errors that may occur nevertheless. The error correcting code in the cpmpact disc system employs not one but two Cross-Interleaved

Reed-Solomon Codes.The decoder corrects all single errors and detects double or triple errors.The probability that the decoder will not detect quadruple or higher multiple errors is only 2^{-19} Detected errors are concealed by linear interpolation between reliable neighbours or by muting the audio signal.The maximum burst length that is guaranteed capable of interpolation is 12304 bits.[14)]

5.Designing for increased immunity.

Noise can enter a control system by conduction or by direct radiation.Radiation has its effect by capacitive or inductive coupling,and conduction through wiring.The amount of filtering is limited by the rate of data transmission.Employing coaxial cables or shielded twisted-pair lines for signal transmission may help.A single point in a system should be chosen and d nated as the system ground point and care should be taken to avoid ground loops.Radiated interference should be avoided by shielding.High threshold logic may avoid some immunity problems. Inductors and bypass capacitors keep high frequency noise on the ac line from entering the power supply.An electrostatic shield between the primary and secondary of the transformer further reduces noise being injected into the power supply.There are numerous books in this field.[15)]Audio amplifiers containing operational amplifiers may be vulnerable to rf interference. The German authorities have put down stringent requirements to avoid this.It has been found experimentally that capacitors "added to the circuit in the right places" can reduce the RFI effects considerably and Goedbloed,Riemens and Stienstra[16)] presented a theoretical background to this observation for amplifiers with negative feedback.

6.Statistical-physical models of electromagnetic interference.

General EM noise environments are classified into three categories with respect to narrow-band receivers:Class A,typically narrower spectrally than the receiver;Class B,where the bandwidth is larger than that of the receiver's front end stages;Class C, a linear combination of A and B.He used the Amplitude Probability Distribution(APD),the probability,that the instantaneous amplitude observed at the receiver's output exceeds some threshold X_o.The noise sources are independent,randomly distributed in space and emit arbitrary waveforms randomly in time, so that the basic statistics are Poisson. The emitted waveforms obey appropiate propagation laws and explicitly include the effects of source and receiver antenna patterns, relative Doppler effects, source distributions in space, and other geometrical factors. The results are highly non-gaussian, but they are analytically tractable and canonical (i.e. the form of the probability structures are essentially invariant of waveform and kinematic and geometric details). For Class A interference, the APD, or exceedance probability depends on three global parameters: A_A = the impulsive index, this is the average number of emission "events" times the mean duration of a typical interfering source emission. $\Gamma_A' \equiv \sigma_G^2 / \Omega_{2A}$ = the ratio of the intensity of the independent gaussian component σ_G^2 of the input interference including received "front-end" noise to the intensity Ω_{2A} of the "impulsive" non-gaussian component. Ω_{2A} = the intensity of the "impulsive component. Furthermore ϵ and ϵ_0 are the normalized envelopes. With these choices, the APD is given by

$$P_1(\varepsilon > \varepsilon_0) = e^{-A_A} \sum_{m=0}^{\infty} \frac{A_A^{\,m}}{m!} \; e^{-\varepsilon_0^2 / 2\hat{\gamma}_{mA}^2}$$

$$0 < \varepsilon_0 < \infty$$

$$2\hat{\gamma}_{mA}^2 = (m/A_A + \Gamma_A')/(1 + \Gamma_A')$$

For Class B there are six global parameters, that again all have a physical meaning. For the rather complex mathematics we refer to Middleton[17]. He notes, that the agreement between theory and experiment is excellent, the canonical nature of the models is demonstrated (the form of the results is invariant of the specific source mechanism, class A and class B are observably and quantitavely different noise types, the parameters can be obtained from approximate experimental data. On the next page we reproduce some of the experimental comparisons with the model.

The purpose of this research was the application to communication problems by Spaulding and Middleton[18]. They consider: Antipodal signals, orthogonal signals, and on-off keyed signals. For the optimum Bayes (Minimum average risk) detection strategy the likelihood ratio for Class A noise is found. The calculation is straightforward but lengthy,and it is usually uneconomical to build such a receiver.Spaulding and Middleton have calculated bounds on the performance of the optimum detector for several cases,but their main effort was directed at finding a locally optimum Bayes detector(LOBD) for small signal and a large number of independent samples

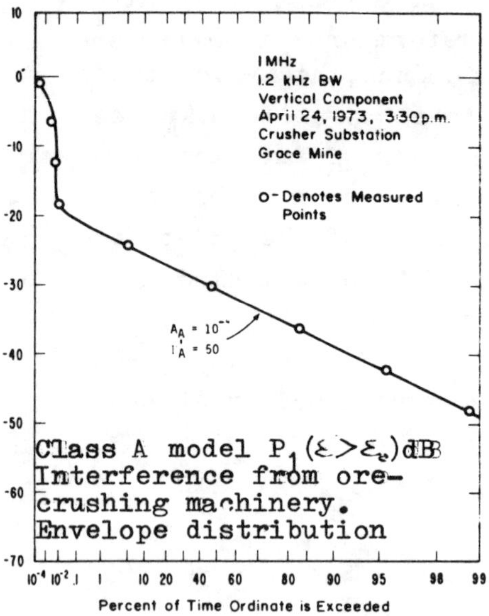

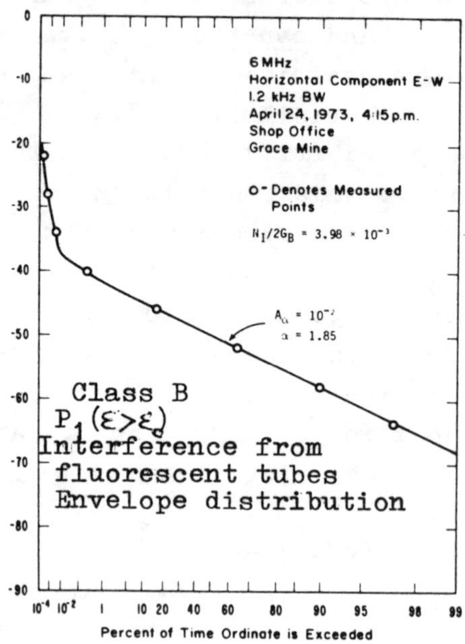

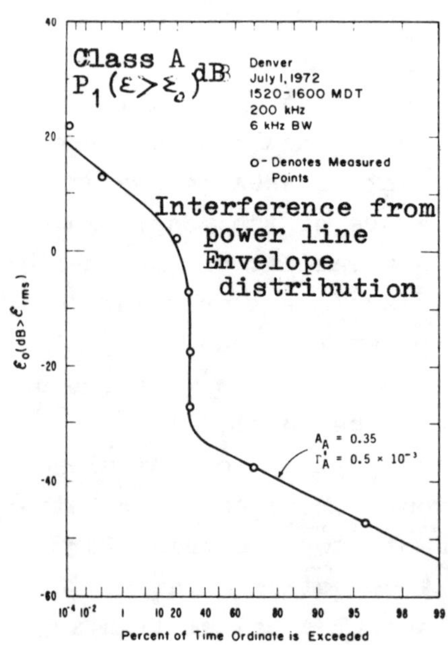

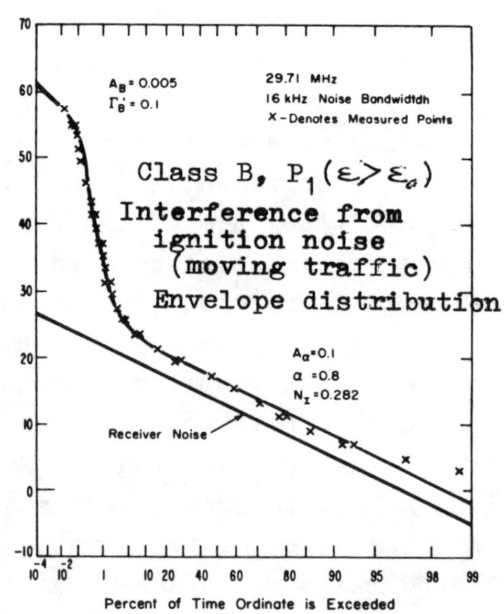

The optimum threshold receiver is the standard Gaussian receiver for coherent signals,preceded by a logarithmic nonlinearity. This receiver must be adaptive,i.e.it is required to adjust itself to changing interference conditions.It must know a priori or estimate the probability distribution.

Berry(19) noted that it was implicit in Middleton's scenario for class A interference,that nearly all potential interference sources are within a relativel small distance range,e.g. the middle 80% of the potential sources of interference to mobile reception should be between 6 and 13 km from the receiver.In an effort to meet this criticism,Middleton[20] extended his earlier work to include additional source distributions in a quasi-canonical model.He agreed that only for sources essentially located at equal distances from the receivers we can expect the strictly canonical results to hold.He found general quasi-canonical forms showing acceptable agreement to related EMI scenarios.

Spaulding[21] got new results,e.g. for diversity reception in atmospheric noise.Middleton[22] further critically examined the problem of developing weak signal detection algorithms for general classes of man-made and natural noise.Spaulding[23] could remove some critical assumptions("sufficiently"small signal level and large time-bandwidth product).This leads to a better estimate of performance in actual situations.A further discussion of space-diversity reception in fading was given by Izzo and Paura[24].

7)Fundamental limits on EMC .

Fundamental and state of the art limits were treated in a survey paper by Showers,Schulz and Lin [25].Apart from other subjects discussed above,they give attention to interference problems in radar and sonar,and to coupling limitations(radiation,induction,transmission line coupling).

REFERENCES.

1)W.F.Chow:Impulse noise reduction circuit for communication receivers.I.R.E.Trans.Veh.Commun.,VC9,May 1960,1-9.

2)W.Gosling:Impulsive noise reduction in radio receivers.The Radio and Electronic Engineer,vol 43,341-347,1973.

3)J.D.Martin:Audio Clipping and Blanking in an AM Receiver,IEEE Trans.Veh.Tech.,VT 29,1980,298-304.

4)G.Hepp.Suppressing interference in FM radio receivers.British Patent Spec.1279756,publ.1972.Suppressing interference in receivers of electrical signals.British Patent Spec.1312935,publ.1973. Circuits for suppressing interference in receivers of electrical signals.German Patent Publ.2051632,publ1971(in german).

5)R.A.Wainwright:On the Potential Advantage of a Smearing-Desmearing Filter Technique in Overcoming Impulse-Noise Problems in Data Systems.IRE Trans.Comm.Systems,1961,362-366.

6)S.R.Al-Araji,A.H.Abdullah:Impulsive Noise Reduction Employing Smear/Desmear Technique.1984 Int.Symp.Electromag.Compatibility. Tokyo,1984,860-866.

7)G.F.M.Beenker,T.A.C.M.Claasen,P.J.Van Gerwen:Design of Smearing Filters for Data Transmission Systems.IEEE Trans.Comm.COM33, 1985,955-963.

8)J.S.Engel:Digital Transmission in the Presence of Impulsive Noise.Bell Syst.Tech.J.,1965,1699-1743.

9)E.J.Baghdady:Linear Cancellation Technique or Suppressing Impulse Noise.IRE Trans Veh.Commun.,VC9,December 1960,1-9.

10)D.E.Watt-Carter,L.K.Wheeler:The Lincompex System for the Protection of H.F.Radio Telephone Circuits.Post Office Elect. Eng.J.vol.59163-167.

11)B.Widrow,J.R.Glover,J.M.Mc Cool,J.Kaunitz,C.S.Williams,R.H. Hearn,J.R.Zeidler,E.Dong Jr,R.C.Goodlin:Adaptive Noise Cancelling: Principles and Applications.Proc.IEEE,vol.63,1975,1692-1716.

12)E.Plotkin,A.Plotkin:An Adaptive Approach to Suppress Powerful Impulsive Interference.Signal Processing 4,1982,25-33.

13)E.I.Plotkin,L.M.Roytman,M.N.S.Swamy:A New Approach to Suppression of Finite-Length Pulse Interference Using Modified Linear Predictor.IEEE Trans.Acoustics,Speech,Signal Processing,ASSP31, 1983,622-629.

14)H.Hoeve,J.Timmermans,L.B.Vries:Error correction and concealment in the Compact Disc system.Philips tech.Rev.vol40,1982,166-173.

15)V.Boaen:Designing logic circuits for high noise immunity. IEEE Spectrum,Jan.1973,53-59.

16)J.J.Goedbloed,K.Riemens,A.J.Stienstra:Increasing the RFI immunity of amplifiers with negative feedback.Proc.EMC Symp.1983,

Zurich,471.476.
17)D.Middleton:Statistical-Physical Models of Electromagnetic interference.IEEE Trans.Electromagnetic Compatibility,EMC 19, 1977,106-127.
18)A.D.Spaulding,D.Middleton:Optimum Reception in an Impulsive Interference Environment-Part I.Coherent Detection,Part II,Incoherent Detection.IEEE Trans.Commun.COM25,1977,910-923,924-934.
19)L.A.Berry:Understanding Middleton's canonical formula for Class A Noise.IEEE Trans.Electromag.Compat.,EMC 23,337-344,1981.
20)D.Middleton:Canonical and Quasi-Canonical Probability Models of Class A Interference.IEEE Trans.Electromag.Compat.,EMC25,76-106,1983.
21)A.D.Spaulding:Optimum Threshold Signal Detection in Broad-Band Impulsive Noise Employing Both Time and Spatial Sampling.IEEE Trans.Comm.,COM29,147-152,1981.
22)D.Middleton:Threshold Detection in Non-Gaussian Interference Environments:Exposition and Interpretation of New Results for EMC Applications.IEEE Trans.Electromag.Compat.,EMC 26,1984,19-28.
23)A.D.Spaulding: Locally Optimum and Suboptimum Detector Performance in a Non-Gaussian Interference Environment.IEEE Trans. Comm.COM 33,1985,509-517.
24)L.Izzo, L.Paura:Asymptotically Optimum Space-Diversity Detection in Non-Gaussian Noise.IEEE Trans.Comm.,COM34,97-103.
25)R.M.Showers,R.B.Schulz,S-Y.Lin:Fundamental Limits on EMC. Proc.IEEE,vol.69,1981,183-195.

COMPATIBILITY OF 144 Kbits ISDN DIGITAL SIGNALS WITH EXISTING SYSTEMS

Z.C. ZHANG

Nanjing Institute of Posts and Telecommunications, NANJING, CHINA

1. INTRODUCTION

Recently, architectures and standards of digital networks for voice and data transmission have been established. These are referred to as Integrated Digital Networks (IDN). The need for different types of digital networks (for telephone, for bulk data, and for bursty data) has stimulated the development of the concept of an Integrated Services Digital Network (ISDN) |1,2|. ISDN provides capabilities for a variety of services using a variety of communications modes. The information rate available to each ISDN subscriber will probably be 144 Kbit/s |3,4|, as resulting from the sum of three digital channels: B1 (64 Kbit/s) + B2 (64 Kbit/s) + D(16 Kbit/s). Channels B1 and B2 will convey the useful information, while channel D is reserved for signalling, and on a time-sharing basis, for packet-switched data transmission |4|.

In this paper we explore some design issues for a 144 Kbit/s digital transmission system. In particular, our attention is focused on the compatibility of this system with pre-existing signals flowing in the same network. In fact, as the new digital system will coexist in the distribution network with conventional systems, like analog telephone and data transmission at various speeds, several design choices (e.g., the type of transmission system - frequency division or burst-mode, or echo cancellation -, of line code, and of transmitted pulse shape) are affected by the compatibility problem. Our goal is to address the compatibility issue by calling the attention to a set of numerical techniques that provide a computationally efficient, as well as accurate, tool for the analysis.

2. DIGITAL-TO-DIGITAL CROSSTALK

Fig.1 shows the model assumed for the analysis of the effects of digital-to-digital crosstalk. The source symbols are first passed through a line encoder; the codes considered here will be pseudoternary (in particular, AMI and MS43), but our analysis can be easily extended to different types of line codes. The resulting signal is filtered by a shaping (or transmitting) filter; then, sent to the transmission line. Addition of thermal noise is accounted for by adding to the signal at the line output a member of a white Gaussian noise process. At the receiver's front end, the signal is processed by a line amplifier and a receiving

J. K. Skwirzynski (ed.), Performance Limits in Communication Theory and Practice, 383–390.

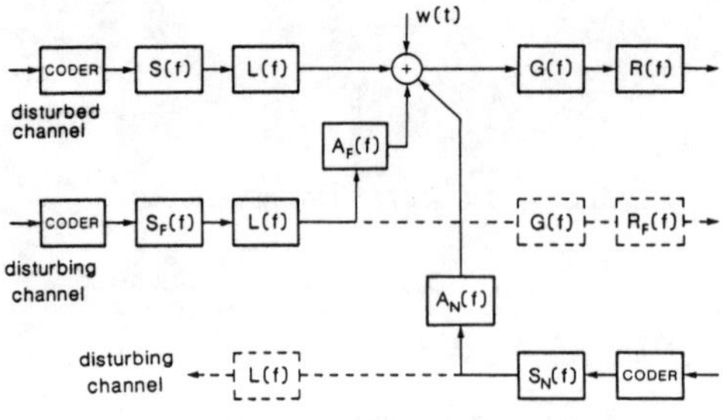

Fig.1

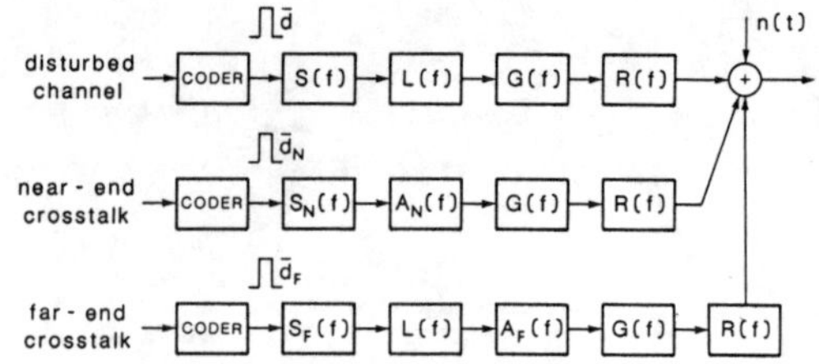

Fig.2

filter. More specifically, in Fig.1 we denote by S(f), L(f), G(f), and R(f) the transfer functions of the transmitting filter, of the line, of the line amplifier, and of the receiving filter, respectively. The noise process w(t) is assumed to have a power spectral density No/2. We consider two disturbing channels giving rise to far-end crosstalk (FEXT) and to near-end crosstalk (NEXT), respectively. They are identified by the subscripts F and N; the linear systems with transfer functions AF(f) and AN(f) account for FEXT and NEXT attenuations.

For the purpose of our computations, the transmission system model is changed into the equivalent form of Fig.2. Here, we also show the amplitudes of the pulses entering the transmission filters in the three channels under consideration. Notice that the noise n(t) entering the system is not white anymore.

To examine the effects of digital-to-digital interference, the standard performance measure is error probability. Thus, we show how it can be computed using an approach which combines simulation with analysis. Observe first that the average symbol error probability P_e *is given by*

$$P_e = \sum_l P_l \, P_e(l) \tag{1}$$

where P_l *denotes the probability that the symbol l is transmitted,* $P_e(l)$ *denotes the probability that the symbol l is received incorrectly, and the summation is extended to the values +1,0,-1 when we are assuming ternary line encoding, and to the values +1,0 when there is no line code (and hence we are considering a "return-to-zero" signal format). Notice that we assume a suboptimum receiver, i.e., one taking decisions on a symbol-by-symbol basis: in other words, in the decision process we take no advantage of the correlation among the transmitted symbols, as it could be done by assuming a Viterbi receiver (see, e.g., |5, Chapter 9|).*

In our model, the received signal is perturbed by additive Gaussian noise, intersymbol interference (ISI), FEXT, and NEXT. If $X_{ISI}(l)$, X_{FEXT}, *and* X_{NEXT} *denote the interference terms corresponding to unit pulse amplitudes (i.e., to* $d=d_F=d_N=1$ *in Fig.2), we can define the interference random variable (RV)*

$$X(l) = X_{ISI}(l) + \frac{d_F}{d} X_{FEXT} \frac{d_N}{d} {}_{NEXT} \tag{2}$$

Thus, if δ_n^2 *denotes the variance of the samples of the noise* $n(t)$, h_o *is the response (at sampling time) of the overall disturbed channel (whose transfer function is* $S(f)L(f)G(f)R(f)$ *) to a unit input pulse, and* $f_{X(l)}(x_l)$ *is the probability density function of the RV* $X(l)$ *defined in (2), having range* $\underline{\underline{X}}(l), l=-1,0,1$, *we have*

$$P_e(0) = \int_{\underline{\underline{X}}(0)} \operatorname{erfc}\left[\frac{d(\frac{h_o}{2} - x_o)}{\sqrt{2}\ \delta_n}\right] f_{X(0)}(X_o)\, dX_o$$

$$P_e(\pm 1) = \frac{1}{2} \int_{\underline{\underline{X}}(\pm 1)} \operatorname{erfc}\left[\frac{d(\frac{h_o}{2} - x_{\pm 1})}{\sqrt{2}\ \delta_n}\right] f_{X(\pm 1)}(X_{\pm 1})\, dX_{\pm 1} \tag{3}$$

As exact evaluation of the integrals appearing in (3) is in general hopelessly complicated, we resort to their computation using the technique of Gauss Quadraturea Rules (see, e.g., Appendix E of |5|). This technique is based on the approximation

$$\int_{\underline{\underline{X}}} \operatorname{erfc}\left[\frac{d(\frac{h_o}{2} - x)}{\sqrt{2}\ \delta_n}\right] f_x(X)\, dX = \sum_{i=1}^{J} w_i \operatorname{erfc}\left[\frac{d(\frac{h_o}{2} - x_i)}{\sqrt{2}\ \delta_n}\right], \tag{4}$$

where the constants x_i, w_i, $i=1,\ldots,J$, *are called the* <u>*abscissas*</u> *and the* <u>*weights*</u> *of the quadrature rule, respectively, and can be determined from the first 2J moments of the RV X. In conclusion, the problem of evaluating the error probability (1) is reduced to the calculation of the parameters* h_o, δ_n^2, *and of the moments.*

Here, the situation is further complicated by the fact that, as $X(l)$ *depends on the transmitted symbol l, so do its moments*

$$M^{(h)}(l) = E\left\{X^h(l)\right\} \tag{5}$$

We shall now describe in some detail the computational steps that we used for the computation of the error probability through (4).

Under the assumption that the three terms appearing in the RHS of (2) are independent, the moments of their sum can be determined if we know the moments of each term. With obvious notations, we have:

$$M^{(h)}(l) = E\left[X_{ISI}(l) + \frac{d_F}{d} X_{NEXT} + \frac{d_N}{d} X_{FEXT}\right]^h$$

$$= \sum_{j=1}^{h} \sum_{i=1}^{j} \binom{h}{j} \binom{j}{i} M_{ISI}^{(i)}(l)\, M_{FEXT}^{(j-i)}\, M_{NEXT}^{(h-j)} \tag{6}$$

The computation of the moments $M_{ISI}^{(h)}(l)$ *of the ISI terms for a correlated symbol sequence, and hence when line codes are used, was considered in |6| by Cariolaro* <u>*et al.*</u>*. Here that method was applied; the response of the overall disturbed channel to a pulse of unit amplitude was determined by simulation, using the TOPSIM III package |7|. The moments of*

FEXT and NEXT were computed using the same techniques, but introducing a random misalignment $\vartheta \cdot T$ between the disturbing and the disturbed signal, with ϑ a continuous RV uniformly distributed in the interval (0,1), and T the signaling period. If $M^{(h)}_{FEXT}(\vartheta)$ and $M^{(h)}_{NEXT}(\vartheta)$ denote the conditional moments of FEXT and NEXT for a given value taken by the RV ϑ, we have

$$M^{(h)}_{i} = \int_{0}^{1} M^{(h)}_{i}(\vartheta)\, d\vartheta , \qquad i = FEXT, NEXT . \tag{7}$$

In turn, the integral (7) can be evaluated using a standard Gauss Quadrature Rule (see, e.g., |8|).

The computation of the Gaussian noise variance δ^2_n is obtained by recalling that n(t) of Fig.2 is the response of a system with transfer function G(f)R(f) to a white Gaussian noise with power spectral density No/2. Thus, evaluation of δ^2_n is based on the evaluation of the equivalent noise bandwidth of that system. This can be done using TOPSIM.

Finally, to the purpose of drawing error probability curves versus E/No, where E denotes the average energy per symbol of the signal at the receiver's front-end, we observe that because of the linearity E is proportional to d and to the duration T of a symbol in the disturbed channel, i.e.,

$$E = d^2 T/\alpha \quad , \tag{8}$$

where α is a constant to be determined by simulation.

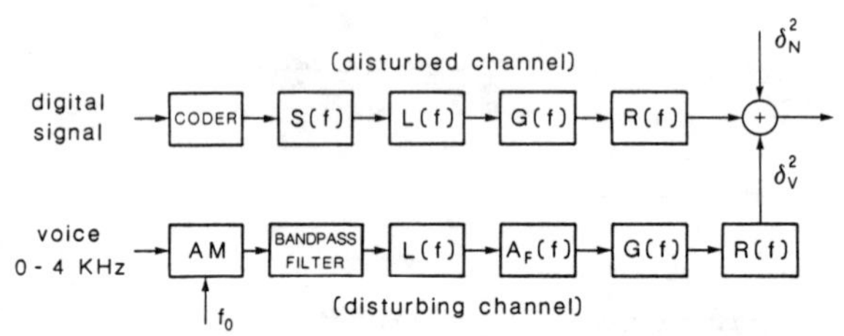

Fig.3

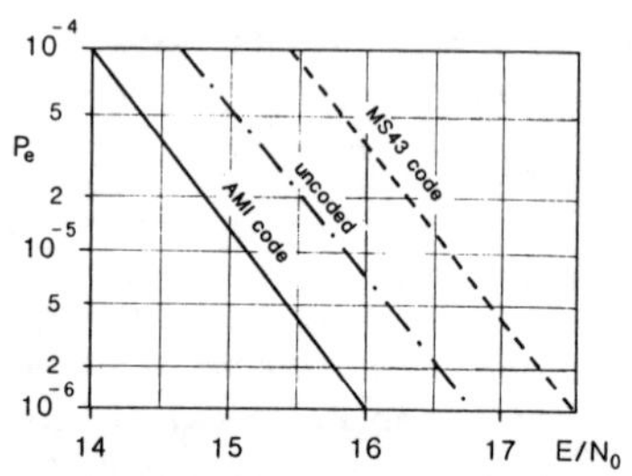

Fig.4

3. VOICE-TO-DIGITAL CROSSTALK

The model assumed for the computation of voice-to-digital crosstalk is shown in Fig.3. Here we consider only one disturbing channel (say, FEXT), but the method presented in this section can be easily extended to encompass more than one. The technique used to evaluate the error probability is similar to that presented in Section 2. In particular, (3) still holds, provided that (2) is changed into

$$X(1) = X_{ISI}(1) + \frac{1}{d} X_V . \tag{9}$$

The new problem here is the computation of the RV X_V in (9), which represents the disturbing voice signal. Our assumption is that the voice *signal can be modeled using a Gaussian probability density function with mean zero and variance δ_V^2, so that the disturbing sample X_V is still Gaussian. Hence, its moments can be evaluated, because all of them depend only on δ_V^2. We note in passing that with these assumptions the effect of more than one voice channel can be easily accounted for, because the disturbing RV in this case would still be Gaussian, with a variance equal to the sum of the variances of the single disturbing voices (provided of course that they are assumed to be independent).*

4. COMPUTATIONAL RESULTS

In this section, we shall present some results obtained along the lines previously described. In particular, we shall examine the degradation of signal-to-noise ratio caused by the different types of impairment considered. A comparison of the degradation suffered by signals with different codes will show the influence of line coding on system performance.

As said previously, we consider digital transmission at a net rate 144 Kbit/s; the actual rate of transmission over the user's line is 160 Kbit/s, resulting from the addition to the information symbols of framing and housekeeping symbols |4|. With reference to Figs. 2 and 3, the system blocks are modeled as follows. S(f) and R(f) are such that their product has a raised-cosine transfer function, with bandwidth 80 KHz (for uncoded transmission or AMI code) or 60 KHz (for the MS43 code) and roll-off coefficent 0.5. S(f) is equal to R(f) times a preshaping term which compensates for the rectangular input pulses. $S_N(f)$ and $S_F(f)$ are assumed to be equal to S(f). The codes considered are AMI and MS43 (see, e.g.,|5|). L(f) is assumed to correspond to 1 Km of 0.6 mm cable; its transfer function is that shown in Fig.16 of |4, p.63|. The line amplifier transfer function G(f) is assumed to compensate exactly for the line attenuation. The transfer functions $A_F(f)$ and $A_N(f)$ of FEXT and NEXT on symmetric-pair cables are modeled as in |4, p.57|, i.e.

$$\begin{aligned} A_F(f) &= -(K_F)_{dB} + \mu_F \log_{10}(f/f_o) + 10\log_{10} D \quad dB \\ A_N(f) &= -(K_N)_{dB} + \mu_N \log_{10}(f/f_o) \quad dB \quad , \end{aligned} \tag{10}$$

where fo=80 KHz, $\mu_F=\mu_N=13.4$ dB (the slopes of the attenuation/frequency characteristics), D is equal to the length (in Km) of the facing lines (here we assume D=5), K_F is the Kilometric FEXT attenuation at frequency fo, and K_N in the NEXT attenuation at frequency fo. Notice that if only K_F and/or K_N change, and the remaining parameters are fixed, the shape of the signal disturbing the useful channel remains the same, with just a variation in its amplitude.

Fig.4 shows the reference error probability curves, i.e., the values of Pe obtained for a signal perturbed only by additive white Gaussian noise. These curves can be obtained as follows. As in (3) there is no interfering term, the error probability has the form

$$P_e = a \; erfc \left(\sqrt{\frac{E}{N_o}} \cdot \sqrt{\alpha} \cdot K \right), \qquad (11)$$

where a is a constant dependent only on the code used, α is the constant defined in (8), and K is a constant independent of the code. In particular, for uncoded transmission we have a=0.5 and α=1/0.474, for AMI code a=0.75 and α=1/0.387, and for MS43 a=0.676 and α=1/0.542.

Fig.5 shows the error probability for transmission disturbed by ISI, NEXT, and FEXT. By choosing the FEXT factor K_F in (10) 10 dB less than the NEXT factor K_N, it was experienced that the effect of FEXT is dominant (compare, for example, the curve labeled ③ and referring to AMI code with ISI, NEXT, and FEXT, with curve ②, in which NEXT is disregarded). Figs. 6-8 show the effect of ISI and FEXT in different situations. In these figures, the dashed lines refer to a situation in which the disturbed and the disturbing channels are time-aligned (and hence the variable ϑ introduced in Section 2 is identically equal to zero). It is seen that time-alignment causes a slight increase of error probability.

Fig.9 shows the loss in signal-to-noise ratio (with respect to the reference curves of Fig.4) suffered from of ISI and FEXT for an error probability 10^{-6}. For comparison's sake, notice that the loss due to ISI alone is 0.32 dB without coding, 0.82 dB with AMI code, and 0.84 dB with MS43 code. It can be seen that the losses of the two coded systems are similar, and larger than those of uncoded transmission.

Fig.10 shows the effect of far-end voice crosstalk on an AMI-coded signal. It is seen that the influence of a baseband (0-4 KHz) voice crosstalk is very small, whereas the influence of carrier (36-40 KHz) voice crosstalk is comparable to that of digital interference. Fig.11 compares the loss in signal-to-noise ratio (with respect to the reference curves of Fig.4) suffered from ISI and voice crosstalk for an error probability 10-6 and different codes. It is seen that the effect of coding is an improved sensitivity to interference.

Finally, Fig.12 compares the effects of different interferences on MS43 and uncoded system.

5. CONCLUSIONS

We have addressed the compatibility issue in the 144 Kbit/s ISDN system, by collecting and exploiting a number of analytical, computational, and simulation techniques that provide, when put together, an efficient and flexible tool for this analysis. Although the examples refer to a specific situation, the application range is very wide, as it can encompass not only different channels, but also different codes and different interfering sources. The numerical examples given should show how guidelines for the design can be obtained.

6. ACKNOWLEDGEMENT

The author is indebted to his colleague Ezio Biglieri of the Dipartimento di Elettronica, Politecnico di Torino, for suggesting him the subject of this investigation.

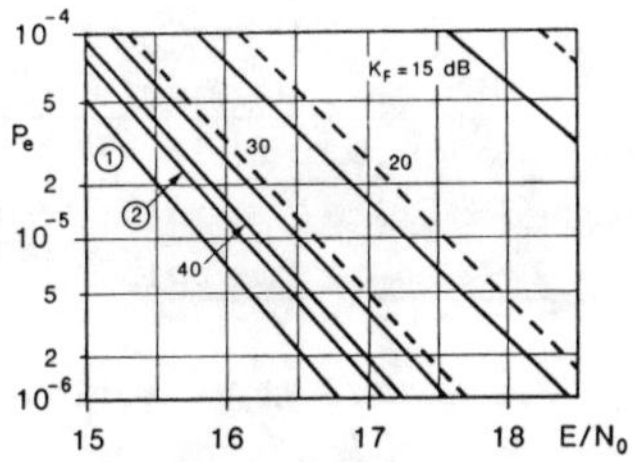

Fig.6 Pe of uncoded

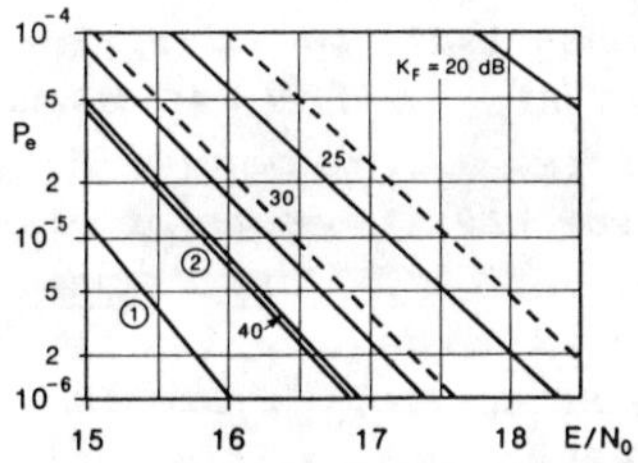

Fif.7 Pe of AMI code

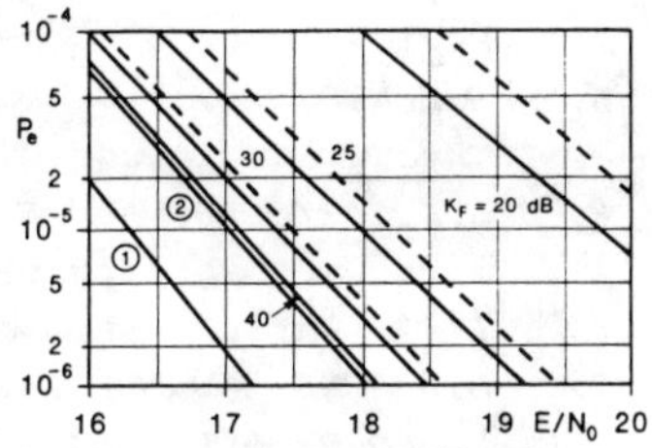

Fig.8 Pe of MS43 code

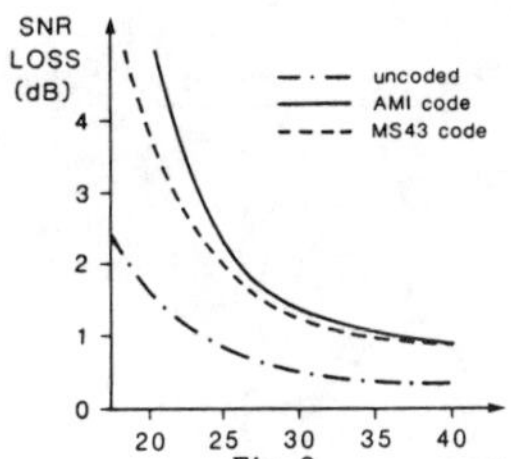

Fig.9

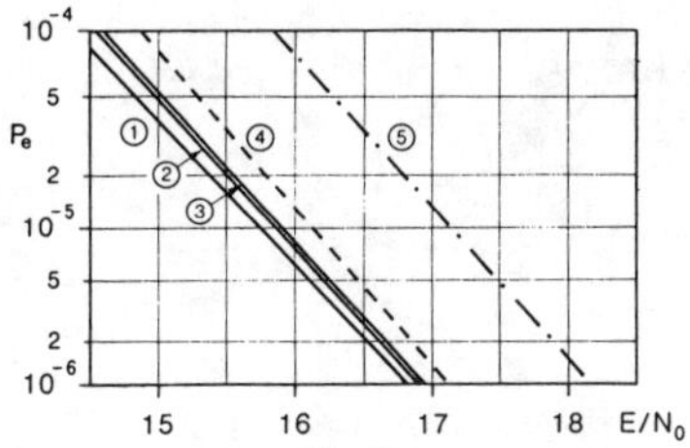

Fig.5

In Fig.5 ① AMI code,Kn=50db
② AMI code,Kf=40db
③ AMI code,Kn=50db and Kf=40db
④ Uncoded ,Kn=50db and Kf=40db
⑤ MS43 code, Kn=50db and Kf=40db

In Fig.6,7,8, ① No FEXT, no ISI
② No FEXT, with ISI

In Fig.10 ① No crosstalk
Solid lines are with 36-40 KHz crosstalk, the dashed line refers to 0-4KHz voice FEXT, Kf=20db

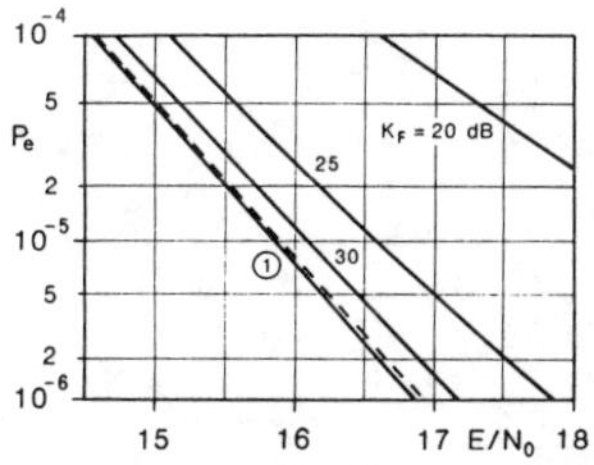

Fig.10

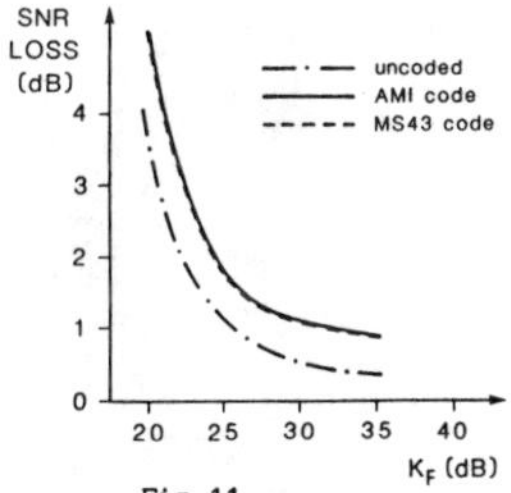

Fig.11

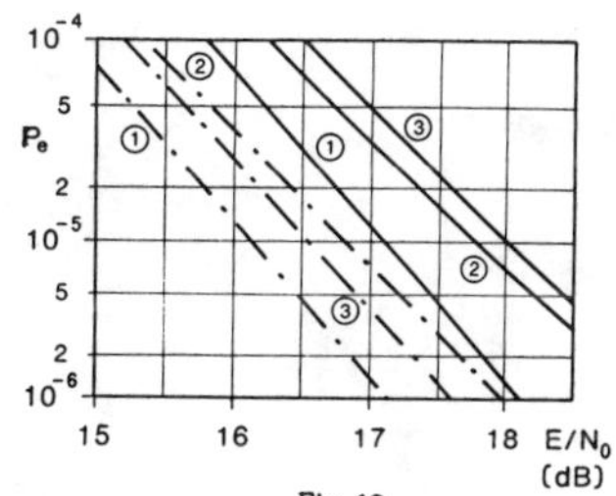

Fig.12

In Fig.12 ① With ISI, nocrosstalk. ② 36-40KHz voice FEXT(Kf=25db) ③ Digital FEXT,Kf=25
Dashed lines: uncoded. Solid lines: MS43 code

REFERENCES

1. ***M.DECINA and A.ROVERI,*** *"Integrated Services Digital Network: Architectures and Protocols", Chapter 2 of: K.FEHER et al., Advanced Digital Communications and Signal Processing, Prentice-Hall, Inc., Englewood Cliffs, NJ, 1986.*
2. ***M.DECINA,*** *"Managing ISDN thourgh international standard activities", IEEE Communications Magazine, vol. 20, no. 5, September 1982, pp. 19-25.*
3. ***P.J.VAN GERWEN, N.A.M.VERHOECKX, and T.A.C.M. CLAASEN,*** *"Design considerations for a 144 kbit/s digital transmission unit for the loca telephone network", IEEE Journal on Selected Areas in Communications, vol. SAC-2, no.2, March 1984, pp. 314-323.*
4. ***L.BELLATO, G.CAPOROTUNDO, and G.GIROLAMI,*** *"144 kbit/s digita transmission on the subscriber loop for ISDN user access", TELETTRA*

Review, no. 36, December 1984, pp. 51-86.

5. **S. BENEDETTO, E. BIGLIERI, and V. CASTELLANI**, *Digital Transmission Theory*. Englewood Cliffs, N.J.ò Prentice-Hall, to be published.

6. **G.L. CARIOLARO and S. PUPOLIN**, "Moments of crrelated digital signals for error probability evaluation", *IEEE Trans. on Inform. Theory*, Vol. IT-21, n.5, September 1975, pp. 558-568.

7. **M. AJMONE MARSAN, S. BENEDETTO, E. BIGLIERI, V. CASTELLANI, M. ELIA, L. LO PRESTI, and M. PENT**, "Digital Simulation of communication systems with TOPSIM III", *IEEE Journal on Selected Areas in Communications*, vol. SAC-2, no.1, January 1984, pp. 29-41.

8. **M. ABRAMOWITZ AND I.A. STEGUN**, *Handbook of Mathematical Functions*. New York: Dover Publ., 1972.

CHANNEL MODELS FOR RANDOM-ACCESS SYSTEMS

J.L. Massey

Institute for Signal and Information Processing
Swiss Federal Institute of Technology
CH-8092 Zurich, Switzerland

1. INTRODUCTION

Our aim in this paper is to examine the ultimate limits on the performance of random-access systems. In broadest terms, a random-access system can be defined as a multiple-access system in which, either by necessity or by choice, it can happen that two or more stations simultaneously attempt transmission over a communications link that can carry only a single message reliably.

Wozencraft and Reiffen [1,P.6] attribute to J.L.Kelly the provocative definition of a channel as that part of the communications system that one is "unwilling or unable to change". Once the channel has been specified, then so has been its ultimate limit or "capacity" to transmit information -- at least in principle. The determination of this capacity may be an extremely formidable task (or a relatively simple one). The devising of practical communications strategies whose performance approaches capacity can be even more challenging.

The logical order of events in communications theory is: (1) specification of the channel model, (2) determination of its capacity, and (3) design of the communications strategy. Historically, however, the last has often been the first. Communications engineers do not always have the luxury to await completion of (1) and (2) before they design their systems. Indeed, communications systems are often designed before anyone has hardly begun to think about a channel model, i.e., to think about what should be considered unchangeable. Such was the case with random-access communications, which began with a particular random-access strategy that formed the cornerstone of the ALOHA system [2].

The ALOHA system was the brainchild of Abramson and his colleagues at the University of Hawaii, who faced the problem of designing a radio network that would link six other stations on the various islands in Hawaii to the computing resources at the main campus near Honolulu [3]. "Even at the beginning [in 1968] of the project, however, it was understood that the intermittent operation typical of interactive computer terminals was a convincing argument against the assignment of point-to-point channels in a conventional frequency-division multiple access (FDMA) or time-division multiple-access (TDMA) manner. Some form of sharing a common communication channel

J. K. Skwirzynski (ed.), Performance Limits in Communication Theory and Practice, 391–402.

resource appeared necessary....The key decision to use the direct form of transmitting user information in a single high speed packet burst...was made at a meeting of the project participants in 1969. The operation of the channel was not well understood at that time... ." [3,p.119]. By the time of the first publication, [2], on the ALOHA system, only a year later, Abramson had already generalized the system concept and had made an intriguing and inventive analysis of its performance, which we now describe.

Abramson considered the situation where infinitely many identical stations share a common radio channel. Each station occasionally obtains messages from an associated message source and each such message requires the same amount of time for transmission. We shall call these messages "packets" and shall call this transmission time one "slot". The totality of new packet arrivals at the stations forms a Poisson process with intensity λ packets/slot. Abramson assumed that a transmitted packet is successfully received unless and only unless its transmission overlaps (completely or partially) that of another packet. Moreover, he assumed that, immediately after transmitting a packet, a station would be informed via a feedback channel as to whether the transmission was successful or not. The ALOHA strategy was that (i) each station transmits a packet immediately upon its arrival and (ii) after an unsuccessful transmission, the station retransmits the packet after some randomly-chosen delay. Abramson postulated (but did not specify) a retransmission strategy such that the starting times of the totality of retransmitted packets formed a Poisson process with intensity λ_r, independent of the arrival process. The starting times of all packet transmissions then form a Poisson process with intensity $\lambda_t = \lambda + \lambda_r$. Consider a particular packet transmission. The probability that it is successful is just the probability that no other packet transmission begins in the two slot interval beginning one slot before the start of transmission of the considered packet and ending with the end of this transmission, namely $\exp(-2\lambda_t)$. Thus, successful packets are received at an average rate of $\lambda_t \exp(-2\lambda_t)$ packets/slot. The stationarity of the retransmission process is tantamount to assuming a "statistical equilibrium" in which new packet arrivals occur at the same average rate as successes, i.e.

$$\lambda = \lambda_t e^{-2\lambda_t}. \tag{1}$$

The maximum λ satisfying (1) occurs when $\lambda_t = 1/2$ packet/slot and the maximum is

$$\max \lambda = \frac{1}{2e} \approx .184 \text{ packets/slot.} \tag{2}$$

Roberts [4] was soon to note that this maximum throughput could be doubled if the stations would slightly delay their transmissions so that the starting times of all packet transmissions were integer multiples of a slot. With this refinement, (1) and (2) become

$$\lambda_t = \lambda_t e^{-t} \tag{3}$$

and

$$\max \lambda = \frac{1}{e} \approx .368 \text{ packets/slot,} \tag{4}$$

where the maximum occurs for $\lambda_t = 1$ packet/slot.

It is now customary to refer to Abramson's original strategy as "pure-ALOHA" and to Roberts' refinement as "slotted-ALOHA". It is also rather common practice to refer to the maximum throughputs 1/(2e) and 1/e as the "capacities" of the pure-ALOHA and slotted-ALOHA "channels", respectively. We would argue that the term "channel" is inappropriate because it is unclear what a system designer is not free to change, and that "capacity" is a characteristic that should be reserved for channels. We prefer to say that 1/(2e) and 1/e are the "maximum throughputs" of the pure-ALOHA and slotted-ALOHA "strategies", respectively.

Abramson's work ignited an explosion of further work on random-access schemes that is still in progress. [As one measure of this activity, I would cite my far-from-exhaustive collection of papers on this subject that now forms a pile about one meter high, most of which dates from the past five years.] There has, however, been scant attention given to channel models for random-access communications. This makes for confusion; throughputs are often compared for schemes that in fact require different channel models. It also impedes progress toward the fundamental understanding of random-access communications.

In what follows, we describe what seem to us to be the most interesting channel models for random-access systems that have appeared in the literature either explicitly or implicitly, and we describe what is known to us about their capacities.

2. COMMON FEATURES OF RANDOM-ACCESS CHANNEL MODELS

In this section, we describe the features that all of the subsequently considered random-access channel models have in common.

2.1 Fixed-length packets

At any given time, a station can be active (i.e., transmitting) or inactive (i.e., silent). When active, a station must transmit a signal of duration one time-unit; the time unit will also be called a "slot". There are Q different such signals or "packet values", where Q is an integer at least 2.

2.2 Propagation delays

A signal sent by station i will arive d_i slots later at the receiver, where d_i is a nonnegative real number.

2.3 Collision-type reception

The received signal identifies the value of a packet when and only when that packet arrives at the receiver without overlapping (completely or partially) any other arriving packet. The received signal also identifies the silent periods when no packet is arriving at the receiver. During reception of packets that overlap other packets, the received signal is independent of the values of any of these "colliding" packets.

3. CAPACITY DEFINITIONS

We shall consider only the "symmetric" case where the stations are fed by identical, but independent, packet-emitting information sources. The task of each station is of course to communicate correctly to the receiver the sequence of "data packets" that this station receives from its source. The following packet-timing prohibiton, abeit somewhat strange, appears to be the key feature that distinguishes random-access channels from other multiple-access channels treated in information theory:

The timing of packet transmissions by a station cannot depend on the values of the data packets to be communicated.

In other words, the randomness used to time access attempts must be independent of the randomness of the data. This excludes a station from conveying information by the pattern of collisions and idle periods that it causes, which seems an undesirable consequence; but it also makes performance independent of the statistics of the data packets, which is a most desirable consequence.

The stations are of course free to code the data packets into transmitted packets however they please.

When the number of stations is finite, say M, we assume that the generation times of data packets by each information source form a Poisson process of intensity λ/M packets/slot. Moreover, all data packets are independent and are equally likely to take on each of the Q possible packet values. We define the capacity then to be

$$C = \sup \lambda \quad \text{(packets/slot)}, \tag{5}$$

where the supremum is taken over all communication strategies such that (i) the receiver can eventually correctly reconstruct the output of each data source and identify its station number from the received signal, and (ii) the backlog of unreconstructed data packets remains bounded with probability one as time increases. Here, by the backlog of unreconstructed data packets at time t, we mean the number B_t of packets that have been generated by the M information sources up to time t but have not yet been reconstructed at the receiver. [When all delays d_i are 0 and when the strategy involves simply transmitting a data packet (at possibly randomly selected times) until it is received correctly, B_t is just the total number of data packets in the station queues at time t plus the number (at most M) in transmission or awaiting retransmission at time t.] Condition (ii) means that

$$\lim_{k\to\infty} \lim_{t\to\infty} \Pr(B_t \leq k) = 1. \tag{6}$$

Because the data packets are independent random variables with entropy $\log_2 Q$ bits, because the packet-timing prohibition implies that the values of these data packets are independent of the received signal except in intervals where uncollided packets are received, and because the entropy of the received signal in these latter intervals can be at most $\log_2 Q$ bits per slot, it follows that

$$0 \leq C \leq 1 \quad \text{(packets/slot)}. \tag{7}$$

We should perhaps stress that the upper bound in (7) depended crucially on the packet-timing prohibition, as it is seldom noted in the literature that such a prohibition applies. It is easy to construct "pathological" channels with $C > 1$ when this prohibition is removed.

When the number of stations is countably infinite, we assume that the totality of generation times of data packets form a Poisson process of intensity λ, and again that the data packets are independent random variables with entropy $\log_2 Q$ bits. Because it would require an infinite amount of information to specify the identity of the station where a data packet was generated, we cannot require such identification at the receiver if we wish to avoid $C \equiv 0$. Moreover, because the probability is 1 that no station will ever have two or more packets to transmit, we only need to consider strategies that deal with at most one packet generation at each station. The only strategies of this type that need be considered are those in which a station transmits its data packet (should its source ever generate one) unchanged in every slot where the station is active. Thus, for this infinite station case, we define capacity to be

$$C_\infty = \sup \lambda \quad \text{(packets/slot)} \tag{8}$$

where the supremum is taken over all such data packet transmission schemes such that (i) each data packet is eventually received collision-free and any subsequent collision-free receptions of this same packet can be recognized as such repetitions from the received signal, and (ii) the backlog of data packets not yet received collision-free remains bounded with probability one as time increases.

It is obvious that

$$0 \leq C_\infty \leq 1 \quad \text{(packets/slot)}. \tag{9}$$

It is interesting to note that the upper bound in (9) holds even if the packet-timing prohibition is removed. In fact, however, we implicitly invoked the packet-timing prohibition in our definition of C_∞ for, without this prohibition, it might be possible for the receiver to determine the value of a data packet without ever receiving that packet collision-free from its transmitting station. Our definition of C_∞ would not be tenable if we had not made the packet-timing prohibition. This subtlety appears to have been overlooked in much of the literature where it is usually assumed with no further justification that every data packet must be transmitted at least once collision-free.

We now make explicit an assumption that has been implicit in both of our capacity definitions, namely that the system begins operation at some finite time, which we take to be $t=0$. This means that we are only interested in communicating the data packets generated by the information sources at times $t \geq 0$, and that no station is active before $t = 0$. Such a system initialization is the usual assumption in information-theoretic analyses.

4. CHANNEL MODELS WITH FINITELY-MANY STATIONS

It should be clear that if there is only one station then $C = 1$ packet/slot. Thus, we assume hereafter that the number M of stations is at least 2.

4.1 Known delays

Suppose that the delay d_i of each station is known in advance to all stations and to the receiver, and let d_{max} be the largest of these delays. For any nonnegative integer j, any station can time a packet transmission so that its leading edge arrives at the receiver at time $d_{max} + j$. Hence, the stations can employ a TDMA strategy in which the received time slot $(d_{max} + Mj + i - 1,\ d_{max} + Mj + i)$ is reserved for station i for $j = 0,1,2,\ldots,\ i = 1,2,\ldots,M$. Each station then sees a noiseless Q-ary channel that he can use once every M slots. By merely transmitting the data packet (if any) at the head of his station queue each time he can use this channel, his queue size will remained bounded with probability one provided only that the arrival rate λ/M satisfies $\lambda/M < 1/M$. Thus, it follows from the definition (5) that

$$C = 1 \quad \text{(packet/slot)}. \tag{10}$$

4.2 Unknown delays

We consider next the channel model in which all the delays d_i are initially unknown to the stations and the receiver. For convenience, let $\underline{d} = (d_1, d_2, \ldots, d_M)$. We now slightly amend our definition (5) of capacity to

$$C = \sup \inf_{\underline{d}} \lambda \tag{11}$$

where the infimum is over all $\underline{d}$ such that $d_i \geq 0$ for $i=1,2,..,$ M and where the supremum is still over all strategies satisfying the two conditions stated after (5) above. What we have essentially done in the definition (11) is to take the worst-case $\underline{d}$ when we evaluate the performance of a particular strategy.

4.2.1 No feedback. We first consider the channel model where there is no feedback from the receiver to the stations so that there is no way that the stations can ever learn $\underline{d}$. This is in fact the model introduced by Massey and Mathys [5] who proved that

$$C = (1 - \frac{1}{M})^{M-1} \quad \text{(packets/slot)}. \tag{12}$$

This capacity decreases monotonically with M and

$$\lim_{M\to\infty} C = \frac{1}{e}, \tag{13}$$

which, of course, invites comparison with the maximum throughput of the slotted-ALOHA strategy. Here, however, the stations cannot time their transmissions to arrive precisely within predefined slots at the receiver -- one might have anticipated an analogy to pure-ALOHA rather than to slotted-ALOHA. Moreover, and more fundamentally, the channel considered here has no feedback, which is the sine qua non of all ALOHA strategies.

In fact, Massey and Mathys [5] considered a different information source model from what we have used here, and they used a somewhat different definition of capacity. Their source model emitted a data packet "on demand" from the station. It is easy to see, however, that the capacity for our model is still the same as theirs. They gave a strategy for every positive integer n that results in each station obtaining $(n-1)\times(M-1)^{M-1}$ uses of a noiseless Q-ary channel every nM^M slots. Thus, in our model, this strategy will keep the station queues bounded with probability one provided that

$$\frac{\lambda}{M} < \frac{(n-1)(M-1)^{M-1}}{nM^M} = \frac{n-1}{n}\,\frac{1}{M}(1-\frac{1}{M})^{M-1} \quad \text{(packets/slot)}.$$

Hence the capacity must be at least as large as the right side of (12). Their converse argument, applied to our model, shows that the time average number of packets received collision-free from a station cannot exceed 1/M times the right side of (12) for at least one station and one choice of $\underline{d}$. Thus, the capacity cannot exceed the right side of (12).

4.2.2 Minimal feedback. We now consider the channel model with unknown delays when there is feedback from the receiver to the stations. The least amount of feedback that suffices to give a capacity different from the no-feedback case is for the receiver to send a special signal at the end of a packet-received collision-free. We assume that this special signal is received at station j d_j slots after its transmission by the receiver. In particular, the station, say i, who transmitted this packet receives the special signal $2d_i + 1$ slots after he began transmission of that packet. The special signal of itself does not identify the transmitting station, and there is silence on the feedback link except when this special signal is sent. We claim for this minimal feedback situation that

$$C = 1 \quad \text{(packet/slot)}. \tag{14}$$

To prove (14), we consider a strategy by which in finite time the stations can learn the delays. The users agree that station 1 will send a packet (whose value is arbitrary) at time $t = 0$, and will retransmit this packet precisely when he receives the special feedback signal. Thus the first two feedback signals are sent exactly $2d_1 + 2$ slots apart and hence received by station j with this same separation for $j = 1,2,...,M$. From this separation, all M stations learn the delay d_j. The users further agree that, as soon as station 2 receives the second special feedback signal, he sends a packet (whose value is arbitrary), and he retransmits this packet precisely when he receives the next special feedback signal. Thus, the third and fourth special feedback signals are received with separation $2d_2 + 2$ slots at all stations, and so all stations learn the delay d_2. Etc. In this manner, at time $t = 4(d_1 + d_2 + .. + d_M) + 3M - d_1$, all stations will have learned all delays. This strategy effectively reduces the problem to the known delay case of Section 4.1, and thus (14) is proved.

Having determined the delays in the above manner, the stations would resort to the TDMA strategy described in Section 4.1 to obtain an overall strategy that would guarantee that the station queues remained bounded with probability one for any arrival rate $\lambda < 1$ packet/slot. It is important to note that this does not imply that true "random-accessing" strategies such as ALOHA (in contrast to the "scheduled-accessing" TDMA strategy) are not useful on this channel model. Suppose that after $\underline{d}$ is determined, the stations agree to time their packet transmissions to arrive in time slots $(j, j + 1)$ for integer j only. They can then employ an appropriate form of the slotted-ALOHA strategy that Tsybakov and Mikhailov [6] have proved will keep the station queues bounded with probability one provided that

$$\lambda < \left(1 - \frac{1}{M}\right)^{M-1} \quad \text{(packets/slot)} \tag{15}$$

but will cause the queues to grow without bound when λ exceeds the right side of (15). [Note that the failure of the special feedback signal to appear $2d_i + 1$ slots after station i transmits a packet informs station i that this packet has suffered a collision -- which is all the feedback information required for an ALOHA strategy.] It might seem a bad trade to exchange the TDMA strategy that accomodates any $\lambda < 1$ for the slotted-ALOHA strategy that, according to (15), can cope with arrival rates no greater than could be accomodated without feedback in the known delay case, according to (12). However, when λ is sufficiently smaller than the right side of (15), the slotted-ALOHA system will yield a much smaller average delay in delivering data packets to the receiver than will the TDMA system. And in fact it was just to reduce this average delay that the ALOHA and later "random-access" algorithms were introduced. It is only when λ lies between the right side of (15) and 1 that one can unequivocally say that the TDMA strategy would be a better choice than the slotted-ALOHA strategy. Even then, there may (depending on the value of λ) be other "random-accessing" strategies that would yield better average packet delay than the "scheduled-accessing" TDMA strategy.

5. CHANNEL MODELS WITH INFINITELY-MANY STATIONS

We now consider the case where the number of stations is countably infinite. To keep the size of this paper within bounds, we consider only the known-delays case in which, moreover, there is a finite upper bound on all delays, say $d_i \leq n$ for some positive integer n. It is easy to see that the best strategies will require stations to time their packet transmissions to arrive in time slots (j, j + 1) for integer j only, because any other strategy, when modified by the necessary additional partial-slot delays to be of this type, yields a new strategy in which a collision occurs only when it would also have occurred in the old strategy. Thus, we need only consider such "slotted-strategies". But having agreed to this, we see that (even with some form of slot feedback) we can always regard the channel as composed of 2n + 1 interleaved channels, each of which has $d_i = 0$ for all stations i = 1,2,3,... . For these reasons, we assume hereafter that all delays are zero and that only slotted-strategies are of interest. Stations can transmit only in slots (j, j + 1), j a nonnegative integer, and packets arrive at the receiver in the same slot as they were sent.

5.1 No feedback

When there is no feedback from the receiver to the stations, it should be clear that a strategy can satisfy condition (i) following (8) only if it provides that a data packet generated at some station is transmitted infinitely often by that station. But this then implies that

$$C_\infty = 0. \tag{16}$$

5.2 Binary feedback

We now assume that there is feedback to the stations from the receiver at the end of each slot (j, j + 1) that informs all stations whether or not there was a collision in that slot. We refer to this collision/no-collision feedback case as binary feedback (but we note that idle/non-idle feedback might also be of interest). Capacity is not presently known for this channel model. All that can now be said is

$$0.4493 < C_\infty \leq 0.568 \quad \text{(packets/slot)}. \tag{17}$$

The lower bound in (17) is due to Studer and Pletscher [7] who, in a student semester project at the Swiss Federal Institute of Technology, showed that a collision-resolution algorithm (CRA) with binary splitting, when used in a random-access protocol of the type first proposed by Gallager [8], resulted in packet backlog that remains bounded with probability one for $\lambda < 0.44937$ packets/slot. The upper bound in (17) is due to Likhanov and Tsybakov [9], who slightly improved a slight improvement by Zhang and Berger [10] of an ingenious bound (0.587) due to Mikhailov and Tsybakov [11].

The idea of a CRA, which is an interesting and appealing alternative to ALOHA strategies, is due independently to Capetanakis [12] and to Tsybakov and Mikhailov [13]. We refer the interested reader to the March 1985 Special Issue on Random-Access Communications of the IEEE Transactions on Information Theory for a thorough coverage of the CRA approach -- and of many other interesting aspects of random-accessing.

5.3 Ternary feedback

By ternary feedback, we shall mean a feedback signal at the end of each slot to all stations that distinguishes the three situations idle/success/collision in that slot. Again for this channel model, capacity is not yet known. All that can be said now is

$$0.48776 \leq C_\infty \leq 0.568 \quad \text{(packets/slot)}. \tag{18}$$

The lower bound reflects the performance of a CRA-based random-access algorithm due to Mosely and Humblet [14] that is the current champion in a fierce battle being waged for the best ternary-feedback algorithm, cf. [15]. The upper bound is again that of [9], which in fact was derived for ternary feedback and thus *a fortiori* is also a bound for the less informative binary feedback. It is a surprising fact that no one has yet proved a capacity upper bound for the binary-feedback model smaller than that for the ternary-feedback model, as it is hard to believe that the two capacities coincide.

5.4 Multiplicity feedback

As our final channel model, we consider the case where the feedback signal at the end of each slot indicates the number of packets that had been transmitted in that slot. This is the only interesting (i.e., $C_\infty > 0$) infinite-station model for which the capacity is known. The result is quite surprising, namely

$$C_\infty = 1 \quad \text{(packet/slot)}, \tag{19}$$

and is due to Pippenger [16]. We can hardly do better here than refer the reader to Pippenger's paper for a description of the clever random-access strategy that he employed to keep the backlog of packets bounded with probability one for any arrival rate $\lambda < 1$.

6. CLOSING REMARKS

We have given rather short shrift to the infinite-station case. Interesting things can be said about the capacity when the strategy is required to be first-come first-serve (FCFS), i.e., when a data packet generated earlier than another packet must be received collision-free before the other packet is received collision-free, cf. [14]. Constraining the time of first transmission of a packet to be in the slot following its generation gives other interesting models, as does constraining the time that a station is permitted to observe the feedback channel. Weakening the constraint on strategies to require only that a fraction θ ($\theta < 1$) of the generated data packets be eventually received collision-free also gives interesting models. [Many of the above modifications are also of interest in the finite-station case.] Allowing "errors" in the forward and/or feedback channel also gives interesting channel models, but the appropriate definitions of capacity are then much modified versions of what we have used in this paper. Our aim has not been to be encyclopedic, but rather suggestive, of the way that precise random-access models can be formulated and analyzed.

REFERENCES

1. Wozencraft JM and Reiffen B: Sequential Decoding. New York: MIT Press and Wiley, 1961.
2. Abramson N: The ALOHA system - Another alternative for computer communications. Proc. Fall Joint Comp.Conf., AFIPS Press, vol. 37, pp.281-285, 1970.
3. Abramson N: Development of the ALOHANET. IEEE Trans. Info. Th., vol. IT-31, pp. 119-123, March 1985.
4. Roberts LG: ALOHA packet system with and without slots and capture. Comp. Comm. Rev., vol. 5, pp. 28-42, April 1975.
5. Massey JL and Mathys P: The collision channel without feedback. IEEE Trans. Info. Th., vol. IT-31, pp. 192-204, March 1985.
6. Tsybakov BS and Mikhailov VA: Ergodicity of slotted ALOHA-system. Probl. Peredachi Inform., vol. 15, pp. 73-87, Oct.-Dec. 1979.
7. Studer P and Pletscher H: Q-ary Part-and-Try Algorithm for Packet Conflict Resolution. To appear in IEEE Trans.Info.Th.
8. Gallager RG: Conflict resolution in random access broadcast networks. Proc. AFOSR Workshop in Comm. Th. and Appl., pp. 74-76, Sept. 1978.
9. Likhanov NB and Tsybakov BS: Private Comm., March 1986.
10. Zhang Z and Berger T: An improved bound for the capacity of the random access channel. Probl. Peredachi Inform., vol. 21, pp. 83-87, Oct.-Dec. 1985.
11. Mikhailov VA and Tsybakov BS: An upper bound to capacity of random multiple access system. Probl. Peredachi Inform., vol. 17, pp. 90-95, Jan.-March 1981.

12. Capetanakis JI: The multiple access broadcast channel. Ph.D. Thesis. MIT, Aug. 1977.
13. Tsybakov BS and Mikhailov VA: Slotted multiaccess packet broadcasting feedback channel. Probl. Peredachi Inform., vol. 14, pp. 32-59, Oct.-Dec. 1978.
14. Mosely J and Humblet PA: A class of efficient contention resolution algorithms for multiple access channels. IEEE Trans. Comm., vol. COM-33, pp. 145-151, Feb. 1985.
15. Verdu S: Computation of the efficiency of the Mosely-Humblet contention resolution algorithm: A simple method. Proc.IEEE, vol. 74, pp. 613-614, April 1986.
16. Pippenger N: Bounds on the performance of protocols for a multiple access broadcast channel. IEEE Trans. Info. Th., vol. IT-27, pp. 145-151, March 1981.

CAPACITY LIMITS FOR MULTIPLE-ACCESS CHANNELS WITHOUT FEEDBACK*

Edward C. van der Meulen

Departement Wiskunde
Katholieke Universiteit Leuven
B-3030 Leuven, Belgium.

ABSTRACT. In this paper several results on the multiple-access channel, which have been obtained by various researchers in information theory around the world during the past decade, are described. The word "multiple-access channel" is to be interpreted here solely from the point of view of multi-user information theory, i.e., in the Shannon-theoretic sense. The recent results are presented through a series of coding theorems and converses, with the emphasis on the statement and interpretation of these results, rather than on proofs. In a previous survey [41] the research advances of several multi-way channels, including the multiple-access channel, were described for the period 1961 - July 1976. The present survey describes research progress on the multiple-access channel since then. However, space limitations have compelled the author to restrict himself to certain topics. Another survey [42], describing the results on the broadcast channel during the period 1976-1980, appeared in the Proceedings of a previous NATO Advanced Study Institute.

1. INTRODUCTION

Within information theory, the communication situation in which two or more users send simultaneously information over a common channel to one receiver was first mentioned by Shannon [35] in connection with his study of two-way communication channels.

A discrete memoryless (d.m.) multiple-access channel (MAC) with two input users, denoted by $(\mathcal{X}_1 \times \mathcal{X}_2, P(y|x_1,x_2), \mathcal{Y})$, consists of two finite input alphabets $\mathcal{X}_1$ and $\mathcal{X}_2$, one finite output alphabet $\mathcal{Y}$, and a transition probability matrix $\{P(y|x_1,x_2)\}$. Here, $P(y|x_1,x_2)$ is the probability that the output symbol y is received, given that the input symbols x_1 and x_2 are transmitted. The problem then is for the two users to communicate simultaneously through the channel as effectively as possible, under various assumptions. The information delivered by the sources to the transmitters may be independent (the classical situation) or correlated. There may be feedback from the receiver to both transmitters, or to only one of them, or no feedback at all. Instead of being discrete, the channel may be continuous, in particular Gaussian. These and other situations have been investigated during the past fifteen years by many workers in information theory. Consideration of these various situations has led to many new intriguing research problems, and has generated a great research

---

* The research for this paper was supported in part by Onderzoeksfonds, Katholieke Universiteit Leuven, Project OT/VI/17-OT/IX/20, and by Project GOA 83/88-53, Ministerie van Wetenschapsbeleid, Brussels, Belgium.

J. K. Skwirzynski (ed.), Performance Limits in Communication Theory and Practice, 403–426.

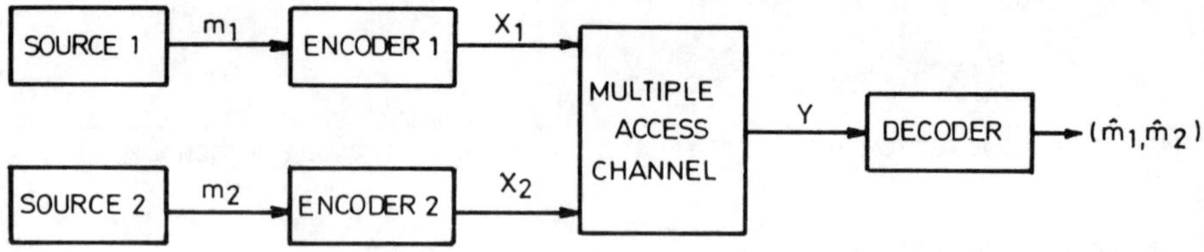

Fig. 1. The multiple-access channel in the classical situation.

activity around the world in multi-user information theory in general, and with respect to the multiple-access channel in particular. The classical configuration of a d.m. MAC with two senders and two independent sources is depicted in Fig. 1. One of the main goals in multi-user channel problems is to characterize the capacity region of the channel under consideration.

The division of this paper is as follows. In Section 2 the basic notation is set forth. Section 3 deals with the d.m. MAC in the classical situation, Section 4 with the d.m. MAC in the Slepian-Wolf setting, and Section 5 treats the d.m. MAC with arbitrarily correlated sources. In Section 6 results on the Gaussian MAC are discussed. Due to space limitations a discussion of the MAC with feedback could not be included here. This topic and other subjects (such as the MAC with cribbing encoders and coding techniques for MAC's) will be described in another paper.

2. BASIC NOTATION

In this paper, sets are denoted by script letters, such as $\mathcal{X}$. In the discrete case, these sets are assumed to be finite, and the cardinality of a set $\mathcal{X}$ is denoted by $\|\mathcal{X}\|$. The cartesian product of two sets $\mathcal{X}$ and $\mathcal{Y}$ is indicated by $\mathcal{X} \times \mathcal{Y}$. To a given finite set $\mathcal{X}$ and a probability distribution (PD) $P(x)$ on $\mathcal{X}$, we make always correspond a random variable (RV) X taking values in $\mathcal{X}$ according to this PD, i.e., $\Pr\{X = x\} = P(x)$. We assume the reader's familiarity with the standard definitions of entropy $H(X)$, conditional entropy $H(X|Y)$, mutual information $I(X;Y)$, and conditional mutual information $I(X;Y|Z)$ for RV's X,Y, and Z. (Cf. [20, pp. 20-21].) Given a d.m. MAC $\{\mathcal{X}_1 \times \mathcal{X}_2, P(y|x_1,x_2), \mathcal{Y}\}$, a PD on $\mathcal{X}_1 \times \mathcal{X}_2$ gives rise to the mutual information functions $I(X_1;Y|X_2)$, $I(X_2;Y|X_1)$, and $I(X_1,X_2;Y)$. In the characterizations of the various achievable rate regions often an auxiliary RV U occurs ranging over a finite set $\mathcal{U}$.

The set of sequences of length n of elements of a set $\mathcal{X}$ is denoted by $\mathcal{X}^n$. The members of $\mathcal{X}^n$ are written as $\underline{x}^n = (x_1,\ldots,x_n)$. A random vector is denoted by underlined, superscripted capitals : $\underline{X}^n = (X_1,\ldots,X_n)$ denotes a random vector of length n. A vector of random vectors is denoted by a superscripted capital underscored with a wavy line : $\utilde{X}^{kn} = (\underline{X}_1^n,\ldots,\underline{X}_k^n)$ represents a vector of k random vectors, each of dimension n. This notation for random vectors holds strictly for the written text. (In the figures, a simpler notation is used. There, a single random vector is not underlined nor superscripted, whereas a vector of random vectors is denoted in the figures by an underlined capital without a superscript, such as $\underline{X}$

(cf. Fig. 13).) When the underlying distribution is unspecified, probabilities are denoted by Pr, such as $\Pr\{U \in \mathcal{A}\}$. When it is given that one deals with the nth power of a certain input PD or a certain channel transition probability function, these probabilities are denoted by P^n, as exemplified by the following. The transmission probabilities of a d.m. MAC $\{\mathcal{X}_1 \times \mathcal{X}_2, P(y|x_1,x_2), \mathcal{Y}\}$ are defined by

$$P^n(\underline{y}^n|\underline{x}_1^n,\underline{x}_2^n) = \prod_{i=1}^{n} P(y_i|x_{1i},x_{2i}) \tag{2.1}$$

for all $\underline{x}_1^n = (x_{11},\ldots,x_{1n}) \in \mathcal{X}_1^n$, $\underline{x}_2^n = (x_{21},\ldots,x_{2n}) \in \mathcal{X}_2^n$, $\underline{y}^n = (y_1,\ldots,y_n) \in \mathcal{Y}^n$, and all positive integers n. Formula (2.1) expresses the memoryless feature of the MAC.

As is customary in information theory, logarithms are in the discrete case taken to the base 2, while in the Gaussian case the natural logarithm is used. Also, h(p) stands for the usual binary entropy function, i.e., $h(p) = -p\log p - (1-p)\log(1-p)$, for $0 \leqslant p \leqslant 1$. The notation $Z \sim \mathcal{N}(0,N)$ means that the RV Z has a normal distribution with mean 0 and variance N. Finally, the convex hull of a set $\mathcal{A}$ is denoted by $co(\mathcal{A})$.

3. THE MULTIPLE-ACCESS CHANNEL IN THE CLASSICAL SITUATION

A d.m. MAC with two input users, denoted by K_{21}, consists of a channel $\{\mathcal{X}_1 \times \mathcal{X}_2, P(y|x_1,x_2), \mathcal{Y}\}$, with transition probabilities given by (2.1). The MAC is said to be in the classical situation if it is used to transmit two separate messages from two independent sources. This communication situation is denoted by (K_{21},I) and shown in Fig. 1. Here, two message sources emit statistically independent messages $m_1 \in \mathcal{M}_1 = \{1,2,\ldots,M_1\}$ and $m_2 \in \mathcal{M}_2 = \{1,2,\ldots,M_2\}$, such that each pair (m_1,m_2) occurs with probability $1/(M_1,M_2)$. Message m_1 is encoded at terminal 1 into the codeword $\underline{x}_1^n(m_1) \in \mathcal{X}_1^n$, and message m_2 is encoded at terminal 2 into the codeword $\underline{x}_2^n(m_2) \in \mathcal{X}_2^n$.

An (n,M_1,M_2,λ)-code for the d.m. MAC in situation (K_{21},I) consists of M_1 codewords $\underline{x}_1^n(m_1) \in \mathcal{X}_1^n$, M_2 codewords $\underline{x}_2^n(m_2) \in \mathcal{X}_2^n$, and M_1M_2 pairwise disjoint decoding sets $B(m_1,m_2)$, such that

$$\frac{1}{M_1M_2} \sum_{\substack{(m_1,m_2)\\ \in \mathcal{M}_1 \times \mathcal{M}_2}} P^n(B(m_1,m_2)|\underline{x}_1^n(m_1),\underline{x}_2^n(m_2)) \geqslant 1 - \lambda . \tag{3.1}$$

A pair (R_1,R_2) of non-negative real numbers is said to be an achievable rate pair for a d.m. MAC in situation (K_{21},I), if, for any $\delta > 0$ and any $\lambda, 0 < \lambda < 1$, there exists, for all sufficiently large n, an (n,M_1,M_2,λ)-code for that MAC such that $\log M_i \geqslant n(R_i - \delta)$, $i = 1,2$. The capacity region $\mathcal{C}(K_{21},I)$ of a d.m. MAC K_{21} is defined as the set of all achievable rate pairs in situation (K_{21},I).

The most important result obtained during the period 1961-1976 was the characterization of the capacity region $\mathcal{C}(K_{21},I)$, which is generally attributed to Ahlswede [1,2] and Liao [28], although a proof of Liao's contribution never appeared in print. This characterization is as follows.

Theorem 3.1. ([1],[2],[28]) : The capacity region of the d.m. MAC in situation (K_{21},I) is given by

$$\mathcal{C}(K_{21},I) = \mathrm{co}\{(R_1,R_2) : 0 \leqslant R_1 \leqslant I(X_1;Y|X_2) \ , \tag{3.2a}$$

$$0 \leqslant R_2 \leqslant I(X_2;Y|X_1) \ , \tag{3.2b}$$

$$R_1 + R_2 \leqslant I(X_1,X_2;Y) \ , \tag{3.2c}$$

$$\text{for some } P(x_1,x_2,y) = P(x_1)P(x_2)P(y|x_1,x_2)\} \ .$$

The total cooperation capacity is defined by

$$C_T = \max_{P(x_1,x_2)} I(X_1,X_2;Y) \ . \tag{3.3}$$

A d.m. MAC is called <u>deterministic</u> if $P(y|x_1,x_2) = 1$ or 0 for all inputs x_1,x_2 and all outputs y. A well-known example of a d.m. MAC K_{21} is that of the deterministic binary erasure MAC, abbreviated <u>deterministic BEMAC</u>. This channel, introduced in [39], has input alphabets $\mathcal{X}_1 = \mathcal{X}_2 = \{0,1\}$, output alphabet $\mathcal{Y} = \{0,1,2\}$, and is defined by the operation $y = x_1 + x_2$. Its capacity region is given by $\mathcal{C}(K_{21},I) = \{(R_1,R_2) : 0 \leqslant R_1 \leqslant 1,$ $0 \leqslant R_2 \leqslant 1,\ R_1 + R_2 \leqslant 1\}$ and shown in Fig. 2.

We now describe some of the main results on the d.m. MAC in situation (K_{21},I) obtained since 1976.

a. Dueck [15] showed that for the d.m. MAC in situation (K_{21},I) the capacity region for <u>maximal error</u> is generally smaller than the corresponding average error capacity region given by Theorem 3.1.

b. Bierbaum and Wallmeier [7] showed that, in the description of $\mathcal{C}(K_{21},I)$ as given by Theorem 3.1., it is indeed necessary to take the convex hull.

c. Dueck [17] proved a <u>strong converse</u> of the coding theorem for the d.m. MAC in situation (K_{21},I). Whereas in information theory a weak converse traditionally means that for small error probability λ and n sufficiently large no code exists with signalling rate above capacity, a strong converse always means that for <u>any</u> λ, $0 < \lambda < 1$, and n sufficiently large no code exists with rate above capacity [45]. For a given λ, $0 < \lambda < 1$, a pair of non-negative real numbers (R_1,R_2) is said to be <u>λ-achievable</u> for a d.m. MAC in situation (K_{21},I) if for any $\delta > 0$ and sufficiently large n there exists an (n,M_1,M_2,λ)-code for that MAC such that $\log M_i \geqslant n(R_i - \delta)$, $i = 1,2$. If we denote for a given λ the set of all λ-achievable rate pairs

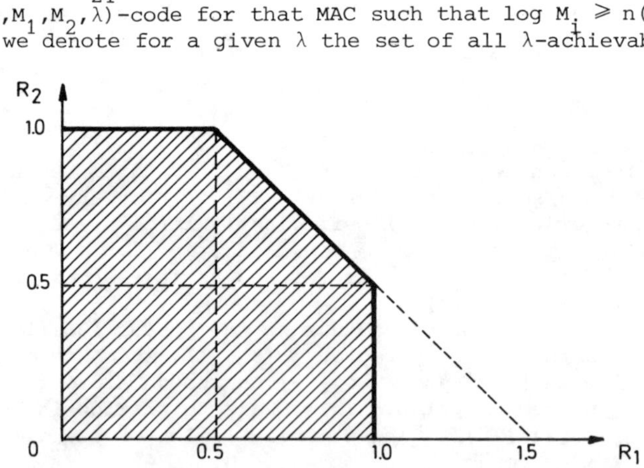

Fig. 2. The capacity region $\mathcal{C}(K_{21},I)$ of the deterministic BEMAC.

by $\mathcal{R}(K_{21},I,\lambda)$, then Dueck's result is the following.

Theorem 3.2. ([17]) : For the d.m. MAC in situation (K_{21},I),

$$\mathcal{R}(K_{21},I,\lambda) = \mathcal{C}(K_{21},I) \text{ for all } \lambda,\ 0 < \lambda < 1 . \tag{3.4}$$

Dueck proved Theorem 3.2 by a method which was called a "wringing technique" by Ahlswede [3] who provided yet another proof of this result.

d. The so-called asynchronous d.m. MAC in situation (K_{21},I) has been investigated by a number of authors [10],[22],[25] and [30]. An illuminating discussion regarding the problems arising when there is no time synchronization is given in [10]. In the literature, various notions of asynchronism have been considered.

(i) The d.m. MAC K_{21} is said to be (fully, totally, or perfectly) synchronous when both senders operate synchronously in the sense that the beginnings of the codewords generated by each sender coincide. This is the classical case (K_{21},I).

(ii) A d.m. MAC K_{21} is said to be quasi-synchronous if the two encoders do not maintain block synchronism while the decoder maintains block synchronism with each encoder. Moreover, it is assumed in this case that bit synchronism is maintained amongst the encoders and decoder. In this situation it is assumed that the beginnings of the two codewords may be shifted relative to each other by an arbitrary interval of duration Δ. The quantity Δ, $\Delta = \ldots,-1,0,1,\ldots$ is called the synchronization shift. The value of the shift is assumed to be unknown at the MAC inputs, but is known at the output. This condition can be guaranteed by transmitting a special synchronization sequence when each sender commences operation [26],[30], also called a preamble sequence [25]. Within the context of a quasi-synchronous MAC it is still possible to distinguish between various situations. Cover et al. [10] considered the case of a quasi-synchronous MAC where the shift is bounded by some constant that is independent of the block length. This case is referred to as "mild asynchronism" in [25]. In general, though, the size of the shift is arbitrary. This situation has been dealt with in [25],[22], and [30]. Hui and Humblet [25] use the terminology "totally asynchronous" in this case.

(iii) Deaett and Wolf [12] considered the situation where not only the assumption of block synchronism but also that of bit synchronism is dropped altogether, both between the encoders and between the decoder and the encoders. We refer to this situation as the non-synchronized MAC. It will not be further discussed in this paper.

We now briefly describe the results on the quasi-synchronous d.m. MAC K_{21} obtained in [10],[22],[25] and [30].

Cover et al. [10] defined the maximum relative delay d as the maximum amount by which the two messages are assumed to be out of synchronization relative to a known or prearranged time. They considered two cases. In the first case it is assumed that d is fixed and known to the receiver, whereas in the second case the receiver's knowledge of d is removed and the delay is allowed to grow to infinity at a certain rate. We denote the capacity region of the quasi-synchronous d.m. MAC in situation (K_{21},I) by $\mathcal{C}(K_{21},I,QS)$. It is shown in [10] that in the first case (d fixed and known to the receiver) the capacity regions of the quasi-synchronous MAC and of the fully synchronous MAC coincide, i.e., $\mathcal{C}_b(K_{21},I,QS) = \mathcal{C}(K_{21},I)$, where the index b indicates that certain conditions are imposed on d, as was done in [10]. In the second case (unbounded delays) Cover et al. [10] allow the delay d and the block length n to tend to infinity in such a manner that $d/n \longrightarrow 0$. Under this assumption it is shown that, by using concatenated codes of increasing block lengths, any rate point (R_1,R_2) in $\mathcal{C}(K_{21},I)$ is still achievable for the quasi-synchronous d.m. MAC K_{21}.

Grigor'ev [22] considered the problem of quasi-synchronous communication over the deterministic BEMAC when the delay can be an arbitrary function of the block length. Grigor'ev found that for a given α, $0 \leq \alpha \leq 0.5$, a rate pair (R_1,R_2) is achievable for the quasi-synchronous deterministic BEMAC in situation (K_{21},I) for all delays k (such that $0 \leq k \leq \alpha n$ or $(1-\alpha)n \leq k \leq n$ for all $n \geq 1$) if (R_1,R_2) belongs to at least one of the following three regions : (i) $0 \leq R_1 < 0.5$, $0 \leq R_2 < 1$; (ii) $0 \leq R_1 < 1$, $0 \leq R_2 < 0.5$; (iii) $0 \leq R_1 < 1$, $0 \leq R_2 < 1$, $R_1 + R_2 < 1.5 - \alpha$. For a given α, $0 \leq \alpha \leq 0.5$, the corresponding achievable rate region is shown in Fig. 3, where it is represented by the shaded area bounded by the broken line ABCEFG.

Poltyrev [30] was the first to determine the actual capacity region of the d.m. quasi-synchronous MAC (situation (K_{21},I,QS)) when the size of the synchronization shift is arbitrary. Let $\mathcal{R}(K_{21},I,QS)$ be the region of pairs given by

$$\mathcal{R}(K_{21},I,QS) = \{(R_1,R_2) : 0 \leq R_1 \leq I(X_1,Y|X_2) , \tag{3.5a}$$

$$0 \leq R_2 \leq I(X_2,Y|X_1) , \tag{3.5b}$$

$$R_1 + R_2 \leq I(X_1,X_2;Y) , \tag{3.5c}$$

$$\text{for some } P(x_1,x_2,y) = P(x_1)P(x_2)P(y|x_1,x_2)\}.$$

Thus, the region $\mathcal{C}(K_{21},I)$, given by (3.2), equals co $\mathcal{R}(K_{21},I,QS)$. Poltyrev showed that for an arbitrary d.m. quasi-synchronous MAC with an arbitrary shift size the capacity region equals $\mathcal{R}(K_{21},I,QS)$. The capacity region of the d.m. quasi-synchronous MAC (K_{21},I) with arbitrary delays differs from that of the synchronous d.m. MAC (K_{21},I) by the fact that in the characterization of $\mathcal{R}(K_{21},I,QS)$ the convex hull operation is absent. Independently, the same result was established by Hui and Humblet [25].

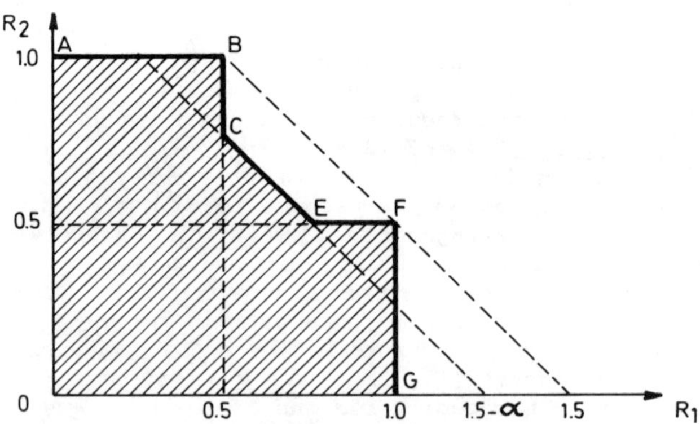

Fig. 3. Achievable region found in [22] for the quasi-synchronous deterministic BEMAC with delays $k \in [0,\alpha n] \cup [(1-\alpha)n,n]$.

If we denote the capacity region of the quasi-synchronous d.m. MAC with arbitrary shift size by $\mathcal{C}_a(K_{21},I,QS)$, we can summarize the main results on asynchronous MAC's as follows.

Theorem 3.3.a ([10]) : When the size of the synchronization shift Δ is bounded, or small relative to the block length n, then the capacity region of the quasi-synchronous d.m. MAC (K_{21},I,QS) is given by

$$\mathcal{C}_b(K_{21},I,QS) = \mathcal{C}(K_{21},I) . \tag{3.6}$$

Theorem 3.3.b ([25],[30]) : When the size of the synchronization shift Δ is arbitrary, $0 \leqslant \Delta \leqslant n$, then the capacity region of the quasi-synchronous d.m. MAC (K_{21},I,QS) is given by

$$\mathcal{C}_a(K_{21},I,QS) = \mathcal{R}(K_{21},I,QS) . \tag{3.7}$$

4. THE MULTIPLE-ACCESS CHANNEL WITH SPECIALLY CORRELATED SOURCES IN THE SENSE OF SLEPIAN AND WOLF

Slepian and Wolf [37] considered a communication situation for the d.m. MAC K_{21} in which the information to be transmitted by the two input users is correlated in a special way. This communication situation is shown in Fig. 4. Here, three message sources emit statistically independent messages $m_0 \in \mathcal{M}_0 = \{1,2,\ldots,M_0\}$, $m_1 \in \mathcal{M}_1 = \{1,2,\ldots,M_1\}$, and $m_2 \in \mathcal{M}_2 = \{1,2,\ldots,M_2\}$, such that each triple (m_0,m_1,m_2) occurs with probability $1/(M_0M_1M_2)$.
Message pair (m_0,m_1) is encoded at terminal 1 by an encoding function f_1 into the codeword $f_1(m_0,m_1) = \underline{x}_1^n(m_0,m_1) \in \mathcal{X}_1^n$. Likewise, message pair (m_0,m_2) is encoded at terminal 2 by an encoding function f_2 into the codeword $f_2(m_0,m_2) = \underline{x}_2^n(m_0,m_2) \in \mathcal{X}_2^n$. Next, the pair $(\underline{x}_1^n(m_0,m_1), \underline{x}_2^n(m_0,m_2))$ is transmitted over the d.m. MAC $\{\mathcal{X}_1 \times \mathcal{X}_2, P(y|x_1,x_2), \mathcal{Y}\}$ using n channel operations, and the decoder must estimate the source triple (m_0,m_1,m_2) based on the received sequence y^n. Thus here, each input terminal attempts to get across one private and one common message to the output terminal. We denote this communication situation by (K_{21},II).

An (n,M_0,M_1,M_2,λ)-code for the d.m. MAC K_{21} in situation (K_{21},II) consists of M_0M_1 codewords $x_1^n(m_0,m_1) \in \mathcal{X}_1^n$, M_0M_2 codewords $x_2^n(m_0,m_2) \in \mathcal{X}_2^n$, and $M_0M_1M_2$ pairwise disjoint decoding sets $B(m_0,m_1,m_2) \subset \mathcal{Y}^n$, such that

$$\frac{1}{M_0M_1M_2} \sum_{\substack{(m_0,m_1,m_2) \\ \in \mathcal{M}_0 \times \mathcal{M}_1 \times \mathcal{M}_2}} P^n(B(m_0,m_1,m_2)|\underline{x}_1^n(m_0,m_1), \underline{x}_2^n(m_0,m_2)) \geqslant 1 - \lambda. \tag{4.1}$$

A triple (R_0,R_1,R_2) of non-negative real numbers is said to be an achievable rate triple for a d.m. MAC in situation (K_{21},II), if, for any $\delta > 0$ and any λ, $0 < \lambda < 1$, there exists, for all sufficiently large n, an (n,M_0,M_1,M_2,λ)-code for this MAC such that $\log M_i \geqslant n(R_i - \delta)$, $i = 0,1,2$. The capacity region $\mathcal{C}(K_{21},II)$ of a d.m. MAC in situation (K_{21},II) is defined as the set of all achievable rate triples in that situation. Slepian and Wolf gave the following characterization of $\mathcal{C}(K_{21},II)$.

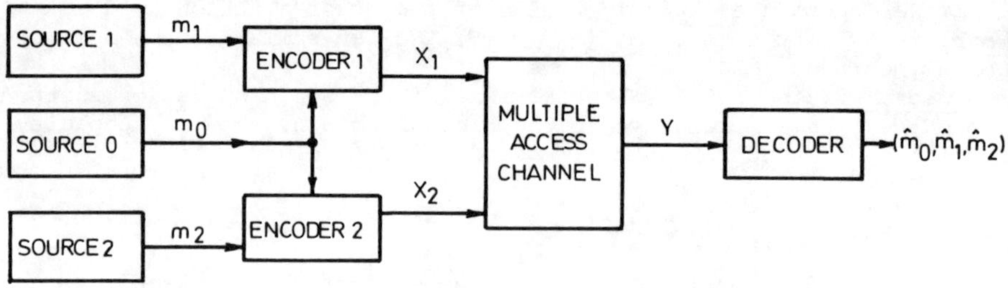

Fig. 4. Slepian-Wolf configuration (K_{21},II) of a d.m. MAC with specially correlated sources.

Theorem 4.1 ([37]) : The capacity region of the d.m. MAC in situation (K_{21},II) is given by

$$\begin{aligned}\mathcal{C}(K_{21},II) = \mathrm{co}\{(R_1,R_2,R_0) \ :\ & 0 \leqslant R_1 \leqslant I(X_1;Y|X_2,U) \ , \\ & 0 \leqslant R_2 \leqslant I(X_2;Y|X_1,U) \ , \\ & R_1 + R_2 \leqslant I(X_1,X_2;Y|U) \ , \\ & 0 \leqslant R_1 + R_2 + R_0 \leqslant I(X_1,X_2;Y) \ , \\ & \text{for some } P(u,x_1,x_2,y) = P(u)P(x_1|u)P(x_2|u)P(y|x_1,x_2)\}. \end{aligned} \tag{4.2}$$

Since 1976 the following specific research progress regarding the d.m. MAC in situation (K_{21},II) was made.

a. Several authors have investigated the problem of upper-bounding $\|\mathcal{U}\|$, and come up with various cardinality bounds. Salehi [32] proved that it suffices to take $\|\mathcal{U}\| \leqslant \min\{\|\mathcal{X}_1\| \cdot \|\mathcal{X}_2\| \ , \ \|\mathcal{Y}\|\}$, while Han [23] proved the weaker bound $\|\mathcal{U}\| \leqslant \|\mathcal{X}_1\| \cdot \|\mathcal{X}_2\| + 3$. Csiszár and Körner [11] and Willems [43] established independently a slightly different formulation of the capacity region $\mathcal{C}(K_{21},II)$. They showed that in the description of the capacity region as given by Theorem 4.1, it is not necessary to take the convex hull. Corresponding with this different formulation of the capacity region, new cardinality bounds arise. Csiszár and Körner proved that $\|\mathcal{U}\| \leqslant \|\mathcal{Y}\| + 3$ can be assumed, while Willems showed that it suffices to take $\|\mathcal{U}\| \leqslant \min\{\|\mathcal{X}_1\| \cdot \|\mathcal{X}_2\| + 2, \ \|\mathcal{Y}\| + 3\}$.

b. (The asymmetric MAC) Haroutunian [24], and then Bassalygo, Pinsker, and Prelov [6], considered the communication situation which results from the Slepian-Wolf configuration if one sets one of the private rates, R_2 say, equal to zero. This leads to a new communication situation, shown in Fig. 5, which is of independent interest. Here, two message sources emit statistically independent messages $m_0 \in \mathcal{M}_0 = \{1,2,\ldots,M_0\}$, and $m_1 \in \mathcal{M}_1 = \{1,2,\ldots,M_1\}$, such that each pair (m_0,m_1) occurs with probability $1/(M_0M_1)$. Message pair (m_0,m_1) is encoded at terminal 1 into a codeword $\underline{x}_1^n(m_0,m_1) \in \mathcal{X}_1^n$, whereas message m_0, the output of source 0, is encoded at terminal 2 into a codeword $\underline{x}_2^n(m_0) \in \mathcal{X}_2^n$. Thus, input terminal 1 attempts to get across the MAC both one common and one private message, whereas terminal 2 transmits only a common message to the output terminal. This communication situation, denoted by (K_{21},III), has been termed the

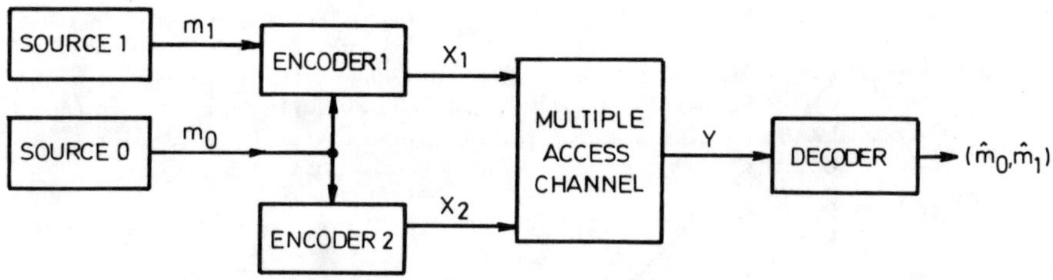

Fig. 5. The d.m. MAC in situation (K_{21},III). (The asymmetric MAC.)

"asymmetric MAC" (abbreviated AMAC) in [14], because of its analogy with the configuration of the asymmetric broadcast channel [11]. At this point we will omit the precise definitions of an (n,M_0,M_1,λ)-code and of a pair (R_0,R_1) of achievable rates for a d.m. MAC in situation (K_{21},III), since these are easily derived from the corresponding definitions in situation (K_{21},II). The capacity region $\mathcal{C}(K_{21},\mathrm{III})$ of a d.m. MAC K_{21} is defined as the set of all achievable rate pairs (R_0,R_1) in situation (K_{21},III). Haroutunian [24] stated a single-letter characterization of the capacity region $\mathcal{C}(K_{21},\mathrm{III})$ and outlined a proof of his result. Bassalygo, Pinsker, and Prelov [6] established the same characterization and gave a more detailed proof. This characterization is stated in the following theorem.

<u>Theorem 4.2</u> ([24],[6]) : The capacity region of a d.m. MAC in situation (K_{21},III) is given by

$$\mathcal{C}(K_{21},\mathrm{III}) = \mathrm{co}\{(R_0,R_1)\ :\ 0 \leqslant R_1 \leqslant I(X_1;Y|X_2)\ , \tag{4.3a}$$

$$0 \leqslant R_0 + R_1 \leqslant I(X_1,X_2;Y)\ , \tag{4.3b}$$

$$\text{for some } P(x_1,x_2,y) = P(x_1,x_2)P(y|x_1,x_2)\}\ .$$

We denote the capacity region of a <u>deterministic</u> d.m. MAC in situation (K_{21},III) by $\mathcal{C}(K_{21},\mathrm{III},\mathfrak{D})$. Bassalygo et al. [6] evaluated $\mathcal{C}(K_{21},\mathrm{III},\mathfrak{D})$ for the deterministic BEMAC and found that in this case the corresponding capacity region is bounded by a curve consisting of three segments, which are respectively specified by the following formulas : (i) $R_0 + R_1 = \log_2 3$, $2/3 \leqslant R_0 \leqslant \log_2 3$; (ii) $R_0 = 1 - \alpha$, $R_1 = h(\alpha)$ for $1/3 \leqslant \alpha \leqslant 1/2$; (iii) $0 \leqslant R_0 \leqslant 1/2$, $R_1 = 1$. This region is shown in Fig. 6, which can be regarded as the counterpart of Fig. 2 for situation (K_{21},III).

Bassalygo et al. [6] were particularly concerned with the problem of the transmission of information over a d.m. MAC in situation (K_{21},III) when it is assumed that the upper bound λ on the average error probability (cf. λ in (4.1)) equals zero. A rate pair (R_0,R_1) is said to be <u>zero-error achievable</u> for a d.m. MAC in situation (K_{21},III), if, for any $\delta > 0$, there exists, for all sufficiently large n, an $(n,M_0,M_1,0)$-code for this MAC such that $\log M_i \geqslant n(R_i - \delta)$, $i = 0,1$. The <u>zero-error capacity region</u> $\mathcal{C}_0(K_{21},\mathrm{III})$ is defined as the set of all zero-error achievable rate pairs (R_0,R_1) in situation (K_{21},III). Let us denote the zero-error capacity

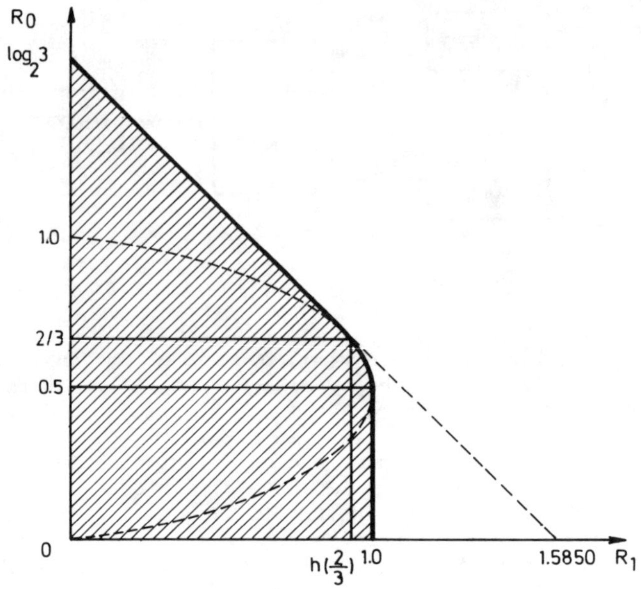

Fig. 6. The capacity region $\mathcal{C}(K_{21},III)$ of the deterministic BEMAC.

region of a deterministic d.m. MAC in situation (K_{21},III) by $\mathcal{C}_0(K_{21},III,\mathfrak{D})$. Then the result by Bassalygo, Pinsker, and Prelov [6] says that $\mathcal{C}_0(K_{21},III,\mathfrak{D}) = \mathcal{C}(K_{21},III,\mathfrak{D})$, i.e., that for a deterministic d.m. MAC in situation (K_{21},III) the zero-error capacity region and the "ordinary" capacity region (as given by Theorem 4.2) coincide. We state their result as a separate theorem.

Theorem 4.3 ([6]) : For a deterministic d.m. MAC one has

$$\mathcal{C}_0(K_{21},III) = \mathcal{C}(K_{21},III), \text{ i.e., } \mathcal{C}_0(K_{21},III,\mathfrak{D}) = \mathcal{C}(K_{21},III,\mathfrak{D}) \ . \tag{4.4}$$

c. Following the methods of Ahlswede and Dueck [4] for the single-user channel, Pokorny and Wallmeier [29] considered codes produced by permutations for the d.m. MAC in situation (K_{21},I), whereas De Bruyn and van der Meulen [14] considered two codeconstructions for the d.m. MAC in situation (K_{21},III), one being a codegeneration by permutations, the other one being an iterative codeconstruction.

d. Ahlswede [1] considered the problem of transmitting information from three independent sources over a d.m. MAC with three input terminals, when each source output is connected to exactly one input terminal, and determined the capacity region in this case. Ulrey [38] continued these investigations and considered the problem of transmitting $t \geqslant 3$ separate messages from t independent sources over a d.m. MAC with t input terminals and also determined the capacity region in that case. We denote a d.m. MAC with $t \geqslant 3$ input terminals by K_{t1} and the situation considered in [38] by (K_{t1},I). In [40] a simplified proof of the converse theorem of [38] is given.

Slepian and Wolf [37] extended their approach (i.e., the setup of situation (K_{21},II)) to the case in which information from seven independent sources is to be transmitted over a d.m. MAC K_{31} and this information, when supplied to the input terminals, is again correlated in a special way. They provided an expression for the capacity region in this case, which however was noted to be incorrect in [31].

Han [23] continued these investigations and considered the general communication situation in which information is to be transmitted from $s \geq 2$ independent sources over a d.m. MAC K_{t1}, such that the source outputs are connected to the input terminals in an arbitrary manner. In order to describe a specific communication situation (within a whole class of possibilities), Han used the notion of an <u>incidence relation</u> $\delta(s,t) = (\delta_1, \delta_2, \ldots, \delta_t)$ between the s sources and the t input terminals, which prescribes for each terminal i ($1 \leq i \leq t$) the subset δ_i of sources observed by that terminal. The entire communication situation thus defined - consisting of the d.m. MAC K_{t1}, the s independent sources, and the incidence relation $\delta(s,t)$ - is here denoted by (K_{t1},$\delta(s,t)$). Following the terminology of [23] we call this communication situation a d.m. MAC with <u>cross observation</u> at the encoders. A typical configuration of this kind is shown in Fig. 7. It includes the situations considered by Slepian and Wolf [37] and Ulrey as special cases.

Han [23] established the capacity region $\mathcal{C}(K_{t1},\delta(s,t))$ of a d.m. MAC K_{t1} in situation (K_{t1},$\delta(s,t)$), for an arbitrary incidence relation $\delta(s,t)$ with s and t finite, by proving a direct part (using a typicality approach) and a weak converse. He showed that his expression for the capacity region includes the expressions obtained by Ahlswede [1] (case (K_{21},I)), Slepian and Wolf [37] (case (K_{21},II)), and Ulrey [38] (case (K_{t1},I)) as special cases. As the capacity region found by Han is notationally too complex to describe here, the reader is referred to [23] for details.

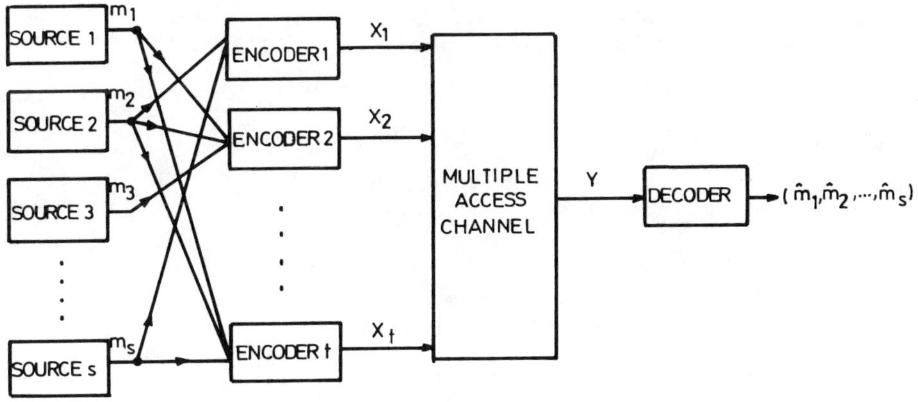

Fig. 7. The d.m. MAC with s sources, t input terminals, and cross observation at the encoders.

Prelov [31] considered the transmission of $t \geqslant 3$ independent sources over a d.m. MAC K_{t1}, when the transmission configuration exhibits a special source hierarchy. Specifically, it is assumed that all message sources are accessible to the first encoder, all sources except the first one are accessible to the second encoder, all sources except the first two are accessible to the third encoder, and so forth. This situation can be regarded as the generalization of the d.m. AMAC (case (K_{21}, III)) to the case of $t \geqslant 3$ sources and terminals. For this situation Prelov obtained results which can be regarded as generalizations of those in [6].

5. THE DISCRETE MEMORYLESS MULTIPLE-ACCESS CHANNEL WITH ARBITRARILY CORRELATED SOURCES

Clearly, the correlation assumed between the messages encoded by the two users in the channel model of Slepian and Wolf [37] is of a special form. In [37] the authors raise the question how to handle more general correlations. Consider thereto the communication situation shown in Fig. 8. On the one hand, there is given a bivariate information source $(\mathcal{U} \times \mathcal{V}, P(u,v))$, putting out independent, identically distributed (i.i.d.) discrete RV's (U_i, V_i), $i = 1,2,\ldots$, according to an arbitrary but fixed PD $\Pr\{U_i = u, V_i = v\} = P(u,v)$, $u \in \mathcal{U}$ and $v \in \mathcal{V}$, where $\mathcal{U}$ and $\mathcal{V}$ are finite sets. On the other hand, there is given a d.m. MAC K_{21} as defined in Section 3. Input terminal 1 observes the sequence of outputs of source $\mathcal{U}$, whereas input terminal 2 observes those of source $\mathcal{V}$. These source output sequences are subsequently encoded into input sequences of the channel, which are then transmitted. At the receiver end the emitted source sequences are to be decoded. The problem now is to determine at what rates the dependent source outputs can be transmitted over the d.m. MAC with arbitrarily small probability of error. In a way, this problem traces back to Shannon [35, Section 14], who raised the problem of transmitting a pair of correlated messages over a two-way channel. The communication situation exhibited in Fig. 8 will be denoted by $(K_{21},(U,V),I)$. The first general treatment of the d.m. MAC K_{21} with arbitrarily correlated sources has been given by Cover, El Gamal, and Salehi [9], who established a sufficient (but not necessary) condition for the reliable transmission of an arbitrarily correlated source over a d.m. MAC K_{21}.

A code of block length n in situation $(K_{21},(U,V),I)$, denoted by $(\phi_1^n, \phi_2^n, \psi^n)$, consists of two encoding functions

$$\phi_1^n : \mathcal{U}^n \longrightarrow \mathcal{X}_1^n , \tag{5.1a}$$

$$\phi_2^n : \mathcal{V}^n \longrightarrow \mathcal{X}_2^n , \tag{5.1b}$$

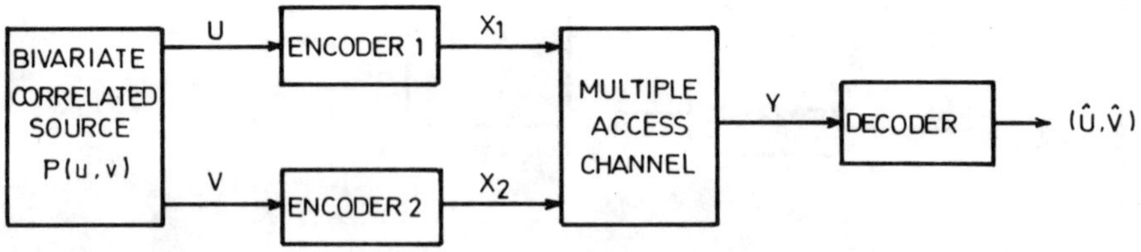

Fig. 8. The d.m. MAC K_{21} with two arbitrarily correlated sources.

mapping source outputs into codewords, and a decoding function

$$\psi^n : \mathcal{Y}^n \longrightarrow \mathcal{U}^n \times \mathcal{V}^n . \tag{5.1c}$$

The probability of error when using this code is given by

$$P_e^n = \Pr\{(\underline{U}^n,\underline{V}^n) \neq \psi^n(\underline{Y}^n)\}$$

$$= \sum_{(\underline{u}^n,\underline{v}^n)} P^n(\underline{u}^n,\underline{v}^n)P^n(\psi^n(\underline{Y}^n) \neq (\underline{u}^n,\underline{v}^n)\,|\,\phi_1^n(\underline{u}^n),\ \phi_2^n(\underline{v}^n)) , \tag{5.2}$$

where the joint probability assignment is given by

$$P^n(\underline{u}^n,\underline{v}^n,\underline{y}^n) = \prod_{i=1}^{n} P(u_i,v_i)P(y_i\,|\,\phi_{1i}(\underline{u}^n),\ \phi_{2i}(\underline{v}^n)) . \tag{5.3}$$

The correlated source pair $(\mathcal{U} \times \mathcal{V},\ P(u,v))$ is said to be <u>reliably transmissible</u> over a d.m. MAC in situation $(K_{21},(U,V),I)$ if for any λ, $0 < \lambda < 1$, and sufficiently large n there exists a code $(\phi_1^n,\phi_2^n,\psi^n)$ for which $P_e^n < \lambda$. The main problem then is to find a simple characterization of the set of all correlated sources that are reliably transmissible over a given d.m. MAC K_{21}.

There are two versions of the achievability theorem of Cover, El Gamal, and Salehi [9], one involving the common part of U and V, and a simpler version omitting the common part.

a. We first state the simpler version of the coding theorem in [9] which assumes that the RV's U and V contain no common part.

<u>Theorem 5.1</u> ([9]) : A correlated source $(\mathcal{U} \times \mathcal{V}\ ,\ P(u,v))$ can be reliably transmitted over a given d.m. MAC K_{21} in situation $(K_{21},(U,V),I)$, if there exist probability mass functions $P(x_1|u)$ and $P(x_2|v)$ such that

$$H(U|V) < I(X_1;Y|X_2,V) , \tag{5.4a}$$

$$H(V|U) < I(X_2;Y|X_1,U) , \tag{5.4b}$$

$$H(U,V) < I(X_1,X_2;Y) , \tag{5.4c}$$

where $P(u,v,x_1,x_2,y) = P(u,v)P(x_1|u)P(x_2|v)P(y|x_1,x_2)$.

Cover et al. [9] noted that applications of Theorem 5.1 yield several known results as special cases : (i) the direct part of the Slepian and Wolf [36] data compression theorem, (ii) the achievability part of the coding theorem for the d.m. MAC with independent sources (situation (K_{21},I)), as proved by Ahlswede [1],[2] and announced by Liao [28] (cf. Theorem 3.1 above), and (iii) the achievability of the total cooperation capacity C_T defined by (3.3).

Cover et al. [9] also gave an example showing that in general the procedure consisting of factorizing the source-channel transmission problem into two separate problems, i.e., Slepian and Wolf data compression applied to the bivariate source $(\mathcal{U} \times \mathcal{V},\ P(u,v))$ followed by the transmission of two independent messages over a d.m. MAC in the sense of situation (K_{21},I), (called the <u>separation principle</u> in [5]), is not optimal, thus motivating the usefulness of Theorem 5.1. The example in [9] involves the deterministic BEMAC defined in Section 3, and the correlated source

$(\mathcal{U} \times \mathcal{V}, P(u,v))$, defined by $\mathcal{U} = \mathcal{V} = \{0,1\}$ and the probability assignment $P(0,0) = P(1,0) = P(1,1) = 1/3$. For this example, plain factorization does not yield reliable transmission at all, but it is easy to see that with the simple choices $X_1 \equiv U$ and $X_2 \equiv V$ errorfree transmission of the source outputs over the MAC is possible.

b. Dueck [16] observed that the construction of Cover, El Gamal, and Salehi [9] is such that the "channel input at time t depends only on the source output at time t". This observation led him to carefully design an instructive counterexample which shows that the coding strategy of [9] for the d.m. MAC with correlated sources is not optimal in the case $(K_{21},(U,V),I)$. The counterexample constructed by Dueck involves the definition of a correlated source $(\mathcal{U} \times \mathcal{V}, P(u,v))$ and a d.m. MAC K_{21} such that the source does not fulfill the conditions of Theorem 5.1 with respect to the MAC. The correlated source presented by Dueck is chosen to be highly asymmetrical and contains no common part. Dueck showed that for this correlated source a strategy like "a large information source output at time t is encoded in several channel inputs" is definitely better than the coding strategy in [9].

c. We next state the second version of the coding theorem of Cover et al. [9], which is applicable when the source RV's do have a common part in the sense of Gács and Körner [19] and Witsenhausen [44]. The <u>common part</u> W of two random variables U and V is defined by finding the largest possible integer $k \geqslant 2$ such that there exist functions f and g, $f: \mathcal{U} \longrightarrow \{1,2,\dots,k\}$ and $g: \mathcal{V} \longrightarrow \{1,2,\dots,k\}$ with $\Pr\{f(U) = i\} > 0$, $\Pr\{g(V) = i\} > 0$, $i = 1,2,\dots,k$, such that $f(U) = g(V)$ with probability one, and then defining $W = f(U) = g(V)$.

<u>Theorem 5.2</u> ([9]) : A correlated source $(\mathcal{U} \times \mathcal{V}, P(u,v))$, having a common part $W = f(U) = g(V)$, can be reliably transmitted over a given d.m. MAC K_{21} in situation $(K_{21},(U,V),I)$ if there exist probability mass functions $P(s)$ on a set $\mathcal{S}$, $P(x_1|u,s)$ on $\mathcal{X}_1$, and $P(x_2|v,s)$ on $\mathcal{X}_2$, such that

$$H(U|V) < I(X_1;Y|X_2,V,S) , \tag{5.5a}$$

$$H(V|U) < I(X_2;Y|X_1,U,S) , \tag{5.5b}$$

$$H(U,V|W) < I(X_1,X_2;Y|W,S) , \tag{5.5c}$$

$$H(U,V) < I(X_1,X_2;Y) , \tag{5.5d}$$

where $P(s,u,v,x_1,x_2,y) = P(s)P(u,v)P(x_1|u,s)P(x_2|v,s)P(y|x_1,x_2)$.

d. Ahlswede and Han [5] continued the investigations of [9] and [16] regarding the d.m. MAC K_{21} with arbitrarily correlated sources. In Section II of their paper, they [5] took a look at the coding theorem in [9] for situation $(K_{21},(U,V),I)$ from the viewpoint of <u>cross observation</u> at the encoders. In doing so, they provided another proof of Theorem 5.2, by first establishing a simple extension of Theorem 5.1 involving three correlated information sources and a d.m. MAC K_{21} with cross observation at the encoders, and then showing that this result contains Theorem 5.2 as a special case.

Consider in this respect the joint source-channel configuration shown in Fig. 9. Here is given a trivariate information source $(\mathcal{U}_1 \times \mathcal{U}_2 \times \mathcal{U}_3, P(u_1,u_2,u_3))$, producing i.i.d. discrete RV's (U_{1i},U_{2i},U_{3i}), $i = 1,2,\dots$, according to an arbitrary but fixed PD $\Pr\{U_{1i} = u_1, U_{2i} = u_2, U_{3i} = u_3\} = P(u_1,u_2,u_3)$. Moreover, there is a given d.m. MAC K_{21}. The channel is

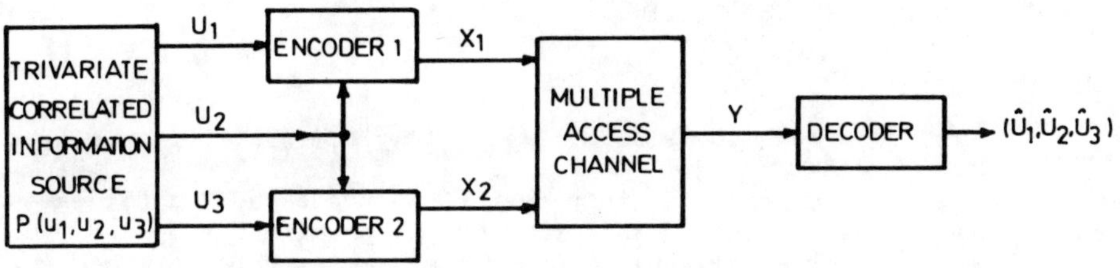

Fig. 9. Trivariate correlated information source and two-user d.m. MAC with cross observation at the encoders.

connected to the source through cross observation at the encoders. Using the notation introduced in Subsection 4.d, we may describe this source-channel configuration by the incidence relation $\delta(3,2) = (\delta_1,\delta_2)$, with $\delta_1 = \{1,2\}$ and $\delta_2 = \{2,3\}$. Encoder 1 observes the sequences of outputs of sources $\mathfrak{U}_1$ and $\mathfrak{U}_2$, and maps these into an input sequence $\underline{X}_1^n$ via an encoding function $\phi_1^n : \mathfrak{U}_1^n \times \mathfrak{U}_2^n \longrightarrow \mathfrak{X}_1^n$. Encoder 2 observes the output sequences of sources $\mathfrak{U}_2$ and $\mathfrak{U}_3$, and maps these into an input sequence $\underline{X}_2^n$ via an encoding function $\psi_2^n : \mathfrak{U}_2^n \times \mathfrak{U}_3^n \longrightarrow \mathfrak{X}_2^n$. These input sequences are subsequently transmitted over the channel. The decoder observes a received output sequence $\underline{Y}^n$ and must reconstruct the three emitted source sequences using a decoding function $\psi^n : \mathcal{Y}^n \longrightarrow \mathfrak{U}_1^n \times \mathfrak{U}_2^n \times \mathfrak{U}_3^n$. The probability of error incurred when using such an encoding and decoding system is defined analogously to (5.2). We denote this communication situation by $(K_{21},(U_1,U_2,U_3),II)$. The correlated source triple $(\mathfrak{U}_1 \times \mathfrak{U}_2 \times \mathfrak{U}_3, P(u_1,u_2,u_3))$ is said to be reliably transmissible over a d.m. MAC in situation $(K_{21},(U_1,U_2,U_3),II)$, if for any λ, $0 < \lambda < 1$, and sufficiently large n there exists a code $(\phi_1^n,\phi_2^n,\psi^n)$ for which the average probability of error P_e^n satisfies $P_e^n < \lambda$.

Ahlswede and Han [5] obtained sufficient conditions in order that an arbitrarily correlated triple source is reliably transmissible over a d.m. MAC in situation $(K_{21},(U_1,U_2,U_3),II)$, thereby generalizing Theorem 5.2 above. We state those conditions in a theorem, which is a reformulation of Theorem 1 in [5]. In this theorem, we assume the sets $\mathcal{S}_i$ to be finite.

Theorem 5.3 ([5]) : A correlated trivariate information source $(\mathfrak{U}_1 \times \mathfrak{U}_2 \times \mathfrak{U}_3, P(u_1,u_2,u_3))$ can be transmitted reliably over a given d.m. MAC K_{21} in situation $(K_{21},(U_1,U_2,U_3),II)$, if there exist random variables S_1,S_2,S_3, conditional probability distributions $P(s_1|u_1)$, $P(s_2|u_2)$, $P(s_3|u_3)$, and functions $f_1 : \mathcal{S}_1 \times \mathcal{S}_2 \longrightarrow \mathfrak{X}_1$, $f_2 : \mathcal{S}_2 \times \mathcal{S}_3 \longrightarrow \mathfrak{X}_2$, such that

$$H(U_1|U_2,U_3) < I(S_1;Y|S_2,S_3,U_2,U_3) \ , \tag{5.6a}$$

$$H(U_2|U_1,U_3) < I(S_2;Y|S_1,S_3,U_1,U_3) \ , \tag{5.6b}$$

$$H(U_3|U_1,U_2) < I(S_3;Y|S_1,S_2,U_1,U_2) \ , \tag{5.6c}$$

$$H(U_1,U_2|U_3) < I(S_1,S_2;Y|S_3,U_3) \ , \tag{5.6d}$$

$$H(U_2,U_3|U_1) < I(S_2,S_3;Y|S_1,U_1) \ , \tag{5.6e}$$

$$H(U_1,U_3|U_2) < I(S_1,S_3;Y|S_2,U_2) \ , \tag{5.6f}$$

$$H(U_1,U_2,U_3) < I(S_1,S_2,S_3;Y) \quad , \tag{5.6g}$$

where $P(u_1,u_2,u_3,s_1,s_2,s_3,y) =$

$$P(u_1,u_2,u_3)P(s_1|u_1)P(s_2|u_2)P(s_3|u_3)P(y|f_1(s_1,s_2),f_2(s_2,s_3)) \; .$$

In the statement and proof of their result, Ahlswede and Han [5] make use of an associated test channel. The proof in [5] parallels the random coding argument used by Cover et al. [9] to prove the simpler version (Theorem 5.1 above) of their coding theorem for situation $(K_{21},(U,V),I)$.

c. To complete this section, we now turn to the communication situation of an asymmetric d.m. MAC K_{21} with dependent sources, i.e., to the case of sending information from an arbitrarily correlated bivariate information source over a d.m. MAC K_{21}, as in the situation (K_{21},III) discussed in Subsection 4.b. This communication situation results from the situation $(K_{21},(U_1,U_2,U_3),II)$ by omitting there the source $\mathcal{U}_3$. The communication configuration is shown separately in Fig. 10. This situation was investigated by De Bruyn, Prelov, and van der Meulen [13], who established necessary and sufficient conditions for reliable transmission in this case. In order to describe these results, we first define the concepts which are needed.

Let be given a bivariate information source $(\mathcal{U}\times\mathcal{V},\ P(u,v))$, emitting i.i.d. discrete RV's (U_i,V_i), $i = 1,2,\ldots$, according to an arbitrary but fixed PD $\Pr\{U_i = u,\ V_i = v\} = P(u,v)$, $u\in\mathcal{U}$, $v\in\mathcal{V}$, where $\mathcal{U}$ and $\mathcal{V}$ are finite sets. Furthermore, let be given a d.m. MAC K_{21}. The encoder at terminal 1 observes the sequences of outputs of both source $\mathcal{U}$ and source $\mathcal{V}$, and maps these into an input sequence $\underline{X}_1^n$ via an encoding function $\phi_1^n : \mathcal{U}^n\times\mathcal{V}^n \longrightarrow \mathcal{X}_1^n$. The encoder at terminal 2 observes only the output sequences of source $\mathcal{V}$, and maps these into an input sequence $\underline{X}_2^n$ via an encoding function $\phi_2^n : \mathcal{V}^n \longrightarrow \mathcal{X}_2^n$. These input sequences are subsequently transmitted over the channel. The decoder observes a received output sequence $\underline{Y}^n$ and must reconstruct the two produced source sequences using a decoding function $\psi^n : \mathcal{Y}^n \longrightarrow \mathcal{U}^n\times\mathcal{V}^n$. The probability of error incurred when using this encoding and decoding system is defined as in (5.2). We denote this communication situation by $(K_{21},(U,V),III)$. The correlated source pair $(\mathcal{U}\times\mathcal{V},\ P(u,v))$ is said to be reliably transmissible over a d.m. MAC in situation $(K_{21},(U,V),III)$, if for any λ, $0 < \lambda < 1$, and sufficiently large n there exists a code $(\phi_1^n,\phi_2^n,\psi^n)$ for which the average error probability P_e^n satisfies $P_e^n < \lambda$. De Bruyn, Prelov, and van der Meulen [13] proved the following result.

Theorem 5.4 ([13]) : A correlated source pair $(\mathcal{U}\times\mathcal{V},\ P(u,v))$ can be transmitted reliably over a given d.m. MAC K_{21} in situation $(K_{21},(U,V),III)$

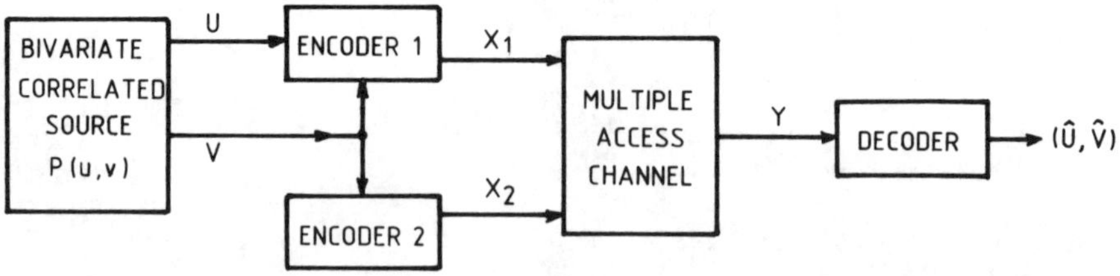

Fig. 10. A d.m. asymmetric MAC K_{21} with arbitrarily correlated sources.

if there exists a probability distribution $P(x_1,x_2)$ on $\mathcal{X}_1 \times \mathcal{X}_2$ such that

$$H(U|V) < I(X_1;Y|X_2) \ , \tag{5.7a}$$

$$H(U,V) < I(X_1,X_2;Y) \ , \tag{5.7b}$$

where $P(x_1,x_2,y) = P(x_1,x_2)\ P(y|x_1,x_2)$.

Conversely, if a correlated source pair $(\mathcal{U} \times \mathcal{V},\ P(u,v))$ can be transmitted reliably over a given d.m. MAC K_{21} in situation $(K_{21},(U,V),III)$, then the following inequalities must be true for some probability distribution $P(x_1,x_2)$:

$$H(U|V) \leq I(X_1;Y|X_2) \ , \tag{5.8a}$$

$$H(U,V) \leq I(X_1,X_2;Y) \ . \tag{5.8b}$$

Thus, Theorem 5.4 provides necessary and sufficient conditions for the reliable transmission of an arbitrarily correlated source pair over a d.m. AMAC, but with no statement when $H(U|V) = I(X_1;Y|X_2)$ or $H(U,V) = I(X_1,X_2;Y)$. This boundary situation should be examined for each source-channel combination separately.

The achievability part of Theorem 5.4 is proven in [13] by applying the Slepian-Wolf coding theorem for correlated sources [36, case 0111], and subsequently the coding theorem for the d.m. MAC in situation (K_{21},III) (Theorem 4.2 above). The converse part is proven by use of Fano's inequality. It is remarkable that for the situation $(K_{21},(U,V),III)$ the separation principle holds, notwithstanding the fact that in the case $(K_{21},(U,V),I)$ it is suboptimal.

In [13] the following example is given as an illustration of Theorem 5.4. Consider again the correlated source $(\mathcal{U} \times \mathcal{V},\ P(u,v))$ defined by $\mathcal{U} = \mathcal{V} = \{0,1\}$ and the assignment $P(0,0) = P(0,1) = P(1,1) = 1/3$, and the deterministic BEMAC defined in Section 3. Suppose we wish to transmit this particular source over this MAC according to situation $(K_{21},(U,V),III)$. There exists no PD $P(x_1,x_2)$ such that the inequalities (5.7) are satisfied for this source-channel combination. However, there does exist a PD $P(x_1,x_2)$ which yields equality in (5.8), viz. the one induced by the encodings $X_1 \equiv U$ and $X_2 \equiv V$. The capacity region $\mathcal{C}(K_{21},III)$ of this d.m. AMAC is shown in Fig. 11 together with the rate region of the given source ([36, case 0111]). The point were the ideal communication take place is also indicated in the figure.

6. THE GAUSSIAN MULTIPLE-ACCESS CHANNEL

In Section 3, 4, and 5 we considered only discrete MAC's in the sense that the input and output alphabets were discrete and the information to be transmitted through the MAC consisted of discrete sequences of symbols. In this section we will discuss MAC's with continuous input and output alphabets, in particular the Gaussian MAC. There are two cases to distinguish : (i) the discrete-time additive white Gaussian noise MAC and (ii) the so-called spectral Gaussian MAC.

The discrete-time additive white Gaussian noise (AWGN) MAC with two input users, input power constraints P_1 and P_2, and noise variance N is defined by the following conditions : (i) at any particular time the operation of the MAC is given by the equation

$$Y = X_1 + X_2 + Z \ , \tag{6.1}$$

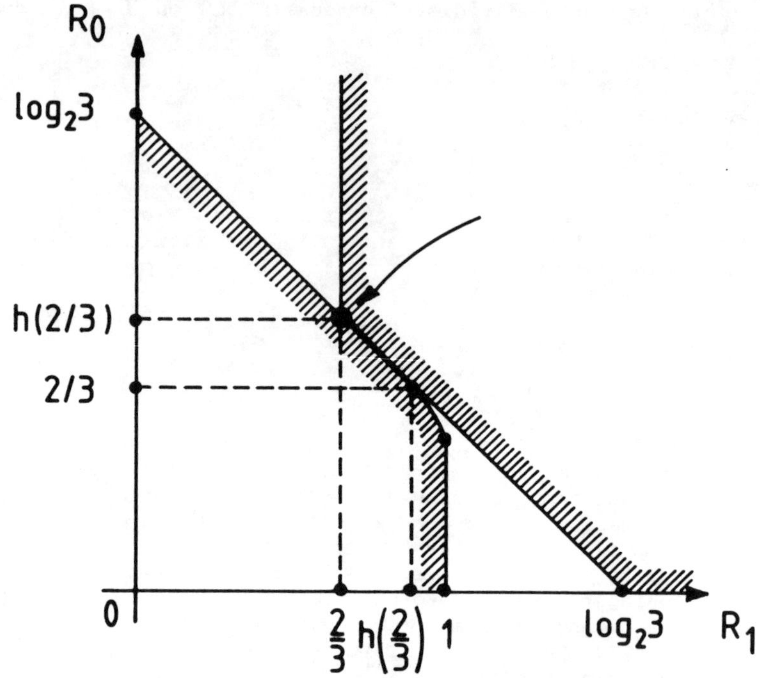

Fig. 11. An optimal point for the transmission of a certain correlated source over the asymmetric BEMAC.

where X_1 and X_2 are the input RV's, Z is a Gaussian noise RV with mean 0 and variance N ($Z \sim \mathcal{N}(0,N)$) independent of (X_1,X_2), and Y is the resulting output RV, all four RV's taking values in the set of real numbers; (ii) when input sequences of length $n \geq 1$ are transmitted over the MAC the channel operation is given by

$$\underline{Y}^n = \underline{X}_1^n + \underline{X}_2^n + \underline{Z}^n , \tag{6.2}$$

where $\underline{X}_1^n = (X_{11},\ldots,X_{1n})$, $\underline{X}_2^n = (X_{21},\ldots,X_{2n})$, $\underline{Z}^n = (Z_1,\ldots,Z_n)$ and $\underline{Y}^n = (Y_1,\ldots,Y_n)$, $\underline{Z}^n$ is independent of $(\underline{X}_1^n,\underline{X}_2^n)$, and $Z_1,\ldots,Z_n$ are i.i.d. RV's with common distribution $\mathcal{N}(0,N)$; and (iii) all input sequences $\underline{x}_1^n = (x_{11},\ldots,x_{1n})$ and $\underline{x}_2^n = (x_{21},\ldots,x_{2n})$ which are to be transmitted must satisfy for any block length n the average power constraints

$$\frac{1}{n}\sum_{t=1}^{n} x_{it}^2 \leq P_i , \qquad i = 1,2 . \tag{6.3}$$

The simplest communication situation of a discrete-time AWGN MAC with two senders is the case where two independent messages are to be transmitted, as in situation (K_{21},I). This communication situation is sketched in Fig. 12. Here the source output m_1 is encoded into the codeword $\underline{X}_1^n(m_1)$ and source output m_2 is encoded into the codeword $\underline{X}_2^n(m_2)$. These codewords are transmitted over the AWGN MAC, whose operation has just been defined, and

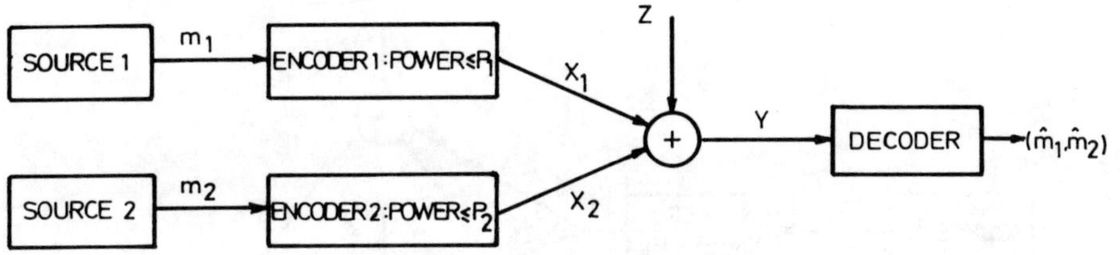

Fig. 12. Discrete-time additive white Gaussian noise MAC with two input users, power constraints P_1 and P_2, and independent noise variable $Z \sim \mathcal{N}(0,N)$. (Situation (AWGN K_{21},N,P_1,P_2,I).)

an output sequence $\underline{Y}^n$ is produced. Applying the decoding rule on this received vector, the source outputs are then to be reconstructed. We denote this communication situation by MAC (AWGN K_{21},N,P_1,P_2,I), or, since it is clear that we are dealing in this paper with the MAC only, by (AWGN K_{21},N,P_1,P_2,I). The formal definitions of a code, error probability, achievable rate pair, and capacity region are the same as the ones given in Section 3 for situation (K_{21},I), except that now discrete probabilities are replaced by probability densities and the constraints (6.3) are imposed on the codewords. We denote the capacity region in this situation by $\mathcal{C}$(AWGN K_{21},N,P_1,P_2,I).

Cover [8] and Wyner [46] established independently an achievable rate region for the MAC (AWGN K_{21},N,P_1,P_2,I). This region can be derived from the characterization of $\mathcal{C}(K_{21},I)$ given in Theorem 3.1 by substituting probability densities for discrete probabilities, taking into account the constraints (6.3), and making use of the Gaussian structure of the channel. However, Cover and Wyner provided direct arguments to prove achievability, based on the well-known properties of the single-input Gaussian channel [34]. A perspicuous proof of the achievable rate region based on the notion of joint typicality is provided in [18], where also the extension of the same result is given to the case of a discrete-time AWGN MAC with $t \geq 2$ independent users. Another direct proof is given in [21].

Although the achievable rate region found by Cover [8] and Wyner [46] is indeed the capacity region, no formal proof of the converse appeared until Keilers [27] gave a proof of the weak converse in his Ph. D. dissertation. For sake of completeness, we combine the results of [8],[27], and [46] into one theorem.

Theorem 6.1 ([8],[27],[46]) : The capacity region of the discrete-time MAC (AWGN K_{21},N,P_1,P_2,I) is given by

$$\mathcal{C}(\text{AWGN } K_{21},N,P_1,P_2,I) = \Big\{(R_1,R_2) \;:\; 0 \leq R_1 \leq \frac{1}{2}\,\ln\left(1+\frac{P_1}{N}\right), \tag{6.4a}$$

$$0 \leq R_2 \leq \frac{1}{2}\,\ln\left(1+\frac{P_2}{N}\right), \tag{6.4b}$$

$$R_1 + R_2 \leq \frac{1}{2}\,\ln\left(1+\frac{P_1+P_2}{N}\right)\Big\} \;. \tag{6.4c}$$

Keilers [27] also considered a k-dimensional discrete-time additive non-white (arbitrarily correlated) Gaussian noise MAC with $m \geq 2$ users, which he

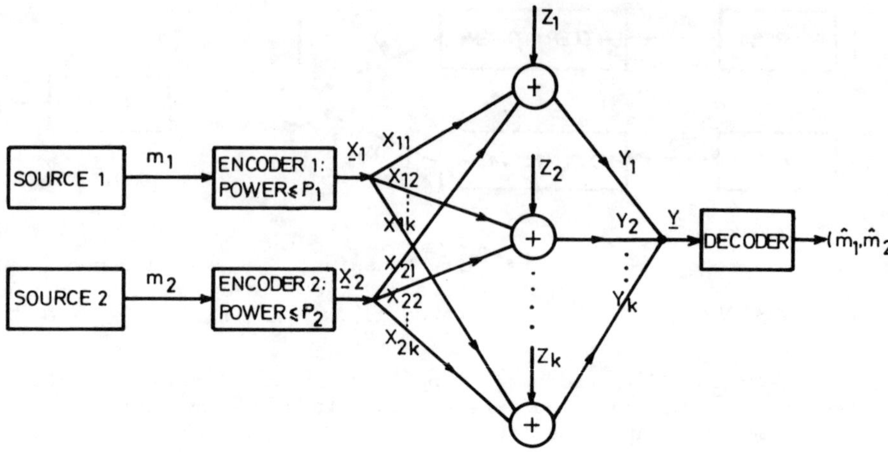

Fig. 13. K-dimensional discrete-time additive non-white Gaussian noise MAC with two input users, power constraints P_1 and P_2, and independent Gaussian noise process $\underline{Z}^k = (Z_1,\ldots,Z_k)$ having an arbitrary covariance matrix Λ_Z; equivalent to a set of k parallel independent additive white Gaussian noise MAC's.

called the spectral Gaussian MAC. This channel is pictured in Fig. 13 for the case of two input users, and can be described as follows. Source output m_i (i = 1,2) is encoded into $\underset{\sim}{X}_i^{kn}(m_i)$, consisting of k n-dimensional vectors $(\underline{X}_{i1}^n,\underline{X}_{i2}^n,\ldots,\underline{X}_{ik}^n)$ of real numbers. At each moment t, t = 1,2,...,n, the k-dimensional vector consisting of the t-th components of the vectors $\underline{X}_{ij}^n$, j = 1,2,...,k, is presented to the channel, i.e., the vector $\underline{X}_{i.t}^k = (X_{i1t},X_{i2t},\ldots,X_{ikt})$. Thus, the codeword $\underset{\sim}{X}_i^{kn}(m_i)$ is fed into the channel as an n-sequence of k-dimensional vectors. Both codewords $\underset{\sim}{X}_1^{kn}(m_1)$ and $\underset{\sim}{X}_2^{kn}(m_2)$ are transmitted, and an output sequence $\underset{\sim}{Y}^{kn}$ is received, upon which the decoder applies its decoding function $\psi^{kn} : \mathcal{Y}^{kn} \rightarrow \mathcal{M}_1 \times \mathcal{M}_2$. The operation of this spectral MAC is statistically defined by the equation $\underline{Y}^k = \underline{X}_1^k + \underline{X}_2^k + \underline{Z}^k$, where X_1 and X_2 are independent, $\underline{X}_i^k$ is the k-th power of X_i with respect to the independent product distribution (i = 1,2), and $\underline{Z}^k$ is a k-dimensional Gaussian random variable (generally not being the product of k independent one-dimensional Gaussian RV's). In this discussion it is further understood that we focus again on the simplest communication situation, viz. the one where two independent messages are transmitted as in situation (K_{21},I), with no cross observation at the encoders. Moreover, we assume that all input sequences $\underline{x}_1^{kn}$ and $\underline{x}_2^{kn}$ satisfy for any block length n the average power constraints

$$\frac{1}{n}\sum_{t=1}^{n}\sum_{j=1}^{k} x_{ijt}^2 \leqslant P_i\,, \qquad i = 1,2. \tag{6.5}$$

In this model of a k-dimensional discrete-time additive non-white Gaussian noise MAC with 2 users it is assumed that at each time instant t (t = 1,2,...,n) the k-dimensional noise vector $\underline{Z}_t^k = (Z_{1t},\ldots,Z_{kt})$ has a k-variate normal distribution with arbitrary covariance matrix Λ_Z, and that the n k-dimensional noise vectors $\underline{Z}_1^k,\ldots,\underline{Z}_n^k$ are statistically independent. The definitions of a code, error probability, achievable rate pair, and capacity region are straightforward for this channel, except that it should be emphasized that each codeword for each user should satisfy the power constraints (6.5). However, as Keilers [27] observed, "each user can apportion power as desired among the k dimensions of the process as long as the total user power is preserved by the sum".

Keilers showed the equivalence of the above model to that of a collection of k parallel independent AWGN MAC's, whereby the noise vector at time instant t, $\underline{Z}_t^k = (Z_{1t},\ldots,Z_{kt})$, consists of k independent Gaussian RV's each with mean zero and respective variances $N_1,\ldots,N_k$, i.e., $Z_{jt} \sim \mathcal{N}(0,N_j)$, $j = 1,\ldots,k$; $t = 1,\ldots,n$. Keilers proved this argument by considering a matrix diagonalization procedure whereby the covariance matrix Λ_Z is transformed into a diagonal covariance matrix Λ_Z' with as principal diagonal elements the variances $N_1,\ldots,N_k$. He argued that the collection of k parallel independent AWGN MAC's thus obtained can serve as an approximation for a continuous-time spectral Gaussian noise MAC, thereby following the approach outlined in Gallager [20, Section 8.5] for reducing the continuous-time AWGN one-way channel to a set of parallel independent AWGN one-way channels. Keilers [27] observed that under the transformation of the original model to that of k parallel independent AWGN MAC's the power constraints (6.5) are preserved, so that the transformed channel is equivalent to the original channel. Thus, achievability theorems and converses can be proved for either channel model, and the extension to the other model is immediate.

As one of his main results, Keilers established the capacity region of a set of k parallel independent AWGN MAC's with $m \geqslant 2$ input users. Following the above line of reasoning, this result can then be carried over to the spectral Gaussian MAC. For simplicity we state Keilers' theorem just for the case of two input users.

Theorem 6.2 ([27]) : The capacity region of a collection of k parallel independent discrete-time AWGN MAC's with two input users, total input powers P_1 and P_2 such that the constaints (6.5) are satisfied, and respective noise variances $N_1,\ldots,N_k$ is given by

$$\mathcal{C} = \text{co}\Big\{(R_1,R_2) \;:\; 0 \leqslant R_1 \leqslant \sum_{j=1}^{k} \frac{1}{2}\ln\left(1+\frac{P_{j1}}{N_j}\right) \tag{6.6a}$$

$$0 \leqslant R_2 \leqslant \sum_{j=1}^{k} \frac{1}{2}\ln\left(1+\frac{P_{j2}}{N_j}\right), \tag{6.6b}$$

$$0 \leqslant R_1 + R_2 \leqslant \sum_{j=1}^{k} \frac{1}{2}\ln\left(1+\frac{P_{j1}+P_{j2}}{N_j}\right), \tag{6.6c}$$

$$\text{such that } 0 \leqslant P_{ji},\ j = 1,\ldots,k;\ i = 1,2;\ \text{and} \sum_{j=1}^{k} P_{ji} = P_i,\ i = 1,2\Big\}.$$

Keilers [27] proved the achievability part of Theorem 6.2 by direct application of the results by Cover [8] and Wyner [46]. He also obtained an alternate form of the achievable rate region in terms of the extreme points which generate it. He furthermore showed that these extreme points are described by a sequential water-fill strategy as in Shannon [34]. Moreover, he proved a weak converse.

ACKNOWLEDGEMENT

The author wishes to thank Kristien De Bruyn for valuable comments on the first draft of this paper and for her assistance with the drafting of the figures.

REFERENCES

[1] R. Ahlswede, "Multi-way communication channels," in Proc. 2nd Int. Symp. Inform. Theory, Tsahkadsor, 1971. Publishing House of the Hungarian Academy of Sciences, pp. 23-52, 1973.

[2] R. Ahlswede, "The capacity region of a channel with two senders and two receivers," Ann. Prob., vol. 2, pp. 805-814, Oct. 1974.

[3] R. Ahlswede, "An elementary proof of the strong converse theorem for the multiple-access channel," J. of Combinatorics, Information & System Sciences, vol. 7, no. 3, 216-230, 1982.

[4] R. Ahlswede and G. Dueck, "Good codes can be produced by a few permutations," IEEE Trans. Inform. Theory, vol. IT-28, no. 3, pp. 430-443, May 1982.

[5] R. Ahlswede and T.S. Han, "On source coding with side information via a multiple-access channel and related problems in multi-user information theory," IEEE Trans. Inform. Theory, vol. IT-29, no. 3, pp. 396-412, May 1983.

[6] L.A. Bassalygo, M.S. Pinsker and V.V. Prelov, "Zero-error capacity for deterministic multiple-access channels with common information," Probl. Pered. Inform., vol. 18, no. 1, pp. 3-11, Jan.-March 1982.

[7] M. Bierbaum and H.M. Wallmeier, "A note on the capacity region of the multiple-access channel," IEEE Trans. Inform. Theory, vol. IT-25, p. 484, July 1979.

[8] T. Cover, "Some advances in broadcast channels," in Advances in Communication Systems, Vol. 4, A. Viterbi, Ed. San Francisco : Academic Press, 1975.

[9] T.M. Cover, A. El Gamal, and M. Salehi, "Multiple access channels with arbitrarily correlated sources," IEEE Trans. Inform. Theory, vol. IT-26, pp. 648-657, Nov. 1980.

[10] T.M. Cover, R.J. McEliece, and E.C. Posner, "Asynchronous multiple-access channel capacity," IEEE Trans. Inform. Theory, vol. IT-27, pp. 409-413, July 1981.

[11] I. Csiszár and J. Körner, Information Theory : Coding Theorems for Discrete Memoryless Systems. New York : Academic Press, 1981.

[12] M.A. Deaett and J.K. Wolf, "Some very simple codes for the nonsynchronized two-user multiple-access adder channel with binary inputs," IEEE Trans. Inform. Theory, vol. IT-24, pp. 634-635, Sept. 1978.

[13] K. De Bruyn, V.V. Prelov, and E.C. van der Meulen, "Reliable transmission of two correlated sources over an asymmetric multiple-access channel". Accepted for publication in IEEE Trans. Inform. Theory.

[14] K. De Bruyn and E.C. van der Meulen, "An iterative code construction and codes generated by permutations for the asymmetric multiple-access channel," IEEE Trans. Inform. Theory, vol. IT-32, pp. 607-617, July 1986.

[15] G. Dueck, "Maximal error capacity regions are smaller than average error capacity regions for multi-user channels," Probl. of Control and Inform. Theory, vol. 7, pp. 11-19, 1978.

[16] G. Dueck, "A note on the multiple access channel with correlated sources," IEEE Trans. Inform. Theory, vol. IT-27, pp. 232-235, March 1981.

[17] G. Dueck, "The strong converse of the coding theorem for the multiple-access channel", J. of Combinatorics, Information & System Sciences, vol. 6, no. 3, pp. 187-196, 1981.

[18] A. El Gamal and T.M. Cover, "Multiple user information theory," Proc. IEEE, vol. 68, pp. 1466-1483, Dec. 1980.

[19] P. Gács and J. Körner, "Common information is far less than mutual information," Probl. of Control and Inform. Theory, vol. 2, pp. 149-162, 1973.
[20] R.G. Gallager, Information Theory and Reliable Communication. New York : Wiley, 1968.
[21] R.G. Gallager, "A perspective on multiaccess channels," IEEE Trans. Inform. Theory, vol. IT-31, pp. 124-142, Mar. 1985.
[22] A.A. Grigor'ev, "Boundary of region of permissible rates for asynchronous communication over a binary summing channel," Probl. Pered. Inform., vol. 15, no. 3, pp. 18-23, July-September, 1979.
[23] T.S. Han, "The capacity region of a general multiple-access channel with certain correlated sources," Inform. Control, vol. 40, pp. 37-60, 1979.
[24] E.A. Haroutunian, "Lower bounds for the error probability of multiple access channels," Probl. Pered. Inform., vol. 11, no. 2, pp. 23-36, April-June, 1975.
[25] J.Y.N. Hui and P.A. Humblet, "The capacity region of the totally asynchronous multiple access channel," IEEE Trans. Inform. Theory, vol. IT-31, pp. 207-216, Mar. 1985.
[26] T. Kasami, S. Lin, V.K. Wei, and S. Yamamura, "Graph theoretic approaches to the code construction for the two-user multiple-access binary adder channel," IEEE Trans. Inform. Theory, vol. IT-29, pp. 114-130, Jan. 1983.
[27] C. Keilers, "The capacity of the spectral Gaussian multiple-access channel," Ph. D. dissertation, Dept. of Electrical Engineering, Stanford University, Stanford, CA, April 1976.
[28] H. Liao, "A coding theorem for multiple access communications," presented at Int. Symp. on Inform. Theory, Asilomar, 1972. Also Ph. D. dissertation, "Multiple Access Channels," Dept. of Elec. Eng., Univ. of Hawaii, 1972.
[29] J. Pokorny and H.M. Wallmeier, "Random coding bound and codes produced by permutations for the multiple-access channel," IEEE Trans. Inform. Theory, vol. IT-31, pp. 741-750, Nov. 1985.
[30] G. Sh. Poltyrev, "Coding in an asynchronous multiple-access channel," Probl. Pered. Inform., vol. 19, no. 3, pp. 12-21, July-Sept., 1983.
[31] V.V. Prelov, "Transmission over a multiple-access channel with a special source hierarchy," Probl. Pered. Inform., vol. 20, no. 4, pp. 3-10, October-December, 1984.
[32] M. Salehi, "Cardinality bounds on auxiliary variables in multiple-user theory via the method of Ahlswede and Körner," Dept. of Statistics, Stanford University, Stanford, CA, Tech. Rep. 33, August 1978.
[33] C.E. Shannon, "A mathematical theory of communication," Bell Syst. Tech. J., vol. 27, pp. 379-423 (Part I), pp. 623-656 (Part II), July and October 1948. Reprinted in Key Papers in the Development of Information Theory, D. Slepian, Ed. New York : IEEE, 1974, pp. 5-29.
[34] C.E. Shannon, "Communication in the presence of noise," in Proc. IRE, vol. 37, pp. 10-21, Jan. 1949. Reprinted in Key Papers in the Development of Information Theory, D. Slepian, Ed. New York : IEEE, 1974, pp. 30-41.
[35] C.E. Shannon, "Two-way communication channels," in Proc. 4th Berkeley Symp. Math. Stat. Prob., vol. 1, pp. 611-644, 1961. Reprinted in Key Papers in the Development of Information Theory, D. Slepian, Ed. New York : IEEE Press, 1974, pp. 339-372.
[36] D. Slepian and J.K. Wolf, "Noiseless coding of correlated information sources," IEEE Trans. Inform. Theory, vol. IT-19, pp. 471-480, July

1973. Reprinted in Key Papers in the Development of Information Theory, D. Slepian, Ed. New York : IEEE, 1974, pp. 450-459.

[37] D. Slepian and J.K. Wolf, "A coding theorem for multiple access channels with correlated sources," Bell Syst. Tech. J., vol. 52, pp. 1037-1076, Sept. 1973.

[38] M.L. Ulrey, "The capacity region of a channel with s senders and r receivers," Inform. Contr., vol. 29, pp. 185-203, 1975.

[39] E.C. van der Meulen, "The discrete memoryless channel with two senders and one receiver," in Proc. 2nd Int. Symp. Inform. Theory, Tsahkadsor, 1971. Publishing House of the Hungarian Academy of Sciences, pp. 103-135, 1973.

[40] E.C. van der Meulen, "On a problem by Ahlswede regarding the capacity region of certain multi-way channels," Inform. Contr., vol. 25, pp. 351-356, 1974.

[41] E.C. van der Meulen, "A survey of multi-way channels in information theory : 1961-1976," IEEE Trans. Inform. Theory, vol. IT-23, pp. 1-37, Jan. 1977.

[42] E.C. van der Meulen, "Recent coding theorems and converses for multi-way channels. Part I : the broadcast channel (1976-1980)," in New Concepts in Multi-User Communication (NATO Advanced Study Institute Series) J.K. Skwyrzinsky, Ed. Alphen aan den Rijn, The Netherlands : Sijthoff & Noordhoff International, 1981, pp. 15-51.

[43] F.M.J. Willems, "Informationtheoretical results for the discrete memoryless multiple access channel," Ph. D. dissertation, Dept. of Mathematics, K.U. Leuven, Leuven, Belgium, October 1982.

[44] H.S. Witsenhausen, "On sequences of pairs of dependent random variables," SIAM J. Appl. Math., vol. 28, pp. 100-113, Jan. 1975.

[45] J. Wolfowitz, Coding Theorems of Information Theory, 3rd Ed. Berlin : Springer, 1978.

[46] A.D. Wyner, "Recent results in the Shannon theory," IEEE Trans. Inform. Theory, vol. IT-20, pp. 2-10, Jan. 1974.

Limits on System Reliability Improvement

W. Kuo

Department of Industrial Engineering

INTRODUCTION

System reliability can be improved through (1) using simple design, (2) using reliable components, or (3) adding redundancy. Frequently, all these techniques are employed during the design stage. While simple design is a matter of art, using reliable components and adding redundancy can be done scientifically. Over the past 30 years, the reliability optimization problem has been studied in an attempt to allocate optimally the redundancy and reliability of system components. System reliability cannot be achieved at the 100% level because the components used in the system are not perfect; if more and better components are used, the system gets voluminous, heavy, and costly. Therefore, the theme of reliability optimization is a tradeoff between reliability (system) and cost, in any acceptable format. The limits of system reliability are also limits of cost.

In reliability optimization problems, the decision variables are the number of redundancies that are integers (redundancy allocation problems), the component reliabilities that are real numbers between 0 and 1 (reliability allocation problems), or a combination of both (reliability-redundancy allocation problems). By these classifications and other concerns, five types of problems have been studied: effort function minimization, reliability allocation, redundancy allocation, reliability-redundancy allocation, and cost minimization. Whenever a

J. K. Skwirzynski (ed.), Performance Limits in Communication Theory and Practice, 427–440.

system is large and complex, optimal solutions become less achievable. This is the limit of computation. In the past, emphasis was on investigating an efficient method to achieve system optimization.

NOTATION

b_i	amount of resource i available
b_i^c	$b_i - \sum_{j=1}^{n} \sum_{k=1}^{x_j} \Delta g_{ij}(k)$
$C_j(r_j)$	cost function of component reliability r_j
C_s	system cost function
g_{ij}	amount of resource i consumed at stage j
G, G_j	effort functions for the system, and associated with component j, respectively
L_{+1}	set of all the stages whose reliabilities can be increased
n	number of stages in the system
R	$(r_1, r_2, \ldots, r_n)$
r_j, q_j	reliability and unreliability of component j
R_j, Q_j	reliability and unreliability of stage j
R_s, Q_s	system reliability and unreliability
$R_{s,min}$	minimal system reliability required
X	$(x_1, x_2, \ldots, x_n)$
x_j	number of components at stage j

DEFINITONS

In this study, system reliability is based on the following definition of a system: A system is composed of one or more stages (or subsystems). A stage, composed of one or more components, is a unique functional unit of the system. Cost functions at different stages are assumed to be additive. The system reliability is the probability of successful operation of a system for a specified period of time under given conditions. It is usually expressed in terms of the reliabilities of both the stages and the components. In evaluating the system reliability, it is necessary to specify system structure, component failure process, and definition of failure.

For a series system, the system is operational only when all the stages are operational. For a parallel system, the system is operational if one or more stages is operational. A general system (or complex system) is a nonparallel, nonseries system whose reliability can be evaluated by probability theory once the system structure is clearly defined. Other types of structures are the parallel-series system and the series-parallel system. In this study, it is assumed that components are independent of each other and the component reliability is deterministic.

The cost functions can be capital investment for high reliability, volume, weight, or a combination of the above factors employed in a system.

MINIMIZATION OF THE EFFORT FUNCTION

Improving the reliability of a system (or component) from a lower level to a higher level requires real effort. Associated with component j is a function $G_j(r_j, r_j')$, which describes the effort to raise reliability from r_j to r_j'. Assume that the functional form of the effort is not available; rather the following properties are agreed upon:

1. application of effort always increases reliability
2. effort required to raise reliability from the current level by a given amount is the same, whether or not this is achieved directly or by a series of separate improvements, however small
3. reliability improvement gets progressively more difficult as reliability increases
4. perfect reliability is unattainable
5. the effort applied to the system is the sum of the efforts applied to the components.

Thus, the optimal policy can achieve the desired reliability of a series system when development effort functions are the same for all components. Work by Albert [1] and Dale and Winterbottom [2] extended and proved that the optimal policy is to increase the reliabilities of the k weakest components to a common value and to leave the remaining components alone if the same effort function is applied to each component. The optimal policy obtained through minimization of the effort function can be used as a valuable managerial tool for assuming desired system reliability.

RELIABILITY ALLOCATION

Another way that the system reliability can be improved is by using reliable components at higher costs. A typical problem of this kind can be formulated as

$$\text{Max } R_s(R)$$

subject to

$$\sum_{j=1}^{n} g_{ij}(r_j) \leq b_i \qquad \forall i$$

Both the functional forms of system reliability and system cost functions can be obtained through fixed system structure and cost structure. The above formulation can be solved by nonlinear programming techniques. Tillman et al. [3] provides a comprehensive introduction.

Another similar problem is to maximize the profit gained by high reliability subject to cost constraints, where the profit is a function of system reliability and system downtime. Apparently, the problem in minimizing the effort function is to achieve the minimal system reliability requirement, whereas the reliability allocation problem is limited by the maximum amount of resources available.

REDUNDANCY ALLOCATION

The redundancy-optimization problem can be formulated as

$$\text{Max } R_s(X|R)$$

subject to

$$\sum_{j=1}^{n} g_{ij}(x_j) \leq b_i \qquad \forall\, i$$

The problem is to determine the number of redundancies at each stage, $X = (x_1, \ldots, x_n)$. The above problem assumes that the component reliabilities, $R = (r_1, \ldots, r_n)$, are given constants.

In the past two decades, reliability optimization techniques have been proposed [3]; these techniques can be classified as the exact method and the iterative method. The exact method, which obtains the solution analytically, generally involves more mathematics and generates a more accurate solution. The Lagrange multiplier with Kuhn-Tucker conditions [4] and dynamic programming [3] are examples of the exact method. The iterative method, which obtains the solution by repeating the algorithm or enumerating the solutions, does not require an extensive mathematical background. The branch-and-bound technique [5] and the heuristic method [6] are examples of the iterative method. Although methods like the branch-and-bound technique and dynamic programming require more computer time, they guarantee global optimality for a variety of problems.

In those methods that are based on differentiation, the decision variables have to be continuous. Earlier studies treat the number of redundancies as real variables [4]. The real number answer is rounded off and the neighboring integer solutions are evaluated. The most feasible solution among the trials is taken as the final solution. This method works well if the problem is simple and the constraints are linear. However, as the problem gets complicated, the rounding off and trial-and-error procedure becomes inefficient and inaccurate. Furthermore, this approach provided no theoretical reasoning.

Depending on the type of structure, failure mode, and constraint functions, the above problem may be converted into different forms. Since 1956, numerous techniques have been proposed to solve a variety of redundancy-optimization problems. Yet, none of them can effectively solve a large-scale general system with multiple, nonlinear constraints. Those proposed techniques are limited to one or more of the following:

1. special type of system configuration, normally the series system
2. special type of constraint function, for example, the linear constraints
3. the dimension of the problem
4. no guarantee for global optimal solution
5. complicated computation
6. nonlinear, noninteger optimization problem; optimal decision variables approximated to integer number through extensive discussions.

Therefore, heuristic redundancy-allocation methods have been suggested. The heuristic method has the properties of simplicity, generality, and efficiency (Kuo et al. [6] and Tillman et al. [3]). Many algorithms have been able to solve the series system (or general system with modification) having multiple nonlinear constraints (e.g., Kohda and Inoue [7]), although global optimality is not yet guaranteed.

Most redundancy-allocation heuristics are based on the following ideas.

1. Initialize each stage with one component.
2. Evaluate sensitivity function to determine the stage to which a redundant component is to be added. (A sensitivity function is defined to consider the tradeoff between the increment of system reliability and the resources consumption, e.g., b_i^c given in the definitions section.)
3. Increase the number of redundancies by one at the proposed stage and repeat Steps 2 and 3 until the constraint is violated.

Although heuristics are the popular approaches, they offer the user little indication of optimality. In addition, heuristics are both inefficient and difficult to justify in solving the reliability-redundancy allocation problem.

RELIABILITY-REDUNDANCY ALLOCATION PROBLEM

Designers must decide whether to use highly reliable components or to add redundancies. For a system like a space shuttle, a system reliability near 1.0 is desirable, yet to minimize the shuttle weight, adding redundancy would be a real burden. On the other hand, in an ordinary industrial product, adding redundancies can be a good solution, since the cost of a reliable component is at least an exponential function of its reliability measure. The following mixed integer reliability optimization problem is formulated to allow flexibility in the decision process for the systems falling between these two extremes:

$$\text{Max } R_s(R,X)$$

subject to

$$\sum_{j=1}^{n} g_{ij} (r_j, x_j) \leq b_i \qquad \forall\, i$$

The $R = (r_1, r_2, \ldots, r_n)$ and $X = (x_1, x_2, \ldots, x_n)$ are to be determined for given g_{ij}'s and b_i's.

The method suggested by Tillman et al. [3] provides a way to optimize the reliability-redundancy allocation problem. Their approach, which combines the heuristic approach by Aggarwal et al. [8] and the Hooke and Jeeves pattern search technique, has suffered numerical instability and low computational efficiency [9].

Lin and Kuo [9] recently completed a comparative study of the heuristic reliability-optimization method. Two sequential search methods and four redundancy-allocation heuristics are combined to solve the mixed-integer reliability optimization for a complex system configuration. A simulation study compares the relative merits of these combinations.

A method of combining the Lagrange multiplier technique with the branch-and-bound technique is proposed by Kuo et al. [10]. The Lagrange multiplier technique quickly reaches an exact, real number solution that is closed to the optimal solution. Then the branch-and-bound method is used to get the integer solution. This method can solve both the redundancy allocation problem and the reliability-redundancy allocation problem. When dealing with the latter problem, only branching and bounding the integer variable is necessary. With the proposed method, a four-stage series system with two linear constraints is illustrated for the redundancy allocation problem, and a five-stage series system with three nonlinear constraints is illustrated for the reliability-redundancy allocation problem. The results showed that the proposed method is better than the previous methods for both the redundancy allocation problem and the mixed-integer-type, reliability-redundancy allocation problem. The proposed method also provides more reasonable explanations when solving reliability optimization problems.

COST MINIMIZATION

The cost minimization problem is similar to the minimization of the effort function problem, except that the cost function is clearly spelled out in cost minimization. Typically, this can be formulated as

$$\text{Min } C_s(R,X)$$

subject to

$$R_s \geq R_{s,min}$$

This is also a mixed-integer-type problem where both the reliability levels and redundancy levels are to be determined. In the past, the redundancy levels were not considered; therefore the problem became an ordinary, reliability-allocation problem, primarily concerned with minimizing cost. Various nonlinear optimization techniques are also available. A multiple-objective optimization problem can be formulated so that the system cost is minimized and the system reliability is maximized.

FUTURE STUDY

For large-scale, reliability-optimization problems an efficient method is always in demand. Many search techniques used have the drawback of being trapped by a local optimum and then not being able to get out of it. Yet some methods that overcome this drawback require extensive computational time and resources. Some cost function is desirable but not readily available; further development in effort function minimization can replenish the cost function modeling. Specifically, an optimal policy is needed if different effort functions are assigned to different components and if the system structure is non-series.

ACKNOWLEDGMENTS

The author acknowledges the grant provided by the NATO Advanced Study Institute through Dr. J. K. Skwirzynski to make the presentation of this paper possible at the 1986 July meeting in Italy. Support from the Engineering Research Institute at Iowa State University is also acknowledged.

REFERENCES

1. Albert, A., "A measure of the effort required to increase reliability," Technical Report No. 43, Applied Mathematics and Statistics Laboratory, Stanford University, Stanford, Calif. 94305, 1958.

2. Dale, C. J., and Winterbottom, A., "Optimal allocation of effort to improve system reliability," IEEE Trans. Reliability, Vol. R-35, pp. 188-191, 1986.

3. Tillman, F. A., Hwang, C. L., and Kuo, Way, Optimization of Systems Reliability, Marcel Dekker, New York, 1985.

4. Misra, K. B., and Ljubojevic, M. D., "Optimal reliability design of a system: a new look," IEEE Trans. Reliability, Vol. R-22, pp. 255-258, 1973.

5. Nakagawa, Y., Nakashima, K., and Hattori, Y., "Optimal reliability allocation by branch-and-bound technique," IEEE Trans. Reliability, Vol. R-27, pp. 31-38, 1978.

6. Kuo, Way, Hwang, C. L., and Tillman, F. A., "A note on heuristic methods in optimal system reliability," IEEE Trans. Reliability, Vol. R-27, pp. 320-324, 1978.

7. Kohda, T., and Inoue, K., "A reliability optimization method for complex systems with the criterion local optimality," IEEE Trans. Reliability, Vol. R-31, pp. 109-111, 1982.

8. Aggarwal, K. K., Gupta, J. S., and Misra, K. B., "A new heuristic criterion for solving a redundancy optimization problem," IEEE Trans. Reliability, Vol. R-24, pp. 86-87, 1975.

9. Lin, H. H., and Kuo, Way, "A comparative study of the heuristic reliability optimization method," Operations Research, 1986 (submitted).

10. Kuo, Way, Lin, H. H., Xu, Z., and Zhang, W., "Reliability optimization with Lagrange multiplier and branch-and-bound," IEEE Trans. Reliability, 1986 (submitted).

LIST OF DELEGATES

Dr. A.A. de Albuquerque, Complexo Instituto Superior Tecnico, Av. Rovisco Pais, 1096 Lisboa. Portugal.
Prof. E. Anarin, Bogazici University, Bebek-Istanbul, Turkey.
Dr. D. Andelman, 27 Ben Yoseph Street, Tel Aviv 69496, Israeal.
Dr. A. Anderman, Commuications System Engineering, 930 Third Street, Santa Monica, CA 90403, U.S.A.
Mr. B.G. Anderson, British Aerospace PLC., Engineering Computer Services Dept. 468, P.O. Box 19, Six Hills Way, Stevenage SG1 2DA, U.K.
Mr. C.P.M.J. Baggen, Phillips Research Laboratories, P.O. Box 80,000, 5600 Eindhoven, The Netherlands.
Prof. I. Bar-David, Department of Electrical Engineering, Technion City, Haifa 32000, Israel.
Prof. V.K. Bhargava, Department of Electrical Engineering, University of Victoria, P.O. 1700, Victoria B.C., Canada V8W 2Y2.
Dr. R.E. Blahut, IBM Owego, N.Y. 13827, U.S.A.
Prof. I.F. Blake, Department of Electrical Engineering, University of Waterloo, Ontario, Canada N2L 3G1.
Mr. P. Bridge, University of Manchester, Department of Electrical Engineering, Dover Street, Manchester M13 9PL, U.K.
Dr. M. Burnashev, Institute of Problems of Information Transmission, USSR Academy of Sciences, Moscow, Ermolovoy Street 19, USSR.
Prof. J.S. Byrnes, Department of Mathematics & Computer Science, University of Massachusetts at Boston, Boston, MA 02125, U.S.A.
Prof. K.W. Cattermole, Fairacre, Gandish Road, East Bergholt C07 6TP, U.K.
Dr. T. Chandrasekeran, Centre for Development of Telematics, Homi Bhabba Road, Bombay-400 005, India.
Mr. Giovanni Cherubini, Department of Electrical Engineering & Computer Sciences, University of California, San Diego, La Jolla, CA 92093, U.S.A.
Mr. F.P. Coakley, Department of Electronic and Electrical Engineering, University of Surrey, Guildford, Surrey GU2 5XH, U.K.
Dr. Brinton A. Cooper III, U.S. Army Ballistic Research Laboratory, Aberdeen Proving Ground, MD 21005, U.S.A.
Mrs. Grace Cooper, 1416 Saratoga Drive, Bel Air, MA 21014, U.S.A.
Prof. T. Cover, Durand, Room 121, University of Stanford, Stanford, CA 94305, U.S.A.
Prof. E. Dalle-Mese, Institute of Electronics & Communications, Via Diotisalvi 2, I-56100 Pisa, Italy.
Dr. R. Dekeyser, Dept. Natuurkunde, Katholieke Universiteit Leuven, Celestijnenlaan 200 D, B-3030 Leuven, Belgiun.
Dr. Y. Desmedt, ESAT Laboratory, Electrical Engineering Department, Katholieke Universiteit Leuven, Kardinaal Mercierlaan 94, 3030 Heverlee, Belgium.
Mr. M. Doroslovacki, Institute "B. Kidric", Department 270, POB 522, 11011 Belgrade, Yugoslavia.
Mr. C.E. Doughty, Room B/0305, Government Communications Headquarters, Oakley, Priors Road, Cheltenham GL52 5AJ, U.K.
Dr. A.M. de Olivera Duarte, Dpto. Electronica e Telecomunicacoes, University of Aveiro, 3800 Aveiro, Portugal.
Mrs. E. Duscombe, 95 Byron Way, Northolt, Middx UB5 6AZ, U.K.
Prof. G. Einarsson, Electrical Engineering Department, University of Lund, 220 07 Lund, Sweden.

Prof. A. Ephremides, Electrical Engineering Department, University of Maryland, College Park, MD 20742, U.S.A.
Dr. F.R. Ergul, The Scientific and Technical Research Council of Turkey, P.K.38, Bahcelievcer, Ankara, Turkey.
Mr. N. Farahati, Engineering Department, Cambridge University, Trumpington Street, Cambridge CB2 1PZ, U.K.
Prof. P.G. Farrell, Department of Electrical Engineering, University of Manchester, Dover Street, Manchester M13 9PL, U.K.
Dr. V. Freitas, Department of Informatics & Control, Centro de Ciencias e Eng. de Sistemas, Universidade de Minho, Largo do Paco, 4719 Braga, Portugal.
Prof. ir. J.H. Geels, Royal (NL) Military Academy, Department of Technical Studies, Telecomlab PFHP 703, P.O. Box 90.154, NL 4800 RG Breda, The Netherlands.
Dr. E. Geraniotis, Electrical Engineering Department, University of Maryland, College Park, MD 20742, U.S.A.
Mr. J.W. Gluck, Electrical Engineering Department, University of Maryland, College Park, MD 20901, U.S.A.
dr. P. Godlewski, Ecole Nationale Superieure des Telecommunications, 75103 Paris, France.
Mr. E.W. Groeneweld, Twente University, P.O. Box 217, 7500 AE Enschede, The Netherlands.
Mr. J.A. Gubner, University of Maryland, Electrical Engineering Department, College Park, MD 20742, U.S.A.
Prof. D. Haccoun, Department of Electrical Engineering, Ecole Polytechnique de Montreal, C.P. 6079, Succ. A, Montreal, Quebec, Canada H3C 3A7.
Mr. G. Hakizimana, Laboratoire CEPHAG INPG/IEG, VA CNRS 346, BP 46, 38406 Saint Martin d'Themes CEDEX, France.
Dr. J.F. Hayes, Electrical Engineering Department, Concordia University, 1445 de Maisonneuve Blvd., West, Montreal, Canada H3G 1M8.
Prof. O. Hirota, Sagami Institute of Technology, Department of Information Science, Tsujido Nishikaigan 1-1-25, Fujisawa 251, Japan.
Dr. B. Honary, Electrical & Electronic Engineering Department, Coventry Polytechnics, Coventry, U.K.
Prof. B. Hughes, Department of Electrical Engineering & Computer Science, The Johns Hopkins University, Barton Hall, Baltimore, MD 21218, U.S.A.
Mr. D.G.W. Ingram, Engineering Department, University of Cambridge, Trumpington Street, Cambridge CB2 1PX, U.K.
Dr. R. Johannesson, Department of Computer Engineering, University of Lund, Box 118, S-221 00 Lund, Sweden.
Mr. B.L. Johnson, The MITRE Corporation, Box 208, M/S E020, Bedford, MA 01730, U.S.A.
Prof. T.J. Jordanides, School of Engineering, California State University, Long Beach, 1250 Belflower Blvd., Long Beach, CA 90840, U.S.A.
Mr. S. Jovanovic, Computer Systems Design Laboratory, "Boris Kidric" Institute, Vinca, POB 522, Beograd, Yugoslavia.
Ms. N. Kalkan, Bogazici Universitesi, Elektrik Muh. Bol., P.K.2, Bebek-Istanbul, Turkey.
Mr. G.A. Karpel, 43 Rother Road, Cove, Farnborough, Hampshire GU14 9LP, U.K.

Mr. O. Katic, Institut "Mihailo Pupin Beograd", Volgina 15, POB 906, 11000 Beograd, Yugoslavia.
Dr. G. Kawas-Kaleh, Ecole Nationale Superrieure des Telecommunications, 46 Rue Barrault, 75634 Paris CEDEX 13, France.
Dr. R.W. Keyes, IBM, P.O. Box 218, Yorktown Heights, N.Y. 10598, U.S.A.
Dr. M. Kimberley, Goldsmiths' College, University of London, Levishm Way, London SE14 6NW, U.K.
Dr. P.J.B. King, Computer Science Department, Herriot-Watt University, 79 Grassmarket, Edinburgh EH1 2HJ, U.K.
Prof. A. Kiper, Department of Computer Sciences, Middle East Technical University, Ankara, Turkey.
Mr. D. Kreit, University College London, Department of Electronic Engineering, Torrington Place, London WC1E 7JE, U.K.
Prof. W. Kuo, Department of Industrial Engineering, Iowa State University, 212 Marston Hall, Ames, Iowa 50011, U.S.A.
Dr. D. Lazic, Faculty of Technical Sciences, V. Vlahovica 3, 2100 Novi Sad, Yugoslavia.
Prof. L.B. Leveitin, College of Engineering, Boston University, 110 Cummington Street, Boston, MA 02215, U.S.A.
Mrs. K. Mallalieu, University College London, Department of Elctrical & Electronic Engineering, Torrington Place, London WC1E 7JE, U.K.
Mr. T. Maseng, Manager, Communications Systems, ELAB, N-7034 Trondheim - NTH, Norway.
Prof. J.L. Massey, Institute for Signal and Information Processing, ETH-Zentrum, CH-8092 Zurich, Switzerland.
Mr. S. Matic, Institute "B. Kidric", Department 270, POB 522, 11011 Beograd, Yugoslavia.
Prof. D.W. McCaughan, Marconi Elactronics Devices Ltd., Lincoln Industrial Park, Doddington Road, Lincoln LN6 3LF, U.K.
Prof. R.J. McEliece, Department of Electrical Engineering (116-81), California Institute of Technology, Pasadena, CA 91125, U.S.A.
Dr. K.D. Mead, Room B/1305, Government Communications Headquarters, Oakley, Priors Road, Cheltenham GL52 5AJ, U.K.
Mr. F. Meier, c/o AUTOPHON AG., Ziegelmattstrasse 1-15, CH-4500 Solothurn, Switzerland.
Prof. D.G. Messerschmitt, University of California, Berkeley, College of Engineering, Department of Electrical Engineering and Computer Science, Berkeley, CA 94720, U.S.A.
Prof. E.C. van der Meulen, Department Wiskunde, Katholieke Universiteit Leuven, Celestijenelaan 200 B, B-3030 Leuven, Belgium.
Prof. J.E. Midwinter, University College London, Torrington Place, London WC1E 7JE, U.K.
Mr. J. Milewski, Warsaw University, Institute of Informatics, P.O. Box 1210, PL-00-901 Warsaw, Poland.
Prof. J.M. Modestino, Electrical, Computer and Systems Engineering Department, Rensselaer Polytechnic Institute, Troy, N.Y. 12180-3590, U.S.A.
Dr. A. Morgul, Electrical Engineering Department, Bogazici University, P.K.2, Bebek, 80815 Istanbul, Turkey.
Prof. J.M.F. Moura, Room 35-213, LIDS, MIT, 77 Massachusetts Avenue, Cambridge, MA 02139, U.S.A.
Dr. R. Nagendra, Ryerson Polytechnical Institute, School of Computer Science, 350 Victoria Street, Toronto, Ontario, Canada M5B 2K3.

Dr. P. Narayan, Electrical Engineering Department, University of Maryland, College Park, MD 20742, U.S.A.
Prof. R. Ottoson, Royal Institute of Technology, S-10044 Stockholm, Sweden.
Mr. E.A. Palo, The MITRE Corporation, Burlington Road, Bedford, MA 01730, U.S.A.
Dr. B. Pehani, Faculty for Electrical Engineering, 6100 Ljubljana, Trzaska 25, Yugoslavia.
Prof. F.C. Piper, 12 Duncan Road, Richmond, Surrey TW9 2JD, U.K.
Dr. M.V. Pitke, Director, Tata Institute of Fundamental Research, Homi Bhabha Road, Bombay 400005, India.
Mr. F. Prodi, V.A. Pisano, Il Pisanello 33, 56100 Pisa, Italy.
Mr. R. Reggianini, Inst. Elettronica e Telecomucatione, Via Diotisalvi 2, I 156100 Pisa, Italy.
Dr. J.B. Riera, ETSI Telecomunicacion, Cmunicacion de Datos, Universidad Politecnica de Madrid, 28040 Madrid, Spain.
Dr. Jose R.F. de Rocha, Department of Electronics and Communications, University of Aveiro, 3800 Aveiro, Portugal.
Prof. Jerome Rothstein, Department of Computer & Information Science, The Ohio State University, 2036 Neil Mall, Columbus, Ohio 43210, U.S.A.
Dr. W. Rupprecht, Universitaet Kaiserlautern, FB Elektrotechnik, Postfach 3049, 6750 Kaiserlautern, Federal German Republic.
Dr. R.S. Safavi-Naimi, Iran Communications Industry, No. 34 Apadana Street, Tehran 15534, Iran.
Prof. J.P.M. Schalkwijk, Department of Electrical Engineering, TH-Eindhoven, Den Dolech 2, 5600 MB Eindhoven, The Netherlands.
Dr. G. Seguin, Royal Military College, Kingston, Canada.
Mr. T. Sekerci, Teletas, R/L Research Laboratories (343), Uramiye 81257, Istanbul, Turkey.
Prof. V. Silindir, Department of Electronic Engineering, University of Hacettepe, Byetepe - Ankara, Turkey.
Mr. A. Sinha, Centre for Development of Telematics, C/O T.I.F.R., Homi Bhabha Road, Bombay 400005, India.
Dr. J.K. Skwirzynski, GEC Research Laboratories, Marconi Research Centre, Great Baddow, Essex CM2 8HN, U.K.
Dr. B.J. Slagsvold, Teledirektoratets Forksningsvdeling, Box 83, 2007 Kjeller, Norway.
Mr. B. Smeets, Department of Computer Engineering, University of Lund, Box 118, S-221 000 Lund, Sweden.
Prof. K. Steiglitz, 130 Longview Drive, Princeton, N.J. 08540, U.S.A.
Mr. B.A. Stuart, 56 Gloucester Avenue, Chelmsford, Essex CM2 9LE, U.K.
Prof. F.L.H.M. Stumpers, Elzentlaan 11, Eindhoven, The Netherlands.
Prof. A.J. Surkan, Department of Computer Science, Ferguson Hall, 115 UNL, Lincoln, Nebraska 68588-0115, U.S.A.
Mjr. M.D. Townsend, USAF EOARD, 223/231 Old Marylebone Road, London NW1 5TH, U.K.
Ms. N. Tugbay, TUBITAK, The Scientific and Technical Research Council of Turkey, Electronics Division, P.O. Box 74, Gebze, Kocaeli, Turkey.
Mr. P. Vanroose, Katholieke Universiteit Leuven, Departement Wiskunde, Celestijnenlaan 200 B, B-3030 Heverlee (Leuven), The Netherlands.
Mr. J.M.M. Verbakel, Room Way 597, Nederlandse Philips Bedryven, P.O. Box 80.000, 5600 IA Eindhoven, The Netherlands.

Dr. S. Verdu, Electrical Engineering Department, Princeton University, Princeton, NJ 08544, U.S.A.
Dr. J. Vinyes, ESTI Telecomunicacion, Comunicacion de Datos, Universidad Politecnica de Madrid, 28040 Madrid, Spain.
Dr. G.G. Weber, Kernforschungszentrum Karlsruhe, IDT, Postfach 3640, D-7500 Karlsruhe, Federal German Republic.
Dr. K. Wesolowski, Technical University of Poznan, The Institute of Electronics, 60-965 Poznan, Poland.
Mr. B.G. West, GEC Research Laboratories, Marconi Rtesearch Centre, Great Baddow, Essex CM2 8HN, U.K.
Mr. P. Wheatley, University College London, Department of Electronic Engineering, Torrington Place, London WClE 7JE, U.K.
Prof. J.K. Wolf, Mail Code S-008, USCD, La Jolla, CA 92093, U.S.A.
Prof. E. Yazgan, Electrical-Electronic Engineering Department, Hacettepe University, Byetepe, Ankara, Turkey.
Mr. E. Yigiter, University of Southern California, Biomedical Engineering Department, Los Angeles, CA 190089-1451, U.S.A.
Mr. G. Zivanovic, "Boris Kidric" Institute for Nuclear Science - Vinca, Computer System Design Laboratory, Dept. 270, PO Box 522, 11001 Belgrade, Yugoslavia.
Mr. Z. Zongcheng, Dipertimento di Elettronica, Politecnico di Torino, Corso Duca degli Abruzzi 24, 10129 Torino, Italy.